Designing Network On-Chip Architectures in the Nanoscale Era

Chapman & Hall/CRC Computational Science Series

SERIES EDITOR

Horst Simon

Associate Laboratory Director, Computing Sciences
Lawrence Berkeley National Laboratory
Berkeley, California, U.S.A.

AIMS AND SCOPE

This series aims to capture new developments and applications in the field of computational science through the publication of a broad range of textbooks, reference works, and handbooks. Books in this series will provide introductory as well as advanced material on mathematical, statistical, and computational methods and techniques, and will present researchers with the latest theories and experimentation. The scope of the series includes, but is not limited to, titles in the areas of scientific computing, parallel and distributed computing, high performance computing, grid computing, cluster computing, heterogeneous computing, quantum computing, and their applications in scientific disciplines such as astrophysics, aeronautics, biology, chemistry, climate modeling, combustion, cosmology, earthquake prediction, imaging, materials, neuroscience, oil exploration, and weather forecasting.

PUBLISHED TITLES

PETASCALE COMPUTING: ALGORITHMS AND APPLICATIONS
Edited by David A. Bader

PROCESS ALGEBRA FOR PARALLEL AND DISTRIBUTED PROCESSING
Edited by Michael Alexander and William Gardner

GRID COMPUTING: TECHNIQUES AND APPLICATIONS
Barry Wilkinson

INTRODUCTION TO CONCURRENCY IN PROGRAMMING LANGUAGES
Matthew J. Sottile, Timothy G. Mattson, and Craig E Rasmussen

INTRODUCTION TO SCHEDULING
Edited by Yves Robert and Frédéric Vivien

SCIENTIFIC DATA MANAGEMENT: CHALLENGES, TECHNOLOGY, AND DEPLOYMENT
Edited by Arie Shoshani and Doron Rotem

COMPUTATIONAL METHODS IN PLASMA PHYSICS
Stephen Jardin

INTRODUCTION TO THE SIMULATION OF DYNAMICS USING SIMULINK®
Michael A. Gray

INTRODUCTION TO HIGH PERFORMANCE COMPUTING FOR SCIENTISTS AND ENGINEERS
Georg Hager and Gerhard Wellein

DESIGNING NETWORK ON-CHIP ARCHITECTURES IN THE NANOSCALE ERA
Edited by Jose Flich and Davide Bertozzi

SCIENTIFIC COMPUTING WITH MULTICORE AND ACCELERATORS
Edited by Jack Dongarra, David A. Bader, and Jakub Kurzak

Chapman & Hall/CRC Computational Science Series

Designing Network On-Chip Architectures in the Nanoscale Era

Edited by

José Flich

Davide Bertozzi

CRC Press
Taylor & Francis Group
Boca Raton London New York

CRC Press is an imprint of the
Taylor & Francis Group, an **informa** business
A CHAPMAN & HALL BOOK

Chapman & Hall/CRC
Taylor & Francis Group
6000 Broken Sound Parkway NW, Suite 300
Boca Raton, FL 33487-2742

Printed in the United States of America on acid-free paper
10 9 8 7 6 5 4 3 2 1

International Standard Book Number: 978-1-4398-3710-8 (Hardback)

Library of Congress Cataloging-in-Publication Data

Designing network on-chip architectures in the nanoscale era / [edited by] Jose Flich, Davide Bertozzi.
p. cm. -- (Chapman & Hall/CRC computational science)
Includes bibliographical references and index.
ISBN 978-1-4398-3710-8 (hardback)
1. Networks on a chip. I. Flich, Jose. II. Bertozzi, Davide.

TK5105.546.D476 2011
621.3815'31--dc22 2010036073

Visit the Taylor & Francis Web site at
http://www.taylorandfrancis.com

and the CRC Press Web site at
http://www.crcpress.com

to María Amparo, Josep, Empar, Miguel, and Encarna...
...for their support and loveliness

to Michela, Giovanni, Giacomo, Gino, Elena, Ivana
and to my CL community in Lugo,
for the exciting journey to Truth
that we are experiencing together

Contents

List of Figures

List of Tables

Foreword

Ten years after the first position papers, the network-on-chip concept is now mainstream. It has been an exciting ride, through the initial skepticism of many industrial players and after a few thousands of published papers. Today, networks-on-chip (NoCs) are an indisputable reality—the communication backbone of virtually all large-scale system-on-chip (SoC) designs in 45 nm and below. Ten years are a relatively short time span from research vision to industrial mainstream. Such a fast evolution has been driven by two converging trends. First, Moore's law has maintained its pace in terms of logic (and storage) density, but it has finally reached hard limits in terms of power consumption and synchronization. Hence, SoCs today heavily rely on multicore parallelism and locally synchronous, globally asynchronous power domains. Second, the complexity of large-scale SoCs combined with the ever-increasing time-to-market pressure have pushed for a strong componentization and modularization of silicon platforms, to reduce design effort through massive reuse combined with hierarchical, divide-and-conquer design flows. NoCs meet all the key requirements imposed by these converging trends—they facilitate the modular construction of heterogeneous multi- and many-core architectures, they provide communication abstractions and services across component boundaries, and they enable the top-down design of highly power-manageable global asynchronous locally synchronous (GALS) architectures. Their fast diffusion in products and roadmaps should therefore hardly be surprising or unexpected.

Looking in retrospect, we could ask ourselves the question: Was today's NoC reality fully encompassed in those early position papers? The (perhaps obvious) answer is "no." Even though the vision was in the right direction, the "devil is in details," and numerous "details"—or, better, key technical insights have been gained through trial-and-error, design experience, and focused research. We now understand much better in how many ways NoC design is unique and profoundly different from off-chip interconnect design. Many naïve architectural assumptions made in the early NoC research papers are now being revised—the key intuition being that the physical fabric of on-chip interconnects is fundamentally different in area, speed, and power from traditional off-chip interconnects. This drives NoC component design into new research ground. In summary, the NoC *concept* is now maturing into NoC *technology*, and this makes an enormous difference in terms of industrial applicability and impact.

This book comes as a timely and welcome addition to the wide spectrum of available NoC literature, as it has been designed with the purpose of describing in a coherent and well-grounded fashion the foundation of NoC technology, above and beyond a simple overview of research ideas and/or design experiences. The book covers in depth architectural and implementation concepts—and gives clear guidelines on how to design the key network components. It provides strong guidance in a research field that is starting to stabilize, it brings "sense and simplicity," and teaches some hard lessons from the design trenches. In addition, the book also covers some of the hottest upcoming research and development trends, such as vertical integration and variation-tolerant design. The editors put enormous effort in orchestrating the content for uniformity and in minimizing overlaps between chapters, while maintaining a solid logical flow. The contributors did an excellent job in helping them. I am personally really impressed with the result. I believe that this book is much more than a collection of chapters, it is a much needed "how-to" guide and an ideal stepping-stone for the next ten years of NoC evolution.

Luca Benini

Università di Bologna & STMicroelectronics

Preface

Chip Multiprocessors (CMPs) are diving into the marketplace very aggressively since the past efforts to speed up processor architectures with techniques that do not modify the basic von Neumann computing model, such as pipelining and superscalar issues, have encountered hard limits. Nowadays, the power consumption of the chip becomes the limiting factor and sets the rules for future CMP systems. As a result, the microprocessor industry is today leading the development of multicore and many-core architectures in the quest for higher performance, but using tens (and even hundreds in the future) of simpler and power-efficient cores instead of a handful of powerful but power-hungry cores. In a multicore chip, as the number of cores increases, efficient communication among them and with off-chip resources becomes key to achieve the intended performance scalability.

This trend has contributed to definitely overcome the skepticism of some system architects to embrace on-chip interconnection networks as a key enabler for effective system integration. Networks-on-chip (NoCs) make performance scalability more a matter of instantiation and connectivity rather than increasing complexity of specific architecture building blocks. NoCs effectively cope with the productivity gap by providing parallelism through the replication of many identical blocks placed each in a tile of a regular array fabric. In practice, they push architecture design styles with strict component orientation. As a consequence, as the number of cores in a CMP increases, the complexity of the NoC keeps constant. The transition to NoCs is today either a well established and unquestionable fact or an inevitable step in product roadmaps.

NoCs have been researched by two different communities. One community steeped in from the field of high-performance interconnects for clusters and massively parallel processors, where the networking paradigm had already been proven to be an extremely effective means of managing parallel communication flows. Indeed, the NoC concept was borrowed from that domain: terms such as topology, switching, routing, and flow control are common to both off-chip and on-chip networks. By distilling the most applicable concepts from the original research domain and by applying them in a way that suits the constraints of silicon technology, NoCs have finally stepped into chip multiprocessors.

The other community evolved from the research on embedded computing platforms, where traditional shared or multilayer buses were poisoning architecture scalability, design predictability, and wire routability. In general, this community was more familiar with the tight resource budgets of systems-on-chip and with the need to support multiple communication protocols at network boundary. This community has initially taken the lead of NoC development for multiprocessor systems-on-chip and, for the cases where target applications are known at design time, has driven a customization effort of NoC architectures for cost-effective realizations.

The degree of interaction between the two communities has grown in time, with exchange of ideas, results and design experiences, thus leading to a fruitful cross-fertilization and to the identification of the most suitable and smart architecture design solutions meeting the constraints of each application domain.

In both cases, NoC architectures have come a long way since their definition as an innovative interconnect concept in early position papers. Early design methods were still too affected by the mind-set of off-chip interconnection network designers and by the trade-offs holding in that domain. However, in the short time frame of ten years researchers have gained awareness of the distinctive features of designing in an on-chip environment and started to shape architectures accordingly. At the same time, the increasing commitment of semiconductor industries with the new on-chip networking paradigm has accelerated this process.

The wide spectrum of available NoC literature mostly reflects this intensive and unrelenting research and development effort. Technical papers from conference proceedings or refereed journals, book chapters, technical reports, and dissemination forums represent a huge amount of material condensing key technical insights, design experiences, and results of focused research. At the same time, there has been the proliferation of dedicated symposiums and workshops (primarily the International Symposium on Networks-on-Chip), as well as special sessions or tracks in more generalist conferences. They aimed at bringing together academic and industrial researchers and developers addressing the issues of NoC-based systems at all levels, from the physical on-chip link level through the network level, and ranging up to system architecture and application software.

Most of the time the resulting material from those events necessarily sets out to bring readers the latest advances in the field of on-chip interconnects for multicores. However, we feel that today we are at a milestone in the development of the NoC concept and something more is needed. The fundamental design principles and implementation methods for NoC architectures tend to stabilize and, as Luca Benini says in his Foreword, the NoC concept has matured into basic NoC technology, which makes an enormous difference in terms of industrial applicability and impact. It is therefore possible to start surveying such technology in a structured and systematic way and gaining unified views on specific topics. This way, the outcome of the research efforts of the last decade can be put at the disposal of different communities, serving the purpose of education, dissemination, and even exploitation initiatives.

This is the ambitious challenge that this book takes on. Its intent is to start structuring knowledge on on-chip interconnection networks and to present the foundations of NoC technology, ideally meeting the needs of NoC designers in both academia and industry. In this direction, we have striven to bring together the treatment of fundamental design issues, the review of alternative design paradigms and techniques, and an overview of the main design tradeoffs always taking the practical perspective. On one hand, for some topics, the book aims at overcoming the chronological order of original technical papers and at presenting the material in logical order. This implied removing overlap between related papers, putting each piece of work in context, and acknowledging logical links between technical contributions. On the other hand, the book intends to serve as a "how-to" guide for the NoC designer, capitalizing on the insights matured in past design experiences. From this viewpoint, the involvement of the editors in the development of much of the presented technology helped to focus each chapter on the key issues that really matter to the designer in real-life NoC design practice.

In spite of this effort, we are aware of the fact that NoC technology is evolving at an unrelenting pace and its development rate is far from peaking. As an increasingly wider range of parallel hardware platforms starts considering NoC architectures as their interconnect backbone, new requirements need to be met by those architectures, thus further pushing them to evolve in new directions. At this time, we view two converging drivers that are shaping the current development effort of NoC architectures. They share an underlying theme: the composition of largely integrated systems out of baseline components is an increasingly challenging process because of a growing degree of *dynamism* and *uncertainty*. To the limit, we could define the ultimate implication of these design constraints as the heterogeneity of NoC architectures in spite of the designer's original intent to design an homogeneous and regular system.

On one hand, the uncertainty conditioning the network "componentization" process stems from the effects of nanoscale physics propagating up the design hierarchy. Controlling manufacturing parameters such as gate oxide thickness, threshold voltage, and effective gate length becomes increasingly difficult as feature sizes shrink. This results in circuit performance and power variability, which impacts yield if not properly addressed by design for manufacturability techniques and by variability-tolerant architecture design methods. In addition, this technology landscape questions the feasibility of a global and synchronous clocking style. This book shares these concerns with NoC designers and selects those technical contributions from the open literature that can provide uncertainty-tolerant architecture design methods and which can tolerate more flexible timing assumptions.

On the other hand, at the opposite side of the design hierarchy, system-level hardware/software (HW/SW) platform management frameworks increasingly drive the constructive NoC composition process by dynamically enforcing network partitioning into arbitrarily shaped regions changing over time.

Each partition might correspond to the performance-power optimal number of processing tiles selected to run a multitask application or to the processing tiles devoted to special purpose functions. Moreover, a partition might be a powered-down region of the network for power management purposes or to mitigate overheating of one or more processing tiles. Heterogeneity of partitions is what makes their composition challenging. It depends on the different number of components per partition, on the shape of the partition, on its interaction with other partitions (especially border effects) and on the different supply voltage and clock speed of each partition. Moreover, some partitions might not even be available just for cross-through connectivity. This book reviews architecture design methods that can support dynamic and flexible partitioning of the network.

Addressing some of the challenges illustrated above is mostly work in progress and only few related and relevant contributions are reported in the open literature. In the absence of much knowledge to structure for these topics, the book nonetheless sets out to structure the work to be done and highlights the main research directions that are being pursued and their expected deliverables. The editors view this ongoing development effort of NoC architectures towards an increased dynamism and flexibility as headed to a new milestone for NoC technology. The book does not intend to miss the opportunity to join this exciting ride, and presents in a coherent and systematic fashion both stabilized and less stabilized topics, with the level of detail enabled by their state-of-the-art, thus hopefully helping understand what is going on in the NoC community and serving as a stepping-stone into future chip multiprocessor architectures.

The intended audience of the book is not limited to NoC designers in industry and academic research groups, but encompasses also faculty and students, even with little prior exposure to interconnection networks. For this purpose, we took the proper course of action to improve understanding of concepts and design techniques. First, Chapter 1 smoothly introduces the reader into the issues addressed by this book starting from the user perception of technology-intensive devices that are today pervasive in everyday life. Second, Chapter 1 also introduces the basic terminology of interconnection networks in a way easily accessible to a broad audience, while preserving accuracy. Third, the chapters of this book have been subdivided into three parts, as hereafter explained. The first part is the one that presents more basic concepts and stabilized design techniques, although selected based on the suitability to tackle the heterogeneity of future platforms. Whenever these chapters go deeper into the details (resulting from actual design experiences), they assume that the uninterested reader in such details can skip them without impacting the understanding of the concepts presented thereafter. This way, we hope that both expert designers and students can profit from this book.

Organization of the Book

The book is structured in three parts to serve as a reference and source of learning, but also as a means of understanding the industrial vision and roadmap and the hottest upcoming research and development trends.

Part I. NoC Technology

The first part includes chapters that deal with the very basic issues of switch architecture and design, topology selection, and routing implementation. These four chapters can be seen as consecutive steps in the design process of the network. In the landscape of nanoscale silicon technologies, such architecture design methods need to prove robust to the sources of heterogeneity and aware of the new clocking challenges. Both constraints are better described in dedicated chapters.

Chapter 1

This chapter introduces the reader to the criticality of interconnection networks for the high performance of computing systems that are pervasive in everyday life. Above all, it raises the awareness of the main challenge that lies ahead for on-chip interconnection network architectures: heterogeneity.

Chapter 2

The very basic component in an on-chip network is the switch. This chapter provides a comprehensive overview of switch architecture design techniques, which are at the core of the main prototypes and relevant design experiences documented in the open literature. Such techniques are quite consolidated as an effect of their experimentation in off-chip networks and of their later review and integration for applicability to on-chip networks.

Chapter 3

The main feature of designing for an on-chip setting is that cross-layer design and optimization becomes an inevitable step for successful designs. In this chapter, implementation of the basic building blocks of a NoC switch at the logic level will be discussed, while pointing out the unique opportunities and strategies for optimizations at this level of the design hierarchy.

Chapter 4

In the general purpose computing domain topology selection has not been perceived as a problem so far, since designers typically opt for a 2-D mesh to stay on the safe side. However, with future systems raising latency and bandwidth scalability concerns, the search for alternative topologies becomes important. This chapter assesses several candidate topologies for large scale multicore systems with awareness of physical design effects and implications.

Chapter 5

The only way to preserve functionality of irregular topologies is by means of topology agnostic routing algorithms. The latter are usually implemented by means of forwarding tables, which scale poorly in delay and area. This chapter reviews the main routing mechanisms for NoCs and focuses on the promising logic-based distributed routing, addressing the challenge to make it flexible in spite of its logic-based implementation.

Chapter 6

In this chapter, architecture design and physical design are tightly interrelated: Today there is little doubt on the fact that a high performance and cost effective NoC can be designed in 45 nm and beyond under a relaxed synchronization assumption. Chapter 6 is therefore devoted to synchronizer-based GALS architectures and design techniques.

Part II. The Industrial Perspective

This part reports selected experiences from the industry and sheds light on their vision and on their roadmap.

Chapter 7

This chapter presents startup Tilera's TILE processor family of multicore processors and the detailed microarchitecture of the replicated mesh networks that interconnect the cores. The chapter also discusses the interfaces of the network to the processor pipeline and the suite of software tools that aid in mapping programs onto the chip.

Chapter 8

This chapter briefly reviews the evolution of on-chip interconnects used in recent Intel products and research prototypes and then discusses new usage models and desired attributes for many-core processors. Following the lessons learned and new insights, the chapter presents the details of a novel on-chip that supports partitioning with performance isolation and the ability to tolerate several faults or irregularities in the topology.

Chapter 9

This chapter illustrates the TRIPS operand (OPN) network, an NoC that interconnects the functional units within the TRIPS processor core. The OPN replaces an operand bypass bus and primary memory system interconnect in a technology scalable manner. The tight integration between the OPN and the processor core elements enables fast operand bypassing across distributed arithmetic logic units (ALUs), providing an opportunity for greater instruction-level concurrency.

Part III. Upcoming Trends

This is the more forward-looking part of the book that covers some of the hottest upcoming research and development trends, such as vertical integration and variation-tolerant design, from an hardware viewpoint, and the efficient implementation of the programming model at the network interface.

Chapter 10

This chapter focuses on NoC interfaces for CMPs and on their architecture design techniques, which are tightly related with the programming model. Therefore, this chapter also addresses some programming model-related issues and relevant synchronization support, with special emphasis on explicit interprocessor communication.

Chapter 11

3D integration is emerging to mitigate the interconnect wire delay problem by stacking active silicon layers. In this chapter, the focus is on various 3-D architectural design options for NoCs and on a novel design that moves from the traditional design options to build a 3-D stacked NoC router.

Chapter 12

This chapter addresses three future challenges for the success of largely integrated multicore architectures, and provides insights into possible strategies to effectively tackle them. They are: traffic characterization in CMPs, congestion management techniques for NoCs, and the impact of process parameter variations on application performance.

Appendix: Switch Models

This appendix illustrates the microarchitectural details of two switch architectures (xpipesLite and gNoC) targeting the multiprocessor system-on-chip and the chip multiprocessor domains, respectively. They serve both as an embodiment of the design methods illustrated throughout the book and as an experimental platform for running tests reported in several book chapters to validate or characterize specific design methods or concepts.

We really wish that the reader, no matter the field he comes from, and no matter the platform he targets for its NoC design, finds the contents of this book as enjoyable and useful as we do.

José Flich
Davide Bertozzi

Acknowledgments

First and foremost, we are very grateful to all the contributors of this book. Forty-three authors contributed twelve chapters and an appendix while guaranteeing their outstanding quality and the harmonization of their technical content; this book was possible thanks to their willingness and dedication. Their commitment and readiness since the early stages of the project were fundamental for its success; therefore, they deserve our deepest appreciation.

The large number of authors enriches this book by offering different yet converging approaches and design experiences in the network-on-chip domain, thus allowing the different contributions to smoothly integrate and complement each other. We are very grateful to Professor José Duato and to Professor Luca Benini, two friends and colleagues of ours. Their advice and forward-looking vision were extremely valuable for the development of a book with timely content that is able to meet the expectations of industry and academia.

We are also very thankful to Randi Cohen (Computer Science Acquisitions Editor at Chapman & Hall/CRC Press) for encouraging us to work on this project and for providing continuous support, from the first draft to the last-minute changes for on-time delivery. Thank you!

Also, we would like to acknowledge the support from other colleagues in Taylor & Francis for the valuable effort they put forth to make all this come true. In particular, our special thanks go to David Tumarkin, who led the way toward a coherent and smooth writing style. Finally, the LaTeX package support from Shashi Kumar (Glyph International) was of the utmost importance to us.

Last, but not least, we would like to thank the PhD students who helped us with the everyday tasks of a long book-editing effort. Alessandro Strano and Antoni Roca are especially worth mentioning for their hard, professional, and mostly hidden work, which has contributed to the overall quality of the final manuscript.

The editors

About the Editors

José Flich received MS and PhD degrees in computer science from the Technical University of Valencia (Universidad Politécnica de Valencia), Spain, in 1994 and 2001, respectively. He joined the Department of Computer Engineering (DISCA), Universidad Politécnica de Valencia, in 1998, where he is currently an associate professor of computer architecture and technology. His research interests are related to high-performance interconnection networks for multiprocessor systems, cluster of workstations, and networks on chip. He has published over 100 papers in peer-reviewed conferences and journals. He has served as program committee member in different conferences, including NOCS, DATE, ICPP, IPDPS, HiPC, CAC, ICPADS, and ISCC, and is an associate editor of *IEEE Transactions on Parallel and Distributed Systems.* He is currently the cochair of the CAC and INA-OCMC workshops. He is the coordinator of the NaNoC FP7 EU-funded project (http://www.nanoc-project.eu).

Davide Bertozzi received his PhD degree from the University of Bologna, Italy, in 2002. He joined the Engineering Department of the University of Ferrara in 2004, where he currently holds an assistant professor position and leads the research group on Multiprocessor Systems-on-Chip. His main expertise is on multicore digital integrated systems, with emphasis on all aspects of system interconnect design (from design technology to physical design issues through architecture design techniques). He has been a visiting researcher at international academic institutions (Stanford University) and semiconductor companies (STMicroelectronics, NEC America, Samsung, NXP Semiconductors). He has been program chair of the main events of the network-on-chip community (Int. Symp. on Networks-on-Chip, Design Automation and Test in Europe Conference – Network-on-Chip Track, NoC tutorial and workshop at HiPEAC 2010). He is member of the editorial board for *IET Computer and Digital Techniques.* He is involved in two EU-funded projects within the FP7 program (Galaxy, on GALS networks-on-chip, and NaNoC, on a NoC design platform) and is an HiPEAC member. Recently, he was funded by the "Future in Research 2008" program of the Italian Government to advance knowledge in the field of photonic interconnection networks.

Contributors

Anant Agarwal
Tilera
USA

Federico Angiolini
iNoCs
Switzerland

Mani Azimi
Intel Labs, Intel Corporation
USA

Liewei Bao
Tilera
USA

Davide Bertozzi
University of Ferrara
Italy

John F. Brown III
Tilera
USA

Donglai Dai
Intel Labs, Intel Corporation
USA

Chita Das
Pennsylvania State University
USA

Reetuparna Das
Intel Labs, Intel Corporation
USA

Giorgos Dimitrakopoulos
University of West Macedonia
Greece

José Duato
Technical University of Valencia
Spain

Soumya Eachempati
Pennsylvania State University
USA

Bruce Edwards
Tilera
USA

José Flich
Technical University of Valencia
Spain

Francisco Gilabert
Technical University of Valencia
Spain

María Engracia Gómez
Technical University of Valencia
Spain

Paul V. Gratz
Texas A&M University
USA

Patrick Griffin
Tilera
USA

Carles Hernández
Technical University of Valencia
Spain

Henry Hoffmann
Tilera
USA

Manolis Katevenis
Foundation for Research and Technology Hellas (ICS-FORTH)

Stamatis Kavadias
Foundation for Research and Technology Hellas (ICS-FORTH)

Stephen W. Keckler
The University of Texas at Austin
USA

Milos Krstic
Innovations for High-Performance Microelectronics (IHP), Germany

Akhilesh Kumar
Intel Labs, Intel Corporation
USA

Mario Lodde
Technical University of Valencia
Spain

Daniele Ludovici
TUDelft
The Netherlands

Matthew Mattina
Tilera
USA

Simone Medardoni
University of Ferrara
Italy

Chyi-Chang Miao
Tilera
USA

Asit K. Mishra
Pennsylvania State University
USA

Vijaykrishnan Narayanan
Pennsylvania State University
USA

Dongkook Park
Intel Labs, Intel Corporation
USA

Vasileios Pavlidis
Integrated Systems Lab (LSI)
EPFL, Lausanne, Switzerland

Dionisios Pnevmatikatos
Foundation for Research and Technology Hellas (ICS-FORTH)

Carl Ramey
Tilera
USA

Antoni Roca
Technical University of Valencia
Spain

Samuel Rodrigo
Technical University of Valencia
Spain

Federico Silla
Technical University of Valencia
Spain

Alessandro Strano
University of Ferrara
Italy

Aniruddha S. Vaidya
Intel Labs, Intel Corporation
USA

David Wentzlaff
Tilera
USA

Yuan Xie
Pennsylvania State University
USA

Part I

NoC Technology

Chapter 1

Introduction to Network Architectures

José Duato

Technical University of Valencia, Spain

Nowadays, computers are a common tool at the office and home. The widespread use of computers has been made possible thanks to the continuously improving performance/cost ratio and, most important of all, to the availability of powerful and useful applications and services. As computers are becoming widespread, the way computers are used is evolving significantly. The widespread availability of Internet access motivated the development of new network-based services (WWW, e-commerce, etc.). As a consequence, the computational burden is moving from desktop computers to large high-performance servers. These servers require powerful processors and fast large-capacity disks. But, what is the role of interconnects in this context? Do we really need high-speed interconnects to provide those services? Many people, especially home users, believe that just increasing the bandwidth of their Internet connection is enough. However, the situation is likely to change very rapidly in the near future, as described in the next sections.

1.1 The Role of Interconnects in the Internet Era

The performance of Internet services—measured by the users as the average response time to their requests—critically depends on the performance of servers and the Internet. Providing low response times is becoming increasingly difficult as the use of Internet services grows because the requests of many users must be concurrently serviced, thus imposing bandwidth requirements on servers and the Internet. Both response time and bandwidth depend on many factors, including the number of concurrent users at a given time, application complexity, computing power of clients and servers, server I/O bandwidth, Internet bandwidth, and bandwidth delivered by Internet Service Providers (ISPs).

Among all those factors, we are interested on analyzing communication requirements. The traditional view of communication when studying computer networks is depicted in Figure 1.1. In this model, the source system is connected to the destination system through some transmission system (e.g., public telephone network (PTN), local area network (LAN), wide area network (WAN), etc.). The main focus is on how to attach the source system to the network and how to establish communication with the destination system. In doing so, communication protocols—which define the communication functionality at different layers—and standardization—which allows systems and protocols implemented by different manufacturers to "understand" each other—play a major role. This traditional model has been described in several excellent books, and it is not the focus of this book.

If we look at how a client accesses an Internet server in more detail, we discover that data are transmitted over several networks, most of which are not covered in traditional books on computer networks. In particular, when accessing remote servers through a WAN, it is very likely that data packets traverse one or more asynchronous transfer mode (ATM) switches and/or Internet Protocol (IP) routers. Those switches and routers require protocols to communicate with other Internet devices and to configure the routing tables used to establish a path through the Internet. A block diagram of the internal structure for a high-performance IP router is shown in Figure 1.2. Incoming packets are stored in the line card buffers and processed by the corresponding network processor, which computes the router output port for each packet from the IP address of the packet destination. The *switch fabric* takes care of switching packets, that is, establishing a path from the input port to the requested output port for each incoming packet. The main bottlenecks inside an IP router are the network processor and the switch fabric. As packet processing is mostly independent from one packet to another, network processor performance can be improved by using either multithreaded processors and/or multiprocessors, or simply by adding more network processors. However, increasing the number of ports and port bandwidth in the switch fabric

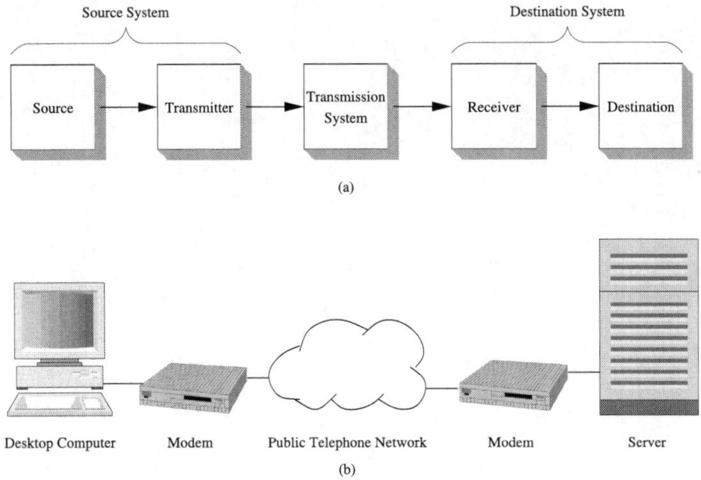

FIGURE 1.1: Classical communication model: (a) system block diagram and (b) example system.

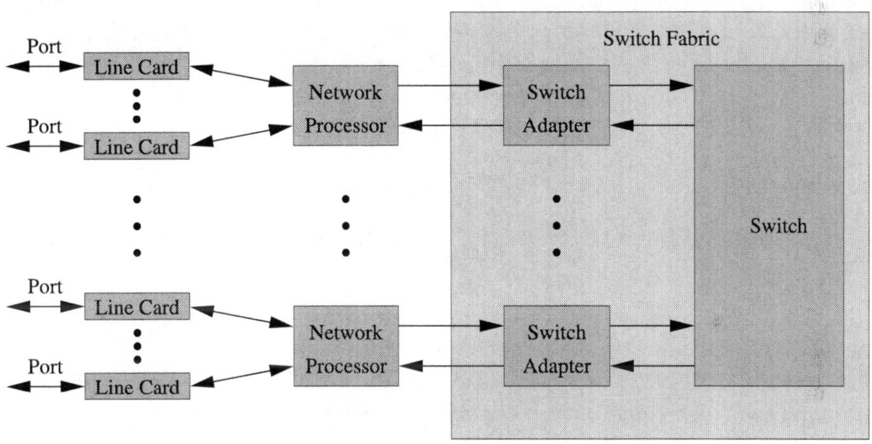

FIGURE 1.2: Block diagram of an IP router.

is more complex since different input/output paths cannot be efficiently decoupled from each other. As VLSI technology only improves by a factor of two every eighteen months (known as Moore's Law), architectural improvements are required in high-speed interconnects to satisfy the increasing bandwidth demand.

Once the user request reaches the destination Internet server, even more high-speed interconnects are needed to process the request with a reasonable delay. Some user requests may require complex database searches and/or building web pages dynamically. Additionally, thousands or even tens of thou-

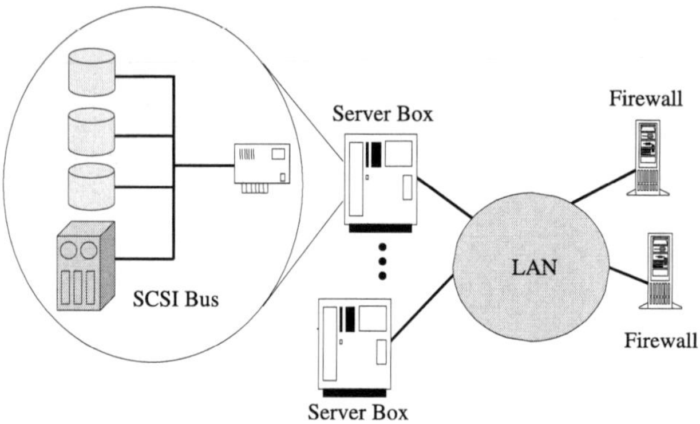

FIGURE 1.3: Traditional server architecture.

sands of clients may be concurrently serviced. In order to provide the required computing power at a relatively low cost, most high-performance servers are currently built around a cluster. Figure 1.3 shows a typical architecture for a traditional server. In this architecture, external requests pass through a firewall and reach a certain server box, where they are processed. Computing power can be easily increased by adding more processors (i.e., more server boxes). However, the main challenges nowadays are solving the I/O bottleneck, mostly produced by I/O buses, and increasing reliability, since the failure of a server box or the LAN path to reach it prevents the access to the attached disks. Current industry trends are toward the use of external reliable disk subsystems (e.g., redundant arrays of inexpensive disks (RAIDs)) and high-speed interconnects linking processors and disk subsystems, usually referred to as *storage area networks (STANs).* [1] Figure 1.4 shows the architecture of a server based on a storage area network. A STAN increases I/O bandwidth with respect to I/O buses and also increases reliability. As an example, Figure 1.5 shows a storage area network in which server boxes are connected to RAIDs through a pair of redundant switches, each one providing an alternative physical path to all the disks. Configurations with hundreds or even thousands of disks are feasible when using STANs for disk access.

But high-speed interconnects are not only required to interconnect processors and disks. Processing a complex request may require executing several processes. Those processes can be executed in parallel in order to reduce response time. Additionally, some processors may require services from other processors (e.g., accessing a database). In these cases, processors have to communicate among them, thus requiring another high-speed interconnect. Figure 1.6 presents a closer view at the internal structure of a server box, showing how CPU boxes are interconnected among them. In most cases, state-of-the-art LAN technology (e.g., Gigabit or 10 Gigabit Ethernet) will suffice. In

[1] We use the short STAN instead of SAN to avoid confusion with system area networks.

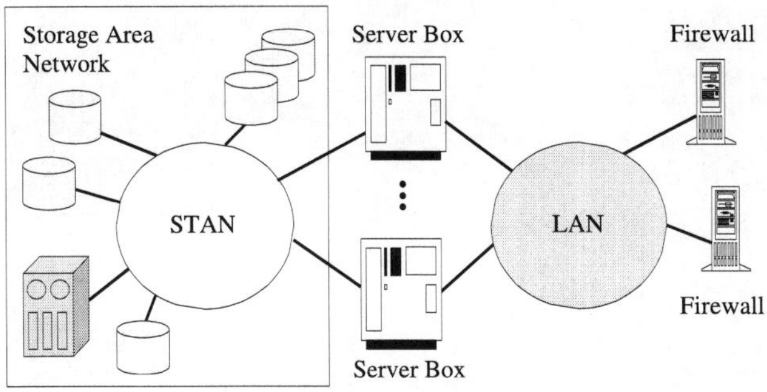

FIGURE 1.4: Server architecture based on a storage area network.

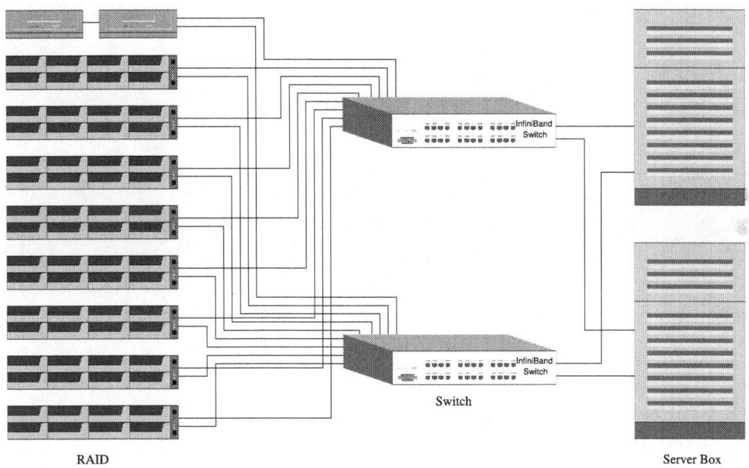

FIGURE 1.5: An example of storage area network with redundant switches.

other cases, more specific interconnects may be required. In particular, some interconnect standards can be used for both I/O accesses as well as for inter-processor communication (e.g., InfiniBand^{TM}), thus unifying the concepts of STAN and LAN into a *system area network (SAN).*

In turn, CPU boxes may be implemented as a shared-memory multiprocessor to increase performance and/or performance density. Dual-processor nodes are very common in current clusters, and multiprocessor configurations with four, and up to eight processor chips per node are frequently used when a higher computing power density is required. Figure 1.7(a) shows a 4-way "blade" format motherboard.

As an example, both AMD and Intel server processors implement an on-chip router and multiple high-speed point-to-point links, allowing several microprocessors to be directly connected to each other, thus forming an interconnection network. This network can be used to access remote memory modules

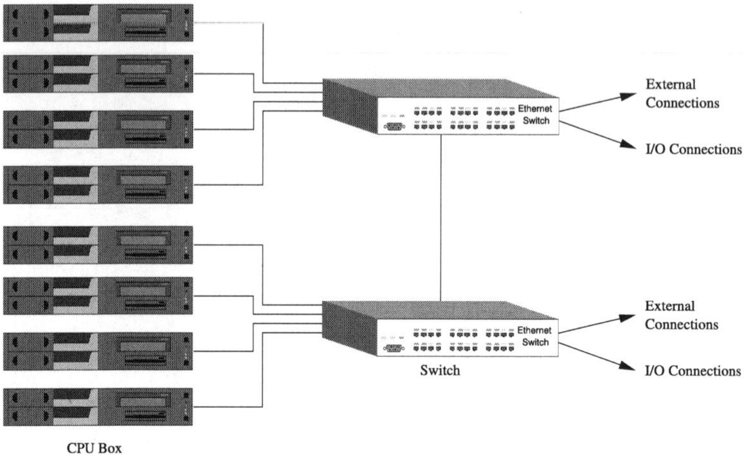

FIGURE 1.6: A closer view at the internal structure of a server box.

as well as for I/O. Figure 1.7(b) shows a typical configuration with four AMD Opteron processors, each of them with multiple Dynamic Random Access Memory (DRAM) modules directly attached to it, and interconnected using HyperTransport™ links. The four open links can be used to access I/O devices through the south bridge chip, as well as standard expansion connectors (e.g., HTX, the standard HyperTransport connector) for plug-in cards. Intel recently introduced a similar technology for the same purposes, referred to as Quick Path Interconnect™ (QPI).

As integration scale increases, more transistors become available and implementing multiprocessors on a chip (chip multiprocessor, or CMP) becomes feasible, thus requiring an on-chip interconnection network, usually referred to as *network on chip* (NoC), to implement fast communication among processors. The interconnection pattern depends on the number of cores and other components (i.e., cache blocks, memory controllers, etc.) that need to be interconnected as well as on their bandwidth requirements. When low bandwidth is required, a bus is usually enough, thus leading to simpler and cheaper systems. When communication bandwidth requirements increase, a crossbar is frequently the best choice. But when the number of components to interconnect exceeds a certain value, a switched network with point-to-point links is usually the best choice, thus delivering good scalability at a reasonable cost. The interconnection pattern can be similar to the one shown in Figure 1.7(b).

Figure 1.8 shows an example of a tiled CMP architecture, where each tile consists of four cores, including the L1 private caches, four shared L2 caches connected to the cores through a local interconnect (e.g., a crossbar), a memory controller to access external memory, a bank of a distributed directory structure to increase the scalability of cache coherence protocols, a set of vias to L3 cache blocks implemented using flip-chip technology, and a router

(a) A 4-way "blade" format server motherboard.

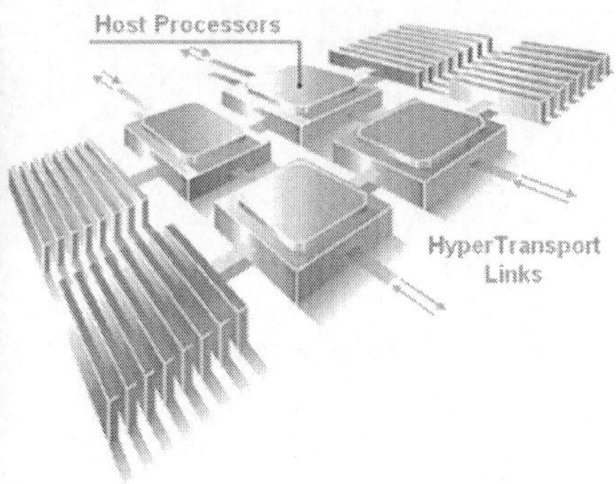

(b) An example of multiprocessor interconnection network.

FIGURE 1.7: Example of a server node and its 3-D block diagram representation (©2007 HyperTransport Technology Consortium, used with permission).

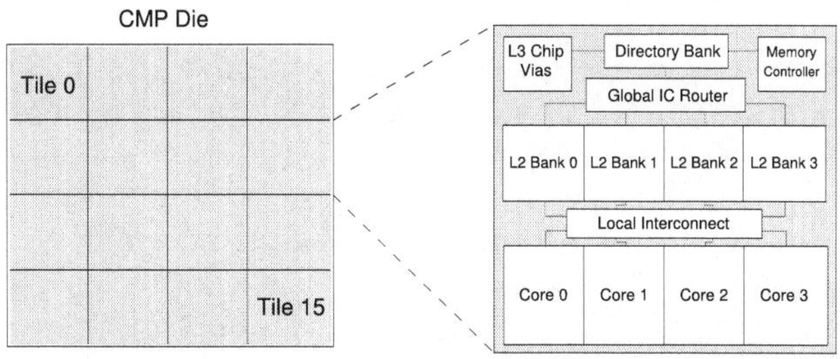

FIGURE 1.8: An example of a tiled CMP architecture.

to interconnect each tile to the remaining ones. The interconnection pattern among the router links can follow different topologies, but most proposals prefer a 2-D mesh since this is the topology that best matches the physical layout of the tiles in the chip.

At a finer level of detail, the architecture of future processors could be a clustered microarchitecture, each cluster consisting of a register file and some arithmetic-logic units (ALUs), again requiring an on-core interconnection network to implement fast communication among clusters. Figure 1.9 shows an example of a processor microarchitecture with four clusters sharing the L1 data cache, including a detailed view of the internal structure of each cluster.

The example about accessing a high-performance server through the Internet has revealed that the performance of future network-based computing systems critically depends on the development of high-speed interconnects able to satisfy the rapidly increasing demands in the server and IP router areas.

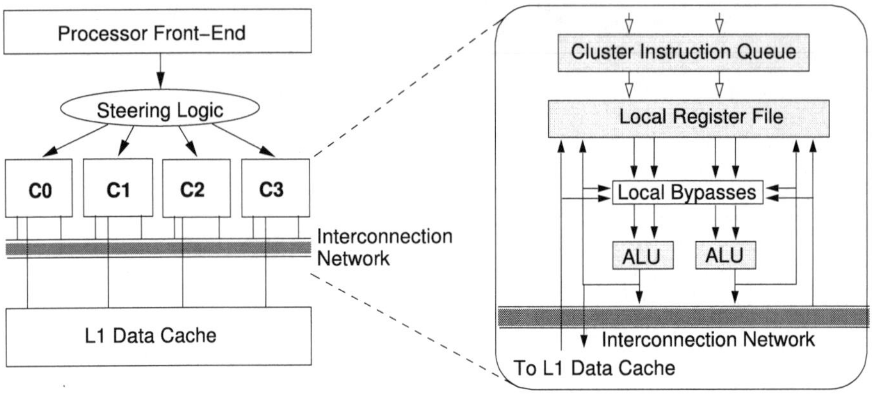

FIGURE 1.9: An example of an on-core network for a clustered microarchitecture.

And there are several other application areas where high-speed interconnects are required.

The different interconnects mentioned above share many common concepts and design issues. However, the particular characteristics and cost of these networks considerably depend on the application area. There are no general solutions. For example, networks that transmit information to distant destinations usually drop packets whenever buffers become full or congestion arises. On the other hand, networks that transmit information across a short distance usually implement flow control to avoid dropping packets and having to retransmit information. This single difference leads to dramatically different solutions for configuring network paths, routing packets, and handling congestion, among others. For some application areas, interconnection networks have been studied in depth for decades. This is the case for telephone networks, computer networks, and backplane interconnects. These networks are covered in many books. However, there are some other application areas that have not been covered as extensively in the existing literature.

An application area that has attracted a lot of interest since multicore processors were accepted as the way to go to keep increasing the computing power without dramatically increasing heat dissipation, is the NoC. Although NoCs were already popular in system-on-chip (SoC) designs, the trend there is to customize the NoC architecture for specific application domains to match the tight resource budgets. The emphasis in SoCs has been more on designing standard on-chip interconnects, like the advanced microcontroller bus architecture advanced extensible interface (AMBA AXI) or the open core protocol (OCP) protocols, able to link intellectual property (IP) blocks from different designers, rather than on delivering top performance.

However, the situation is different for multicore processors. The need to interconnect an increasing number of cores, cache blocks, and memory controllers, with low latency and high bandwidth regardless of the number of cores, has made NoC design a very challenging task.

Initial NoC designs and academic proposals took solutions from the interconnection network designs proposed in the '80s for multicomputers, the message-passing multiprocessors of those days. Those designs were taken as a reference because the most critical design constraints are very similar in both application areas. In particular, in both cases there is a need to implement compact and fast routers that deliver packets with low latency. In the case for multicomputers, the goal was to integrate each router into a single chip, also implementing the routing algorithm in hardware. These constraints led to the invention of flit-level flow control and wormhole switching, two techniques that when combined together allowed the implementation of very compact routers that required very small buffers, thus fitting into a single chip. Even more important, wormhole switching combined with a hardwired routing unit allowed packets to be routed without having to store them in the main memory at intermediate nodes, thus enabling the implementation of network links with bandwidth similar to memory bandwidth, instead of being a small fraction of it, as in packet switching.

Additionally, the need to implement NoC wire layouts in a flat piece of silicon led to the selection of two-dimensional topologies, the 2-D mesh being the preferred one due to its perfect fit with a 2-D layout. For a different reason, low-dimensional topologies were also the preferred ones in the '80s, evolving over the years from a 2-D mesh to a 3-D torus. The most popular routing algorithm has been dimension-order routing (DOR), due to its simplicity and guaranteed absence of deadlock. So, practically all initial NoC designs are based on a 2-D mesh topology, and compact routers implementing flit-level flow control, wormhole switching, and a hardwired DOR algorithm.

Although the lack of standards for high-performance NoCs has enabled the quick adoption of solutions previously developed for a different application area, distributed shared-memory multiprocessors—the architecture of choice for multicore processors—require even higher network bandwidth and lower latency than multicomputers, thus pushing the development of novel solutions specifically devised for NoCs. Even more important, the continuously evolving VLSI technology is starting to approach its own limits. This manifests in an increasing set of limitations and side effects that need to be addressed. Among them, it is worth citing the manufacturing defects, which affect an increasingly large transistor and wire count as integration scale increases, and manufacturing variability, which is going to increase to the point where clocking all the circuits at the frequency dictated by the slowest device is going to deliver very poor performance.

The above-mentioned limitations have triggered new research projects that attempt to develop innovative logic designs and architectural solutions to address them and reduce their impact to the very minimum. In particular, different fault tolerance techniques have been proposed to address manufacturing defects, and several strategies have been developed to clock different chip areas at different clock frequencies and/or disable the slowest components. Most, if not all, of those techniques and strategies introduce irregularities in an otherwise regular and homogeneous design.

Moreover, when related design constraints and market trends are analyzed, it turns out that there is a quite large set of sources for heterogeneity, which require carefully designed solutions to address them efficiently, or at least take them into account. Therefore, there is a need for structuring the various concepts related to heterogeneity in NoCs as well as the solutions that have been devised to address it and deliver power-efficient, cost-effective designs. In this book, we take on this challenge and present, in a structured way, the basic underlying concepts as well as the most interesting solutions currently implemented or proposed in the literature to address heterogeneity in NoCs.

1.2 Basic Network Architecture

The architecture of the network can be significantly different depending on the number of devices that need to communicate and the design constraints. Communication between two devices can be easily achieved by using a point-to-point link. When more than two devices have to exchange information, there are basically two options: i) using a shared transmission medium linking all the devices, or ii) using a switch to dynamically establish connections between pairs of devices upon request. The former approach is simpler and has been traditionally used to interconnect a small number of devices (e.g., the typical buses found in every computer system), but it has some limitations that will become more significant as technology advances. On the other hand, switches are more complex but provide a much more effective way to interconnect several devices, usually being preferred when designing high-speed interconnects.

When the number of devices to be connected exceeds a certain value that depends on the VLSI technology available at a given time, it becomes infeasible to interconnect all of them using a single switch. In this case, it is possible to design networks with multiple switches. These networks are usually based on point-to-point links to connect switches and devices to each other, although a few proposals exist based on shared-medium links, or even on special links with one source and several destination devices. In networks with multiple switches, messages sent by devices attached to the network may require crossing several switches before reaching their destination.[2]

Finally, the communication services provided by the switches may not be

[2]The analysis of the three cases mentioned above (communication between two devices, communication through a single switch, and communication through a network of switches) roughly corresponds to the physical layer, switching layer, and routing layer, respectively, as defined in Chapter 2 of [85].

the appropriate ones for the rest of the system. In this case, devices can be attached to the switch fabric through a network interface, which provides the required services. The complexity of network interfaces varies significantly from one system to another. All of the above interconnect options are briefly analyzed in the next sections, which also introduce the main services provided by network interfaces. The rest of the book is mostly devoted to analyze these issues in more detail.

1.2.1 Direct Communication between Two Devices

Direct communication between two devices is usually implemented by using a point-to-point link. Communicating devices usually have independent clocks and data transmission between those devices is usually performed asynchronously. This implies that some protocol must be implemented among communicating devices to make sure that communication is correctly performed. In general, this protocol allows the transmitter to notify data transmission, also allowing the receiver to notify correct reception and buffer availability.

In order to achieve high performance, link bandwidth should be as high as possible. Link bandwidth is the product of the link clock frequency times the link width. Advances in VLSI technology allow faster clock rates. However, data do not propagate instantaneously through a link. Therefore, in order to use the maximum link clock frequency allowed by a given technology, it may be necessary to inject new data into a link before previously injected data reach the other end of the link. This is usually referred to as *channel pipelining*. In this case, separate electrical signals propagate through the wires (or light propagates through fiber) like waves, without merging with each other, although there is some signal attenuation that limits the applicability of this strategy. Depending on the clock frequency and the link length, a few bits to several packets can be on the same link at a given time. When channel pipelining is used, links should be *full-duplex*, thus allowing simultaneous transmission in both directions. A *half-duplex* connection would be very inefficient since it would be necessary to empty the link before changing the direction of data transfer. Channel pipelining has been traditionally used in computer networks for long-distance communication.

Assuming that the transmitter stores data to be transmitted into a single physical or logical FIFO (first in, first out) buffer queue and that the receiver also stores received data into a FIFO queue, it is necessary to make sure that the capacity of the reception queue is not exceeded. This can be achieved by implementing a *link-level flow control* protocol, which takes care of notifying buffer availability at the receiver side to the transmitter. A *flow control unit* is the amount of data on which flow control is applied. Buffer space is measured in flow control units.

Data are buffered at the reception side before being processed. There are several issues related to data buffering, generically referred to as *buffer management*. First, flow control information regarding the availability of additional buffer space can be returned to the transmitter when a flow control unit has been completely consumed/forwarded, or as soon as it begins to be consumed/forwarded. The latter option reduces delay, but it is more complex to implement. Second, the current trend consists of implementing queues in random access memory (RAM), usually static RAM (SRAM). Therefore, some control circuits are required to properly allocate memory for the queue and keep the list of memory locations forming it. The flexibility provided by implementing queues in RAM allows other policies different from FIFO (like, for instance, dynamically allocated multiqueues (DAMQs)).

In many cases, several flows of information need to be multiplexed onto a single link. Assuming that each flow is stored into a different physical or logical FIFO queue both at the transmitter and receiver sides, it is necessary to extend the flow control protocol to properly identify each flow, store its data into the corresponding buffer queue, and properly implement flow control in each queue. In this case, buffer management becomes more complex. Queue selection at the receiver side can be based on explicit information provided by the transmitter (usually identifying the queue from which data have been transmitted), dynamically computed at the receiver side with the goal of either optimizing the use of resources or reducing queue waiting time, or based on information provided by the message being received (usually indicating the destination of that message). Also, in the case of implementing queues in RAM memory, control circuits have to properly allocate memory for the different queues and keep the list of memory locations forming each queue.

Moreover, when several flows of information are multiplexed onto a single link, a *transmission scheduler* is required at the transmission side to properly select the flow for which a flow control unit will be transmitted next. This scheduler should guarantee a fair use of resources, thus preventing *starvation* of any flow. Also, a *reception scheduler* may be required at the receiver side to schedule the return of flow control information if several flow control units from different queues can be consumed/forwarded at the same time.

The transmission scheduler may or may not implement *priorities* when selecting a flow control unit from a given flow for the next transmission. Moreover, some flows of information require a guaranteed bandwidth and/or a guaranteed bound on the delay experienced by each flow control unit or groups of them. These guarantees are usually referred to as *quality of service*, or *QoS*. In this case, the transmission scheduler becomes much more complex, since it has to provide QoS guarantees to some flows and, at the same time, it has to prevent starvation of the remaining flows. Obviously, it would be impossible to guarantee QoS if the total bandwidth demanded by flows requiring QoS guarantees exceeded link bandwidth. An *admission control* protocol is usually required to prevent bandwidth oversubscription.

Figure 1.10 shows a logical block diagram including most of the elements described above. The link controllers implement flow control. A possible implementation using RAM memory for the buffer queues is shown in Figure 1.11. The buffer managers contain lists of pointers to keep track of the mem-

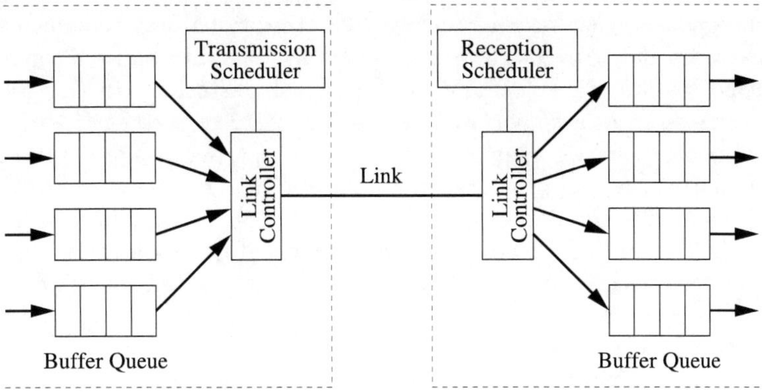

FIGURE 1.10: Components required for data transmission over a point-to-point link.

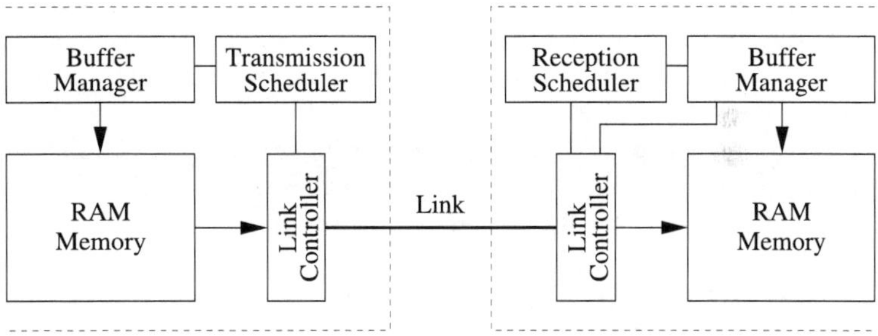

FIGURE 1.11: Implementation of buffer queues using RAM memory.

ory locations associated with each buffer queue, also maintaining a list of free memory locations. Buffer managers also generate memory addresses to store incoming messages.

1.2.2 Communication through a Single Bus

The simplest way to establish communication among several devices consists of using a bus. Besides its simplicity, a bus features two important characteristics. First, it guarantees in-order delivery of all the messages transmitted through it. Second, a bus allows each device to broadcast information to all the devices attached to it without any extra delay with respect to transmitting information to a single destination device.

However, despite its advantages, a bus is not suitable for high-performance communication. The main reason for this is that bus bandwidth must be shared among all the devices communicating through that bus. When the number of devices attached to a bus increases, it becomes a bottleneck since bus bandwidth does not increase linearly with the number of attached devices.

As a consequence, buses are not scalable. The traditional way to enhance performance when buses are used to access memory (e.g., in small-scale symmetric multiprocessors (SMPs)) consists of splitting bus cycles. In this scheme, usually referred to as a *split-transaction bus,* instead of keeping the bus busy during a memory access, memory requests (reads or writes) are transmitted through the bus, releasing it immediately after the transmission. When the memory completes the requested operation, another message is transmitted through the bus, returning either the requested data or an acknowledgment. In general, this scheme consists of using the bus just to transmit data, releasing it immediately after each transmission, and thus minimizing bandwidth consumption for each transmission. However, this optimization does not solve the scalability problem, and therefore it is not enough when the number of devices attached to the bus is large.

Besides limited bandwidth, buses have some additional undesirable features that translate into higher message latency and power consumption. First, signal propagation time is slower than in point-to-point links because devices attached to the bus add capacitance to the bus wires. Second, this additional capacitance increases power consumption with respect to point-to-point links, even if only one device needs to receive the transmitted data. Third, bus wires must reach all the devices attached to it, usually being much longer than point-to-point links in a switch-based system of the same size. This is important since longer wires imply longer transmission latency. Fourth, arbitration among transmission requests is required before data transmission in order to guarantee that a single device is driving the bus at a given time. Even if arbitration is overlapped with the transmission of data from the previous winner, arbitration adds delay for each particular data transmission.

Additionally, buses have reliability problems. In particular, if a device fails in such a way that one or more bus signals are stuck at a certain voltage, the remaining devices may not be able to communicate through the bus. In summary, buses offer limited bandwidth, low reliability, and relatively high latency and power consumption. Therefore, in general, buses are not a good candidate for high-speed interconnects. Because of this, they will not be covered in this book.

1.2.3 Communication through a Single Switch

When a bus does not provide enough bandwidth to communicate between a set of devices, higher performance can be achieved by interconnecting those devices through a switch. A *switch* is a digital circuit designed to interconnect several devices. A switch usually implements a set of input and output *ports* connected to external links, a set of buffers to temporarily store flow control units (or *flits*, for short), an *internal switch*,[3] and some control circuits provid-

[3]Using the words "switch" and "internal switch" to refer to different concepts may produce some confusion. Unfortunately, there is no agreement on the name given to some

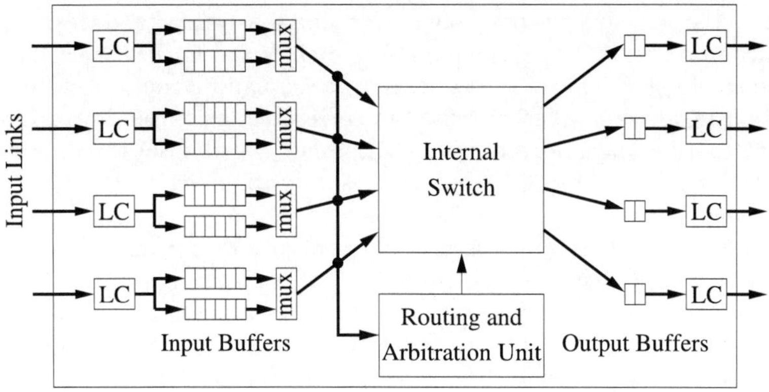

FIGURE 1.12: Basic switch architecture.

ing some functionality that will be described below. The internal organization of a switch is usually referred to as *switch architecture.* Figure 1.12 shows a block diagram for a basic switch architecture. In this diagram, several input buffer queues are associated with each input port. Certainly, this is neither the only option nor the most efficient one. However, this switch architecture is quite popular and will be useful in discussing the relevant switch design issues. An overview of basic switch design issues is given below.

The basic switch behavior is as follows. When a flow control unit[4] is received at some input port, the buffer manager determines the buffer in which it should be stored. Meanwhile, the link controller performs the actions associated with flow control. In order to forward each flow control unit, it is necessary to establish a connection through the internal switch to the appropriate output port. However, this cannot be done without having information on the requested output port. One or more flow control units are usually grouped into larger units, usually referred to as *packets.* Packets contain a header with destination information. That header is sent from the corresponding input buffer to the routing and arbitration unit, which computes the appropriate output port for that packet and requests that port. In case of simultaneous

devices. In particular, devices used in interconnects for parallel computers are usually referred to as "routers", and devices with similar architecture and functionality are referred to as "switches" in the computer communications community. Additionally, there is the problem that there are three levels in the hierarchy, and only two words to distinguish between different components. For instance, an InfiniBand^{TM} network may consist of several subnets interconnected by routers. Additionally, each subnet may consist of several switches interconnecting devices attached to the subnet. Finally, each switch may implement an internal switch to establish connections between input and output links. We will use this terminology since it is the most extensively used practice.

[4]We prefer to use the term "flow control unit" instead of "packet" because it is more general. A packet is a flow control unit only for certain switching techniques. On the other hand, we do not use the term "flit," which is correct in this context, because people usually associate flits with a particular switching technique (i.e., wormhole).

requests to the same output port, only one request is granted and the routing and arbitration unit instructs the internal switch to set up the corresponding path. Then, the flow control units for that packet can be transmitted to the output buffers, and from there through the corresponding output link. In summary, the routing and arbitration unit dynamically establishes paths across the internal switch according to the requests from incoming packets, possibly selecting among contending requests.

The *switching technique* determines when the internal switch is set and the time at which packet components (usually, flow control units) can be transferred along the corresponding path. Switching techniques are coupled with flow control for the synchronized transfer of flow control units through the input links, the internal switch, and the output links in forwarding packets from source to destination. Also, as mentioned in Section 1.2.1, flow control is tightly coupled with buffer management, which determines how switch buffers are granted and released, and as a result determine how packets are handled when blocked due to contention. Different switching techniques differ in decisions made in each of these areas, and in their relative timing, i.e., when one operation can be initiated relative to the occurrence of the other.

Besides deciding when a path through the internal switch will be set for a given packet, it is also necessary to decide which path will be set. This operation is known as *routing*, and is performed by the routing and arbitration unit. The simplest way to implement it in a single switch consists of directly encoding the required output port in the packet header. This issue becomes important in networks with multiple switches, and thus, it will be analyzed in the next section.

As mentioned above, several packets may request the same output port at the same time, thus producing *contention*. Two mechanisms are required to handle contention. First, when contention occurs, an *arbiter*[5] is required to fairly select among contending requests. The winner will be granted the requested port and the loser will have to wait, thus being blocked. Second, buffer management together with flow control are needed to guarantee that no piece of the blocked packet is lost. The blocked packet is buffered and flow control prevents buffer overflow. Contention may have very bad consequences. It not only delays packets losing arbitration, but it also causes those blocked packets to, in turn, block other packets, thus producing a cumulative effect. This is especially serious when some blocked packets in a FIFO queue block subsequent packets in the same queue for which their requested output

[5]Again, there is no agreement on the name given to this component. It is usually referred to as an "arbiter" in the area of interconnects for parallel computers, and as a "scheduler" in the area of computer communications. Indeed, there is a subtle difference between both concepts. An arbiter simply selects a winner among contending requests. Therefore, it follows a greedy strategy with respect to resource assignment. A scheduler delivers transmission schedulings with respect to some goal. It may leave a resource free for a while even if there are requests for it (e.g., when the goal is minimizing some QoS metric, like jitter). We will try to use the appropriate word whenever possible. However, most people do not make any difference between both concepts.

port is free. This situation is referred to as *head-of-line (HOL) blocking.* In this case, link bandwidth is wasted, and may lead to significant performance degradation.

As in the case for data transmission over a single link, different information flows may have different priorities. Some of them may even require QoS guarantees. The solution, again, follows the same guidelines. Some admission control protocol must be used to prevent bandwidth oversubscription, thus guaranteeing that there will be enough bandwidth along the entire path for all the information flows requiring QoS guarantees. Additionally, schedulers are required at each contention point to make sure that every packet receives the appropriate treatment according to its QoS requirements.

1.2.4 Networks with Multiple Switches

When the number of devices to be connected exceeds a certain value that depends on the VLSI technology available at a given time, it becomes infeasible to interconnect all of them using a single switch. For this case, it is possible to design networks with multiple switches. These networks are usually based on point-to-point links to connect switches and devices to each other. In networks with multiple switches, it is at least necessary to specify the interconnection pattern between switches and how to select a path from the source device to the destination device. However, it may not be as simple to do this as it seems. As interconnects grow in size, some problems become more critical and need to be addressed (e.g., performance and reliability). Also, some additional services, like QoS guarantees, become more complex to implement efficiently. An overview of the main design issues related to networks with multiple switches is presented in this section.

The interconnection pattern between switches defines the *topology* of the network. Components can be arranged in several ways. One way is for a single device to be attached to each switch, which is interconnected to other switches following some (usually regular) connection pattern. This arrangement is usually known as a *direct network* because the one-to-one correspondence between devices and switches allows the integration of the switch into the same board or chip as the device, and thus, those boards or chips are directly interconnected between them. Another way consists of arranging switches in several stages, each one connected to the next stage following some regular interconnection pattern, and attaching devices only to the first and last switch stages. This arrangement is usually known as a *multistage interconnection network.* In general, networks where only a subset of switches have devices attached to them are referred to as *indirect networks* because switches are usually arranged into external cabinets, thus allowing the interconnection of devices indirectly through a set of switches, usually referred to as *switch fabric.* In both cases, but especially in the case of indirect networks, the interconnection pattern between switches may not follow any regular pattern. Such networks are usually referred to as *irregular networks.* In general, we can view networks using switched point-to-point links as a set of interconnected switches, each one being connected to zero, one, or more devices. Direct networks correspond to the case where every switch is connected to a single device. *Crossbar* networks correspond to the case where there is a single switch connected to all the devices. Multistage interconnection networks correspond to the case where switches are arranged into several stages and the switches in intermediate stages are not connected to any device. Irregular networks correspond to the general case.

The interconnection of several switches raises new problems. First, a suitable topology for the network must be selected. Sometimes, the topology is decided at design time. In some other cases, the goal is to provide enough flexibility to the end user to configure the desired topology. In some application areas (e.g., supercomputers), it is also important to determine how to physically arrange all the network components into chips, boards, and cabinets, so that links are as short as possible and the number of links crossing each chip/board/cabinet boundary is as small as possible. This arrangement, usually known as *component layout*, is constrained by many factors, usually including chip/connector pin count, board size, thermal dissipation, etc. Selecting an appropriate network topology is crucial to obtain a good component layout.

In networks with more than one switch, packets usually have to cross several switches before reaching their destination. Therefore, it is necessary to establish a path for each packet so that it can reach its destination as fast as possible. The algorithm to compute such a path is referred to as *routing algorithm.* This algorithm can be computed in a distributed way (step by step at the routing and arbitration units in the switches along the path traveled by a packet), in a centralized way for each packet (at the source device or its associated network interface), or in a centralized way for all the network (at a network manager). When the algorithm is not computed in a distributed way, *routing (forwarding) table* (also referred to as *forwarding tables*) are usually used at every switch and/or network interface to store the precomputed paths. Regardless of where the routing algorithm is computed, every switch must contain the circuits to compute or the memory to store enough information to route every incoming packet. Packets must also store destination information in their header. In some networks, the packet header encodes the entire path for that packet. In this case, the switch just has to extract path information from the packet header and use it to set the internal path accordingly.

Two problems associated with routing are deadlock and livelock. A *deadlock* occurs when some packets cannot advance toward their destination because the buffers requested by them are permanently full. All the packets involved in a deadlocked configuration are blocked forever, each one requesting buffers occupied by other packets and occupying buffers requested by other packets. A *livelock* occurs when some packets are routed forever, never reaching their destination. In this case, a packet travels around its destination node, never reaching it because the channels required to do so are occupied by other packets. Deadlock and livelock can be avoided by carefully designing the routing algorithm. There exist well-known techniques for doing so. This always implies imposing some routing constraints (restrictions), so that certain paths cannot be used to reach certain destinations. These routing constraints may reduce performance. Alternatively, deadlocks can be detected and recovered from when they occur. However, this strategy also introduces some overhead that may affect performance.

In general, there exist multiple paths from each source device to each destination device in a network with multiple switches.[6] However, the routing algorithm may provide a single path from each source to each destination (referred to as *deterministic routing*). This strategy usually leads to faster routing but contention is higher and the switch fabric usually saturates at a lower average load. Altenatively, those paths can be used either to avoid congested regions or to provide service in the presence of failures. *Load balancing* strategies aim at distributing traffic among alternative paths so that link utilization is balanced (i.e., similar for different links), thus avoiding or delaying the occurrence of *hot-spots* in the network. A particular (and perhaps the most effective) way to implement load balancing consists of gathering information about traffic in the network and selecting the path for each packet according to resource availability, thus minimizing the occurrence of hot-spots. This strategy is known as *adaptive routing*, and most of the times uses only local information about resource utilization, thus minimizing overhead. The main problem introduced by adaptive routing is *out-of-order* (OOO) delivery of packets, which is not acceptable in some application areas.

Sometimes, the same information must be delivered to several destinations. This is known as the *multicast routing* problem. In some other cases, several devices must send some information to the same device, which has to combine all the received messages. This communication operation is referred to as *reduction.* These communication primitives are generically referred to as *collective communication operations.* Besides the two communication operations mentioned above, there exist many other collective communication operations. Some of them are particular cases. For example, a particular case of the multicast routing problem is *broadcast*, in which the same information must be delivered to all the devices attached to the network. Some other operations are combinations of simpler operations. For instance, a frequently used communication operation involving several devices is the *barrier*, and is used to synchronize those devices. It consists of a reduction (in which every participating device notifies that it reached the barrier to a single coordinating device) followed by a broadcast to all the participating devices (notifying that the barrier has been reached).

The multicast routing problem can be solved by sending independent copies of the same packets to every destination. Although this solution works, it introduces a significant overhead in the network. In most cases, it is pos-

[6]There are exceptions for this (e.g., a network with a tree topology).

sible to arrange communication in such a way that some packet destinations collaborate in the distribution of packet copies by forwarding those packets to other destinations, which in turn may also forward copies to other destinations until all the destinations are reached. This strategy, known as *software-based multicast routing*, requires devices receiving copies of the packets to be able to forward them and improves performance significantly. When this is not possible or higher performance is required, it is possible to implement hardware support for multicast routing within the switches. Besides providing simultaneous paths from each input port to several output ports in the internal switch, this strategy requires an appropriate encoding of all the packet destinations in the packet header. Multicast routing also introduces new potential deadlock scenarios.

As in the case for multicast, reduction can also be implemented by sending independent packets from every source device to the common destination. Again, this introduces a significant overhead in the network, possibly creating a hot-spot. It is also possible to arrange communication in such a way that some source devices send packets to other source devices, which in turn combine the received packets with their own information and transmit a new packet either to another source device or to the final destination. This procedure, known as *software-based reduction*, continues until the destination device has received packets containing information from all the source devices. Finally, when the latter strategy is not possible or higher performance is required, it is possible to implement hardware support for reduction within the switches. However, as reduction implies waiting for several packets and combining them, large buffers as well as some specific logic may be required. Therefore, when reduction is implemented in hardware, it is usually constrained to small packets and simple combining operations.

Some problems are not specific to switch fabrics with multiple switches but become more critical and difficult to solve as the size of the fabric increases. One such a problem is contention. Frequent contention produces congestion, and this congestion may propagate through the switch fabric, forming a *congestion tree* that propagates from the congested area to the source devices that are transmitting packets toward (or through) that area. This is due to packets blocking when they request buffer resources occupied by other blocked packets. This situation only occurs when the switch fabric or some fraction of it saturates. Therefore, it will only represent a serious problem if the average load injected into the switch fabric usually represents a large percentage of its capacity. In such a case, congestion trees may significantly degrade network performance and some solution is required. The traditional solution to this problem consists of detecting congestion, notifying that situation to the source devices, and limiting injection at those devices until congestion vanishes. This strategy is referred to as *congestion control*, and can be implemented in very different ways depending on where congestion is detected, how this information is propagated to the source devices, and how injection is limited at the sources.

Another related problem that is also aggravated when the switch fabric becomes larger is providing QoS support. Effectively, when the network size increases, paths through the switch fabric must be shared by a larger number of packet flows. Although there exist many more alternative paths in the switch fabric, their use may be restricted by the routing algorithm in order to avoid deadlock. Therefore, in networks with multiple switches, it becomes even more necessary to use some admission control protocol to make sure that there is enough bandwidth available for every accepted connection along the entire path it will use.

Reliability may also become more critical when the switch fabric becomes larger. Effectively, the probability of a component failure increases with the number of components. Therefore, interconnects should be designed in such a way that a single component failure does not bring the entire network down. The traditional approaches to address the reliability problem consist of either implementing redundant components (in some commercial products, the entire switch fabric is replicated) that are switched in when a failure is detected, or implementing a *fault-tolerant routing algorithm*, which is able to route packets toward their destination in the presence of a few failures in the interconnect (provided that the destination device did not fail). Fault-tolerant routing is implemented by taking advantage of the alternative paths provided by the topology. This strategy is usually cheaper than replicating network components but, obviously, performance degrades when some switches and/or links fail. Fault-tolerant routing may introduce new deadlock scenarios. In all the cases, efficient and reliable *fault detection* techniques are required.

More recently, another approach has been proposed to address network failures. Some application areas require providing enough flexibility so that the end user could define the required network topology, and even modify it dynamically (hot addition and replacement of network components and devices). This flexibility requires some *network configuration protocol* to discover the topology of the switch fabric, compute the routing tables, and download them to the switches and/or network interfaces. However, there is no significant difference between the failure of a network component and the hot removal of the same component. Therefore, the same network configuration protocols can be used to implement *network reconfiguration* in case of failure. This reconfiguration can be done either by first stopping all the traffic through the network (*static reconfiguration*) or while packets continue being routed (*dynamic reconfiguration*). The main problem that arises in the latter case is that old and new routing tables coexist at different switches and/or network interfaces during routing table update, usually introducing new deadlock scenarios.

Finally, as mentioned above, some application areas require providing enough flexibility so that the end user could define the required network topology. This flexibility is usually provided by using configurable switches and/or network interfaces (e.g., switches with programmable routing tables to define the routing algorithm, programmable arbitration tables to define the behavior of link schedulers providing QoS guarantees, etc.) and some generic solutions (e.g, a generic routing algorithm for which it is possible to compute routing tables for every topology). However, the use of configurable network components implies that some initialization phase is required before the network is fully operational. Among other things, this initialization phase must implement the network configuration protocol. Some maintenance operations may also be required. All these operations are usually referred to as *network management* and can be performed by a dedicated *network manager*, or implemented in a distributed way across network interfaces and/or switches.

1.2.5 Services Provided by the Network Interface

The *network interface* is the component that establishes the connection between devices attached to the network to the switch fabric itself. Network interfaces significantly vary from one application to another. In some cases, it simply consists of a pair of queues (or even a pair of registers), which store messages to be transmitted and received messages, respectively. In some other cases, it is a sophisticated component containing several general-purpose and specialized processors, several tens of megabytes of RAM memory, several direct memory access (DMA) devices, interfaces to some standard system bus and network links, and more. Therefore, the functionality and services provided by the network interface drastically vary from one system to another. This section briefly describes the main functions that are implemented in some systems. This functionality can be classified into four groups:

- Support for data transmission and reception. This is the most basic functionality and it is implemented in almost every network interface. At least a pair of queues are required to store messages to be transmitted and messages that have been received, respectively. However, some network interfaces provide support for many queues, usually implemented in RAM memory. This is the case for IP routers, where a different queue is usually used to store messages destined to different output ports, thus reducing HOL blocking. Also, this is the case in most network interface cards that are typically used to attach a computing node to a cluster. Besides support for message buffering, network interfaces may also implement one or more DMA devices, thus being able to read data to be transmitted and write received data from/to some external memory (e.g, the main memory of computing nodes attached to some interconnect to form a cluster). In case of direct networks, the network interface is usually integrated in every switch. However, network interfaces are usually connected to the switch fabric through a link in an indirect network. In this case, link-level flow control should be implemented in the network interface. Network interfaces may also provide support for splitting messages into fixed-size packets.
- Support for services implemented in the switch fabric. The network inter-

face usually cooperates with the switch fabric to provide communication services. It can help to implement routing, load balancing, reliable transmission, QoS, network configuration, etc. The distribution of functions between the network interface and the switch fabric drastically varies from one system to another. For example, routing for a fixed topology can be implemented without network interface support by using a distributed algorithm implemented in switch hardware. On the other extreme, the entire packet path can be computed (or read from a routing table) at the network interface, including it in the packet header. In this case, each switch simply reads and removes from the header the next piece of the path, and selects the indicated output port. Hybrid approaches are also possible. For instance, one or more network interfaces compute the paths during network initialization and download those paths to the routing tables implemented in the switches.

Network interfaces may also implement load balancing techniques. This can be done by distributing packets for a given destination among alternative paths existing in the switch fabric. This distribution may or may not take into account congestion in those paths.

Also, network interfaces usually help to improve transmission reliability by implementing some end-to-end protocol. These protocols traditionally work by returning some acknowledgment every time a packet is correctly received, taking care of retransmitting a packet when a negative acknowledgment is returned or when no acknowledgment is returned after a certain timeout. Correct reception is usually checked by computing and adding a cyclic redundancy check (CRC) to each packet at the source, and checking it at the destination. Network interfaces may implement CRC computation and checking in hardware for faster operation.

When QoS is implemented in the interconnect, the network interface usually plays a fundamental role. Network interfaces usually implement admission control protocols, either in a centralized or distributed way. Network interfaces should also schedule packet transmission, taking care of guaranteeing that no connection will consume more bandwidth than allocated to it.

Finally, network interfaces may implement all (or most of) the functionality required for network configuration in those application areas where the topology is defined by the end user, and/or may dynamically change. In general, network management functions can be implemented in a centralized or distributed way in the network interfaces.

- Support for additional services. The communication services provided by an interconnect can be improved by adding support for different programming models (e.g., message-passing and remote writes), implementing some operations in the processors located in the network interface, etc. For example, in addition to traditional message transmission, the network interface may signal an interrupt when a message is received, may provide support for remote write operations by specifying in the packet header the address of a memory location in the destination network interface, or may execute a message handler whose address is contained in the packet header upon packet reception. The network interface may also execute in one of its processors part of the functions required to implement some software-based collective communication operations.

- Interface to standard system buses. Network interfaces also act as bridges between network links and some standard system buses, where I/O devices are usually plugged into. Functionality varies from one system to another, but in general, support is provided for programmed and DMA data transfers, for interrupts, for mapping network interface memory, and for downloading code to be executed in the network interface processor(s).

1.3 Heterogeneity

Most theoretical studies of interconnection networks assume homogeneous systems with regular topologies. Homogeneity simplifies the design of topologies, routing and load balancing strategies. It also allows the use of a single switch design as a building block for implementing larger switch fabrics. As the number of cores per chip increases, buses are becoming a bottleneck, and researchers started to consider switched networks with point-to-point links in order to make communication bandwidth more scalable. In doing so, the initial proposals for networks on chip (NoCs) inherited many properties of the interconnection networks that were proposed in the '80s. The reason for this is that they share a common goal: minimizing packet latency while devoting the minimal amount of resources to the network. In the case for the interconnection networks developed in the '80s, the goal was to implement single-chip routers. In the case for NoCs, a goal is to minimize silicon area requirements.

Therefore, it is not surprising that the initial proposals for NoCs are based on wormhole switching, two-dimensional meshes, and dimension-order routing (DOR). Wormhole switching delivers low latency with very small buffers, and hence, with small silicon area and power consumption. DOR is easy to implement with a finite-state machine, and leads to very compact and fast routers. Finally, 2-D meshes not only were considered the optimal topology for wormhole networks in the '80s, they also match the 2-D layout of current chips, enabling the lowest communication latency among neighbors and minimizing wiring complexity. Interestingly enough, some theoretical studies in the '80s concluded that the 2-D mesh is the optimal topology under the assumption of constant bisection bandwidth but such a constraint never became true in practice. However, it may become true for NoCs, therefore emphasizing the use of 2-D topologies.

As a consequence, most of the work presented in this book lies around a 2-D mesh topology, since it is the current trend both in academia and industry. In Chapter 4 we review the set of suitable topologies for NoCs where different alternatives evolve from a 2-D mesh. In Chapter 5 we analyze the implementation of routing algorithms for 2-D mesh topologies and their variations. Examples from real products and test chips also lay around the 2-D mesh topology. So, we present some examples like Tilera's chips (Chapter 7), the latest Intel test chips (Chapter 8), and the TRIPS Operand Network (Chapter 9).

1.4 Sources of Heterogeneity

Although it is widely assumed that a 2-D mesh topology (and its derivatives) is a good choice for NoCs in CMP systems, we need to be aware of emerging challenges that may break the homogeneity such topologies offer. There are many sources that may affect the final topology and the NoC system must be designed to deal with them. Those sources may break the homogeneity, thus ending up in a NoC that is no longer regular but irregular (thus, being heterogeneous).[7] In the following sections we identify potential issues that may end up rendering the NoC heterogenous.

1.4.1 Architectural Sources

Interconnects for current chips have significantly more complex requirements than just minimizing latency and power consumption. First of all, cores need to be connected to the external components (e.g., DRAM modules), and most current processors already integrate one or more on-chip memory controllers. Even if the number of memory controllers is increased as the number of cores increases, there will always be a smaller amount of memory controllers than processing cores or cache blocks. As an example, current graphics processing units (GPUs) integrate several hundred cores and less than ten memory controllers. This introduces heterogeneity in the topology of the interconnection network, which must properly interconnect those devices. Moreover, it also introduces asymmetry in the traffic patterns since memory controllers are usually located near the chip edges. Additionally, since memory bandwidth

[7] It has to be stated that in this book we define the term heterogeneity from the point of view of the topology of the network, and not from the point of view of the types of components that are connected through the on-chip network.

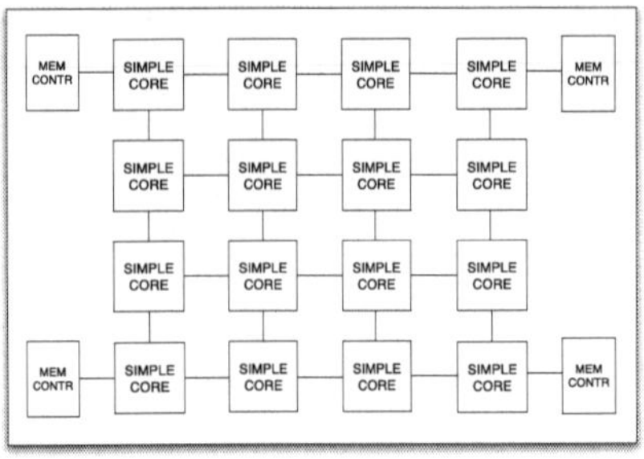

(a) Different number of memory controllers

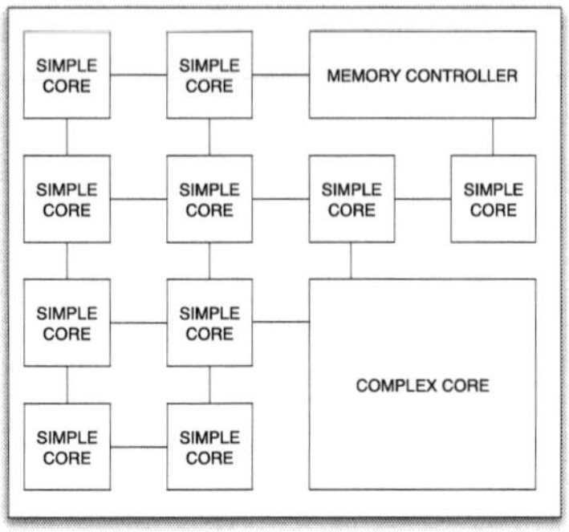

(b) Cores with different sizes

FIGURE 1.13: Different examples of heterogeneity sources.

will become a scarce resource, traffic destined to memory controllers will very likely produce congestion within the NoC, no matter how overdimensioned it is. Figure 1.13(a) shows an example where four memory controllers are laid out in the chip at the corners, thus forcing memory traffic to concentrate.

Another potential source of heterogeneity for the NoC is the fact that the devices attached to it have different functionality. In addition to processing cores, there are also cache blocks, which are either shared by all the cores or by subsets of them. Those cache blocks will likely differ in size and shape from the cores, possibly introducing irregularities in the topology of the NoC. Even if caches and cores have a similar size, they will generate different traffic patterns that may recommend some asymmetry in the bandwidth of the different links, or even the use of separate networks for different purposes. For example, a design may implement a 2-D mesh topology for the transmission of cache lines and a binary tree (or even a fat tree) for the ordered transmission of coherence commands. This kind of solutions have already been implemented in off-chip networks (e.g., Sun E10000), and therefore, are not unexpected for NoCs as well.

Although the number of cores per chip is increasing at a steady rate, many applications are still sequential. Therefore, manufacturers are wondering whether increasing the number of cores per chip for the desktop, laptop, and mobile markets is a good idea. In this situation, the best way of using the increasingly higher number of transistors per chip that will become available is by integrating more functionality into the processor chip. The next large component that will be integrated is the GPU, at least for application areas that do not require top graphics performance (e.g., all the desktop and laptop applications except for gaming and engineering design). Both Intel and AMD announced plans for this kind of integration. As a consequence, they need to figure out how to effectively use those GPUs in the server market so as to reuse CPU designs. The solution to this problem has already been provided by NVIDIA, and consists of using the GPU as an accelerator for numerical applications. High-performance computing (HPC) platforms and datacenters will make good use of those accelerators, not only to drastically increase computing power at a relatively low cost, but also to dramatically enhance the Flops/watt ratio. Again, the implementation of a GPU into the processor chip will introduce a significant amount of heterogeneity, due to both the different size of this component and the very different traffic requirements with respect to general-purpose processing cores.

Another way of increasing the Flops/watt ratio is by replacing a small number of complex out-of-order cores with deep pipelines by a large number of simple in-order cores with shallow pipelines. However, manufacturers did not make that move because sequential applications would run much slower, and therefore, the number of sales for multicore chips would have dropped dramatically. A possible approach for increasing the Flops/watt ratio while being able to run sequential applications very fast consists of implementing a small number of complex cores (e.g., one or two) and a large number of simple cores, possibly with the same instruction set architecture (ISA). This is the approach followed by the Cell processor, although in this case the ISA is different for simple and complex cores, thus making it more difficult to generate code for this processor. Again, this strategy constitutes another source of heterogeneity, not only because of the different size of the different kinds of cores, which will force the use of irregular topologies, but also because of the different traffic requirements. In Figure 1.13(b) we have an example where the chip has cores with different sizes, thus leading to a nonregular mesh structure.

1.4.2 Technology Sources

In addition to the above-mentioned architectural sources of heterogeneity, manufacturing processes will also force designers to adopt solutions that will end up introducing even more heterogeneity. One of the problems is that, as integration scale increases, the number of manufacturing defects (see Figure 1.14(a)) is expected to increase. Therefore, yield will drop dramatically unless future designs incorporate support for fault tolerance. Fortunately, interconnection networks can implement relatively cheap solutions to increase fault tolerance by using the alternative paths provided by the network topology. Such a use, however, introduces asymmetries in the way links can be used, both to avoid deadlocks and also because fault-free regions will have more alternative paths, and therefore, will be less heavily loaded. A detailed thorough analysis and solution is provided in Chapter 5.

Another source of heterogeneity is the expected increase in manufacturing process variability. Up to now, chips are tested to determine the clock frequency at which they can safely run, and therefore, clock frequency is determined by the slowest devices in the chip. This is acceptable because variability is still relatively small. As process variability increases, the former approach becomes less interesting, and researchers are proposing different ways to allow different parts of the chip to run at different speeds. When those techniques

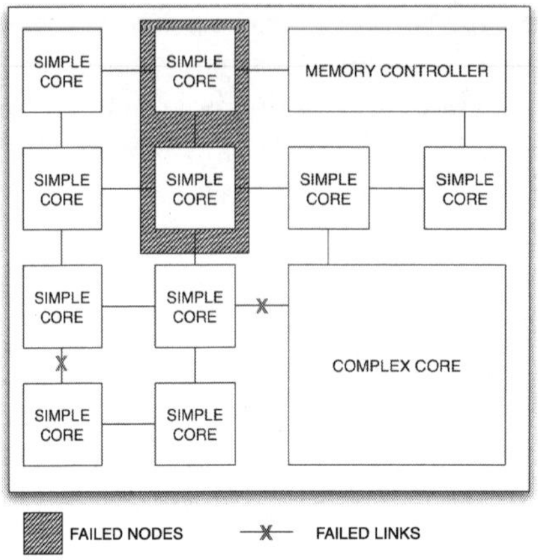

(a) Faulty components due to manufacturing defects

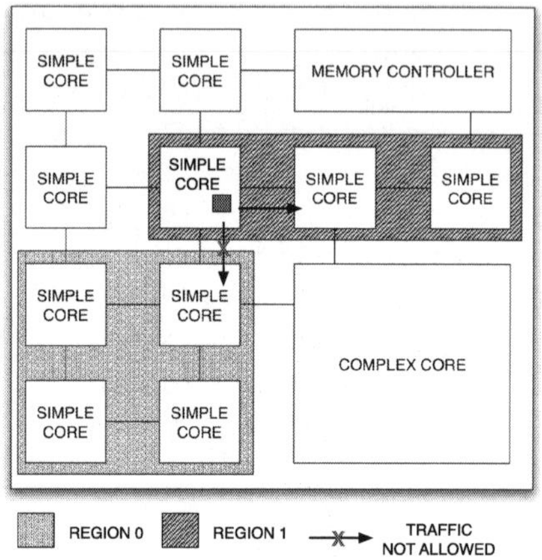

(b) On-chip virtualization

FIGURE 1.14: Different examples of heterogeneity sources.

are applied to a NoC, they result in links and/or switches that require a variable number of clock cycles to transmit information, depending on where they are located. Synchronization is described in Chapter 6 and process variability in Chapter 12.

It is also predicted that VLSI technology will reach a point in which it will be feasible to integrate more transistors as long as they are not all active at the same time. Future processor chips will implement sophisticated temperature controllers able to dynamically adjust clock frequency for independent clock domains, thus introducing functional heterogeneity even in completely homogeneous systems. But this will affect performance guarantees for different virtual machines (see the discussion in the next section). Moreover, pipelined switching techniques like wormhole and virtual cut-through perform quite poorly when some links and/or switches in the path are slower than others, because traffic accumulates at the buffers located before entering the slower regions and may even produce congestion. So, this problem needs to be addressed in order not to waste bandwidth.

The temperature problem will be aggravated with the introduction of 3-D stacking technology. Effectively, 3-D stacking is considered to be the most promising technology to alleviate the memory bandwidth problem of future multicore chips. By stacking multiple DRAM chips together with a multicore chip, it will be possible to drastically reduce the pressure on external memory bandwidth as well as memory access latency. However, those stacked chips will need to dissipate heat, and that heat must go through the already very hot, temperature-throttled multicore chip. In addition to this, 3-D stacking is an important source of heterogeneity. Not only chips in the stack will be different. They will also have very different communication requirements. Moreover, communication with each chip will be significantly faster than communication from chip to chip in the stack, due to the much larger number of wires in the metal layers with respect to the number of vias between chips. 3-D stacking is addressed in Chapter 11.

1.4.3 Usage Model Sources

Finally, the usage model is also a source of heterogeneity. Current market trends, including outsourcing of IT services, have led to the massive adoption of virtualization as a solution to the problem of running applications from different customers in the same computer while guaranteeing security and resource availability. Moreover, in order to optimize utilization, resources are dynamically assigned to different virtual machines according to customer application requirements. As a beneficial side effect, virtualization splits a given computer into smaller chunks, thus reducing the severity of the problem of developing parallel code. Providing efficient virtualization support at the chip level (see Figure 1.14(b)) is not trivial because splitting the set of cores into smaller disjoint subsets is not enough. It is also necessary to guarantee that each partition forms a region whose internal traffic does not interfere with traffic from other regions. This implies that regions should be formed by a contiguous set of nodes and that the routing algorithm is properly defined, which is not trivial because certain paths must be forbidden to avoid deadlocks. Moreover, each region should contain cores and caches in order to be as independent as possible from each other region, which again has some implications on the component layout and introduces heterogeneity at a finer granularity. Anyway, shared caches will introduce some interference among regions. Even if all the cache levels within the chip were private, memory controllers must be shared. This also implies that memory controllers must be an integral part of each and every region, thus introducing even more heterogeneity in the definition of regions. Region definition at the routing level is introduced in Chapter 5.

Finally, there are application areas in which the application to run is fixed (e.g., some embedded devices). In those cases, it is very likely that the application generates nonuniform traffic, sometimes even using only a subset of the links in the NoC. In those cases, efficiently supporting heterogeneity can lead to significant savings in silicon area and power consumption. For instance, an application-specific design may implement only the links and switches that are really needed. Also, different links may provide different bandwidth, according to the application requirements.

Analyzing the heterogeneity problems described above is one of the main goals of the present book. Indeed, most of the chapters will deal in one way or another with the heterogeneity problem.

Chapter 2

Switch Architecture

Giorgos Dimitrakopoulos

University of West Macedonia, Greece

Davide Bertozzi

University of Ferrara, Italy

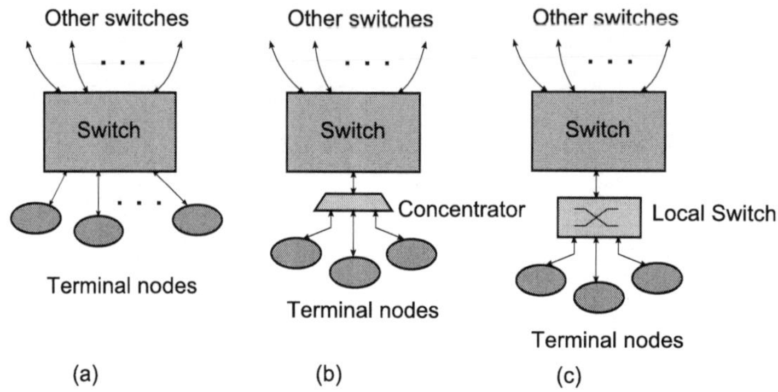

FIGURE 2.1: The connection of the switches to the terminal nodes of the network. (a) By offering them individual input-output ports or sharing a common link of the switch, (b) via a single multiplexer, and (c) by attaching them to an independent local switch.

2.1 Introduction

The switches (or routers) are the basic building blocks of the interconnection network. Their design critically affects the performance of the whole network both in terms of throughput and latency. Switches are connected, through their links, to other switches or terminal nodes. Their function is to make routing decisions and to forward packets that arrive through the incoming links to the proper outgoing links.

Each switch consists of a number of input and output ports that are separated into two classes. It is common to denote the total number of input-output ports of the switch as *switch radix*. The first class of ports is utilized for connecting the switch to the neighbor switches of the network that communicate via direct links, while the second class is dedicated to the terminal nodes [69].

In the straightforward implementation each switch is connected to one or more terminal nodes (see Figure 2.1(a)) via direct bidirectional connections. However, the more the number of connected terminals to the same switch the higher the switch radix and the less the maximum operating speed after physical synthesis [17]. *Concentration* allows the connection of many terminal nodes to the same switch without necessarily increasing the switch radix since local terminal nodes can share a common input/output port of the switch, as shown in Figure 2.1(b). Of course, the multiplexing of the terminal nodes on a common port reduces the available bandwidth that they can utilize.

Taking concentration one step further the local terminal nodes can communicate directly via an additional local switch, as shown in Figure 2.1(c). This local switch would still offer at least one path to the main switch of the

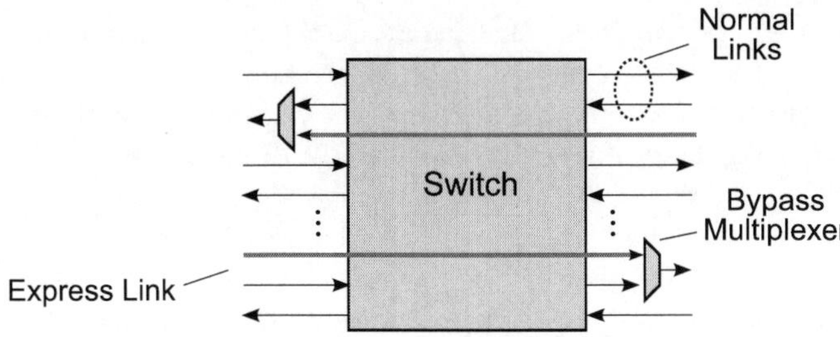

FIGURE 2.2: Connecting express links to switches via bypass paths.

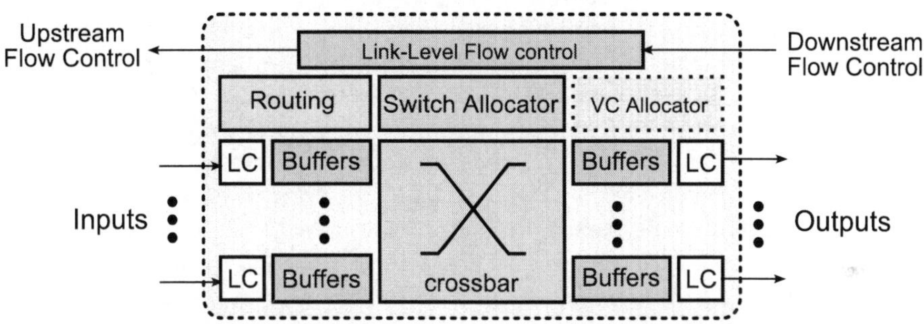

FIGURE 2.3: The main components of a switch.

network similar to the multiplexer of Figure 2.1(b). However, in this case, all traffic between terminal nodes passes through the local switch and does not interfere with global network traffic. In many cases, this local switch can even be a simple local bus that connects this local cluster of nodes.

There are cases where certain links arriving or leaving the switch have a higher priority than the rest. A characteristic example of such cases are the express channels [62]. These express links participate in a preferred network path that offers lower latency than the normal links. Therefore, they don't connect to the switch as normal links or terminal nodes but they enjoy instead a faster bypass path. When express links are activated the operation of the switch is bypassed and a direct connection between bypass input/output links is achieved taking priority over normal traffic. An example of switch bypassing is shown in Figure 2.2. The express links connect directly to the appropriate output in a single cycle via the output multiplexers added outside the switch.

2.1.1 What Are the Main Parts of the Switch?

Switches follow roughly the architecture shown in Figure 2.3. The incoming flits are first received by the link controllers (LC) and stored in input buffers. The flow control logic is responsible for communicating buffer availability among neighbor switches, while routing logic unwraps incoming packets' headers and determines their output destination. The inputs that are allowed to send their data over the crossbar are determined by the switch allocator that resolves all conflicting requests for the same outputs. Optionally, the flits that cross the crossbar may be stored in output buffers. In switches that support virtual channels (VC) a VC allocator is also needed that selects which output VC the input flits would use when leaving the switch.

The configuration decided by the switch allocator and performed by the crossbar must meet certain constraints. An input can be connected to at most one output for unicast traffic or several outputs for multicast traffic. At the same time, an output can be connected to at most one input for either unicast or multicast traffic. The crossbar can efficiently satisfy these requirements due to its fully nonblocking nature. In an application-specific environment, when some turns of the network are disallowed, reduced-connectivity crossbars suffice. In principle, for Networks-on-Chip (NoCs) simpler switch elements inside the switch, such as Benes networks, are rarely considered due to their blocking features (see Chapter 4). The datapath of the switch consists of the crossbar and the buffers while the switch/VC allocator, the routing logic, as well as flow control, logic participate in the control path of the design.

In the rest of the chapter we present in detail the organization of state-of-the-art switch architectures and discuss the implications arising from their VLSI (Very Large Scale Integration) implementation. The chapter's content evolves in a step-by-step fashion beginning from the basic switch microarchitectural alternatives and moving gradually to high-performance design choices. In parallel, besides the analysis of well-known techniques, unexplored alternatives are discussed.

2.2 How Packets Move between Switches?

The switching policy determines how the packets flow in the network and which network resources, like buffers and links, they can utilize over time. The main goal of a switching mechanism is to provide a fair access to these resources and to resolve output contention without unnecessary blocking of ready-to-go packets.

In *store-and-forward* switching each node waits until an entire packet has been received before forwarding any part of the packet to the next node. This approach entails large delays at each node of the network even at low traffic. An efficient solution to the aggravated delay was given by *virtual cut-through* [150]. With virtual cut-through, a packet's header begins to leave the switch even if the entire packet has not been received yet at the input port of the switch. Therefore, a packet's switch traversal is overlapped in time with packet reception and significant delay is thus saved. Crossbar connections are allocated on a per-packet basis and once allocated to a packet's header are kept

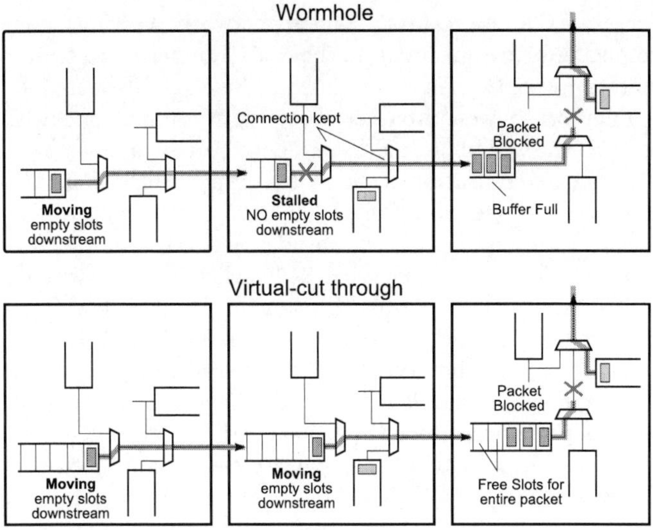

FIGURE 2.4: A blocked packet when using virtual cut-through and wormhole switching.

until the tail of the packet passes the crossbar. Also, before a packet starts to flow to the downstream node, the available buffer space for the whole packet should have been allocated. In this way, it is guaranteed that even if the head of the packet stops at the downstream switch there will be enough space to accommodate the in-flight part of the packet.

Alternatively, *wormhole* switching applies flow control at the flit level rather than at the packet level [66]. Other than that, the wormhole attains packet-wise switching and routing properties, that resemble virtual cut-through.[1] Even though a crossbar connection is maintained for all flits of a packet, every separate flit must first secure space in the downstream switch before moving forward. Thus, switch buffers do not need the capacity for "large" packets, as in packet-wise switching; instead a packet can span multiple switches along its path, consuming buffering resources from multiple switches.

There are significant differences in the performance of wormhole and virtual cut-through under different traffic loads [290]. Wormhole switching requires fewer buffers than virtual cut-through, but its maximum throughput is relatively limited and depends on the distance a packet has to travel in the network. The reason for this throughput loss is simple. Packets that travel more hops, stall more links when blocked, which is the cause of a snowball

[1]Switching and routing decisions can be also performed at the flit level but require additional header bits at each flit, which increases the bandwidth lost to maintain the additional header information. This would be equivalent to applying virtual cut-through to single-flit packets.

effect that creates contention throughout the network. At heavy loads, virtual-cut through outperforms wormhole at the cost of buffering in-transit packets. In implementation constrained environments this may offset the performance gains. The difference in how a packet is stored in the network in the case of blocking between the wormhole and virtual cut-through is shown in Figure 2.4.

An interesting generalization of the switching policies, called *hybrid switching*, that lies between the wormhole and virtual cut-through has been presented in [290]. The hybrid approach decides whether to buffer (like virtual cut-through) or stall (like wormhole) blocked packets at each switch depending on the load of the network. Following the same path, layered switching [186], a technique inspired from hybrid switching, chooses dynamically on how many consecutive flits to apply the operations of a wormhole switch. Therefore, if the group size equals the packet size then it behaves like virtual cut-through, while for smaller group sizes tends to behave like the wormhole.

The efficiency of any switching mechanism can be improved by the introduction of virtual channels. With virtual channels the allocation of buffers are decoupled from the allocation of links and thus performance is improved by allowing packets to bypass stalled packets. Virtual channels are discussed in Section 2.4, after having presented in detail the organization of a baseline wormhole-based switch in Section 2.3.[2]

2.2.1 Link-level Flow Control

A switching mechanism cooperates on a cycle-by-cycle basis with link-level flow control (or backpressure) to guarantee the lossless nature of the network. Link-level flow control protocol provides a closed-feedback loop to control the flow of data from a sender to a receiver. The main objective of such protocols is to regulate the flow of data from sender to receiver in such a way that the receiver's buffer is not forced to either drop packets due to lack of space (overflow) or idle unnecessarily (underflow).

One choice for link-level flow control is the use of credits. Credits keep track of the number of buffer slots available at the next hop.[3] No flit is allowed to leave the switch unless it has the available credits. Credits are decremented when a new flit leaves the switch and incremented when a buffer slot at the downstream node is released. Incrementing the credit counter is practically equivalent to returning the credits to their owner. Credit return is propagated to the upstream switch either with additional wires, which is the simplest and the preferred option in the case of on-chip networks, or embedded inside the packets that travel to the opposite direction (piggybacking).

On the other hand, stop&go signaling involves a signal between adjacent switches that is asserted (go) or de-asserted (stop) in order to inform the upstream switch when downstream buffers are full (or almost full). On the

[2]Although virtual channels can be built on top of any switching policy, in this chapter, we present them as an extension to wormhole operating at the flit level.

[3]The slots can be packets or flits depending on the flow control policy.

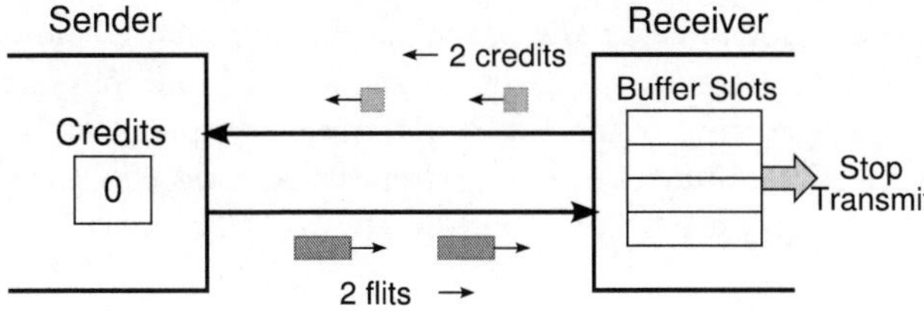

FIGURE 2.5: A graphical representation of the minimum amount of buffering required by credit-based flow control in order to sustain full throughput.

"go" case the upstream switch is free to send as many flits it wants, while on the "stop" case, every transmission should end. The stop signal is not sent when the downstream buffers are already full, but a little earlier so as to cover the flits that are already in-flight. In the same manner, we should not wait for the downstream buffers to empty before resuming to the go case. This is performed earlier as long as the number of available slots in the buffer pass a predetermined threshold.

Every time that a new credit or a "go" signal is sent upstream, it causes the re-initiation of the transmission from the sender and the arrival of a new flit at the receiver. The time that passes between the departure of the first credit and the reception of a new flit at the downstream switch is called the round-trip time (RTT) and determines the maximum throughput that the flow control policy can support. In order to achieve full throughput, such as the transmission of 1 flit per cycle, we should guarantee that at the downstream switch there are enough buffer slots to cover the RTT frame. This scenario is shown in Figure 2.5, where the downstream switch is suddenly blocked when 2 flits and 2 credits are already in flight. The 2 flits will fill up 2 buffer positions, while the 2 credits will notify the upstream switch to send 2 more flits before stopping transmission. Thus, when the link between switches is pipelined in two stages as in the case shown in Figure 2.5, 4 buffer positions are required to support sending a flit every cycle. For credit-based flow control the minimum number of buffer size that is required equals the product of the RTT and the link's bandwidth. On the contrary, the minimum number of buffers required by the stop&go mechanism is twice than that of credit-based flow control.

When the NoC uses pipelined links between switches, multiple flits can be in flight. Instead of letting these pipeline registers increase the RTT, we can transform them to elastic buffers [209], [164]. An elastic buffer accepts and transmits flow control signals informing its neighbors on its state (full or empty). Therefore, adding elastic buffers to a link results into the distribution of the input buffer of the downstream switch across the link.

The efficiency of the flow control mechanism can be improved by speculatively transmitting flits to the downstream switch even if no free buffer space is guaranteed [212]. At low loads, where latency matters, this speculation is not expected to lead to packet losses. However, at high loads speculation should be avoided and a traditional credit-based mechanism should be re-initiated. Although these techniques offer significant reduction in latency, especially for bursty traffic, the hardware overhead for handling misspeculation may not fit the strict design constraints of a NoC. Alternatively, we could try to reserve the next hop buffer space before actual flit arrival to the switch [250].

2.3 Baseline Wormhole Switch

In wormhole switching data movement is performed on a per-flit basis, while routing and switch allocation decisions are taken once for the head flit and applied to the remaining flits of the packet. The application of a wormhole relies on the existence of a single queue at each input of the switch. Each input queue is exclusively used (or left idle) by flits of the same packet.[4] Once a packet, actually its leading, head flit (head-of-line, HOL), wins switch allocation, the established crossbar connection is maintained for all flits of the packet; effectively, the downstream input queue also becomes populated by flits of the same packet, as no flit from other input queues (thus from other packets) can intervene between the flits of the connected packet. In that sense, the wormhole behaves similar to packet-wise networks, even though its unit of buffering and flow control is a packet segment.

When a head flit arrives at one input of the switch five distinct tasks need to take place: Buffering, packet decoding and routing computation, switch allocation, crossbar, and link traversal. On the contrary, when a body or a tail flit reaches the switch only a subset of the tasks is required. For example, routing computation and switch allocation are omitted since they are inherited directly from the decisions taken for the head flit of the packet. A graphical representation of the tasks devoted to each flit type along with their dependencies is shown in Figure 2.6(a).

As we will shortly see, some tasks may be overlapped in time because there is no dependency between them or they can be completely bypassed depending on the status of the switch and the load injected to the network. When ultralow latency is required, even if some tasks are interdependent, they are executed speculatively in parallel, in order to increase performance for the average case that speculation is correct.

If the input buffer is empty, then buffering can be bypassed, and the packet's header can be sent to the routing logic that decides to which output each flit should be forwarded. If this is not the case the flit is buffered and waits until it reaches the front position of the input queue. In an alternative

[4]This is not always true. If an input queue is large enough, it may store flits from more than one packet. However, in a wormhole, the input port can only forward flits from the packets that has exclusive access to the head-of-line position.

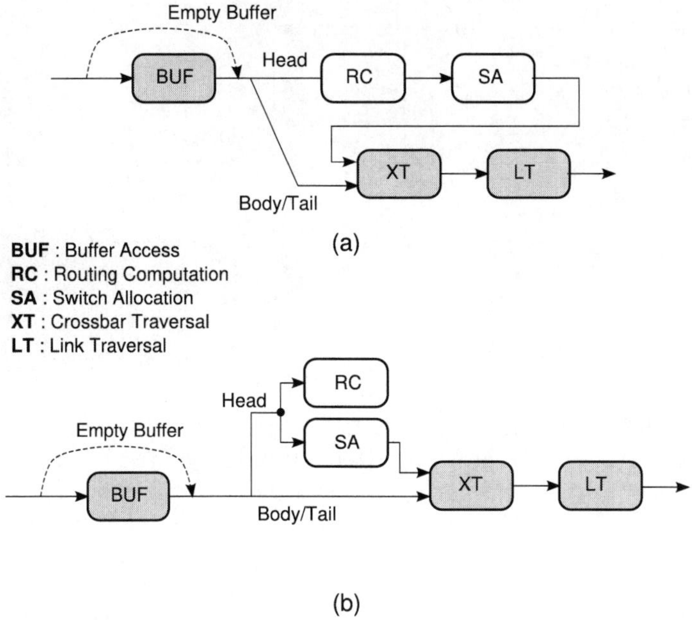

FIGURE 2.6: The main tasks of a wormhole-based switch: (a) Naive ordering of tasks and (b) ordering with reduced dependencies that favors parallel task execution.

scenario the head flits can carry already decoded information concerning routing computation. Decoding and routing computation can be prepared before the flit leaves the previous switch in order to use them in the current switch. This technique is called *lookahead routing* and was first employed in the Silicon Graphics Spider switch [97] and extended to adaptive routing algorithms in [327]. With lookahead routing, routing computation is performed independently from the other tasks since it depends only on the tasks of the next switch. The implementation of lookahead routing is shown in Figure 2.7.

The head-of-line (HOL) flit of the queue requests access to the crossbar via the switch allocator that receives requests from all inputs and assigns the available output ports to the requestors. Which output ports are actually available is determined by the flow control information returned by the downstream switch. The outputs that have available buffer slots at the other end of the link can be connected to any eligible input. Care should be taken so as to avoid the scenario that switch allocation is performed prior to buffer space allocation in the downstream switch, allowing the switch allocator to select a flit that cannot move due to lack of buffer space, thus leaving an output needlessly idle.

The actual transfer of flits from one switch to the next is performed by the last two tasks: crossbar and link traversal. Both tasks are straightforward

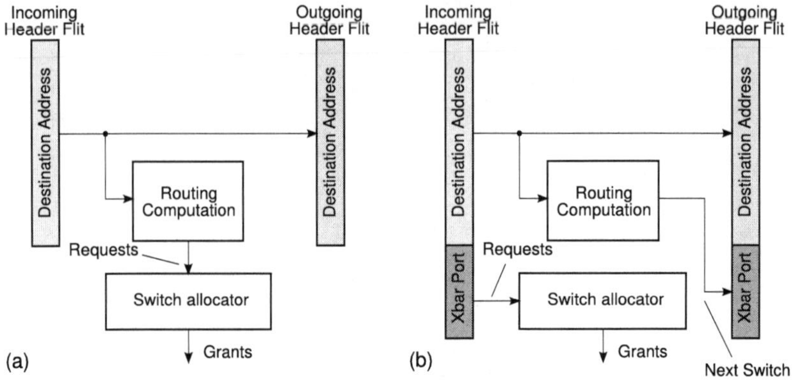

FIGURE 2.7: The dependency between routing logic and switch allocation: (a) Routing computation prior to switch allocation—a serial dependency. (b) The implementation of dependency-free lookahead routing where selected output port is encoded in the head flit and used directly in the next switch (from [327] © 1999 IEEE).

to implement since they involve only crossbar and wire connections. Packets that "won" switch allocation in the current switch may nevertheless stop, when their head flit fails to establish the next crossbar connection at the downstream switch. Flow control information does not return fast enough, and the flits of the packet are blocked as they wait for available downstream buffer slots in order to move on. Effectively, the acquired connection at the current crossbar is wasted and crossbar input and output slots are underutilized. This is not a problem when the switched packet is congested, but its a clear performance deficit considering that the congested packet deprives crossbar access to other packets that come from the same crossbar input, or go to the same output, since those other packets may not be congested. This unnecessary blocking is removed by virtual channels as described in Section 2.4.

An optimized version of the task dependencies is depicted in Figure 2.6(b) where as many tasks as possible are executed in parallel. The execution of all tasks in a single cycle is not easy even for this optimized version, when considering a fast clock cycle of 15–25 FO4.[5] To shorten the clock cycle pipelining needs to be employed, even if this choice increases the zero-load latency of the switch, i.e., the number of cycles a flit has to spend in an empty network in order to travel from a given source to a given destination. A reasonable pipelined organization of a wormhole switch is shown in Figure 2.8. The pipeline consists of 4 stages; 3 dedicated to the switch and one to link traversal. Head flits

[5]The FO4 delay metric equals to the delay of an inverter that drives four equally-sized inverters, and it is used since it provides in some sense a technology independent way to express the delay of a circuit [340]. For example, in 65 nm under typical conditions 1 FO4 $\sim$ 30 ps.

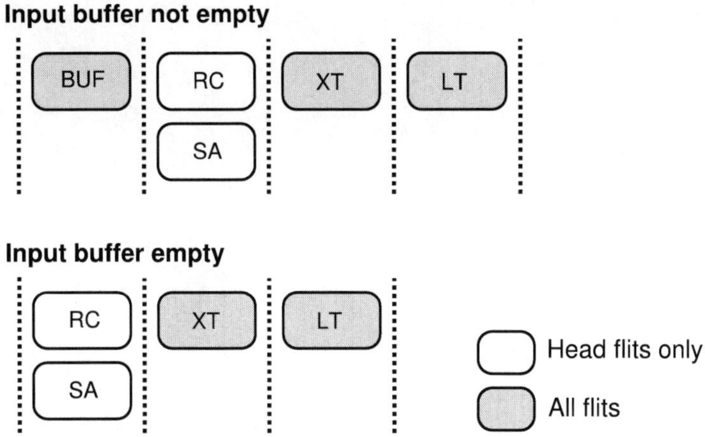

FIGURE 2.8: The pipelined organization of a latency-optimized wormhole switch. Body/tail flits bypass the pipeline stages that are devoted to tasks related only to head flits.

see a deeper pipeline than body/tail flits since they are involved in routing computation and switch allocation. In all cases, when input buffers are empty incoming flits move directly to switch allocation (head flits) or to crossbar traversal (body/tail flits) depending on their type. If the designer wishes to reduce the zero-load latency (s)he can merge any of the consecutive pipeline stages, increasing the clock cycle inevitably.

The pipelined organization allows the designer to decouple switch allocation from crossbar traversal, with the possible risk of performance loss under certain circumstances. This decoupling is practically an application of precomputing the switch-allocation decisions ahead of time and can be implemented in two ways.[6]

The first approach was proposed by [225] and allows switch allocation to happen in parallel to crossbar traversal. This technique allows the switch allocator to compute a grant vector depending on the requests of each cycle. However, those grants do not drive the multiplexers of the crossbar, as in the typical case, but instead they are stored and act as grant-enable signals for the next cycle. Concurrently, the crossbar is driven by the request signals that are first masked with the grant-enable signals that correspond to the granted requests of previous cycles.

The second method, proposed by [202], allows bypassing the switch allocation stage and guiding the HOL flits directly to the crossbar assuming that the crossbar connections have been precomputed earlier by the prediction logic. If

[6]Both techniques have been proposed in the context of virtual-channel (VC) switches. However, the main idea behind the two approaches is more generic and is equally applicable to VC-less wormhole-based switches.

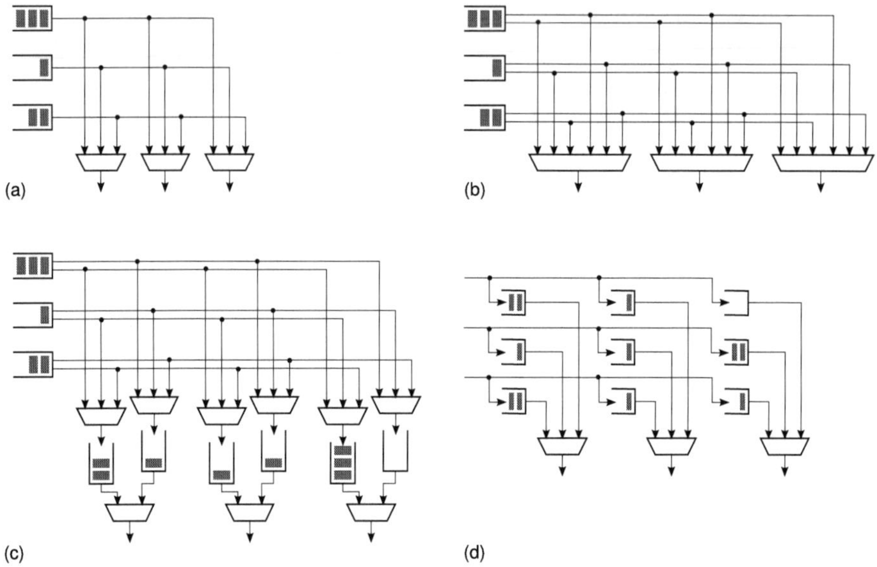

FIGURE 2.9: Buffer organization in an input-queued wormhole switch: (a) Baseline input-queued switch, (b) with input speedup 2, (c) with both input and output speedup, and (d) crosspoint queuing.

the prediction is correct the task of switch allocation is avoided and one cycle is saved per switch. In the opposite case, the normal switch pipeline is executed without any additional overhead. The prediction logic forecasts which outputs will be used by future incoming packets and reserves accordingly future crossbar connection slots. The predictions are based on the characteristics of the topology and the routing algorithm as well as the network's traffic at runtime.

2.3.1 Crossbar and Buffer Organization

Switch buffer architecture is closely related to the organization of the crossbar. Various alternatives can be designed based on the placement of buffers in the switch [53]. We will present the various alternatives for a wormhole switch using as an example a 3×3 switch.

Figure 2.9(a) shows the simplest buffering architecture, called *input queuing*, where a single queue is placed in front of each input of the crossbar. The input queues should be able to perform both a dequeue (read) and an enqueue (write) operation per cycle if the cut-through operation is supported. Hence, two separate ports are needed—one for reading and another for writing. Their bandwidth is proportional to the link bandwidth and not to the number of links. This requirement is fairly easy to satisfy irrespective of the way the input queues are actually implemented, either using SRAM (Static Random Access Memory) macros or using latches and multiplexers in a logic-synthesis-based environment.

With simple input queuing, every clock cycle, only the head flit of each input queue is eligible to request an output port. If more than one head flits request the same output, then only one of them will win the competition and the rest will have to wait. This condition blocks all other packets behind the HOL that are destined for another output port that may stay idle. This problem is known as *head-of-line blocking* and is the cause of switch throughput saturation for random traffic at roughly 59% of the available link capacity [143], [40]. Under bursty traffic, with bursts larger than the input buffers the saturation is observed at lower loads [63].

The performance of input queuing can be improved by enhancing the number of input/output ports of the crossbar or by using more than one crossbar network. In this way, the throughput of the crossbar is higher than that of the incoming links. For example, Figure 2.9(b) shows an example of a switch where each input can use twice the number of crossbar's input ports [219]. In order to support this *input speedup*, the input queues should support more than one read ports in order to feed the crossbars every cycle. High-port requirement needs custom memory solutions when SRAM blocks are employed for the implementation of the input queues. In a standard-cell ASIC (Application-Specific Integrated Circuit) environment adding more read ports is more straightforward to implement, since it is equivalent to adding more multiplexers and address decoders to the basic latch-based structure.

The crossbar's throughput can be further enhanced by introducing *output speedup*. This form of speedup allows each output to receive 2 flits per cycle, as shown in Figure 2.9(c) where both input and output speedup is utilized. To balance the difference in throughput between the switch and the links additional queues should be added at each output, supported also by internal flow control between input and output buffers. The increase in throughput can be equivalently achieved by utilizing a crossbar that runs twice as fast as the links of the network (speedup in time). However, increasing the speed of the crossbar is not as scalable as the use of a crossbar with the higher number of input/output ports (speedup in space).

To reach maximum throughput, one can go to crosspoint queueing; there is a single queue for each input-output link pair as depicted in Figure 2.9(d). Every outgoing link can now be kept busy, by selecting and forwarding a packet to it, independent of what the other links do. This architecture achieves optimal link utilization, but has the disadvantage of needing many buffers and flow control information separately for each crosspoint.

2.3.1.1 High Throughput Buffering Alternatives

Output queuing associates a buffer with each output port that must be able to accept, in the worst case, flits arriving simultaneously from all inputs [123].

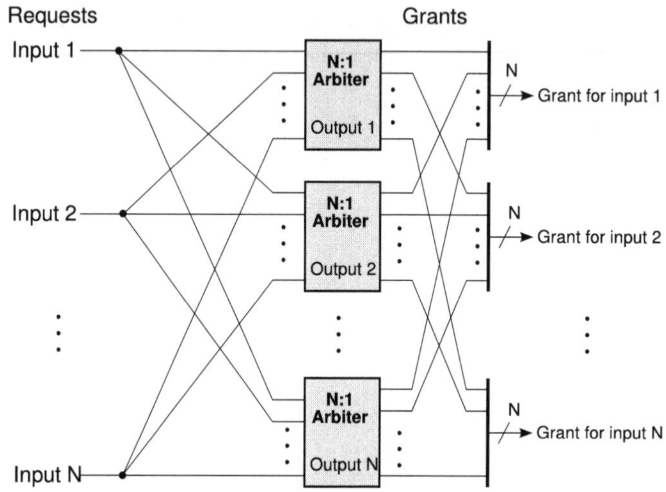

FIGURE 2.10: The organization of the switch allocator in the case of a wormhole input-queued switch.

Each output queue actually unifies all the queues that appear in one column of the crosspoint queue architecture. This alternative represents a switch with full output speedup. This merging improves buffer memory utilization without sacrificing throughput. However, the number of write ports that each output queue must support equals the number of input ports, which is a very hard constraint to satisfy even in small low-radix switches.

The many-port memory problem of output queueing can be also solved by the pipelined memory organization [146]. This pipelined organization is actually a form of shared buffering where only one large buffer is used for the entire switch. This shared buffer must have a throughput equal to the aggregate throughput of all input and output links. This requirement can be partially satisfied by utilizing an internal datapath that is N times wider than the width of the links. However, this approach is only applicable to NoCs with links of small width and not in cases that the links are 512 bits-wide or more.

Finally, a simplified form of output queuing that has not yet been explored in the NoC context is the knockout switch [346]. In this organization, each output port is associated with a number of parallel buffer memories, say k. When the number of input ports that send flits to the same output is smaller than k then all inputs are satisfied at once. When there are more than k flits for the same output, the extra flits should be either dropped (as in baseline knockout) or not acknowledged to the upstream switch and re-transmitted in a later cycle. The upstream switch should be smart enough to keep a valid copy of the transmitted flit until its receipt is acknowledged. Only a small value of k suffices for significant performance, thus partially alleviating the high write-port problem of output queuing.

2.3.2 Switch Allocation Organization

The switch allocator accepts the requests from each input and decides which inputs to grant in order to produce a valid connection pattern for the crossbar. Each input possibly holds packets for different outputs. From those packets only the head flit of each input is eligible to request an output port. Therefore, each input port generates one request that is broadcasted to all outputs. At each output an arbiter is required that will receive the corresponding requests and will grant only one of them. Therefore, the switch allocator in the case of the baseline wormhole switch is constructed using a single arbiter per output of the switch, which decides independently which input to serve. The arbiter's decision is kept until all flits of the same packet pass through the corresponding output port. This organization is shown in Figure 2.10.

2.4 Virtual Channels

In terms of performance, the introduction of virtual channels (VCs) overcomes the shortcomings of the baseline wormhole switch by improving the effective statistical multiplexing on network's links [63]. In practice, adding virtual channels to an interconnection network is analogous to adding lanes to a street network. VC-less networks, either wormhole or virtual cut-through, are street networks composed of one-lane streets. In such networks, a single blocked packet (car) blocks all following packets. Adding virtual channels to the network removes this constraint and allows otherwise blocked packets to continue moving by just turning to an empty (less congested) lane of the same street. Since the additional lanes are virtually existent their implementation involves the time multiplexing of the packets that belong to different lanes (virtual channels) on the same physical channel.

Briefly, virtual channels behave similar to having multiple wormhole channels present in parallel. However, adding extra lanes (virtual channel) to each link does not add bandwidth to the physical channel. It just enables better sharing of the physical channel by different flows. In this context, certain mechanisms built upon virtual channels suggest that incoming packets should be assigned to virtual channels according to their destination [229] or to the network's condition on the path towards their destination [145]. In this way, congested traffic, or congested parts of the network, can be isolated and let packets that move to quiet regions to continue their travel uninterrupted, as shown in Figure 2.11.

Besides performance improvement, virtual channels are used for a variety of other purposes. Initially, as described in detail in Chapter 5 virtual channels were introduced for deadlock avoidance [64]. A cyclic network can be made deadlock-free by restricting routing so that there are no cycles in the

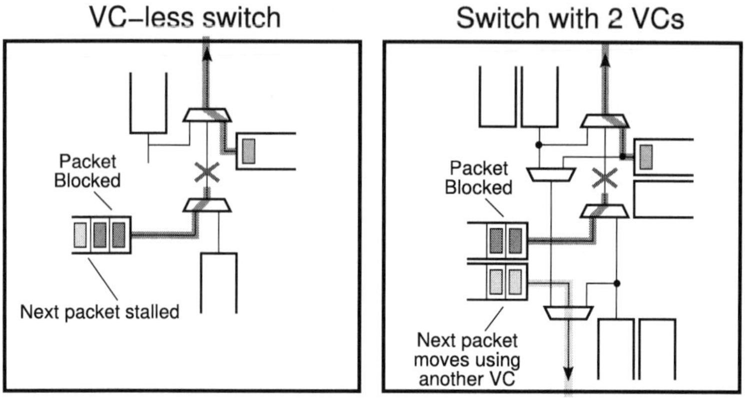

FIGURE 2.11: Virtual channels remove unnecessary blocking.

channel dependency graph. The channels can be thought as the resources of the network that are assigned to distinct virtual channels. The transition between these resources in the packet's routing path is being restricted in order to enforce a partial order of resource acquisition, which practically removes cyclic dependencies [82]. In a similar manner, different types of packets, like requests and reply packets, can be assigned to disjoint sets of virtual channels to prevent protocol-level deadlock that may appear at the terminal nodes of the network.

2.4.1 Basic Operations of a Switch with Virtual Channels

To divide a physical channel into V virtual channels, the input queues at every crossbar input need to be separated into as many independent queues as the number of virtual channels. These multiple input queues are now named virtual channels and each one of them is reserved on a per-packet basis. These virtual channels maintain control information that is computed only once per packet. The leading (head) flit of the packet sets up this information during *virtual-channel allocation* and its followers body flits find this information ready upon arrival at the corresponding VC structure. Tail flits when leaving the switch release the allocated virtual channel. To support the multiple independent queues link-level flow control is also augmented and includes separate information, either credits or stop&go signals, per virtual channel.

In a switch with virtual channels, crossbar connections are not maintained for the entire duration of the packet, but are instead set up and torn down on a flit-by-flit basis. All body flits of a packet use the virtual channel allocated by the packet's header and fight for a flit buffer on the same VC as well as access to the crossbar. Thus, even when a head flit fails to establish a connection to its output, flits from other packets may still use the crossbar. In this way, body flits may be interleaved over the shared channel with flits from other packets,

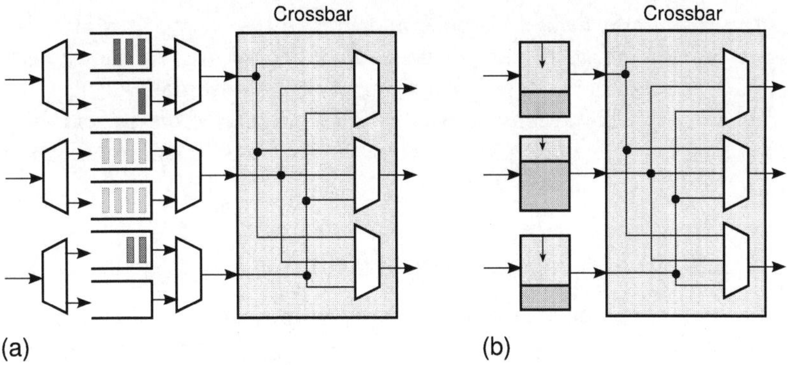

FIGURE 2.12: Buffer organization for a virtual-channel switch. Virtual channels allocated (a) to separate banks and (b) to a shared memory space.

but have a virtual channel exclusively reserved for them. This increased degree of multiplexing can in principle increase network utilization; however, it may increase packet delay at low loads by introducing gaps between the arrival of consecutive flits of the same packet at the sink VC.

Flits from different VCs of the same input can compete independently for a crossbar connection. This increases the complexity of the switch allocator, but mitigates the needless blocking within the present crossbar. Virtual channels allow the upstream switch to use a second, free lane (a VC with buffer space available), when a "first" packet is blocked downstream and the switch has run out of downstream buffer slots for the VC of the blocked packet.[7]

2.4.2 Enhanced Crossbar and Buffer Organization

Every virtual channel requires a separate queue at each input of the switch. All architectures presented in Figure 2.9 for the case of wormhole switches can be adopted after splitting the buffers into as many banks as the number of virtual channels. An example for two virtual channels is shown in Figure 2.12. Notice in Figure 2.12 that the complexity of the crossbar as well as the number of ports of the input buffers is equal to that of the original input queued switch. This is achieved by sharing an input port of the crossbar between all virtual channels via a local multiplexer. This is enough for the case of on-chip switches, since the rate of incoming traffic from each VC does not justify the allocation of a separate port of the crossbar to each one of them. However, this structural choice increases significantly the complexity of switch allocation relative to the baseline input-queued wormhole switch.

The input queues can be either implemented using separate memory banks

[7]This is one good reason why each virtual channel should have private buffer space allocated; to avoid buffer hogging. Otherwise, the buffer space per input can be shared among virtual channels, with all sharing benefits that one can imagine.

where the maximum available space allocated for each VC is statically determined or can utilize a monolithic memory structure, as shown in Figure 2.12(b), where independent VCs are implemented internally using a linked-list structure [312]. The second alternative allows dynamic sharing of the space allocated for each VC, while in very aggressive switch designs even the number of VCs per input can change at runtime [230].

2.4.3 Switch Allocation Organization

The switch allocator is responsible for determining in each cycle the connections between the input and the output ports of the switch. Since now the input buffers are organized in multiple-independent queues, each input can send multiple requests per clock cycle.

2.4.3.1 Centralized Switch Allocators

A centralized switch allocator, such as the wavefront arbiter [311],[8] or the 2-D round-robin [174], handles all input requests at the same time and produces a global schedule for the whole crossbar. The schedule is maximal when it is not possible to insert a new input/output connection, without altering some of the connections that already exist in it. In general, centralized switch allocators produce maximal schedules and do not leave any outputs unoccupied if there is a request for them. The main drawback of the commonly used centralized schedulers is their delay.

The simplest form of a centralized switch allocator with N input/output ports receives directly the requests of all input VCs, that is $N \times V$ in total, and grants only N of them; one for each output. The scheduler in this case is asymmetric since it shares N outputs among $N \times V$ requests. In this way, we may end up granting two or more VCs from the same input that share a common crossbar port but have flits for different outputs. To avoid this conflict the granted VCs of the same input pass from an additional $V : 1$ arbiter that selects only one of them.

Of course, this per-input local arbitration stage (local $V : 1$ arbiter) could have been done at the beginning of switch allocation. Then, each input would transmit only one request to a $N \times N$ centralized allocator that again would produce one grant for each output. This grant can be consumed directly by the input port since the eligible VC is preselected.

Following this analysis, it is evident that both techniques besides the centralized allocator need an additional local arbiter for resolving output contention. In the baseline implementation this arbiter lies in the critical path of the switch allocator. However, based on the observation of [22] such a constraint can be removed, if we rely on a centralized allocator that gives to each

[8]Please, note that the wavefront arbiter, although called an "arbiter," is a complete switch allocator. The term arbiter refers to a circuit that accepts many requests and grants only one of them.

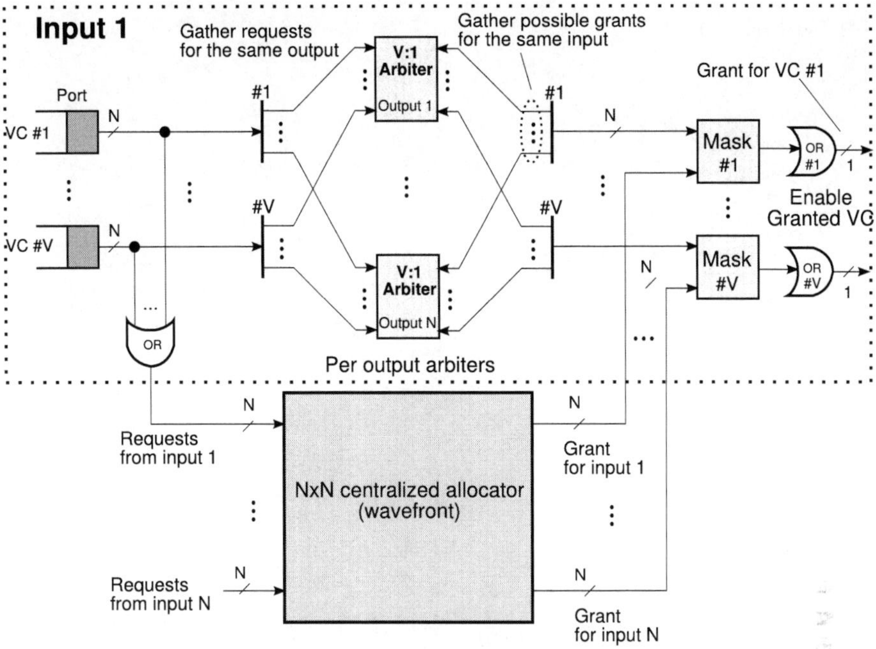

FIGURE 2.13: Centralized switch allocators for virtual-channel switches.

input at most one grant. No grants are given to the same input for two distinct outputs; this constraint is already satisfied by the wavefront allocator. This faster alternative is depicted in Figure 2.13.

At first, each input requests all outputs for which it has HOL flits irrespective of the input VCs that they belong to. The requests are received by a $N \times N$ centralized allocator that grants N requests, one for each output. Therefore, at this point we know the output to which each input will be connected to. The only thing that remains to be selected is the appropriate VC from each input. The good news is that this choice can be performed in parallel to the centralized allocator, using N $V:1$ arbiters. Each arbiter corresponds to an output and selects one candidate VC for this output. At the end, from all derived (output, VC) pairs at each input, only one would be chosen according to the schedule of the centralized allocator.

2.4.3.2 Separable Switch Allocators

Separable switch allocation is organized in two phases since both a per-output and a per-input arbitration step is needed [11]. An example organization is shown in Figure 2.14. In the first case, each input is eligible to send to the outputs only one request (Figure 2.14(a)). In order to decide which request to send, each input arbitrates locally among the requests of each VC.

On the contrary, in the case of output-first allocation all VCs are allowed

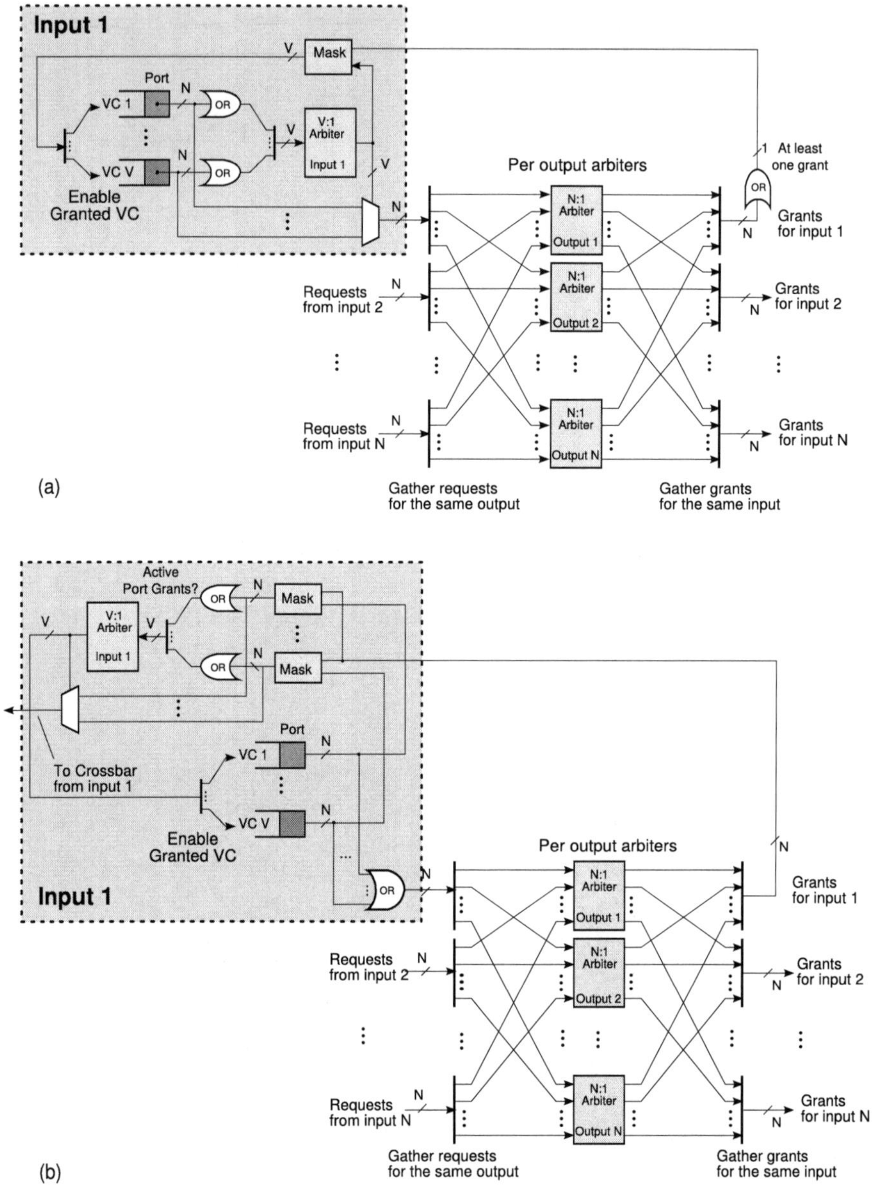

FIGURE 2.14: Separable switch allocators in the case of switches with virtual channels. (a) Input-first and (b) output-first switch allocators.

to forward their requests to the output arbiters (Figure 2.14(b)). In this way, it is possible that two or more VCs of the same input will receive a grant from different outputs. However, only one of them is allowed to pass its data to the crossbar. Therefore, a local arbitration needs to take place again that will resolve the conflict.[9]

In either form of separable switch allocation, in order to ensure fairness and deterministic service guarantees the priority vector of any arbiter should be updated only if the grant that it produced is also accepted in the second arbitration stage [204]. Appropriate priority selection can also increase switch allocator's throughput by biasing the input/output pairs that correspond to heavily backlogged flows [265], [4]. In this way, flits of the same packet (or flows in general) are kept together as much as possible. Low-latency approaches that follow this principle, that are also easy to implement in the context of NoCs, have been presented in [172] and [54].

Even though the arbiters operate independently, their eventual outcomes in switch allocation are very much dependent. For example, when two or more input arbiters request the same output simultaneously, a conflict occurs. Only one input will be matched, while others will be left idling. There is nothing better these idling inputs could have done if they had HOL flits only for the requested output. However, it would have been much better for them and for the collective throughput of the switch if they had tried to bind to some other output. Such conflicts are unavoidable, when every port has an opportunity of only one request in every clock cycle. In order to improve their efficiency, separable switch allocators have two generic options. They can either (a) try to "desynchronize" their bindings, so that each input (output) requests a different output (input) on every new scheduling cycle, or (b) they can attempt multiple requests per port, per scheduling operation [11, 204].

Desynchronization is hard to achieve in the context of NoC since it requires the addition of as many independent queues per input as the number of output ports [204]. This is either prohibitive, or it may lead to very shallow buffers that will destroy throughput.

On the other hand the execution of multiple scheduling iterations is a more scalable solution. At each scheduling iteration, the input-output pairs that remain unmatched retry to find an available output port. In this way, the set of already matched input-output pairs is augmented at each iteration. Iterative crossbar scheduling converges to a maximal schedule. Even if iterations clearly improve performance they prolong significantly scheduling time, which increases linearly with the number of iterations. Thus, with no doubt a multi-iterations switch allocation would definitely exceed the tight clock cycle budget of NoC switches. Pipelining between iterations is a viable alternative [205], however a more scalable solution is desirable.

---

[9]In both cases, the local arbitration between either the requests in the case of input-first allocation or the grants in the case of output-first allocation can be avoided, if the multiple-queues per input have direct access to an input port of the crossbar. If this is the case, the switch allocation operation is again reduced to one bigger arbiter per output.

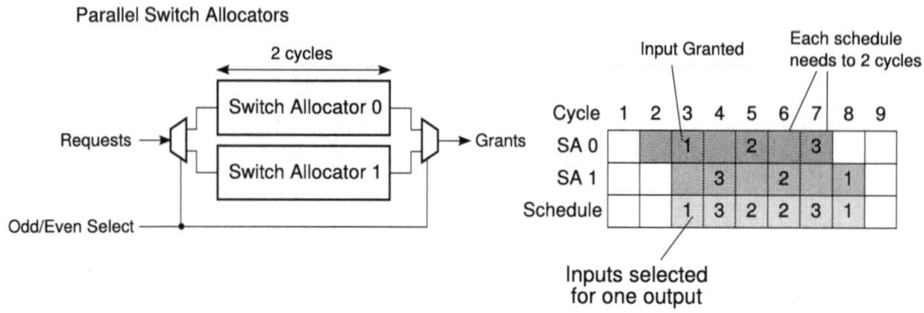

FIGURE 2.15: An example of the application of parallel switch allocators when each switch allocator performs 2 iterations.

2.4.3.3 Parallel Switch Allocators

In order to compensate for large scheduling delays, we can resort to parallel switch allocators [213], [211]. The basic idea of parallel allocation is to use multiple instances of a switch allocator and operate these instances in parallel. In this way, we can produce crossbar schedules at the desired clock rate, even if each switch allocator requires multiple clock cycles to return a schedule. To achieve this, the crossbar schedule utilized at each cycle will be provided by a different allocator. Therefore, to achieve one matching per clock cycle when allowing each allocator to perform m iterations we simply need m parallel allocators.

If we employ two parallel switch allocators, as shown in Figure 2.15, we can devote the odd cycles of the switch to the first one and the even cycles to the other. In this way, each allocator is allowed to perform two scheduling per-input (output) and per-output (input) iterations since its decisions are only needed once every two cycles. This time division yields a collective behavior that looks similar to that of a single, monolithic switch allocator, which would be responsible for all clock cycles. For example, if output 1 is decided by the first scheduler to serve inputs 1, 2, and 3 in odd clock cycles while the same output, as decided by the second scheduler, serves inputs 3, 2, and 1, during the even cycles the resulting time schedule, shown in Figure 2.15, is both throughput efficient and approximately fair.

One central issue that needs to be properly managed is the reaction of an input queue that stores only one flit. If it requests an output on cycle t, the result will come back from the first allocator on time $t + 2$. Thus, on time $t + 1$ the input does not know which is the wisest path to follow; to repeat the request to the second allocator on cycle $t + 1$ hoping that this action would increase the probability to receive a grant in the following two cycles (now both switch allocators are requested) or to hold the request waiting first for the result of the first switch allocator. In the first case, the input may be matched by both switch allocators. Then one of the two grants should be discarded thus possibly leaving an output idle. In the second case, the first switch allocator may not grant the input, which will remain idle for at least 2 more cycles before receiving a grant.

2.4.4 Virtual-channel Allocation

Each flit before leaving the switch and transferred to an input port of the downstream switch, should have been first allocated a virtual channel at the downstream switch. Of course the movement will actually happen when there is a free buffer slot at the corresponding downstream VC. The operation of assigning a VC of a downstream switch to a packet is called *virtual-channel allocation.*

In the most general case, during virtual-channel allocation, the head flits that are buffered at any input VC must compete for an available downstream VC, which belongs to the output port that is selected by the routing function [251]. Thus, the VC allocator performs a matching between $N \times V$ requestors and $N \times V$ resources, subject to the constraint that any output VC is requested only by input VCs that need to be forwarded to the same output port. This allocation problem can be solved either by using a centralized approach, or by adopting a VC separable allocator with separate per-input and per-output arbiters.

2.4.4.1 Centralized Virtual-channel Allocation

In a centralized virtual-channel allocator the requests of each input VC are expanded to a $N \times V$-wide request vector depending on the output port selected by the packet and the choices enforced by the routing logic, as shown in Figure 2.16. The wide request vector cannot have more than V active requests, while all requests concern VCs of the same output. After a matching between input and output VCs is developed, the $N \times V$ grant signals coming from all output VCs are gathered per-input VC. The OR function at each input VC just detects whether at least one of the V returning lines corresponds to a grant.

2.4.4.2 Separable Virtual-channel Allocation

In this case, the organization of a virtual-channel allocator follows roughly the same structure as the separable switch allocator. Each VC of each input port tries to get access to a specific output port. Since more than one VC is associated with each output port, each input VC should decide whether it should request all of them at once[10] or if it should select only a subset of the available output VCs. Its decision is guided solely by the routing function. In any case, the design of the virtual-channel allocator requires one arbiter per

[10]Requesting all the available VCs of each output port practically should reduce the requests to one request per output port. Then it is a job of the output port to return at least one available VC.

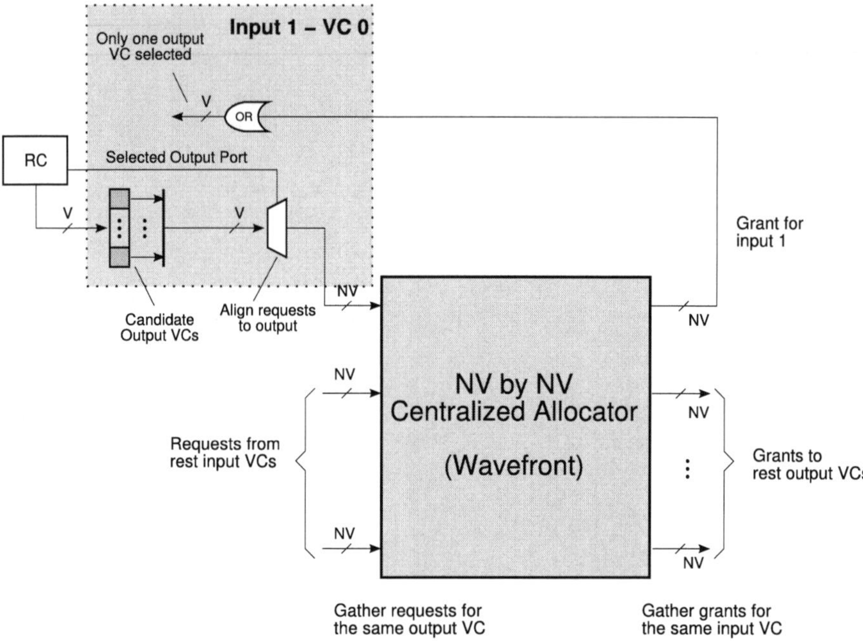

FIGURE 2.16: Centralized virtual-channel allocator.

VC per output port. Thus, $N \times V$ arbiters are required in total at the outputs, with each one accepting $N \times V$ requests.

Depending on how many output VCs each input VC is allowed to request two alternatives are possible. In the case of input first allocation, each input VC selects only one of the available VCs returned by the routing function for the selected output, as shown in Figure 2.17(a). Then, it is up to the output arbiters to grant one of the requesting input VCs. In the symmetric case of output-first allocation, shown in Figure 2.17(b), each input VC requests all available VCs that belong to the output selected by the routing function. Since each input VC asks for more than one output VC it is possible that it receives a grant from more than one output arbiter. Thus, an additional stage of arbitration is needed to decide which grant to finally accept.

When separable VC allocation returns poor matchings, we could choose to perform multiple iterations as in the case of switch allocation. Nevertheless, as shown in [22], this option is not needed since the matching quality of the VC allocator does not determine the network performance of the switch.

2.4.4.3 How Many Output VCs Should Each Input VC Request?

We have already discussed that virtual channels apart from improving performance actually may serve several other purposes. As noticed in [22], the set of available VCs that exist in a system should not be treated in a

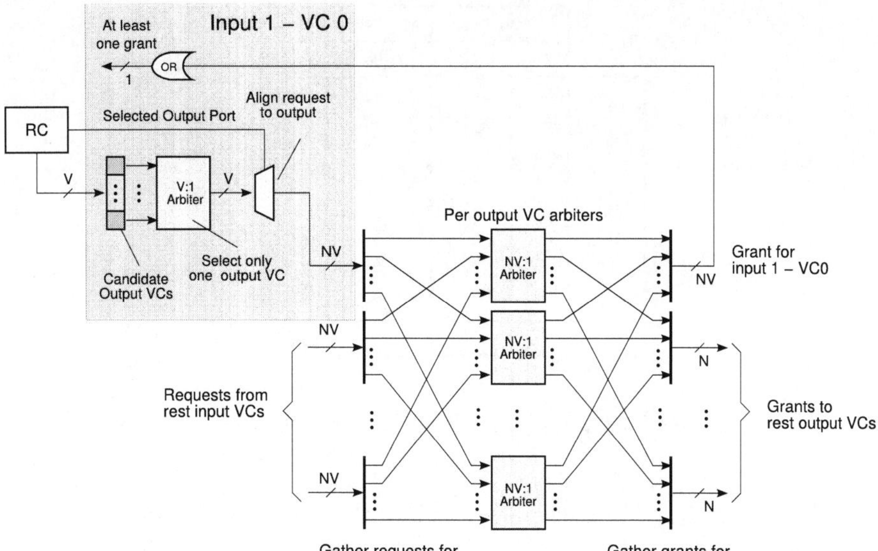

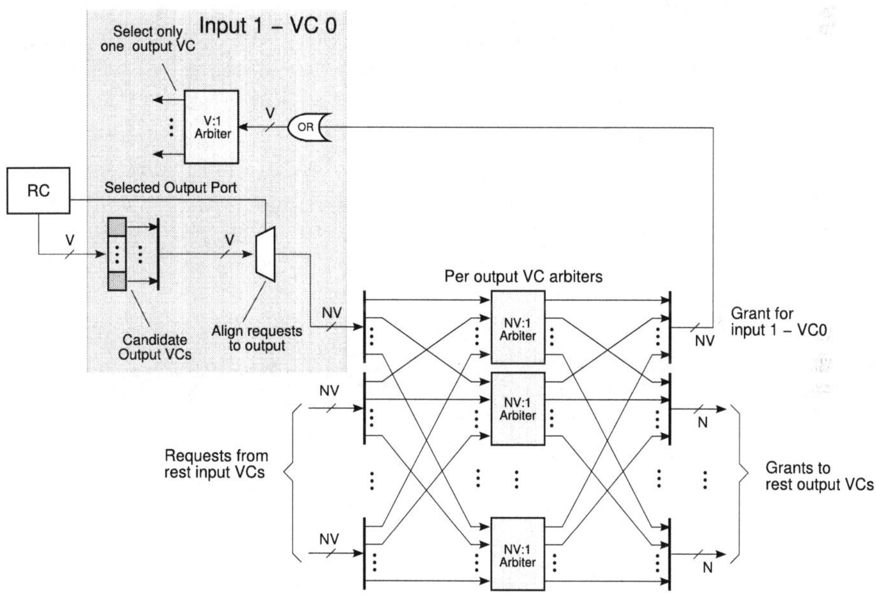

FIGURE 2.17: Separable virtual channel allocators: (a) input-first and (b) output-first VC allocation.

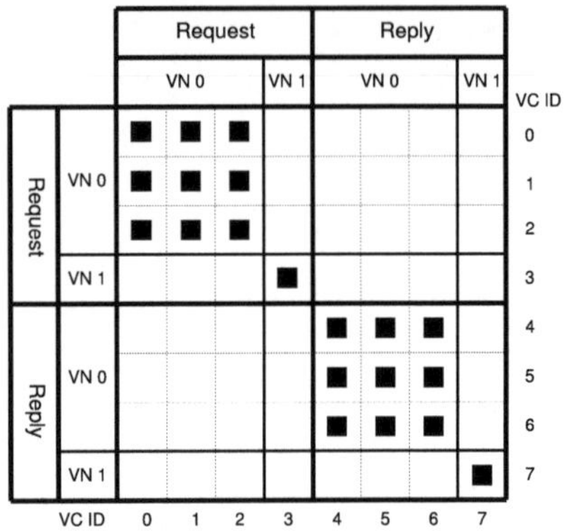

FIGURE 2.18: The VC-to-VC mapping table.

unified manner by the virtual-channel allocator but instead they should be partitioned according to their role. The total number of VCs that exist in a system depends on the number of message classes M that flow in the network, the number of resource classes R used for deadlock avoidance, and the number of unreserved VCs used for performance improvement, say C. Therefore, there is no need for the virtual-channel allocator to try to match all $N \times V$ input VCs to all available $N \times V$ output VCs, since some of the combinations of input-output VC pairs actually never occur.[11]

The pairs that should be omitted by virtual-channel allocation depend on the actual class that the corresponding VCs belong to. For example, packets are assigned to message classes depending on the purpose that each type of message fulfills in the system, like request and reply packets. This message class remains unchanged throughout the packet's lifetime. In this way, protocol-level deadlock that may arise when messages of different classes are mixed in the network is also avoided. Thus, since packets would never change their message class, there is no reason for them to request a VC at an output port that belongs to a different message class.

Based on this observation and the assignment of VCs to different classes, we can remove any redundancy in the virtual-channel allocator by constraining the set of candidate output VCs that an input VC can generate requests for. For example, a $NV \times NV$ virtual-channel allocator that handles M separate message classes, with each one assigned to a different set of VCs, can be implemented by M smaller and completely independent $(NV/M) \times (NV/M)$ virtual-channel allocators. Each smaller allocator would handle only the allocation of VCs inside each message class, while the number of inputs and outputs of the allocator's arbiters would be reduced by a factor of M, thus reducing significantly their delay.

[11]In the generic case only the bit vector supplied by the routing logic is used to constrain the range of allowable VCs at runtime.

The design of a virtual-channel allocator should be followed by a VC-to-VC assignment map as the one shown in Figure 2.18. The purpose of this map is twofold. First, it clearly depicts the role of each VC, the message or resource class (virtual network–VN) it belongs to, as well as the number of VCs assigned per class in order to improve the performance of the network. In this case VCs are asymmetrically assigned to classes with VN0 allowed to use 3 VCs, while VN1 is allowed only one. Second, the assignment map clearly identifies all the available input-output VC matchings. The black squares in the map show which output VCs are eligible by the input VCs. In this way, any redundant connections in the large $NV \times NV$ VC allocator are removed and significant delay and area are saved from the implementation.

When the black squares in the VC-to-VC map cover all output VCs of a given message or resource class then the VC allocator requests don't need to specify individual candidate VCs within the class, but instead they can select the class as a whole. The C per-output-VC arbiters can be replaced by a new circuit [79] that returns all the granted output VCs at once. An alternative way is to put all free VC identifiers at each output port in a queue and return the head of the queue during virtual-channel allocation, thus avoiding the need to arbitrate among multiple free VCs [225], [172].

2.4.5 Low-latency Pipelined VC Switch Organization

Having described the implementation of the main components of a virtual-channel-based switch, we will discuss in this section their interaction and the overall switch organization.

Similar to a VC-less wormhole switch, when a head flit arrives to a virtual-channel switch needs to perform the following 6 tasks: buffering, packet header decoding and routing computation, virtual-channel allocation, switch allocation, crossbar traversal, and link traversal. In the case of body and tail flits, routing computation and virtual-channel allocation are omitted, inheriting the decisions taken by the head flit. In virtual-channel switches, switch allocation settings are not passed from the head flits to the remaining flits of the packet, and every flit competes on its own for gaining access to an output port of the crossbar. In this case, compared with VC-less switches the number of tasks is increased by one due to the introduction of virtual-channel allocation. Also, now, some tasks like switch allocation have increased complexity. Besides that, the dependencies between the tasks remain almost the same and they can be executed as shown in Figure 2.19.

In the task dependency graph of Figure 2.19, VC allocation (VA) and switch allocation (SA) are performed sequentially. This guarantees that only packets that have successfully obtained an output VC from the VC allocator can make requests for their desired output channel [126]. However, with this task order VCs may be assigned to packets that subsequently will lose in switch competition; this may happen for instance when the flits encounter significant input crossbar contention. An example of this organization for a

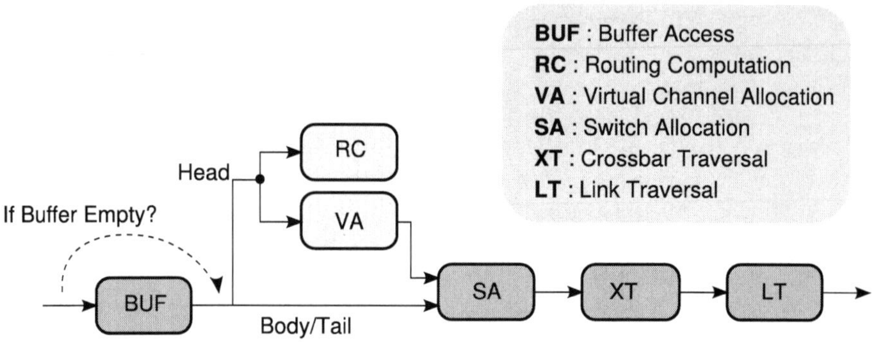

FIGURE 2.19: The task dependencies of a baseline virtual-channel switch.

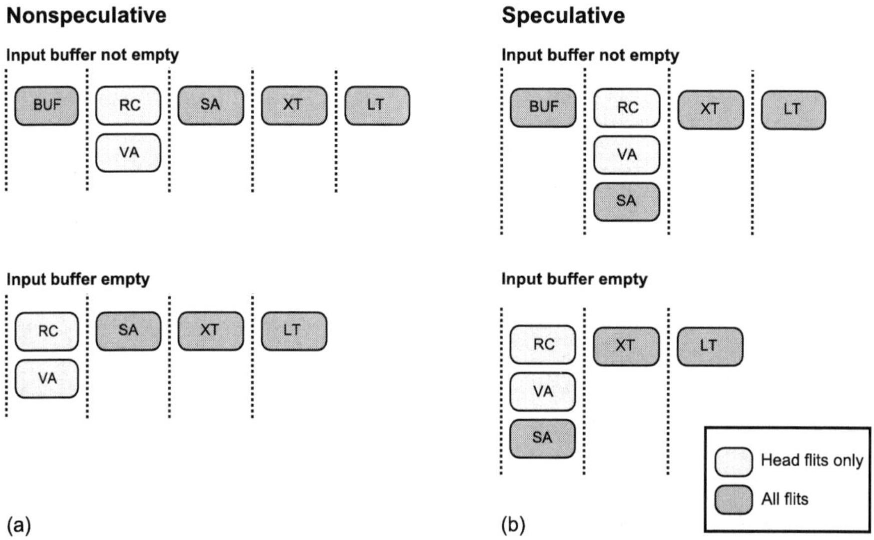

FIGURE 2.20: The pipelined implementations of a virtual-channel switch: (a) The organization of a nonspeculative pipelined VC switch and (b) its speculative counterpart.

pipelined virtual-channel switch is shown in Figure 2.20(a). Alternatively, the order of VC allocation and switch allocation can be swapped after taking care that the flit that has gained access to the crossbar will find at least one free VC at the corresponding output [172].

Nevertheless, when aiming at the reduction of latency, switch allocation and VC allocation should be performed in parallel for the head flits [251], [127]. This is only possible via speculation. In this case, the head flits bid for crossbar access at the same time they request an output VC (see Figure 2.20(b)). This type of speculation may waste crossbar grants when VC allocation fails.

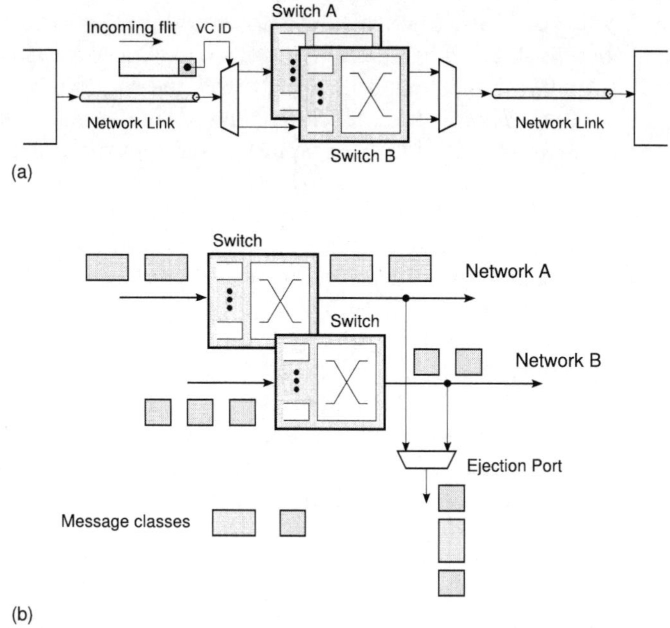

FIGURE 2.21: Virtual-channel switch implemented with separate components: (a) Virtual channels share the network's links but use separate switches and (b) completely separate networks are used by each virtual channel that are merged only at the ejection port of the network interface.

To avoid performance loss, two separate switch allocators are used. The first handles speculative requests that come from flits that have not yet gained an output VC, and the other deals with nonspeculative requests, which refers to packets with an already allocated output VC. Nonspeculative grants are prioritized over speculative ones. Speculative grants are discarded when nonspeculative grants match the same input-output pair.

Finally, please notice that similarly to a VC-less wormhole switch, when performing switch allocation we should guarantee that there are available buffer slots of the corresponding output VC. Especially, in the case of speculative requests, additional care is required so as to ensure that besides allocating an output VC to a granted input VC, the corresponding output VC has the required buffer space to receive a new flit.

2.4.5.1 Virtual-channels in Separate Networks

Up to now the only structural difference between a virtual-channel switch and a VC-less wormhole switch is the separation of the buffers to parallel buffer lanes and the addition of a virtual-channel allocator. As described though in Section 2.4.4.3, depending on the class that the VC corresponds, not all VC-to-VC mappings are possible. If we take into account only the VCs of a given message class, then, instead of just simplifying the VC allocator, we can build a simpler switch that is designed specifically for this message class.

At each node of the network we replicate as many simpler switches as the number of the message classes [101]. A graphical representation of this operation is shown in Figure 2.21(a). Replicated switches share the same physical input and output ports, similar to conventional virtual-channel switches, but with the main difference that, in the new implementation, VCs that belong to different message classes have their own access to a replicated crossbar and switch allocation units. The control path of each subswitch is smaller since it handles only V/M virtual channels per physical link, while the datapath remains roughly the same. The only simplification appears at the buffers where only V/M parallel lanes needs to be supported in each subswitch.

Extending this methodology, each message class can be implemented in a completely separate network, as done in the Tilera processor [339] (see Chapter 7) and represented graphically in Figure 2.21(b). In this case, both the switches and the links are replicated and practically the traffic of each message class does not interfere at any point with the traffic of the other message classes. The only point of convergence between packets of different message classes is the ejection point, where the terminal node should serialize the simultaneously arriving packets from different subnetworks.

2.5 Modular Switch Construction

Under certain circumstances, from either an architectural or an implementation perspective, it is beneficial to implement the switch in a modular fashion. The modularity of the implementation is achieved by employing a hierarchical switch construction, based on *port slicing*. At each hierarchy level, the input and output ports of the switch are grouped together and handled by smaller switches. To complete the design, the intermediate results of each subswitch are merged together using additional multiplexers at the upper levels of hierarchy [156].

The modular construction of a $N \times N$ switch is based on the separation of the inputs and the outputs into distinct groups. Assume that we separate the inputs into k groups of N/k inputs each and the outputs into m groups with each group consisting of N/m outputs, respectively. Each input group broadcasts its signals to a row of m smaller subswitches that have N/k input ports same as the number of inputs per group, and N/m output ports, equal to the number of outputs per group. Thus, there are k such rows of subswitches in total, leading to a 2-D array of km subswitches. Each subswitch handles only the traffic that flows between a certain input-output group. Hence, at each column of the array, there are k subswitches that all want to connect

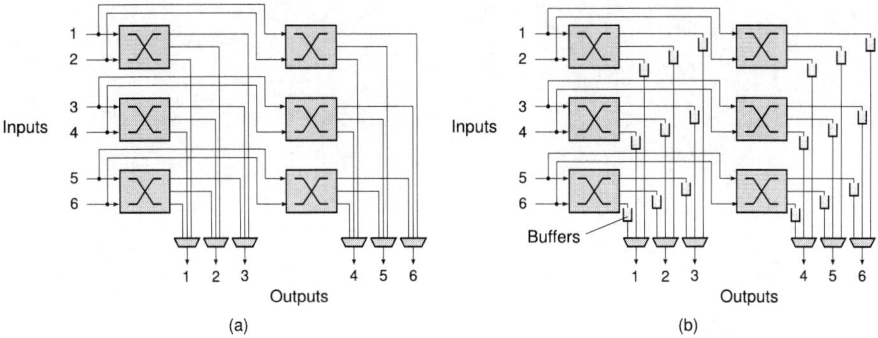

FIGURE 2.22: Hierarchical switch construction using port slicing. (a) An example of a 6×6 switch with 3 input port slices and 2 output port slices as well as (b) its buffered version that decouples switch allocation at each subswitch from the arbitration of the merging multiplexers at each output.

to the same output group. Selecting the correct output of the k subswitches requires N/m additional k-to-1 multiplexers at each output group.

An example application of port slicing on a 6×6 switch that is built hierarchically after slicing the input ports to $k = 3$ groups and the outputs to $m = 2$ groups is shown in Figure 2.22(a). The switch now consists of 6 smaller subswitches of 2 inputs and 3 outputs. Also, 3 additional 3-to-1 multiplexers are needed at each output group that merge the intermediate results that appear on each column of subswitch array.

Although the switch is hierarchically constructed, the switch allocation logic should still handle all available input-output connections, even if it can be built from smaller allocators following the modular structure of the switch. If we want each partition to act independently as far as the scheduling decisions are concerned, we should add intermediate buffers at the outputs of each subswitch. In this way the traffic that leaves each subswitch is decoupled from the traffic of the other subswitches and the scheduling decisions at the merging points are handled by independent arbiters. This decoupling is depicted in Figure 2.22(b).

The most common application of hierarchical switch organization is *dimension slicing*, where the ports are grouped according to the network dimension they belong. Each group is handled by smaller switches that handle only the traffic that remains on the same network direction. The traffic that does not belong to the same dimension and corresponds to the turns imposed by the routing algorithm is handled by additional links that interconnect the per-dimension subswitches. An example of a dimensioned-sliced switch is shown in Figure 2.23 for the case of 2-D mesh. Also, it is possible to introduce intermediate buffers to decouple the x-dimension from the y-dimension.

The main argument in favor of this kind of port slicing is that, since the per-dimension switches are smaller and perform a limited set of functions, they

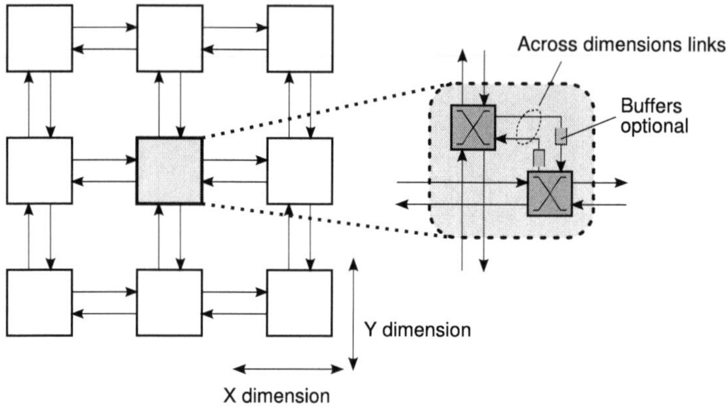

FIGURE 2.23: An application of port slicing: Dimension slicing groups input-output ports according to the network dimension they connect to.

can offer also lower latency than a switch that handles all input-output ports at once. Therefore, when such an approach is used in concert with a routing algorithm that keeps a packet on the same dimension for as long as possible, then the total latency is reduced, even if extra cycles are required when a packet turns and changes dimensions. The design of dimension-sliced switches was taken one step further by Kim in [154] after following techniques similar to Tilera's switches [339] and the Alpha 21064 router [223]. In this design, switch allocation is almost eliminated by providing priority to packets that are in flight and continue to travel in the same dimension, without sacrificing fairness with proper modification of the flow control mechanism.

Dimension slicing should not be considered as a special case of switch re-organization but as a simple subcase of the generic port slicing presented in Figure 2.22. The only difference compared to Figure 2.22 is that instead of allowing all possible connections between ports, we impose to ports that belong to different dimensions to have reduced interconnectivity as shown in Figure 2.23. Similarly, concentration, shown in Figure 2.1(c), is also just an application of port slicing. In this case, the ports of the switch are sliced in two groups, one for the terminal ports and the second for the network ports, which are handled by two separate subswitches that communicate only via a single link.

Finally, another form of switch slicing is *bit* or *channel slicing*. In this case, the datapath of the switch is sliced to independent datapaths of smaller width. This operation is equivalent to spreading parts of each flit to multiple smaller switches. For example, Figure 2.24 shows how a 32-bit switch datapath is split into 4 smaller switches that each one handles 8-bit inputs. Although, this organization sounds fairly easy to implement, it still has some vague points, with respect to the control logic that need to be clarified. One approach suggests that the control logic should remain unified and all its decisions to be

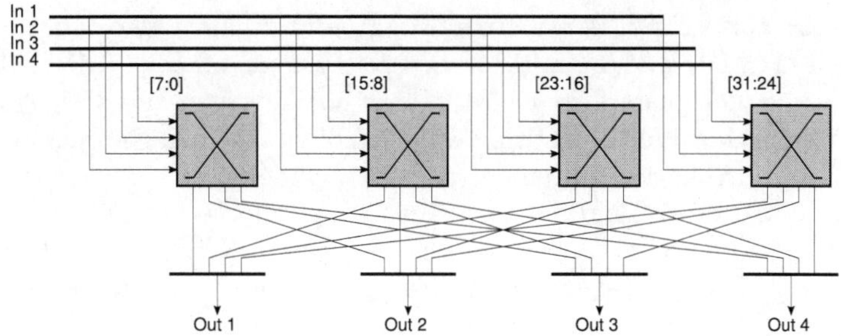

FIGURE 2.24: Application of bit slicing on a 4 × 4 switch with a 32-bit datapath.

broadcasted to each subswitch. Of course this technique requires that the head flits, which are possibly segmented to smaller words, should be first gathered and then handled by the unified control logic. In a different perspective the control logic is copied at each subswitch allowing each one of them to operate independently. However, in terms of hardware implementation, this may not always be the best choice, since the smaller the width of the datapaths the larger the portion dedicated to control logic at each subswitch and thus the larger the overhead.

Originally, bit slicing was used in order to divide a network switch into multiple chips since the area of a single chip did not suffice to fit the complete switch. With current levels of integration this reason no longer exists except for ultrademanding cases. Besides area constraints, bit slicing was employed in TinyTera [205] to handle the excessive power budget that a unified switch would require, mainly due to the power consumption of the high-speed IO links. Unlike off-chip networks, in an on-chip setting these constraints are not expected to show up. However, even in simpler NoCs, when packets of different width travel in the network, the parts of the channels that do not participate in the traversal of narrow flits can be switched off in order to reduce unnecessary power consumption. In such cases, bit slicing may still be a useful technique.

2.6 Conclusions

This chapter tried to categorize and present the most efficient switch architectures in a concise manner so that the similarities and the differences between various design alternatives are appropriately highlighted. Also, several unexplored design techniques have been shortly discussed taking also into account their true VLSI implementation cost.

Switch architectures are constantly evolving with the goal to achieve the tight latency/throughput constraints of high-performance chip multiprocessing systems (CMP) and/or multiprocessors systems-on-chip (MPSoCs). At the same time, designers face a host of new challenges in the design of NoC switches that lie beyond the natural evolution of an increasing number of cores and higher performance requirements. These include power management, synchronization and error handling, as well as adaptivity to both manufacturing variations or application characteristics. Therefore, even more sophisticated architectures are expected in the future that would be versatile enough to offer efficient solutions in this new demanding environment.

Acknowledgment

This work was supported by the project NaNoC (project label 248972), which is funded by the European Commission within the Research Programme FP7.

Chapter 3

Logic-level Implementation of Basic Switch Components

Giorgos Dimitrakopoulos
University of West Macedonia, Greece

3.1 Introduction

The main components of a switch are the buffers, the crossbar, and the routing logic as well as the switch and virtual-channel (VC) allocator. In this chapter we explore the various alternatives for the logic-level implementation of these components and discuss pros and cons of each available option. A single optimum solution does not exist for any of the main switch components and the designer is responsible for selecting the appropriate solution that best fits the area-delay-energy constraints of the specific switch implementation. In any case, a complete design space exploration is required that examines at the same time the various architectural-level solutions analyzed in Chapter 2 along with their implementation complexity. We hope that this chapter acts as

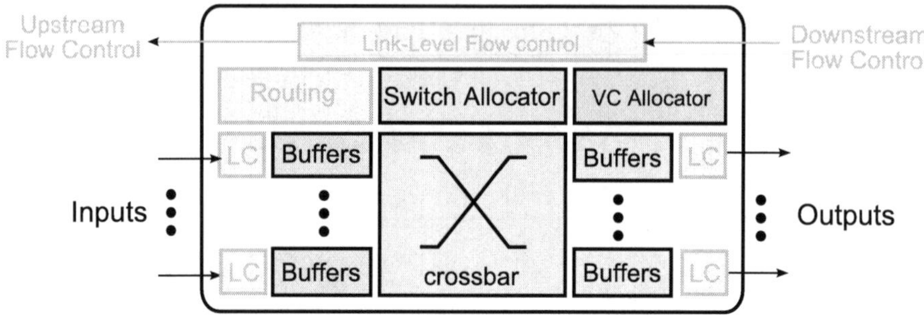

FIGURE 3.1: The main parts of the switch. This chapters deals with the implementation of the highlighted components: the buffers, the crossbar, and the switch/VC allocator.

a guide to the designer providing several alternatives for the implementation of the basic switch components.

This chapter focuses on the design of the buffers, the crossbar, and the switch/VC allocators as shown graphically in Figure 3.1. The implementation aspects of the routing logic and any synchronization primitives used inside the link controllers of the switch are presented in detail in the following chapters. In particular Chapter 5 discusses the various alternatives for the design and implementation of routing logic, while Chapter 6 deals with the switch-to-switch synchronization issues and the design of the corresponding link controllers.

We start our discussion with the datapath of the switch that is the buffers and the crossbar and continue with the components used to build the switch and virtual-channel allocators.

3.2 Buffer Memory

We have two options for the implementation of the memories at the inputs of the switch. The first one is to statically partition the memory per VC or to share the buffer space among all VCs. The second alternative is more efficient both in terms of performance and buffer utilization, assuming that additional care is taken to avoid buffer starvation of lightly utilized VCs. Any other architecture presented is actually a mix of these two basic techniques.

Statically partitioned buffers are the easiest to implement. In the baseline case that a single read and write port is needed, a simple SRAM macro or an array of registers with decoders and multiplexers suffice. An example of ASIC (Application-Specific Integrated Circuit) register-based implementation of an input buffer is shown in Figure 3.2(a). Of course, when only one port is avail-

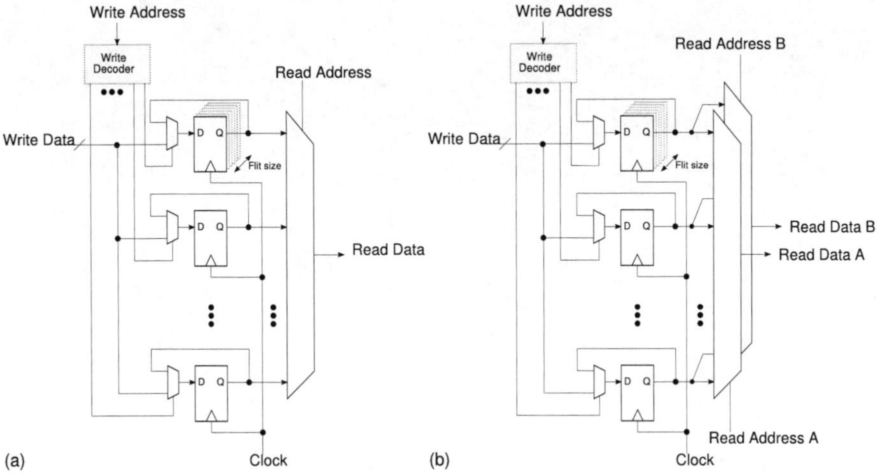

FIGURE 3.2: Statically partitioned buffer organization of 4 VCs for the case of an ASIC register-based implementation: (a) One write and one read port and (b) adding an additional read port by an extra read multiplexer.

able for both reading and writing, the two operations need to be serialized and performed in different cycles one after the other. In the case of SRAM (Static Random Access Memory), if more read ports are required multiple banks need to be present, [1] while in the case of register-based memories the addition of extra output multiplexers suffice for providing the needed amount of read ports. This organization for the case of an ASIC register-based implementation is shown in Figure 3.2(b) where one additional read port has been added to the baseline input buffer.

The number of write ports need to be increased only in the case of high-throughput memory organizations as output queuing. In this case, making the memory k times wider than the input word size and writing new data into the memory in batches of k words offers equal throughput to a memory with k independent write ports. The burden in this case moves to appropriately managing the data written to the wider memory.

A shared buffer per input is probably the most interesting approach and offers many potentials. All the queues that are present per input are stored inside a common memory array and they are handled by a set of linked lists [312],[230]. Each linked list follows the storage of each packet in the array, while separate pointers link to the start and the end of each queue. Empty cells do not remain unstructured but they are organized in a free list. These pointer structures are stored in independent and smaller memory arrays, best implemented with latches, that can be accessed fast, removing the high-port

[1]We assume that we are not able to add extra read ports to the SRAM macro. This is only viable in a full custom design where we design the input memories from scratch.

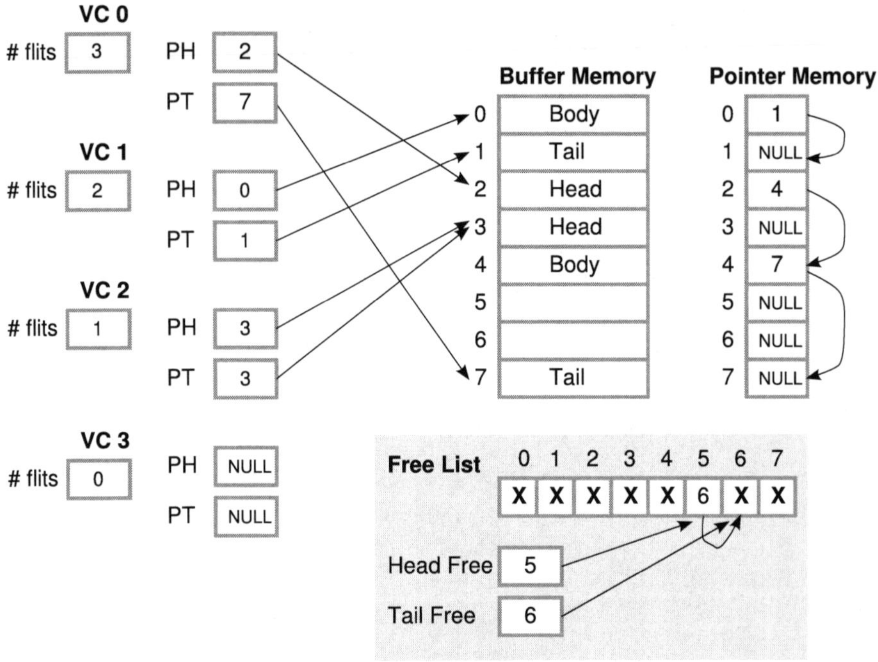

FIGURE 3.3: The shared buffer organization of four VCs.

requirement of the main buffer memory. In the following we will describe the organization of shared buffer that handles four virtual channels each one holding one packet of arbitrary length. Although this organization may not be the most complex case that linked lists can handle, it is a good representative of a shared buffer organization in the case of on-chip networks.

Every VC is associated with two pointers, PH (Header Pointer) and PT (Tail Pointer), that point to the position of the head and the tail flits of the packet, respectively. In case of single-flit packets the pointers refer to the same position. Also, when the head flit of the packet has departed from the switch, PH points to the first of the following body flits, while an uninitialized pointer just contains the symbol NULL. Also, an additional counter may be added to each VC that counts the number of flits of the packet stored in the VC, which can also guide the generation of full and empty flags required by the link-level flow control mechanism. The pointers required for the link-list implementation of each VC are stored in a separate memory block called the *pointer memory*. According to the example organization shown in Figure 3.3, VC 0 stores a 3 flit packet. Its head and tail pointers, PH(0) and PT(0), point to positions 2 and 7 of the buffer memory, while the intermediate body flit lies in position 4. This is depicted by the pointer stored in position 2 of the pointer memory. Also, the free list consists of 2 cells that lie in positions 5 and 6.

Enqueue NewFlit at VC #k

```
FreeSlot = Head(FreeList)              // Get an empty slot from the Free List
BufferMemory[FreeSlot] <– New Flit     // New Flit put in buffer
PointerMemory[PT[k]] <– FreeSlot       // Connect previous tail of VC k to new flit
PT[k] <– FreeSlot                      // Update tail pointer of VC k
PointerMemory[FreeSlot] <– NULL        // Make new flit the last flit of the queue
... // Update the Head pointer of the Free List
```

Dequeue OutputFlit from VC #k

```
Head <– PH[k]                          // Get the head pointer of VC k
OutputFlit <– BufferMemory[Head]       // Get flit from the head position
Next <– PointerMemory[Head]            // Connect next pointer to old head
PH[k] <– Next                          // Make new head the next flit int the queue
PointerMemory[Head] <– NULL            // Release old head flit from list of VC k
... // Return position Head to the Free List
```

FIGURE 3.4: The operations involved in the (en)dequeue operations of a shared buffer virtual-channel organization.

The input buffer is needed to enqueue at most one new flit per cycle. In brief, one buffer slot is released from the free list and the tail pointer of the corresponding VC moves to the appropriate position. The steps involved in the enqueue operation at VC k assuming that the free list has available slots, are shown in Figure 3.4. Similarly, for the case of a dequeue operation from VC k, the flit that is pointed by the PH pointer is removed from the list and its position is returned to the free list. In order for the list to remain connected, the head flit pointer of the corresponding VC should be updated and link to the flit following the one that has just been dequeued. These steps are summarized in Figure 3.4.

3.3 Crossbar Design Options

The crossbar is responsible for connecting the inputs to the outputs of the switch. The crossbar is a nonblocking network in the sense that an unused input can always connect to an unused output without destroying the connections of other input/output pairs. The connection realized by the crossbar is determined by the switch allocator. In a single-cycle implementation the decisions of the switch allocator are given to the crossbar in the same cycle. In this configuration the delay of both components should be aggressively optimized in order to satisfy the tight delay constraints. Alternatively, in a pipelined

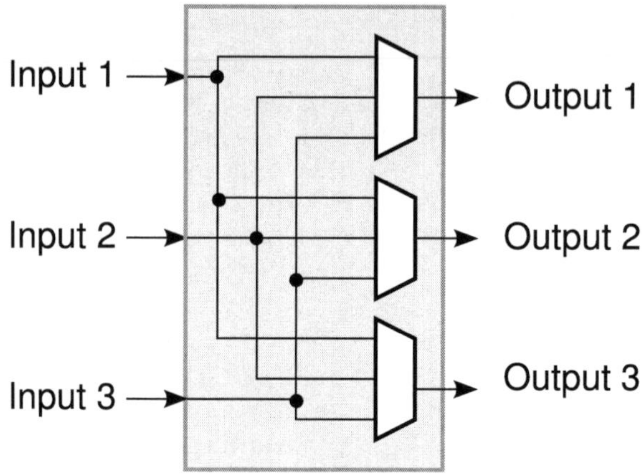

FIGURE 3.5: A 3 × 3 crossbar implemented using a multiplexer at each output.

implementation the crossbar (crossbar traversal stage) and the switch allocator operate in different cycles. The only difference with respect to single-cycle implementations is that the output of the switch allocator is at first latched before being given to the multiplexer. Thus, in the pipelined implementation, even if the switch allocator and the crossbar do not contribute directly as a pair to the delay of the switch, the choices made for one of them directly affects the other. This argument will be clarified in the following paragraphs.

3.3.1 Crossbar Logic

The crossbar is commonly built with a single multiplexer per output, as shown in Figure 3.5. All inputs are connected directly to all output multiplexers. Therefore, any connection is possible between inputs and outputs without any restrictions.

The multiplexers are controlled by their select lines, which are actually connected to the grant signals of the corresponding arbiters either directly, in a single-cycle implementation, or via registers in the case of a pipelined switch. The circuit-level implementation of the multiplexer restricts the format of the grant decision. Four different 4-to-1 multiplexer implementations are shown in Figure 3.6. In the cases shown in Figs. 3.6(a)–(c) the grant decision is encoded in one-hot form and the multiplexer is implemented either using (a) an AND-OR tree of gates, (b) a parallel combination of transmission gates, or (c) tristate buffers. One-hot encoding utilizes a separate line for each input, where the single 1 denotes, which input has won the grant. On the contrary, in Figure 3.6(d) the multiplexer is implemented as a tree of 2-to-1 multiplexers. In this case, the select lines (grant signals) that configure the paths of each level of the tree are encoded in weighted binary representation. Therefore, in

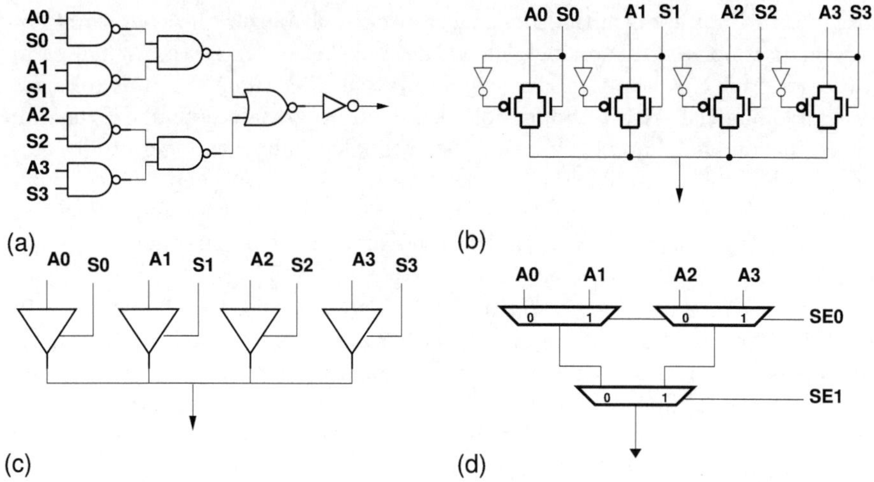

FIGURE 3.6: Possible circuit-level implementations of a 4-input multiplexer.

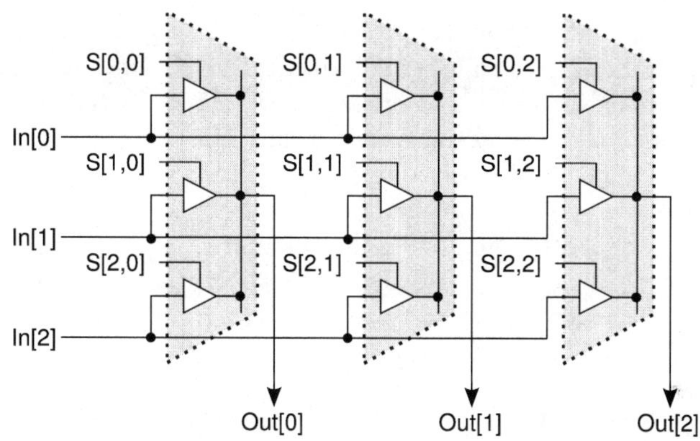

FIGURE 3.7: 2-D implementation of a 3 × 3 multiplexer-based crossbar.

order not to limit the designer's choices concerning the implementation of the multiplexer, the outputs of arbiters, i.e., the grant signals, should cover a wide range of encodings.

Of course, we could design arbiters that produce the grant signals in one form only, either in one-hot or weighted binary and translate it to the required format using either an encoder (one-hot to weighted binary) or a decoder (weighted binary to one-hot) before driving the multiplexer. Although this is a viable choice, it increases the delay significantly.

Differently from Figure 3.5, crossbar designs are often presented as a 2-D array of transistors that short the input and output lines in order to achieve a connection. Practically, as shown in Figure 3.7, this approach is not different from Figure 3.5, after choosing for the multiplexer the tristate buffer implementation. If the multiplexers are actually placed in a rectangular form is a matter of the layout of the whole switch. The signals $S[i,j]$ are used for controlling the tristate buffers of each column of the crossbar. In general, depending on the characteristics of the technology and the logic family employed, different multiplexer implementations may be the best choice [340]. Nevertheless a widely applicable principle is that for very wide multiplexers a tree of smaller multiplexers should be used that may combine more than two inputs at each level of the tree.

Finally, concerning the various circuit-level options, we should keep in mind that transmission-gate implementations are not supported in a standard-cell based design flow, while tristate buffers should be avoided since they may create testability problems if no care is taken [340]. In a standard-cell logic-synthesis based environment the implementation shown in Figure 3.6(a) and 3.6(d) are the best choices.

3.3.2 Crossbar Floorplan

It will be proved in Chapter 4 that when going through a mainstream physical synthesis toolflow, high radix crossbars used inside NoC switches become prohibitively expensive (area- and critical path-wise). To the limit, physical design convergence becomes even infeasible since the physical implementation tool is not able to simultaneously enforce the timing constraint, control routing congestion, and avoid crosstalk. This results in thousands of layout rule violations that even a 50% increase of floorplan area or a 50% decrease of the target frequency are not able to counter. However, hand-crafting the crossbar floorplan (whenever affordable) provides unprecedented opportunities to successfully infer crossbars with a large number of input/outputs. This is the discovery of the work in [245](©[2010] IEEE), which shows that by proper guidance of the EDA (Electronic Design Automation) tools through an algorithmic placement of the circuit, very high radix crossbars are feasible, which can be in fact very cheap both in terms of area and delay. In this section based on the results of [245] we will categorize the various alternatives for placing the crossbar inside the switch and analyze the characteristics of each approach.

We assume that the switch follows a tiled implementation where each tile is dedicated to serve one input-output port. The tiled organization is modular and scalable, while it represents most of the industrial switch implementations presented in open literature, like the one presented in Chapter 8. Each tile consists of the corresponding input buffers, the per-input or per-output arbiters used in switch and VC allocation logic, as well as link controllers and possibly dedicated per-input routing logic. The buffers consume the majority of the tile's area and consist mostly of SRAM blocks, plus a small area for port control, while nonbuffer areas correspond to the above-mentioned blocks. The physical layout of SRAM blocks already utilizes some of the lower wiring

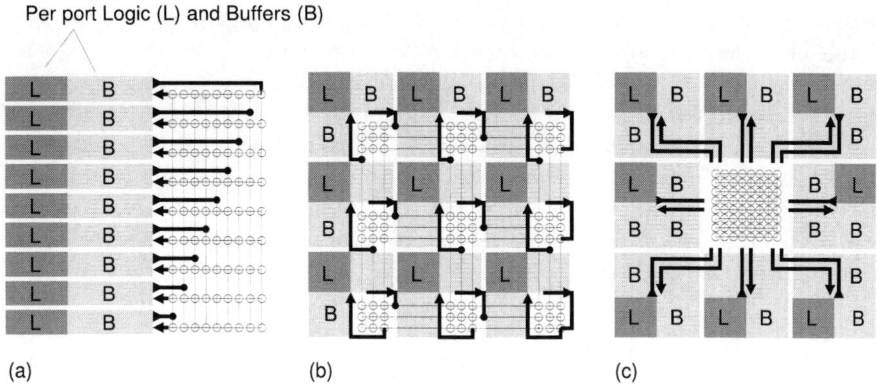

FIGURE 3.8: Crossbar possible floorplans: (a) column of tiles using side crossbar, (b) matrix of tiles using distributed crossbar, (c) matrix of tiles using centralized crossbar (from [245] ©2010 IEEE).

metal layers of the chip. In the reference implementation of [245] SRAM blocks utilize only the four lower of nine wiring metal layers. The remaining five upper metal layers are not obstructed and they are available for routing in both directions, thus implementing the interswitch wiring connections.

Based on the placement of the tiles three different alternatives for the placement of the crossbar inside the switch are identified. The three design options are depicted graphically in Figure 3.8.

Figure 3.8(a) shows the first floorplan considered. The tiles are arranged in a single column, with the crosspoints to each of them lying on its side. While being simple, this floorplan is problematic when N is large because it implies skinny tiles.

Figure 3.8(b) depicts a more scalable floorplan. The tiles are Γ-shaped with input buffers per port occupying at least the two-thirds of the tile and the remaining area is given to other per-port switch components. The tiles are placed in a $\sqrt{N} \times \sqrt{N}$ matrix where each tile owns a crossbar input line spanning its row and a crossbar output line spanning its column. Thus, each tile also embraces the $\sqrt{N} \times \sqrt{N}$ crosspoints between the $\sqrt{N}$ (horizontal) crossbar input lines that are owned by the tiles in its row and the $\sqrt{N}$ (vertical) crossbar output lines that are owned by the tiles in its column.

Providing space for the crossbar lines can be achieved in two ways. The traditional way is to leave wiring channels in both directions between the port tiles. This approach has considerable overhead. Alternatively, as shown in Figure 3.8(b), the crossbar lines can be routed above the input buffers. The second approach can be considered as an enriched mesh, where each tile uses a subswitch with more crosspoints and links.

Figure 3.8(c) shows an alternative two-dimensional floorplan. The tiles are now square and the buffers occupy at least three-fourths of their area. The crosspoints are concentrated in the central region of the matrix replacing some of the tiles. Hence, the crossbar lines are shorter, as all the crosspoint logic is kept in close proximity, while the wires connecting the tiles with the crossbar are extended to reach the center. Essentially, instead of routing over the tiles the crossbar lines as done in Figure 3.8(b), the links that connect the tiles to the crossbar are routed. The over-the-tile link density can be balanced if the tiles are divided in four isometric groups and the tiles of each group are routed to a distinct edge of the crossbar region. In this way, the link density varies from its lowest at the corners of the switch, to its highest at the periphery of the crossbar.

Depending on the characteristics of the technology and the switch radix either the distributed or the centralized crossbar are equally good solutions. Which one offers the best possible implementation should be verified by actual circuit implementations.

3.4 Switch/VC Allocators

According to Chapter 2 the implementation of switch and VC allocators can take two forms. The first one is the centralized approach where one circuit is responsible for deriving the complete input-output matching. The second approach, which is more scalable, relies on separate per-input and per-output arbiters that operate independently and reach an agreement (input-output matching) after communicating their intermediate decisions. In this section, we will discuss first the implementation of centralized allocators and then we will move to an in-depth presentation of variable-priority arbiters that are the core of separable switch and VC allocators.

3.4.1 Centralized Allocators

Centralized allocators do not rely on independent per-input (output) and per-output (input) arbitration stages but resolve any output contention at once using a single circuit. The most practical and widely agreed centralized allocator is the wavefront allocator that was presented in [311].

The wavefront allocator represents the set of input requests as a 2-D matrix, where the rows correspond to inputs while the columns represent the outputs. For example, when input i wants to connect to output j it asserts the request $R[i, j]$. The wavefront allocator generates a matching by starting at an initial diagonal and granting all requests that fall under it. For each request that is granted, all requests that belong to the same row or column are discarded. The allocation proceeds in the same fashion with the next diagonal, wrapping around at the end of each row and column, until all diagonals are examined. Please note that granting requests that belong to the same diagonal

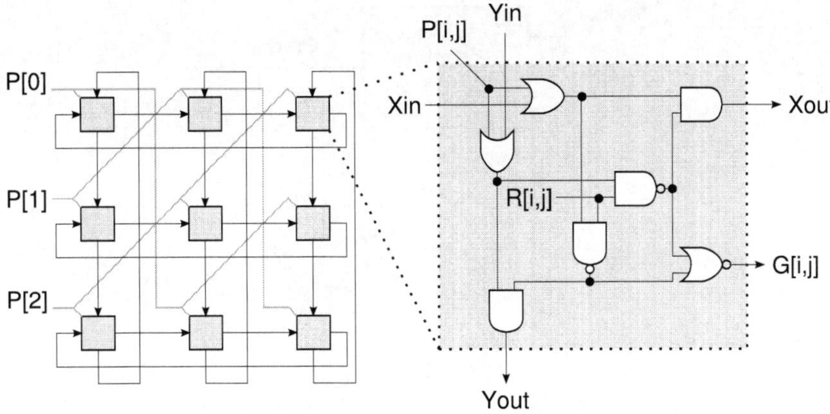

FIGURE 3.9: The implementation of a 3×3 wavefront allocator.

can be performed without any restrictions since they correspond to distinct input-output pairs. A problem would appear only if the allocator granted more than one request of the same row (same input) or the same column (same output), respectively.

Wavefront allocator is implemented as a 2-D array of simple nodes as shown in Figure 3.9. Due to this structure the critical path grows linearly with the number of inputs. The main drawback of wavefront allocator is the wrapped-around connections that are inherent to its operation and create combinational loops that cannot be handled by synthesis tools. In order to derive a loop free implementation, the designer could use an exhaustive technique, where the wavefront allocator is replicated N times, and each allocator assumes a different diagonal of the matrix as the one with the highest priority. At the end, only one of the multiple instances of the wavefront allocators would have made the correct decision. Selecting the appropriate one is done according to the actual priority. Alternatively, a very interesting technique for breaking the loops has been proposed in [133]. Each loop is first unrolled leading to a $2N \times 2N$ array of nodes, where, at each cycle, only the appropriate $N \times N$ submatrix is chosen based on the highest priority diagonal.

Finally, the NoC switch implementations can benefit from other centralized allocators that have been proposed for the case of Internet-wide network routers. A characteristic example of a simplified centralized allocator that can achieve high-throughput and offer high-speed implementation is the ALOOFA allocator [324]. ALOOFA is based on an approximation of LOOFA (Longest output first algorithm) that favors heavily backlogged input queues (similar algorithms have been also proposed in [324]). The ALOOFA approach removes any complex sorting circuit involved in the selection of the input with the largest number of waiting flits, and allows a regular implementation following the 2-D array principle of the wavefront allocator.

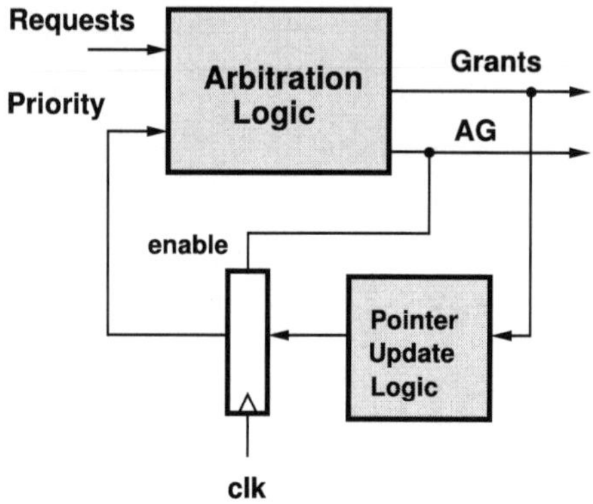

FIGURE 3.10: The general structure of a variable-priority arbiter.

3.4.2 Arbiter Design

The arbiter is responsible for resolving conflicting requests for the same resource and granting only one of them. Besides its basic functionality, the arbiter should guarantee in some form the fair allocation of the shared resource to the inputs. Such guarantee should grant first the request with the highest priority. To allow a fair allocation of the resources and to achieve high overall system performance, we should be able to change the priority of the inputs dynamically. The way the priority changes is part of the policy employed by the system's scheduling algorithm. For example, the most widely used round-robin policy dictates that the request served in the current cycle gets the lowest priority for the next arbitration cycle. Therefore, in order for the arbiter to be practical, it should support variable priorities and this extra feature should add the minimum possible delay overhead.

The block diagram of a general Variable Priority Arbiter (VPA) is shown in Figure 3.10. It consists of two parts; the arbitration logic, which is a combinational logic block that decides which request to grant based on the current state of the priority vector, and the pointer update logic that decides according to the current grant vector, which input to promote. In other cases, the pointer update logic can be avoided and set the priority directly by an external source (processor or network interface).

Due to the variable priority nature of the arbiter, the search for the winning request should begin from the position with the highest priority and continue in a circular manner until all requests are examined. For example, if the ith request has the highest priority in the case of an N-input arbiter then the arbiter should search circularly to positions $i+1$, $i+2$, ..., $N-1$, 0,1, ..., i, until it finds the first active request. If after scanning all inputs the circuit does not find an active request it de-asserts the AG (Any Grant) signal that declares that no input was granted. This circular search complicates the design

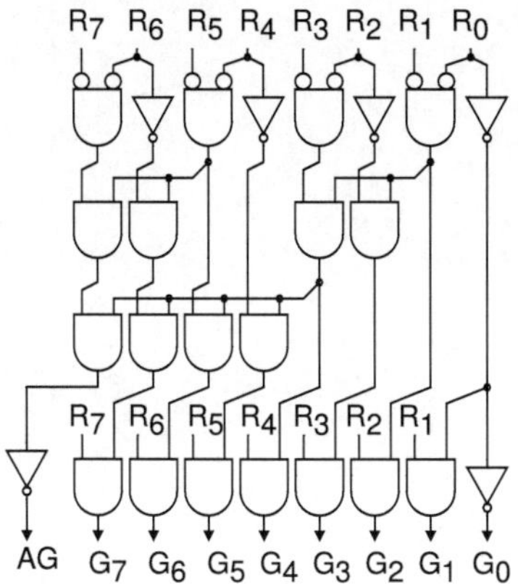

FIGURE 3.11: The parallel-prefix implementation of the 8-input FPA (priority encoder). The grant signals are one-hot coded.

of VPAs. Nevertheless, several elegant solutions have been presented so far and will be analyzed in the following paragraphs.

The commonly used variable priority arbiters can be categorized in three major groups. The arbiters designed using priority encoders belong to the first group. In the second group, we find the arbiters designed utilizing carry-lookahead structures, while in the last group we put the matrix arbiter [69]. The more scalable arbiter in terms of delay belongs to the carry-lookahead-based category and was introduced in [80].

3.4.2.1 Priority-encoding-based VPA

The VPAs of this category utilize as their main block the fixed priority arbiter (FPA). The FPA does not allow any dynamic priority change and always position 0 has the highest priority and position $N-1$ the lowest. The FPA produces its grant signals G and an additional flag AG (Any Grant) that denotes whether at least one input request was granted [340]. A grant is given to the ith request when $R_i = 1$ and no other requests exist to any position with an index smaller than i.

Depending on the encoding chosen for the grant signals the FPA can take two forms. In case that the grant signals are required to be in one-hot form, each G_i is computed via the well-known priority encoding relation:

$$G_i = R_i \cdot \overline{R}_{i-1} \cdot \ldots \cdot \overline{R}_1 \cdot \overline{R}_0$$

where $\cdot$ represents the boolean-AND operation and $\overline{R}_i$ denotes the complement of R_i. Similarly,

$$AG = R_{N-1} + \ldots + R_1 + R_0 = \overline{\overline{R}_{N-1} \cdot \ldots \cdot \overline{R}_1 \cdot \overline{R}_0}$$

showing that at least one request line is asserted. Assuming 2-input gates, the

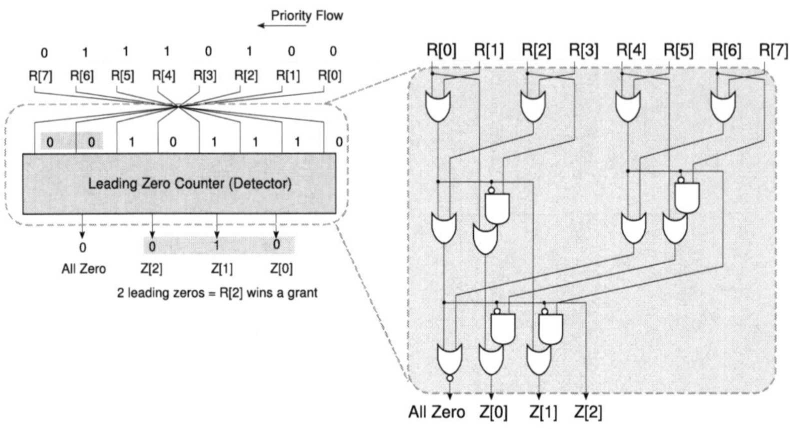

FIGURE 3.12: An 8-input FPA utilizing a leading-zero counter. The grant signals are coded in a weighted binary representation.

output of the FPA is computed in the best case in $\log_2 n + 1$ logic stages, using regular parallel prefix structures as the one depicted in Figure 3.11 for an 8-input FPA. When using AND gates with more inputs it is possible under certain circumstances to get faster circuits.

When the FPA returns its decision in the weighted binary representation, then the equivalent circuit to a priority encoder is called the *leading-zero counter* (or *leading-zero detector*). The leading-zero counter computes the position of the leading digit (1 in our case) by counting the number of zeroes that appear in front of it. The best leading-zero counters have been presented in [81] and can be built hierarchically as shown in Figure 3.12.

The two circuit forms can be used interchangeably depending on the desirable encoding for the grant signals. Therefore, in the following when we refer to FPA we mean either of them. The designs that are already built around the priority-encoding FPA can be transformed by replacing the priority encoders with the corresponding leading-zero counters.

In order to support variable priorities the arbiters utilize either more FPAs and a selection mechanism, or additional circuits that help them mimic the behavior of a variable-priority arbiter. Three design options are presented in the following paragraphs: the exhaustive, the Rotate-Encode-Rotate, and the dual-path approach [113].

Exhaustive: A N-input exhaustive VPA utilizes N independent FPAs. Since each FPA considers the request connected to input 0 to be the one with the highest priority, in order for the exhaustive VPA to work properly, it should present to each FPA a different permutation of the request lines. So, as shown in Figure 3.13(a), FPA0 receives requests R_3, R_2, R_1, R_0, while FPA1 gets R_0, R_3, R_2, R_1. In this design, the FPAs produce N different intermediate grant decisions. Selecting the correct one depends on the value of

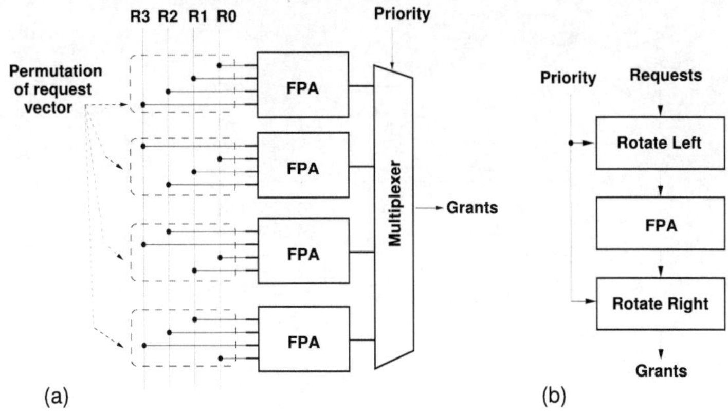

FIGURE 3.13: The exhaustive and the rotate-encode-rotate variable priority arbiters.

the priority vector. The main problem with this design is its area overhead and the presence of the multiplexer at the output, which increases the delay significantly compared to the delay of a single FPA. A similar approach has been proposed in [289]. The only difference is that instead of the output multiplexer the final grant decision is computed by a tree of OR gates and the whole design follows a hierarchical structure, which does not offer much to energy-delay performance of the circuit.

Rotate-Encode-Rotate: The rotate-encode-rotate VPA is practically the "folded" version of the exhaustive VPA. Instead of utilizing N independent FPAs that receive a different rotated version of the input requests, a rotator is utilized, as shown in Figure 3.13(b), that rotates the input requests so as to bring the highest priority request to position 0 of the FPA, i.e., its highest priority input. The number of rotations depends on the priority vector. The rotator following the FPA restores back the grant to its original position by shifting to the opposite direction. The main drawback of this approach is its delay overhead, which is roughly equal to more than 3× the delay of a single FPA.

Dual-Path: One of the most efficient VPAs is the dual-path arbiter that employs two FPAs in parallel. In order to work properly it assumes that the priority signal is thermometer coded; when the ith position has the highest priority then the $i + 1$ less significant bits of P are equal to 1 while the rest $n - i - 1$ most significant bits are equal to 0. The block diagram of the dual-path arbiter is shown in Figure 3.14. The upper FPA is used to search for a winning request from the position denoted by the priority vector P, i.e., $Ppos$, up to position $N - 1$. It does not cycle back to input 0, even if it could not find a request among the inputs $Ppos \dots N - 1$. In order not to allow the upper FPA to search in positions $0 \dots Ppos - 1$ its requests are masked with the thermometer coded priority vector P. The lower FPA is driven by the

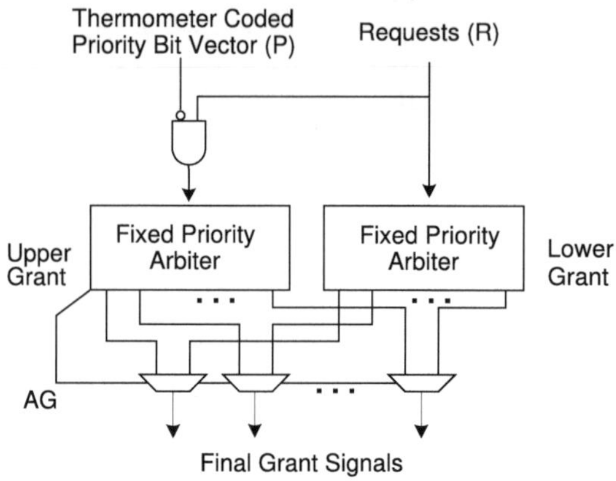

FIGURE 3.14: The block diagram of the dual-path VPA.

original request lines and searches for a winning request among all positions assuming that position 0 has the highest priority. The two arbitration phases work in parallel and only one of them has computed the correct grant vector.

The selection between the two outputs is performed by employing a simple rule. If there are no requests in the range $Ppos \ldots N-1$, the correct output is the same as the output of a lower FPA. If there is a request in the range $Ppos \ldots N-1$, then the correct output is given by the output of the upper FPA. Differentiating between the two cases is performed by using the AG signal of the upper FPA. The dual-path arbiter requires 2 more stages of logic compared to the FPA, while the AG line of the upper FPA needs to drive N logic gates in order to perform the selection procedure.

3.4.2.2 Comment on the Design of Large Priority Encoders

The design of arbiters with many inputs has long been considered a hard problem in terms of delay and various heuristics have been presented to deal with it. One interesting alternative is the hierarchical arbiter. In this approach, the requests are not treated all at once but they are partitioned into P groups. Each group contains N/P requests and is handled separately by a N/P-input arbiter. Inside each group only one (if any) request is granted by the corresponding arbiter. Then, at the next level of arbitration the grant of only one of the available P groups is selected. This is performed by a second arbiter that receives the OR of the outputs of the first arbiters and selects only one of them. The decision of the second arbiter masks accordingly the first set of grants in order to produce the final N-bit wide grant vector, as shown in Figure 3.15.

Alternatively, instead of ORing the grants of the first level of arbiters, which is practically done faster via the AG signals, and then feeding them to the second arbiter as done in Figure 3.15, we can equivalently OR the

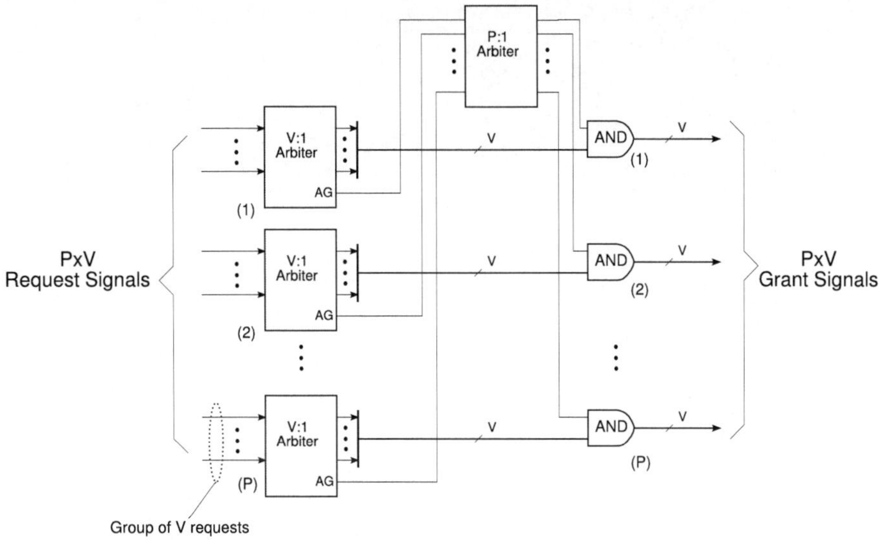

FIGURE 3.15: A hierarchical implementation of a wide arbiter.

requests of the first level of arbiters and give the result to the corresponding P-input arbiter. This arbiter would select a group of requests instead of a group of grants, thus saving some delay, without changing the functionality of the circuit [225].

These hierarchical techniques are very interesting from an algorithmic point of view. However, they don't offer any true delay benefit when compared to the parallel-prefix organization of the fixed-priority arbiter. In that case the delay of the arbiter grows logarithmically with the number of inputs. In the base case, where only 2-input gates are used, when doubling the number of inputs of the arbiter the delay is increased by only 1 logic level. Look for example at the implementation of a 32-input FPA in Figure 3.16. Therefore, partitioning is both a redundant approach and inefficient. Also, even when a smaller area is required one can resort to sparse parallel-prefix solutions as the one shown in Figure 3.17, where both delay and area efficiency are balanced. Sparse structures precompute the grant signals in small blocks of 4 inputs using simple ripple-carry-like structures, while the final grants are selected by a sparse priority encoder. Similar approaches can be adopted when the FPA is implemented with a leading-zero counter in the place of a priority encoder.

3.4.2.3 Carry-lookahead-based VPAs

Another class of VPAs are built using carry-lookahead-like structures. In this case the highest priority is declared using a priority vector P that is encoded in one-hot form. As in the case of FPA a request is granted when no other higher priority input has an active request. The main characteristic

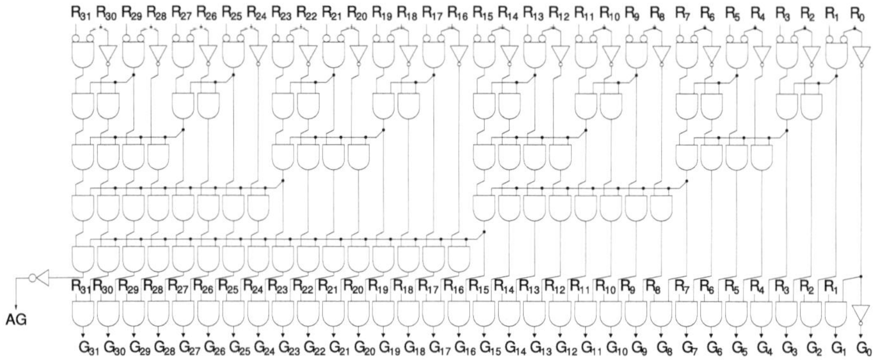

FIGURE 3.16: The parallel-prefix implementation of the 32-input FPA (priority encoder).

of the carry-lookahead-based VPAs is that they don't require multiple copies of the same circuit and they inherently handle the circular nature of priority propagation.

In this section we will present the round-robin arbiters introduced in [80]. The arbiter gives a grant to the position with an active request that has at the same time the highest priority. To encode that the ith position has the highest priority we introduce a new signal, named X_i. When $X_i = 1$ it means that the ith request has the highest priority to win a grant. Therefore, grant signal G_i is asserted when both X_i and R_i are equal to 1:

$$G_i = R_i \cdot X_i$$

Now we need to explore the case that activates signal X_i. At first, the ith request gets the highest priority when bit P_i of the priority vector is set. Practically, the position indexed by the P vector is the place where the priority is generated. This priority can be transferred to the most-significant positions, step-by-step, only when the positions in between do not have an active request and thus cannot produce a grant. Therefore, position i can get the priority from its neighbor $i-1$, only when $R_{i-1} = 0$. Merging the two cases, we can say that X_i is set when either $P_i = 1$ or when $R_{i-1} = 0$ and the incoming priority transfer X_{i-1} is activated. Expressing this result in boolean algebra leads to the following relation

$$X_i = P_i + \overline{R}_{i-1} \cdot X_{i-1}$$

In a variable priority arbiter the search for the winning position should continue in a circular manner after all positions are examined. Therefore, the priority transfer signal X_{N-1}, out of the most-significant bit position, should be fed back to position 0, $X_{N-1} = X_{-1}$, in order to guarantee the cyclic transfer of the priority. Please notice that if the priority transfer X_{N-1} is connected

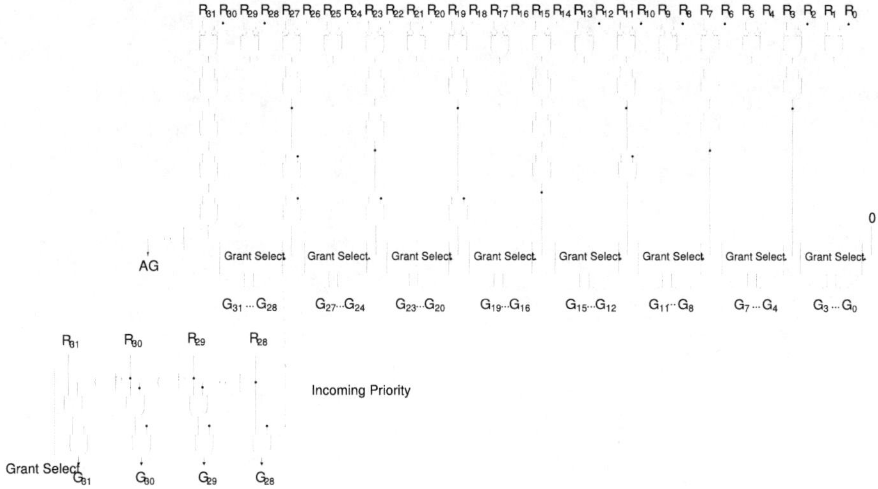

FIGURE 3.17: The sparse implementation of the 32-input FPA that saves a considerable amount of area.

directly back to position 0, a combinational loop is created, that cannot be treated efficiently by commercial static timing analysis tools and may create testability problems.

However, the cyclic operation of arbitration can be treated differently based on the following observation. The recursive definition of X_i has exactly the same form as the well-known carry lookahead equation $c_i = g_i + p_i \cdot c_{i-1}$, where in place of the carry generate bit g_i we have here the priority signal P_i (called *priority generate*), and instead of the carry propagate bit p_i we use the inverted request signal $\overline{R}_i$ (called *priority propagate*). In fact, the computation of the priority transfer bits is equivalent to an end-around carry operation, where the carry-out signal should be fed back to the carry-in position.

Multiple solutions exist to the end-around carry problem of adders that can be applied to the proposed recursive arbitration formula. For example, the combinational loop can be avoided if we utilize two carry computation units where the carry-out (X_{N-1}) will be used as a carry in (X_{-1}) for the next. This method introduces significant redundancy and should be avoided. Alternatively, the end-around carry can be virtually re-inserted in the carry computation unit by employing a carry increment block as shown in Figure 3.18(a) for the case of a 4-input arbiter.

Additionally, when high speed is required, priority transfer can be computed directly from the request and priority input bits utilizing a butterfly like carry-lookahead structure. The operation of such efficient solutions relies on fully unrolling and simplifying the recursive relation that describes the

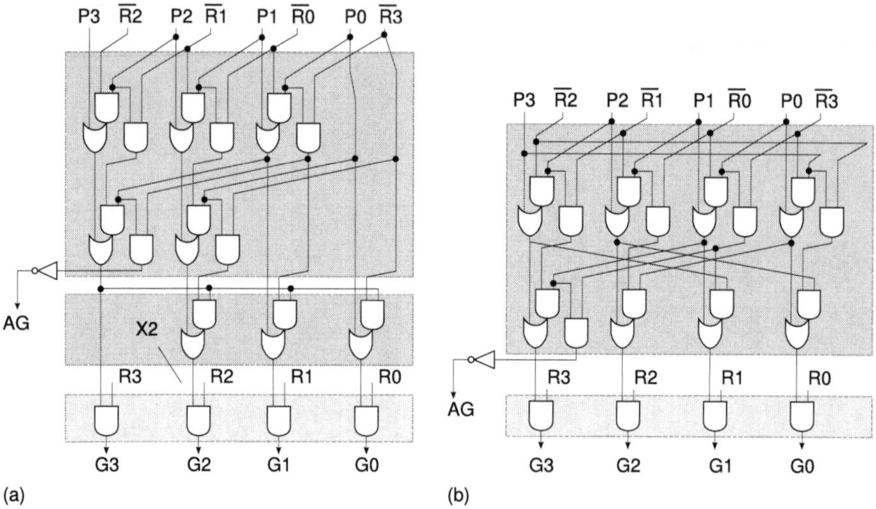

FIGURE 3.18: The implementation of a 4-input VPA using: (a) the smart end-around carry approach and (b) the faster approach that transfers the priority inside the carry-lookahead unit.

priority transfer signal X_i:

$$X_i = P_i + \sum_{j=0}^{n-2} \left(\prod_{k=j+1}^{n-1} \overline{R}_{|k+1|_N} \right) P_{|i+j+1|_N}$$

where $|y|_n$ denotes the operation $y \mod n$. For example, in the case of a 4-input VPA the priority transfer signals and the final grant signals are given by the following fully-unrolled relations.

$$G_0 = R_0 \cdot X_0 = R_0 \cdot (P_0 + \overline{R}_3 \cdot P_3 + \overline{R}_3 \cdot \overline{R}_2 \cdot P_2 + \overline{R}_3 \cdot \overline{R}_2 \cdot \overline{R}_1 \cdot P_1)$$
$$G_1 = R_1 \cdot X_1 = R_1 \cdot (P_1 + \overline{R}_0 \cdot P_0 + \overline{R}_0 \cdot \overline{R}_3 \cdot P_3 + \overline{R}_0 \cdot \overline{R}_3 \cdot \overline{R}_2 \cdot P_2)$$
$$G_2 = R_2 \cdot X_2 = R_2 \cdot (P_2 + \overline{R}_1 \cdot P_1 + \overline{R}_1 \cdot \overline{R}_0 \cdot P_0 + \overline{R}_1 \cdot \overline{R}_0 \cdot \overline{R}_3 \cdot P_3)$$
$$G_3 = R_3 \cdot X_3 = R_3 \cdot (P_3 + \overline{R}_2 \cdot P_2 + \overline{R}_2 \cdot \overline{R}_1 \cdot P_1 + \overline{R}_2 \cdot \overline{R}_1 \cdot \overline{R}_0 \cdot P_0)$$

An implementation of this faster VPA for the case of 4 inputs is shown in Figure 3.18(b). The grant vector is computed in exactly $\log_2 N + 1$ logic levels, as in ordinary FPA. Also, no large fanout line is required, since the cyclic nature of the priority transfer is performed inside the carry-lookahead tree. The only drawback of the proposed circuit are the long lines inside the priority transfer computation unit. Although the extra capacitance added by these lines degrades by a small percentage the delay of the circuit, the overall circuit is significantly faster than the most efficient previous implementations. For large number of inputs sparse carry-lookahead structures similar to the

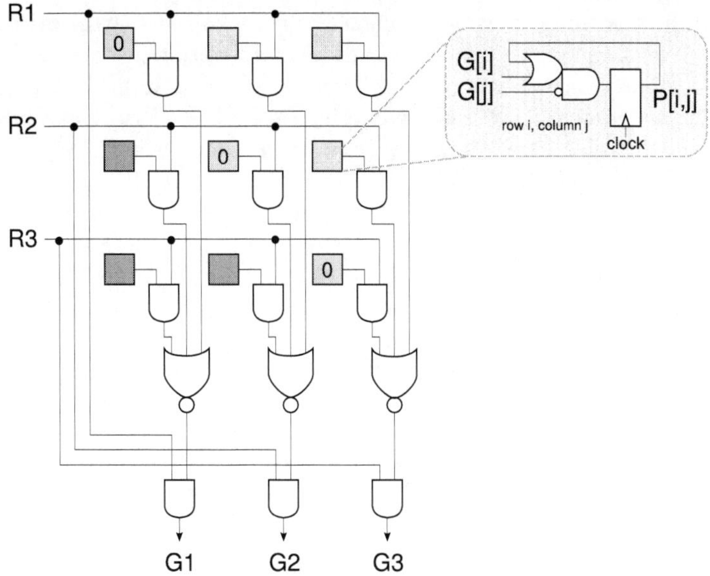

FIGURE 3.19: The implementation of a 3-input matrix arbiter. The priority cells of the main diagonal can be omitted.

ones presented in Figure 3.17 can be used. Of course, alternatively, the new boolean equations for the grant signals can be fed to any logic synthesis tool and derive some equally efficient netlists.

3.4.2.4 Matrix Arbiter

The matrix arbiter is another form of variable priority arbiter that implements the Least Recently Served policy [69]. The matrix arbiter stores more priority bits than simpler VPAs, and it is able to handle more complex relations between the priorities of each input.

As deduced from its name, the matrix arbiter uses a $N \times N$ matrix of bits to store the priority values. Each matrix element $M[i,j]$ of the ith row and jth column records the priority between the pair of inputs i and j. When $M[i,j] = 1$ then requester i has a higher priority than requester j. To reflect the priority of i over j, the symmetric matrix element $M[j,i]$ should be set equal to 0. Also, the elements of the diagonal $M[i,i]$ have no physical meaning. Thus, only the $n(n-1)/2$ elements of the priority matrix are actually needed. Care should be taken with the priority values since the possibility of a deadlock exists. For example in the case of a 3-input matrix arbiter with $M[0,1] = M[1,2] = M[2,0] = 1$ a circular priority dependency is produced, which blocks the matrix arbiter from producing any grant.

According to the operation of the matrix arbiter a requester will receive a grant if no other higher priority requester is bidding for the same resource.

Assume at first the case that all requests are active at each arbitration cycle. Then, if at least one 1 exists on column j of the priority matrix then the request of input j cannot be granted. This holds since an activated priority on column j, means that at least one input, say k, has a priority over j since $M[k, j] = 1$. Therefore, the request of input k should be granted over input j. This condition for column j can be identified by ORing all bits of the same column and nullifying the corresponding grant (jth output). However, in the general case not all requests are active. Therefore, in this case, the request from each input is ANDed first with all the priority bits that belong to the same row. In this case, an active request keeps all priority bits of the same row alive, while an inactive request on input k erases every priority of input k over other inputs by zeroing all $M[k, *]$ bits. A 3-input matrix arbiter is shown in Figure 3.19.

Once the ith requester is granted, its priority is updated and set to be the lowest among all requestors. This is performed at first by clearing all bits of the ith row, e.g., setting $M[i, *]$ to 0, and second by setting the bits of the ith column, e.g., $M[*, i] = 1$, so that all other requests will have higher priority over request i. The implementation of the priority update mechanism is also shown in Figure 3.19. The only drawback of this technique is the high fanout of the grant signals that have to drive both the outputs of the circuit as well as the priority update module.

3.5 Conclusions

The implementation chosen for the main switch components directly reflects on the complexity of the whole switch. The architectural parameters such as buffer depth, link width, and switch radix may critically determine the performance of the interconnection network but the implementation details of the buffers, the crossbar, and/or the arbiters determine the maximum delay of the switches as well as their area and power characteristics. Therefore, the study of the logic-level implementation of the main switch components is crucial when designing high-end switches under strict performance and implementation-related constraints.

Acknowledgment

I would like to thank Giorgos Passas for his contribution to the crossbar floorplan section.

Chapter 4

Topology Exploration

Francisco Gilabert

Technical University of Valencia, Spain

Daniele Ludovici

TUDelft, The Netherlands

María Engracia Gómez

Technical University of Valencia, Spain

Davide Bertozzi

University of Ferrara, Italy

4.1 Introduction

The network topology is a key factor for the performance and cost of any Network-on-Chip (NoC) design. As anticipated in Chapter 1, the actual trend for chip multiprocessors is to choose a 2-D mesh because of its regularity, its suitability for the bidimensional silicon surface, and for the better predictability of its electrical parameters. However, as the number of end nodes in future on-chip networks increases, this might not be the best choice, since the 2-D mesh is well-known to incur bandwidth and latency scalability issues.

This raises the need to search for substitute topologies and to assess their properties by means of a topology exploration framework for use in the early design stage [28]. The design space of network topologies has been extensively explored in the domain of off-chip interconnection networks and some promising connectivity patterns can be evaluated for an on-chip setting too (e.g., k-ary n-mesh or k-ary n-tree topologies). However, the reader should keep in mind that the on-chip environment has radically different constraints than the off-chip one, especially when it comes to channel characteristics, wiring cost, signaling techniques, and synchronization mechanisms.

In the off-chip domain, a significant cost arises from communication channels, huge pads, and the adoption of expensive connectors, cables, and optics. On the other hand, in the on-chip landscape most of the cost is due to silicon area and power, which is typically dominated by storage resources (buffering). Wire density is certainly higher than in the off-chip domain and is not constrained by pin limitations.

Regarding *channel characteristics*, on-chip wires are relatively short and latency is comparable with logic delay. At the same time, large bandwidth can be easily achieved due to optimized links possibly integrating logic and/or buffers along them. Vice versa, latency is a critical metric for off-chip links.

The gap between the constraints driving the design of on-chip vs off-chip interconnection networks (and hence the gap between the final network topology selected for use in each domain) is widening even more as an effect of the relentless pace of technology scaling to the nanoscale regime. New physical effects come into play and may either degrade performance/power in an unpredictable way or even affect feasibility of specific connectivity patterns or architecture design techniques.

In particular, the reverse scaling of the delay of on-chip interconnects as an effect of their shrinking cross-section area is making the performance of designs targeting sub-45 nm silicon technologies increasingly interconnect-dominated. This effect becomes more and more severe at each technology node and tends to widen the gap between post-synthesis and post-place&route performance figures and even to move critical path delays from logic blocks to long global wires.

Performance and even feasibility of topologies for on-chip networks is extremely sensitive to the performance of on-chip interconnects and to similar subtle design issues, with deep implications on the network architecture as well. In an on-chip setting, topologies must in fact match the 2-D silicon surface, while off-chip realizations are dictated by board/rack organization. The 2-D mapping constraint raises implementation issues such as wire crossings, wires of uneven length or the decrease of switch operating frequency with the number of Input/Output (I/O) ports. As an ultimate consequence, topologies borrowed from off-chip networks should be reassessed in the on-chip environment and validated against the design pitfalls in this domain.

This is the motivation that is at the core of this chapter. We are aware that many regular topologies feature better abstract properties (e.g., diameter, bisection bandwidth) than a 2-D mesh, however their silicon implementation is very challenging. The objective of this chapter is to quantify to which extent their inherently better abstract properties are impacted by the degradation effects of the physical implementation on nanoscale silicon technologies.

4.2 The High-level View

Network topologies can be classified as *direct* or *indirect*. In *direct* topologies, each end node is directly connected to the network since the end node also includes a tightly coupled switching fabric for connection to its neighbor network nodes. The most common direct topologies in the open literature are orthogonal, since switches can be arranged in an orthogonal n-dimensional space, in such a way that every link produces a displacement in one dimension (see Figure 4.1(a)). In those topologies, routing of messages is simple and easy to implement.

In *indirect* topologies, the end node accesses the network indirectly through a link that is connected to a switch, in turn connected to the network. In *indirect* topologies, switches can be classified into two general groups: the group of switches that connect to end nodes and other switches, and the group that connects only to other switches. Among indirect topologies, the most

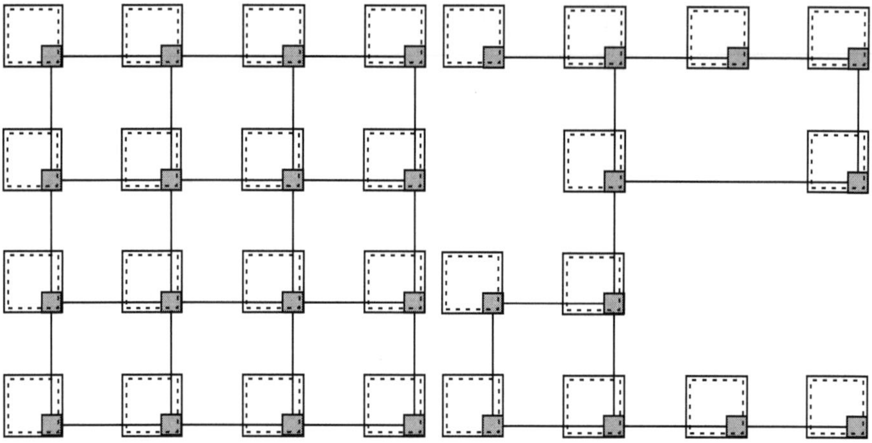

(a) Regular (b) Irregular

FIGURE 4.1: Example of direct topologies.

scalable ones are multistage networks, in which end nodes are interconnected through a number of intermediate switch stages (see Figure 4.2(a)).

On the other hand, there are two main subclasses among both, direct and indirect categories: *regular* and *irregular* (see Figures 4.1 and 4.2). In *regular* topologies there is a regular predefined pattern that defines the way links connect switches with other switches and/or end nodes, while *irregular* topologies have no predefined interconnection pattern. Usually, regular topologies show better scalability than irregular ones, although their main advantage is topology re-usability and reduced design time. While regular topologies are most suited for general purpose designs [316], irregular topologies are typically used in the embedded computing domain, where NoCs are customized for the communication load of a specific application or set of applications that are known at design time [26].

Although regular topologies are the typical choice for high-performance chip multiprocessors, there are additional factors that may break the topology regularity assumption at run time. Power management events or permanent and transient faults are examples such factors. It is therefore essential to design the on-chip network in such a way that it can preserve correct operation and high performance in the presence of such runtime events, a challenge that is addressed especially in Chapters 5 and 6 of this book.

4.2.1 High-level Topology Properties

As the number of topologies that can be considered for a NoC design increases, the need to predict the capabilities of each topology arises. Several properties exist that help to predict topology characteristics and properties

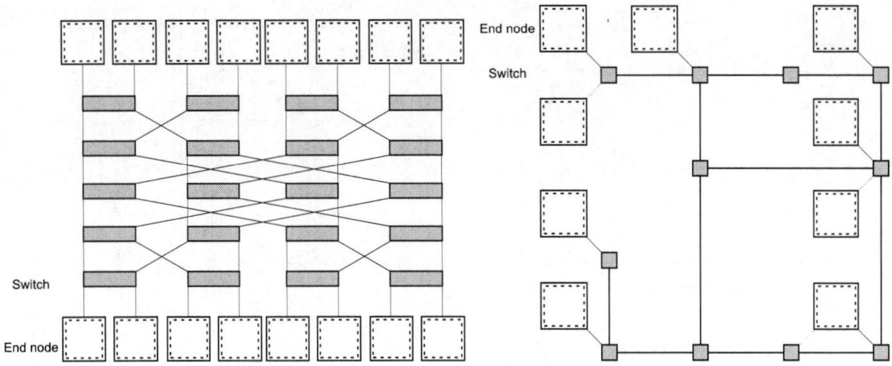

(a) Regular (b) Irregular

FIGURE 4.2: Example of indirect topologies.

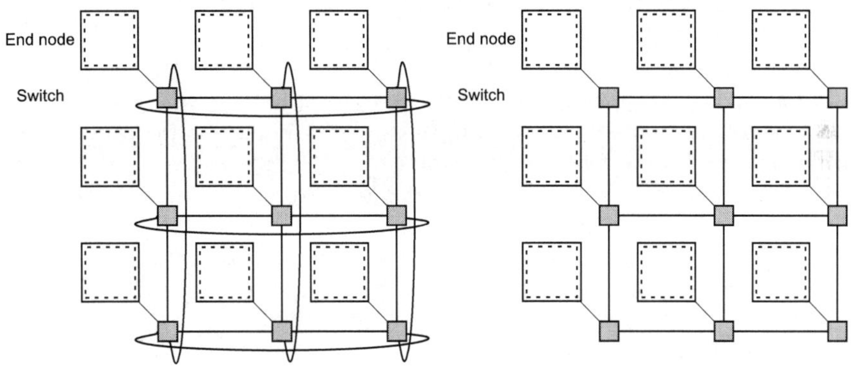

(a) Symmetric (b) Nonsymmetric

FIGURE 4.3: Example of symmetric and nonsymmetric topologies.

from a theoretical point of view. The rest of this section enumerates and defines some of the most common high level properties utilized in abstract topology analysis and comparison.

Symmetry. A topology is symmetric when the network looks the same from every switch. Figure 4.3(a) and Figure 4.3(b) show different topologies: only the first one has the symmetric property. Although symmetric topologies may provide abundant communication paths between any two end nodes, taking advantage of this effect while providing deadlock free routing is usually a complex task, due to the overabundance of cycles.

Switch degree. It is defined as the total number of input/output ports of a switch. The operating frequency of a switch and its area requirements are strongly related to this property: the higher the switch degree, the lower the switch maximum operating frequency, and the higher its area cost.

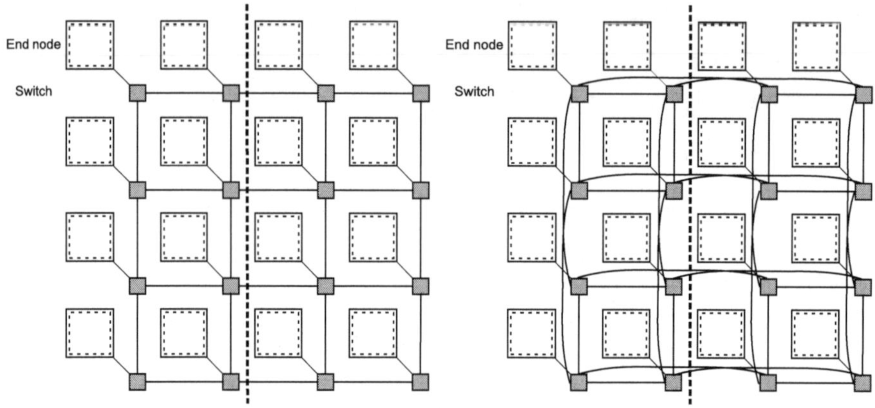

(a) 4 × 4 Mesh (b) 4-Hypercube

FIGURE 4.4: Examples of topology bisection bandwidth.

Homogeneity. A topology is homogeneous if all of its switches have the same degree, that is, the same number of ports. Figure 4.3(a) and Figure 4.3(b) show two different topologies, where only the first one is homogeneous. In the NoC environment, nonhomogeneous topologies may present switches with different maximum operating frequency, since it strongly depends on the switch degree. This may force potentially faster switches to work at the speed of the slowest ones, or to use techniques to support switches with different operating frequencies in the same NoC. Although homogeneous topologies are more modular and in principle easier to implement, homogeneity is a desirable property but not an indispensable one. As an example, the reader should consider that the widely used 2-D mesh is not homogeneous. This property can in fact be tolerated not to lose other important properties of a topology such as the minimization of communication resources and/or the simplification of the connectivity pattern.

Bisection bandwidth. It is defined as the smallest aggregated bandwidth of all the pairs obtained by dividing the topology into two equal-size halves. It is a common measure of theoretical network performance: the higher the bisection bandwidth, the better the topology is suited to cope with high traffic loads. Figure 4.4 shows two examples of bisection bandwidth (depicted as a dotted line).

Hop count. It is defined as the maximum number of switches that must be traversed in the topology in order to travel between any two end nodes by means of minimal routes. The higher the value of this property, the longer messages take to reach their destinations and the higher the probability of collision with other messages. Figure 4.4 shows two regular direct topologies with the same number of end nodes. While it takes 7 hops to travel from top-left to bottom-right in the 4x4 mesh, this number is only 5 in the 4-hypercube.

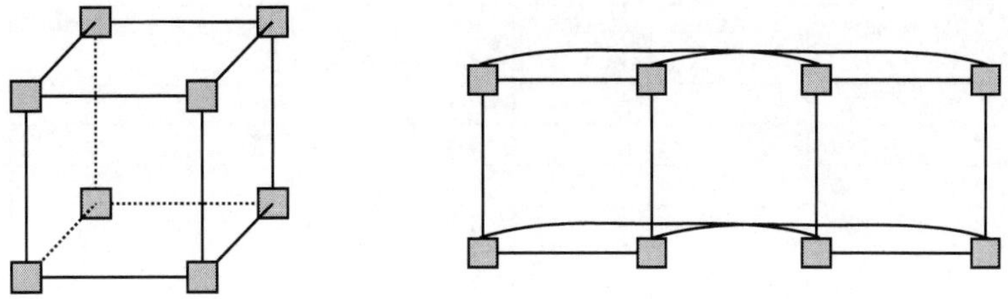

(a) Conceptual (b) Projected into a 2-D plane

FIGURE 4.5: Representations of a 3-hypercube.

Diameter. Notice that multicycle links may distort the classical hop count metric, as the hop count of a path may differ too much from the real cost in cycles for traversing this path. In order to have a more accurate parameter, the diameter of a topology is defined as the maximum distance between any two cores in cycles. This property is completely dependent on the physical implementation of the topology. For example, Figure 4.5 shows a conceptual 3-hypercube and its projection into a bidimensional plane. Of interest, longer links are the candidates to become multicycle, therefore, although both representations of the same topology have the same hop count, they may end up presenting a different diameter. This is a classical example of a theoretical metric that does not hold when the topology is laid out on silicon because it needs some engineering effort to be effective.

Connectivity. The connectivity is defined as the minimum number of network links that have to be disconnected in order to prevent an end node from sending and/or receiving messages. This property indicates the maximum number of network link failures that a topology can tolerate without isolating any end node.

Total number of switches. It is defined as the total number of switches required to fully interconnect all the end nodes with a particular topology. This property is usually related to the number of end nodes that need to be interconnected.

Total number of links. It is defined as the number of unidirectional network links required to fully interconnect all end nodes of a certain topology. As links are not a critical resource in NoCs, this is not an interesting property for NoC designers but the delay of each link is a key factor. In fact, as it will be discussed in the physical implementation section, a long link may require several pipeline stages to be inserted in order to keep the whole topology frequency above a certain threshold, thus affecting its link delay. Furthermore, since the real link delay is dependent on the topology mapping strategy and the technology library used, it is very difficult to provide an accurate estimation of this parameter in the early design stage. On the other hand, this property is still a good implementation cost indicator as it is directly related to the

Topology	Switches	Nodes/ switch	End nodes	Max. switch degree	Symm.	Homog.
k-ary n-mesh	k^n	m	mk^n	2n+m	No	No
k-ary 2-mesh	k^2	m	mk^2	4+m	No	No
2-ary n-mesh	2^n	m	$m2^n$	n+m	Yes	Yes
k-ary n-cube	k^n	m	mk^n	2n+m	Yes	Yes
k-ary l-rec	k^l	m	mk^l	k+m	No	No
k-ary n-tree	nk^{n-1}	0 or k	k^n	2k	No	No
k-ary n-fly	nk^{n-1}	0 or k	k^n	k	No	Yes
k-C-mesh	k^2	m	mk^2	4+m	No	Yes/No
k-ary n-flat	k^{n-1}	k	k^n	(n-1)(k-1)+k	Yes	Yes

Topology	Unidirectional networks links	Bisection bandwidth	Hop count	Connect.
k-ary n-mesh	$2n(k-1)k^{n-1}$	$2k^{n-1}$	n(k-1)+1	n
k-ary 2-mesh	4(k-1)k	2k	2k-1	2
2-ary n-mesh	$n2^n$	2^n	n+1	n
k-ary n-cube	$2nk^n$	$4k^{n-1}$	n(k/2)+1	2n
k-ary l-rec	k^{l+1}-k	$k^2/2$	2^l	k-1
k-ary n-tree	$2k^n(n-1)$	k^n	2n-1	k
k-ary n-fly	$k^n(n-1)$	$k^n/2$	n	1
k-C-mesh	$4k^2$	4k	k+1	4
k-ary n-flat	$(n-1)(k-1)k^{n-1}$	$k^n/2$	n	(n-1)(k-1)

TABLE 4.1: High-level properties of analyzed topologies.

total number of ports of a topology. The total number of ports of a topology is equal to the total number of unidirectional network links plus the total number of end nodes. Notice that to perform a comparative analysis of the implementation cost of topologies with an equal number of end nodes, only the number of unidirectional network (switch–to–switch) links varies between topologies, as the end node-to-switch links are a constant in this case. For this reason, from now on, we will refer only to the *total number of networks links.*

The properties listed above perform generic but inaccurate topology comparisons in terms of both, performance and cost, since their definition abstracts away the trade-offs of the physical implementation on a specific silicon technology. While this makes sense for a rough estimate of abstract system level properties, this chapter intends to warn the designer against selecting the topology for its on-chip network based only on these considerations.

4.2.2 Topologies for NoCs

As mentioned, the 2-D mesh is the most common topology for general purpose NoCs in the open literature. It is also the common solution for the latest commercial products and industrial prototypes, like the Tilera multicore processor family [339] and the Intel 80-core Polaris chip [57]. However,

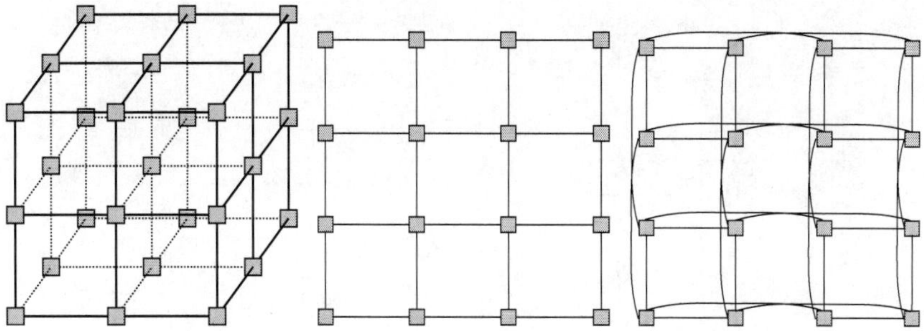

(a) 3-ary 3-mesh (b) 4-ary 2-mesh (c) 2-ary 4-mesh

FIGURE 4.6: Switch distribution of several mesh configurations.

this trend is expected to change in the near future due to 2-D mesh scalability limitations. Alternative topologies are either optimizations or modifications of a 2-D mesh, like the C-mesh [17], or based on radically different connectivity patterns, like WK-Recursive [264] or Fat-Tree [188] topologies. This section presents some of the most promising regular topologies that have been recently proposed for NoCs in the open literature. Table 4.1 summarizes the properties of all the presented topologies, as well as those of their most common subtypes (if any).

4.2.2.1 Mesh

Although commonly depicted as a direct topology, a mesh can be implemented as both, direct and indirect. In the former case, every end node is directly connected to other nodes. In the latter case, every node is connected to the network through a link and a switch. Regardless of their type, meshes are always orthogonal.

A generic mesh is defined by three parameters: the number of dimensions n, the number of end nodes attached to each switch m, and a n-tuple $\langle k_1, k_2, ...k_n \rangle$, in which each k_i defines the number of switches in dimension i. This nomenclature is usually represented as: $k_1 \times k_2 \times \cdots \times k_n$. If the value of k is different across dimensions, the topology becomes asymmetric, thus worsening the topology hop count and its bisection bandwidth. For this reason it is common to consider a subset of meshes in which the value of k is constant for all the dimensions, known as k-ary n-mesh.

A k-ary n-mesh has k^n switches distributed across n dimensions with k switches in each dimension. Each switch may have up to m end nodes attached, thus interconnecting up to mk^n end nodes. Switch degree increases with the number of dimensions and depends on switch placement inside the topology. The maximum degree is found in switches located in the center of the mesh, with $2n + m$ ports. The hop count is $n(k - 1) + 1$, while the total

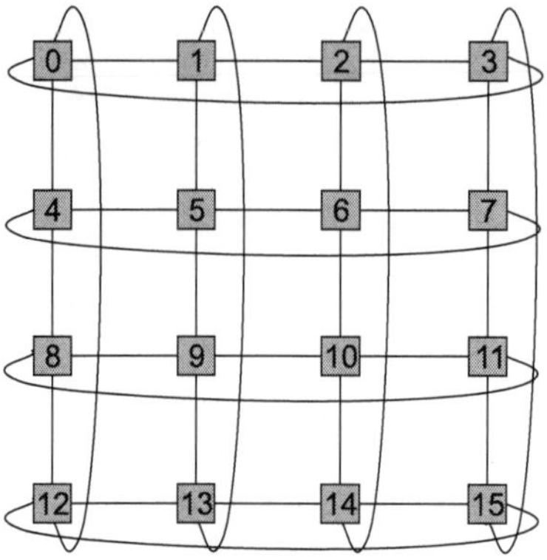

(a) 4-ary 2-cube

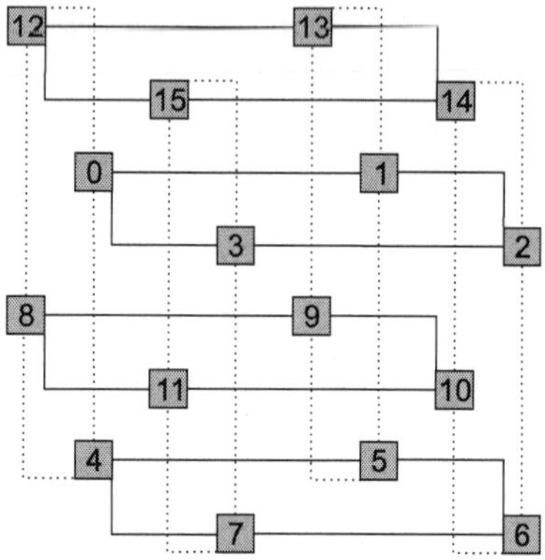

(b) 4-ary 2-cube folded

FIGURE 4.7: Switch distribution of several torus configurations.

number of unidirectional network links is $2n(k-1)k^{n-1}$. Finally, the bisection bandwidth is $2k^{n-1}$ unidirectional links, and the topology connectivity is n links, corresponding to the number of links of switches located at the corners. As a general rule, a k-ary n-mesh is neither symmetric nor homogeneous, with the exception mentioned below. Figure 4.6 shows several examples of the switch distribution in k-ary n-mesh topologies. Notice that each switch may have several end nodes attached (although they are not shown).

There are two subsets of meshes that require special attention: *hypercubes* and *2-D meshes.* A 2-D mesh (see Figure 4.6(b)) is a mesh with two dimensions, so it can be represented as k-ary 2-mesh. On the other hand, a hypercube (see Figure 4.6(c)) is a 2-ary n-mesh, that is a mesh in which the number of switches in each dimension is always 2, thus they may also be defined as n-hypercubes. Hypercubes are the only subset of k-ary n-mesh that is both, symmetric and homogeneous, with a switch degree of $n+m$, regardless of switch placement.

Although the 2-D mesh is the most common topology for NoCs, meshes with more than two dimensions are less common in the open literature. While a 2-D mesh perfectly matches the layout of the end nodes, other meshes present a more complex layout, introducing long links and high-degree switches. Anyway, the work in [155] introduces a new topology that in some cases is equivalent to a hypercube with several end nodes attached to each switch, while the work in [102] proposes the use of hypercubes in NoCs. Moreover, 3-D technologies are becoming a real possibility for multicore designs [337], favoring the use of topologies with more complex wiring patterns. For example, a 3-D mesh (see Figure 4.6(a)) may have the optimal wiring pattern for a 3-D environment.

4.2.2.2 Torus

A generic torus is constructed by adding wrap-around links to the equivalent mesh and thus they can be defined by similar parameters. Thus, tori are orthogonal topologies that are usually implemented as a direct topology. But, like in meshes, it is possible to get the switch outside the node, converting a torus into an indirect topology. For example, Figure 4.7(a) shows the switch distribution of the torus resulting from the addition of wrap-around links to the 2-D mesh in Figure 4.6(b). A torus can be defined by the number of dimensions (n), the number of end nodes attached to each switch (m), and the n-tuple that defines the number of switches per dimension ($\langle k_1, k_2, ...k_n \rangle$). As in the case of the mesh, a torus with varying number of switches per dimension may not be the best choice.

For this reason, it is usual to consider k-ary n-cubes, a subset of torus in which the number of switches per dimension (k) is constant. A k-ary n-cube has k^n switches distributed across n dimensions with k switches in each dimension, with up to m end nodes attached to each switch, thus interconnecting up to mk^n end nodes. Switch degree increases with the number of dimensions and is constant for all the switches: $2n+m$ ports. The hop count is $n(k/2)+1$, while the total number of unidirectional network links is $2nk^n$. The bisection bandwidth is $4k^{n-1}$ unidirectional links, and the topology connectivity is $2n$ links. Finally, a k-ary n-cube is symmetric and homogeneous.

Considering the same values of k and n, a torus lowers the hop count and increases the bisection bandwidth and the connectivity over a mesh. Therefore, it provides better performance and fault tolerance, but at the cost of a slightly higher number of links Also, the wrap-around links of a torus may present overlength issues. A common way to solve this problem is to use a folded torus, in which the nodes are re-allocated over the floorplan, obtaining an equivalent topology with reduced wrap-around link length at the cost of increasing the length of the other links. Figure 4.7(b) shows the folded version of the 4-ary 2-cube shown in Figure 4.7(a).

Although not as common as 2-D meshes, tori have been proposed in several NoC frameworks. The Proteo NoC [291] employs a ring topology, and rings can be defined as a k-ary 1-cubes. Also, 2-D folded tori have been proposed for their use in NoCs [68, 216, 173, 267]. Besides rings and 2-D tori, other topologies have been proposed. However, they are very difficult to find even in the open literature due to their complexity and expected layout intricacy.

4.2.2.3 WK-Recursive

WK-Recursive topologies were first proposed in [75], and are regular direct nonorthogonal topologies, constructed by recursively replicating a basic structure called *virtual node*. The first virtual node is constructed by connecting k nodes in a fully connected structure, leaving k links free. Like in meshes and tori, it is possible to take the switch out of the node, thus converting the WK-Recursive topology into an indirect topology. In this case, m end nodes

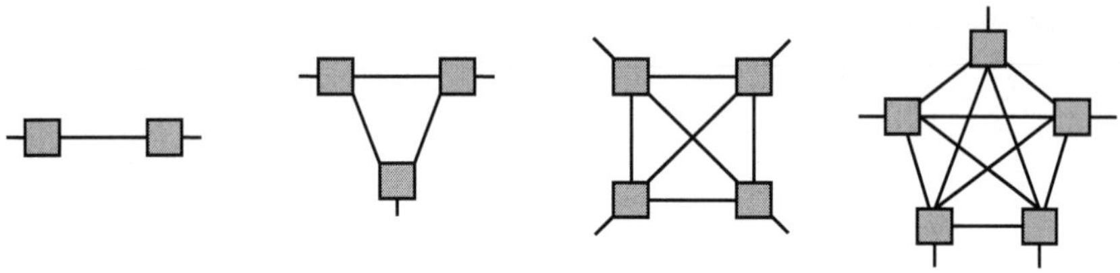

(a) Degree 2 (b) Degree 3 (c) Degree 4 (d) Degree 5

FIGURE 4.8: Switch distribution in WK-Recursive first-level virtual nodes.

could be attached to each switch. Hence, a virtual node is virtually similar to a switch with k free ports and mk end nodes attached to it. Figure 4.8 shows the switch distribution of several first level virtual nodes. Following the same strategy, k virtual nodes can be used to define a second level virtual node with degree k. In fact, a level l virtual node is constructed by using k virtual nodes of level $(l-1)$. According to this definition, a WK-Recursive topology is defined by two parameters: the virtual node k and the expansion level l. The most common nomenclature of this topology in the open literature is to use the pair (k,l) but, in order to provide an homogeneous nomenclature among all the topologies described in this chapter, from now on, this family of topologies will be denoted as k-ary l-rec. Figure 4.9 shows several WK-Recursive topologies for different values of k and l.

A k-ary l-rec has k^l switches, with m end nodes attached to each switch, thus interconnecting mk^l end nodes. The switch degree is directly defined by virtual node degree and depends on the switch placement inside the topology, being the maximum switch degree $k+m$ ports. The hop count is 2^l, while the total number of unidirectional network links is $k^{l+1}-k$. The bisection bandwidth is $k^2/2$ unidirectional links, and the topology connectivity is $k-1$ links. Finally, a k-ary l-rec is neither symmetric nor homogeneous. Notice that in the case of k-ary 1-rec configurations, the maximum switch degree is $k-1+m$ ports and the topology becomes symmetric and homogeneous.

Although it is difficult to provide efficient deadlock-free routing in a k-ary l-rec [264], this topology has attracted the attention of NoC researches [264, 307], due to some appealing properties. First, 4-ary l-rec has similar layout than 2-D mesh with the same number of end nodes per switch. Second, the switch degree is exactly the same in both topologies. Third, the number of switches is exactly the same in both topologies, although the number of links is slightly higher in the WK-Recursive topology. Finally, a WK-Recursive topology has a much lower hop count than its equivalent 2-D mesh. Anyway, the improvements of 4-ary l-rec topologies over 2-D meshes come at a cost, as their bisection bandwidth scales worse than in a 2-D mesh.

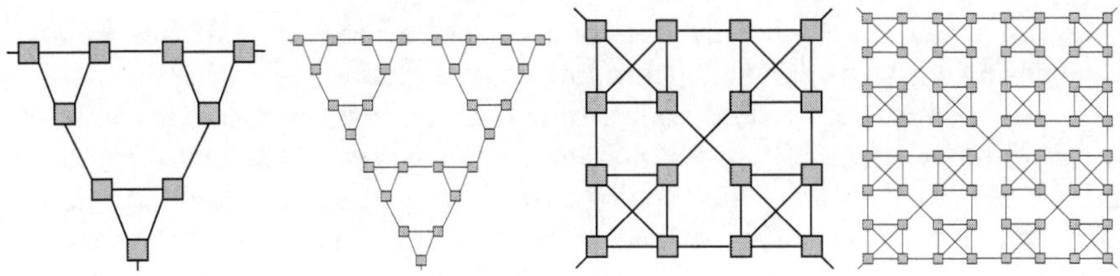

(a) 3-ary 2-rec (b) 3-ary 3-rec (c) 4-ary 2-rec (d) 4-ary 3-rec

FIGURE 4.9: Switch distribution of several WK-Recursive configurations.

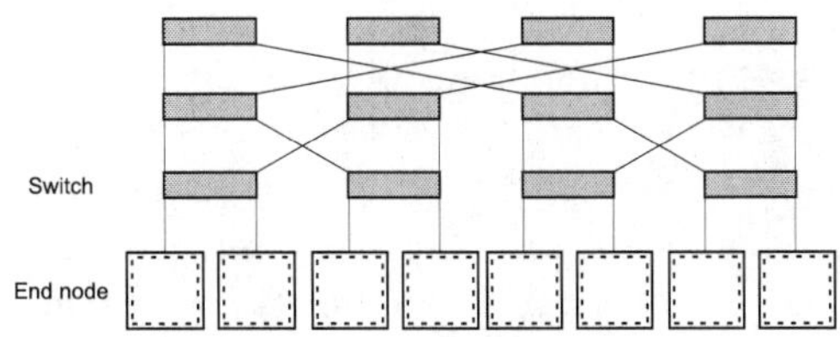

(a) 2-ary 3-tree

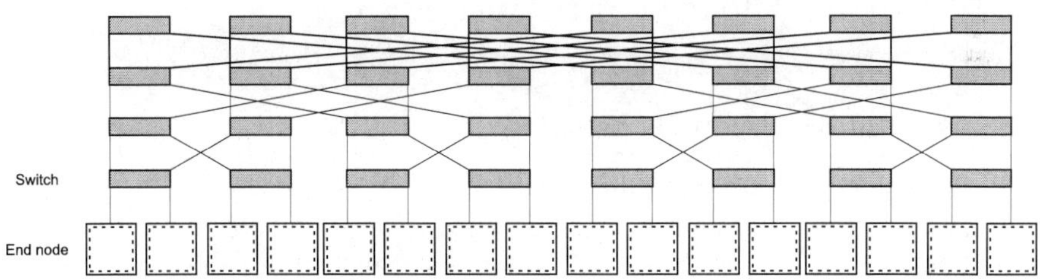

(b) 2-ary 4-tree

FIGURE 4.10: Examples of fat-tree configurations.

4.2.2.4 Fat-tree

Fat-trees are regular indirect topologies based on complete trees. The difference between a fat-tree and a complete tree is that a fat-tree gets thicker (offers more bandwidth) near the root, thus making switch degree higher the closer the switch is placed to the root. If the switch degree becomes too high, the physical implementation may become infeasible due to an unrealistically low operating frequency and/or cost. For this reason, some alternative implementations have been proposed in order to use switches of fixed degree.

Among those alternatives, one of the most common subclasses are butterfly fat-tree or folded butterfly, also known as k-ary n-trees [252]. In this notation a fat-tree is defined by two parameters: the number of stages n and the switch degree k. A k-ary n-tree belongs to the bidirectional Multistage Interconnection Network family (bMIN), and thus switches are organized in stages, in which a switch only has connections to the next and the previous stage according to a regular pattern, and only switches belonging to the first stage have connections to end nodes. In particular, a k-ary n-tree is composed by nk^{n-1} switches distributed in n stages. Each switch in the first stage has k end nodes attached, thus the topology provides connection to k^n end nodes. Each switch employs k bidirectional links to connect with the next stage and k bidirectional links to connect to the previous stage, with the exception of switches belonging to the first and last stages. Switches in the last stage have a switch degree of k, while all the other switches have a degree of $2k$. The hop count is $2n - 1$, while the total number of unidirectional network links is $2k^n(n - 1)$. The bisection bandwidth is k^n unidirectional links, and the topology connectivity is k links. Finally, a k-ary n-tree is neither symmetric nor homogeneous. Figure 4.10 shows several configurations of k-ary n-trees. Notice that unlike the previous topologies, in this case the number of end nodes connected by the topology is strictly fixed.

Due to their low hop count and high bisection bandwidth, fat-trees have been proposed as topologies for NoC aiming at low latency communication, like the works presented in [241, 5]. Anyway, k-ary n-trees require a much higher number of switches and links than an equivalent mesh with the same number of end nodes.

4.2.2.5 Unidirectional Topologies

Most topologies proposed for NoC in the open literature employ bidirectional links. But, in resource-constrained environments like NoCs, unidirectional topologies may be a better cost-performance trade-off in the design space. A unidirectional topology requires one-way ports and links, so the number of links and switch ports of the topology can be smaller than in a 2-D mesh with an equal number of end nodes. This translates into important area and power consumption savings, but there are several issues that question the viability of these topologies in NoCs. For example, simple unidirectional topologies only offer a unique path between any two end nodes, making them highly vulnerable to link failure, as there is no way to avoid those faulty links. Although it is possible to use more complex unidirectional topologies to overcome this limitation, this is likely to increase the cost of the topology, thus offsetting the low-cost advantage of unidirectional topologies.

There are two kinds of unidirectional topologies that are well known in the open literature: rings and multistage. Although unidirectional rings are not very popular due to their severe scalability issues, unidirectional Multistage Interconnection Networks (uMINs) have some properties that make them appealing as NoC topologies. For example, the work in [12] proposes the use of the butterfly topology as a means to provide high-performance communications with an affordable implementation cost, while the work in [188] proposes the use of uMINs in order to reduce the implementation cost of NoCs.

In a uMIN, switches are organized in stages, and each stage only connects

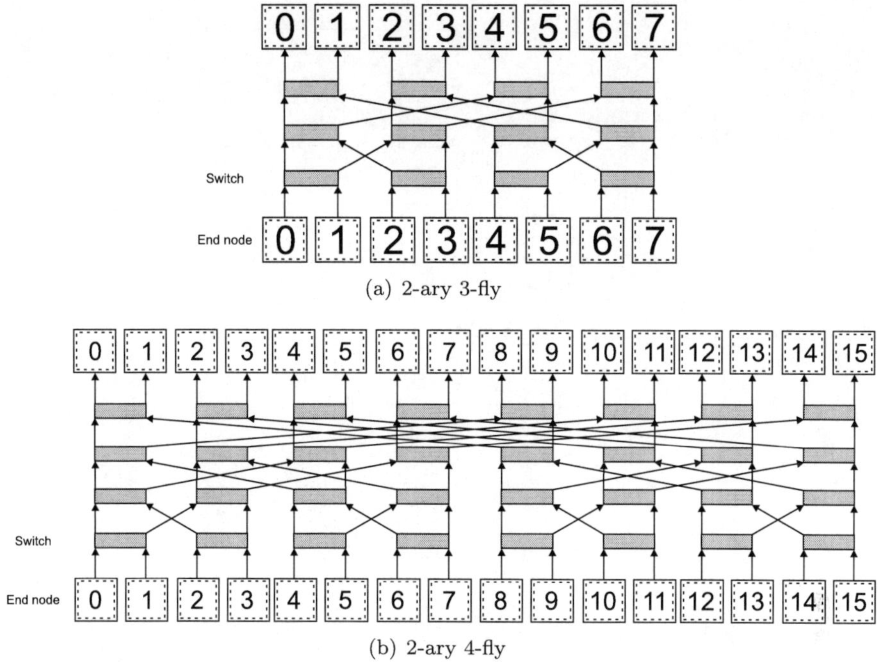

FIGURE 4.11: Examples of butterfly configurations.

to the next stage following a regular pattern, with the exception of the first and the last stage, that have connections to the end nodes. In those topologies, communications are one-way: Messages are always delivered from the first stage to the last stage. An uMIN is characterized by the pattern defining how stages interconnect with each other. There are several common patterns, and one of the most common patterns is the butterfly, which is also used to build the k-ary n-tree topology. A k-ary n-fly is a regular indirect topology, composed by nk^{n-1} switches distributed in n stages. The topology provides connection to k^n end nodes. Switch degree is constant and equal to k, while the hop count is n, and the total number of unidirectional network links is $k^n(n-1)$. The bisection bandwidth is $k^n/2$ unidirectional links. Due to the existence of a single path between any pair of end nodes, the topology does not tolerate any link failure, thus its connectivity is 1 unidirectional link. Finally, a k-ary n-fly is homogeneous, but it is not symmetric.

Figure 4.11 shows several configurations of k-ary n-fly. As can be observed, at the first stage, each switch has k end nodes attached only to send messages through the topology, while at the last stage, each switch has the same k end nodes attached only to receive messages from the topology.

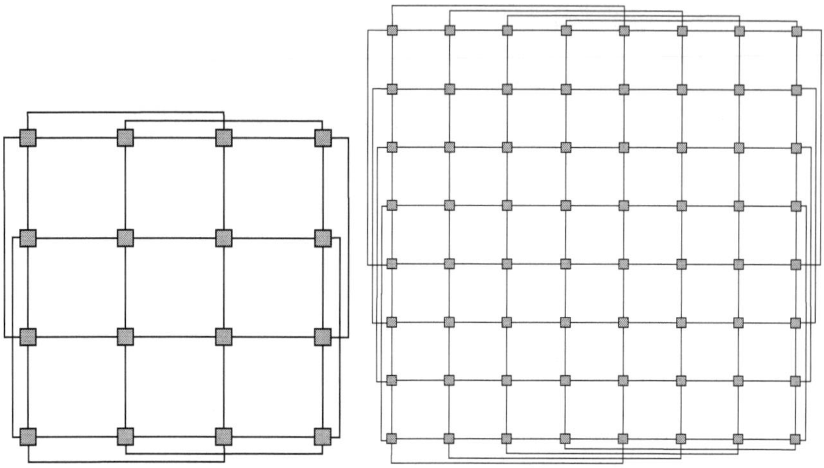

(a) 4 C-mesh (b) 8 C-mesh

FIGURE 4.12: Switch distribution of several C-mesh configurations.

4.2.2.6 Concentrated Mesh

The Concentrated Mesh (or C-mesh) topology was first proposed in [17]. This topology is an evolution of a 2-D mesh, and aims at solving 2-D mesh scalability issues while keeping its advantages.

In [17], a C-mesh is built around a 4×4 2-D mesh with four cores attached to each switch. This way, a lower hop count with respect to that of a classical 2-D mesh, with one core per switch, can be achieved. However, this comes at the cost of a diminished bisection bandwidth. For this reason, the adoption of express links is often used to improve C-mesh bisection bandwidth. Express links are added only along the perimeter of the network, thus taking advantage of the switches that have a lower degree than the maximum one. This keeps the same switch degree for the elements at topology border. Although the main goal of using express links is to increase the bisection bandwidth, they may be also used to greatly reduce the hop count as they effectively represent alternative routes. Unfortunately, this opportunity comes with a cost, i.e., deadlock may more easily become an issue. Therefore, the provision of deadlock-free routing algorithms is a more complex task with respect to the original mesh. For this reason, in the original proposal, express links are used only to advance messages that travel along the perimeter of the topology a distance equal or higher than the switches that this express link overtakes. Albeit the use of the express links considered in [17] is limited, those express links can be used in a similar way to the wrap-around links of a torus, regardless of whether the message is traveling through the perimeter of not. From a pure topology point of view, a C-mesh can be seen as a classical 2-D mesh with

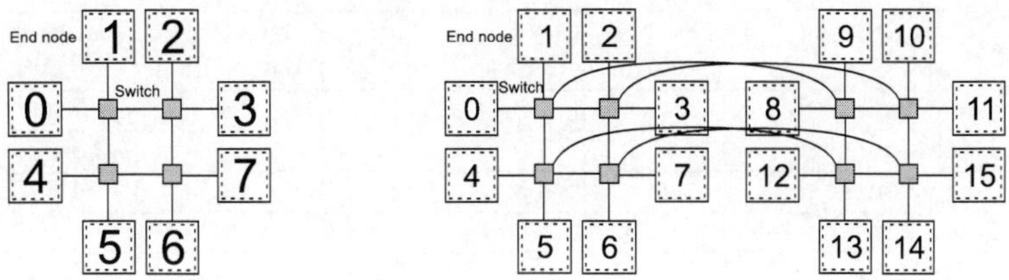

(a) 2-ary 3-flat

(b) 2-ary 4-flat

FIGURE 4.13: Examples of flattened butterfly configurations.

express links, regardless of the number of cores attached to each switch. Figure 4.12 shows several examples of switch distribution in generalized C-meshes, in which each switch can connect to a parameterizable number cores.

A C-mesh is defined by the same parameters of a conventional 2-D mesh, so it can be defined as a k-C-mesh, with k^2 switches distributed across 2 dimensions and k switches in each dimension, with m end nodes attached to each switch, interconnecting a total of mk^2 end nodes. Express links connect switches from the same dimension that are located at the perimeter of the network and are $k/2$ hops distant from each other, thus configurations with a number of switches per dimensions (k) of less than three, are infeasible in this topology. The maximum switch degree is equal to $4 + m$ ports, and is constant only when k is an even number. The hop count is $k + 1$, while the total number of unidirectional network links is $4k^2$. The bisection bandwidth is $4k$ unidirectional links, and the topology connectivity is 4 links. Finally, a k-C-mesh is not symmetric, and it is homogeneous only for even values of k.

4.2.2.7 Flattened Butterfly

The flattened butterfly topology was first proposed as a NoC topology in [155], as a solution for cost-efficient high-degree networks. The topology is an evolution of the butterfly topology, and it is formed by flattening the switches in each column of a butterfly while keeping the same interswitch connections. The resulting topology is a regular orthogonal topology that provides the same bisection bandwidth of a butterfly and a lower hop count, at the cost of a greatly increased switch degree. Notice than the switch degree greatly increases with both original butterfly parameters: switch degree (k) and number of stages (n), thus the real performance achievable is extremely sensitive to the network architecture employed in the final implementation.

A flattened butterfly may be defined as a k-ary n-flat, with k end nodes attached to each switch. In this way, a flattened butterfly can be designed as a k-ary n-flat, with k^{n-1} switches distributed across $n - 1$ dimensions with k switches in each dimension, interconnecting a total of k^n end nodes. Switch degree is constant and increases with the number of dimensions, being equal

Topology	Sw	Nodes/ switch	Max. degree	Unidir. links	Bisection bandwidth	Hop count	Connect.
4-ary 2-mesh	16	1	5	48	8	7	2
2-ary 4-mesh	16	1	5	64	16	5	4
2-ary 3-mesh	8	2	5	24	8	4	3
2-ary 2-mesh	4	4	6	8	4	3	2
4-ary 2-cube	16	1	5	64	16	5	4
4-ary 2-rec	16	1	5	60	8	4	3
8-ary 1-rec	8	2	8	56	32	2	7
2-ary 4-tree	32	0 or 2	4	96	16	7	2
4-ary 2-tree	8	0 or 4	8	32	16	3	4
2-ary 4-fly	32	0 or 2	2	48	8	4	1
4-ary 2-fly	8	0 or 4	4	16	8	2	1
4-ary 2-flat	4	4	7	12	8	2	3
4-C-mesh	16	1	5	64	16	5	4

TABLE 4.2: High-level parameters of topologies for 16-end nodes.

to $(n-1)(k-1)+k$ ports. The hop count is n, while the total number of unidirectional network links is $(n-1)(k-1)k^{n-1}$. The bisection bandwidth is $k^n/2$ unidirectional links, and the topology connectivity is $(n-1)(k-1)$ links. Finally, a k-ary n-flat is both symmetric and homogeneous. Figure 4.13 shows the flattened butterfly configurations obtained from flattening the butterfly configurations presented in Figure 4.11.

Notice that this topology may overlap with previous topologies in several cases. First, as can be observed in Figure 4.13, when flattening a 2-ary n-fly, the resulting topology will be identical to a 2-ary $(n-1)$-mesh with two end nodes attached to each switch. Second, when flattening a butterfly of two stages, the resulting topology is equivalent to the basic virtual node of the WK-Recursive of degree k with k end nodes attached to each switch. In particular, a k-ary 2-flat will be equivalent to a k-ary 1-rec with k end nodes attached to each switch, for any value of k. Finally, a k-ary n-flat is completely equivalent to a generalized hypercube [29] of degree k and $n-1$ dimensions, with k end nodes attached to each switch.

4.2.3 High-level Topology Exploration

It is possible to utilize the properties exposed above to perform a high-level comparison of topology capabilites. This analysis may be employed in the early stages of a system design in order to select the subset of the most promising topology candidates. This section presents an example of high-level topology exploration for two systems with different sizes: 16- and 64-end nodes. The maximum number of end nodes attached to each switch in those topologies has been limited to four, not to limit too much the maximum operating speed of the switches. Also, the placement of end nodes around the switches in those cases would not be a trivial task, since as the length of the injection/ejection links grows, the NoC would suffer from a significant drop in performance [103].

Table 4.2 summarizes the values of the parameters of all 16 end nodes configurations considered for each topology. For such small systems, there are only a small number of different configurations to consider. As can be seen, apart from the classical 2-D mesh, several hypercubes with different concentration degrees of end nodes at each switch are considered. On the other hand, only one torus configuration is considered. Although it is possible to concentrate end nodes in a torus, for this particular system size, the resulting torus topologies would have a k value of 2, becoming very similar to the hypercubes already considered in the table. Also, two WK-Recursive configurations with different concentration degrees of end nodes on switches are considered. In the case of the C-mesh topologies, there is only a single valid configuration, as configurations with a k value lower than three are not valid in this topology. Regarding MINs, two kinds of fat-tree and butterfly topologies are considered, as well as the respective flattened versions of the fat-trees, with two exceptions. First, the 2-ary 4-flat configuration is completely equivalent to the 2-ary 3-mesh with two end nodes attached to each switch. Finally, when flattening the 4-ary 2-fly solution the resulting topology overlaps with a 4-ary 1-rec solution with four end nodes attached to each switch.

As can be observed, the 8-ary 1-rec solution shows the highest bisection bandwidth, making this configuration the best candidate for high traffic loads. It also provides the minimum possible number of hops, while requiring the highest switch degree of all the considered solutions. This fact questions the achievable maximum operating frequency of its physical implementation. On the other hand, from a low-latency (low number of hops) viewpoint there is not a clear winner. There are three solutions that can reach any destination in only two hops: 4-ary 2-fly, 8-ary 1-rec and 4-ary 2-flat. The 4-ary 2-fly solution seems the best choice: It requires a similar number of switches with a lower degree than the 8-ary 1-rec solution, with a lower number of links as well. Unfortunately, it has an important drawback associated with its connectivity. In fact, with a connectivity of only a single unidirectional link, this solution is highly sensitive to faulty links. Regarding the 4-ary 2-flat, although it offers a smaller bisection bandwidth than the other two solutions, it has the lowest number of switches and links with respect to the other two low-latency solutions. However, there is an issue that further complicates the comparison in this case: the switch degree. While the 8-ary 1-rec and the 4-ary 2-flat topologies require high-degree switches, the 4-ary 2-fly solution has a maximum switch degree of 4. Finally, the 2-ary 4-mesh, 4-ary 2-cube, 4-ary 2-tree, and 4-C-mesh solutions appear to be equivalent topologies providing the second highest bisection bandwidth with an acceptable number of hops. Among them, the only topology that shows a clear inconvenient is the 4-ary 2-tree, that has a switch degree as large as the 8-ary 1-rec.

Overall, the high-level analysis for 16 end node systems shows that the best topology is the 8-ary 1-rec solution. This topology has a low number of

Topology	Sw	Nodes/ switch	Max. degree	Unidir. links	Bisection bandwidth	Hop count	Connect.
8-ary 2-mesh	64	1	5	224	16	15	2
4-ary 3-mesh	64	1	7	288	32	10	3
4-ary 2-mesh	16	4	8	48	8	7	2
2-ary 6-mesh	64	1	7	384	64	7	6
2-ary 5-mesh	32	2	7	160	32	6	5
2-ary 4-mesh	16	4	8	64	16	5	4
8-ary 2-cube	64	1	5	256	32	9	4
4-ary 3-cube	64	1	7	384	64	7	6
4-ary 2-cube	16	4	8	64	16	5	4
4-ary 3-rec	64	1	5	252	8	8	3
4-ary 2-rec	16	4	8	60	8	4	3
8-ary 2-rec	64	1	9	504	32	4	7
2-ary 6-tree	192	0 or 2	4	640	64	11	2
4-ary 3-tree	48	0 or 4	8	256	64	5	4
2-ary 6-fly	192	0 or 2	2	320	32	6	1
4-ary 3-fly	48	0 or 4	4	128	32	3	1
8-ary 2-fly	16	0 or 8	8	64	32	2	1
4-ary 3-flat	16	4	10	96	32	3	6
8-C-mesh	64	1	5	256	32	9	4
4-C-mesh	16	4	8	64	16	5	4

TABLE 4.3: High-level parameters of topologies for 64-end nodes.

switches and links, while providing a high connectivity, the highest bisection bandwidth and the lowest number of hops. As anticipated, the only parameter that may arise some concerns is its switch degree, as will be more clear after reading Section 4.4.2. In case the 8-ary 1-rec topology proves inefficient for physical implementation, the best solution might be one of the following solutions: 2-ary 4-mesh, 4-ary 2-cube, or 4-C-mesh.

Regarding the 64 end node system analysis, Table 4.3 summarizes the properties of all 64 end node configurations considered for each family of topologies. In this case, due to the increased system size, there are more topology configurations to be considered, even when excluding overlapped topologies. Notice that there are several configurations that are excluded from the analysis due to an overly high switch degree. Those configurations are the 8-ary 2-flat, 8-ary 2-tree, and the 16-ary 1-rec with 4 end nodes concentrated in each switch, thus resulting in a switch degree of 16, 17, and 19 respectively. Unless full custom solutions are conceived, the implementation of such high-degree switches with a standard Application Specific Integrated Circuit (ASIC) flow is infeasible (see Section 4.4.2).

For this system size, the solutions best suited for high traffic loads are the 2-ary 6-mesh, 4-ary 3-cube, 2-ary 6-tree and 4-ary 3-tree. Among those, the 2-ary 6-tree topology has the highest hop count (11 hops), making it the worst candidate for latency sensitive systems and/or applications. Also, it requires the highest amount of resources: 192 switches of degree 4 and 640 unidirectional links. Regarding the 4-ary 3-tree, it has an acceptable number of hops (5 hops), while requiring a moderate amount of resources: 48 switches of degree 8 and 256 unidirectional links. On the other hand, the 2-ary 6-mesh and 4-ary 3-cube solutions are similar from a high-level point of view, as both require 64 switches of degree 7 and 384 unidirectional links, that is more switches and links than the 4-ary 3-tree solution, with only one port less in each switch. They also have a higher number of hops than the 4-ary 3-tree topology: 7 hops against 5. So, the best configuration for high traffic loads in a 64 end node system is the 4-ary 3-tree, as it presents the lower hardware resource requirements and number of hops among all the amenable solutions for high communication loads.

On the contrary, from a low-latency point of view, the best solution is the 8-ary 2-fly, which needs only 2 hops to reach any end node. This solution employs 16 switches of degree 8 and only 64 links, providing a bisection bandwidth of 32 undirectional links. Its only drawback is its low connectivity, common to all butterfly topologies.

Overall, for bandwidth-intensive systems the 4-ary 3-tree solution seems to be the best option, while for latency-sensitive systems the 8-ary 2-fly topology seems the most promising one. However, the 8-ary 2-fly offers the second best bisection bandwidth, becoming the overall best solution unless the highest bisection bandwidth is really required. Unfortunately, it is not suited for systems that require some degree of fault tolerance and/or a variety of paths between any pair of end nodes. In this case, the 4-ary 3-flat solution provides the same bisection bandwidth than the 8-ary 2-fly, while increasing the number of hops and switch degree. The specific requirements of the system designer and the features of the network architecture can push one solution instead of another, as there is in general no clearly winning topology but a trade-off between many factors has to be evaluated.

4.3 Physical Design Pitfalls

The question that this chapter is now going to address is the following: Do the results of the high-level theoretical analysis still hold when mapping a topology onto silicon? In order to make the answer to this question more clear, this chapter will point out several physical design pitfalls that are not always easily recognizable at first glance.

A common approach to capture physical design implications of a topology in the early design stages is by means of pencil-and-paper floorplanning considerations. Unfortunately, a number of assumptions are often made that are not verified on actual layouts.

For instance, with this approach, the same length is typically assumed for

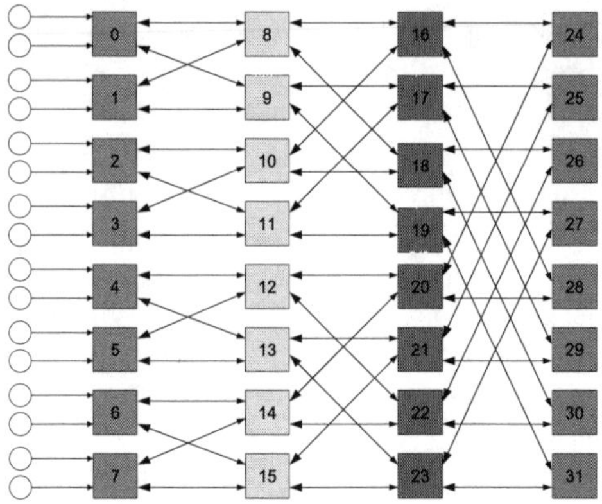

FIGURE 4.14: High-level sketch of a 2-*ary* 4-*tree* (fat-tree).

all the links in the topology, thus neglecting the operating principles of physical routing tools and the constraints of a 2-D layout. The example in Figure 4.14 depicts the high-level connectivity pattern of a 2-*ary* 4-*tree* topology (commonly known as fat-free). If we look at a possible layout realization of that connectivity pattern (see Figure 4.15, where only some links are illustrated), it is clear that the physical view is radically different from the abstract representation and it may not be intuitive at all. First of all, the width/height of a core is much larger than that of a simple network switch. This is a simple yet layout-constraining condition that is often neglected when devising a possible topology floorplan. In this case, the floorplanning strategy was to minimize wirelength between consecutive switch stages. For this reason, cores are clustered in groups of four and the connected switches (of the first and second stage) are placed in the middle of each cluster. The third switch stage is split into 2 subgroups and placed between the upper and lower clusters. Each subgroup serves its relative counterpart from the first and second stage. The last switch stage is located in the center of the chip. The presented layout exhibits equalized wirelengths between the second, the third, and the last stage of switches. However, it is easy to imagine a case where this result does not hold. When an asymmetric Intellectual Property (IP) core (i.e., with a form factor different from 1) is used, the wire length regularity of the topology is broken since a different length between wires in the horizontal and in the vertical directions has to be expected. For the same reason, wires in the first and in the second dimension of a 2-D mesh may not have the same length when the IP core is not square but rather rectangular.

In a more complex topology, such as torus or C-mesh, the utilization of express links comes at a considerable price. In fact, as an express link turns out to be a very long link to be routed, it is likely to limit the overall operating

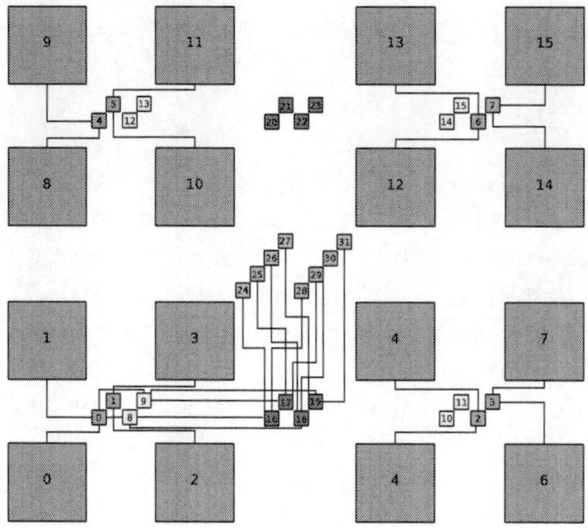

FIGURE 4.15: Floorplanning of a 2-*ary* 4-*tree* (fat-tree).

frequency of the topology. Obviously, there are well-known methods to solve this problem, namely repeater insertion and/or link pipelining. However, they come with an area cost and in some cases they induce an ever higher overhead in the architecture. As an example, let us recall the need to match the number of buffer slots in a link receiver to the round trip latency for the sake of correct flow control and of maximum throughput.

Another design pitfall concerns the actual topology implementation. In fact, a network may require different switches to be implemented. For instance, in the 2-D mesh topology, there are typically three different types of switches: 4×4, 5×5, 6×6 (assuming that an end node requires two ports towards the switch it is connected to). Of course, the switch degree depends on the number of connections required by the position of the switch itself in the mesh. A switch in the corner is a $4{\times}4$ and is potentially the fastest of the entire network. However, the final operating frequency of the topology is set by the switch with the highest degree in combination with the longest link in the topology. Therefore, these border switches will end up being synthesized at a lower speed than the maximum one they could achieve, resulting in lower area. An early area estimation framework neglecting these effects in nonhomogeneous topologies would be highly misleading.

A design pitfall at a lower level of abstraction concerns physical routing. Sometimes, a topology is used to interconnect different IP blocks provided by various vendors. The IP block could be provided as a nonroutable hard-macro, thus forcing the physical designer to reserve physical routing channels for the routing of NoC links. Instead, whenever the IP block is provided as a soft-macro, over-the-cell routing may become feasible, thus giving rise to more compact layouts. However, even in this case routing on top of IP blocks may not be convenient for the intricacy of inferring link repeaters in the same fence of the IP block.

4.4 16-tile Systems

This section will provide more solid evidence of the design pitfalls illustrated above by means of a couple of case studies where intuitive physical design assumptions are put in discussion when considering the actual layout constraints. In the first case study, topologies for a 16-tile system will be laid out on silicon. In the second one, richer connectivity topologies for 64-tile systems will be considered, although they will be evaluated by means of a modeling framework of physical effects inspired by the first case study. All physical synthesis experiments of this chapter have been carried out by means of a commercial synthesis toolflow on an industrial 65 nm technology library.

4.4.1 Total Wire Length

The way link design techniques impact topology quality is twofold. On the one hand, the longest link in the topology determines its maximum achievable speed. Therefore, cutting down on the delay of that link is beneficial for the whole topology. On the other hand, the cost for performance boosting of the critical link becomes relevant for the topology only when there are more such critical links and they account for a significant fraction of the total topology wire length.

In the experiment that follows, the impact of switch-to-switch wires on total topology wire length will be assessed. Moreover, it will be demonstrated that drawing a conclusion on this based on abstract considerations on the topology connectivity pattern is highly misleading. An interesting family of regular NoC topology, offering representative design points for our analysis, consists of the k-ary n-mesh topologies.

As explained in Section 4.2.2, an optimization of this topology consists of connecting more end nodes per switch while keeping the overall number of end nodes the same. This way, the total number of switches is reduced, hence resulting in a lower number of hops, but the bisection bandwidth is reduced as well. In this analysis, the focus is on 16-end node systems, hence we consider a 4-ary 2-mesh (referred to as 2-D mesh from now on) with one end node per switch and a 2-ary 2-mesh with 4 end nodes per switch. We denote this latter solution as the *concentrated 2-D mesh.* Table 4.4 recalls some distinctive properties of the studied topologies.

Both the 2-D mesh and its concentrated counterpart feature 2 dimensions and homogeneous interswitch wire lengths. However, such wire length will not be the same in the two topologies due to floorplan constraints reflecting a well-known trade-off for these topologies: spread-around topologies on one hand (large number of low-radix switches) as opposed to more concentrated ones with a small number of high-radix switches *placed further apart* to optimize connectivity with a large number of end nodes. In fact, these latter have to

Topology	4-ary 2-mesh	2-ary 4-mesh	2-ary 2-mesh
Max. Arity	6	6	10
Link Length	1	1,2	1
Switches	16	16	4
Max. Hops	7	5	3
Bisection bandwidth	4	8	2
End node per Switch	1	1	4

TABLE 4.4: Topologies under test.

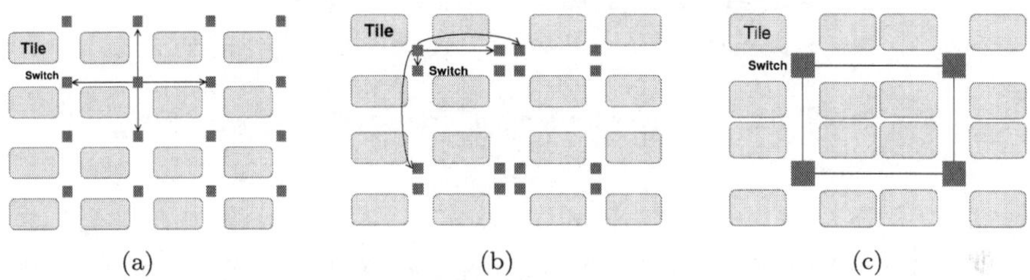

FIGURE 4.16: Floorplan directives for (a) the 2-D mesh, (b) the 4-hypercube and (c) the 2-ary 2-mesh.

be placed around the switches they have to be connected to, thus separating the switches in space.

A different design point is represented by hypercubes. For 16-end node systems, a 2-ary 4-mesh (4-hypercube) can be obtained from the 2-D mesh by increasing the number of dimensions and by reducing the number of switches in each dimension. Interestingly, the maximum switch degree stays the same, but the 4-hypercube is homogeneous unlike the 2-D mesh. Unfortunately, this comes at the cost of links with uneven length. In fact, in a mesh with more than two dimensions, the links used to connect the dimensions greater than two are longer, and this holds for 50% of the 4-hypercube interswitch links.

Theoretically, the length of a link of the dimension t is generally assumed to be $k^{(d-2)/2}$, where d is equal to t if t is an even number and d is equal to $t+1$ if it is an odd number. Unfortunately, placement and layout constraints put this picture in discussion, thus making it very difficult to predict the impact that the cost of a specific link performance boosting technique might have on the cost metrics of the entire topology.

Let us now consider the floorplanning directives given for topology synthesis reported in Figure 4.16(a), 4.16(b) and 4.16(c).

The asymmetric end node size plays in favor of the 4-hypercube wiring, since the length of the horizontal and of the vertical express links turns out to be comparable with that of horizontal wires in the 2-D mesh. This latter

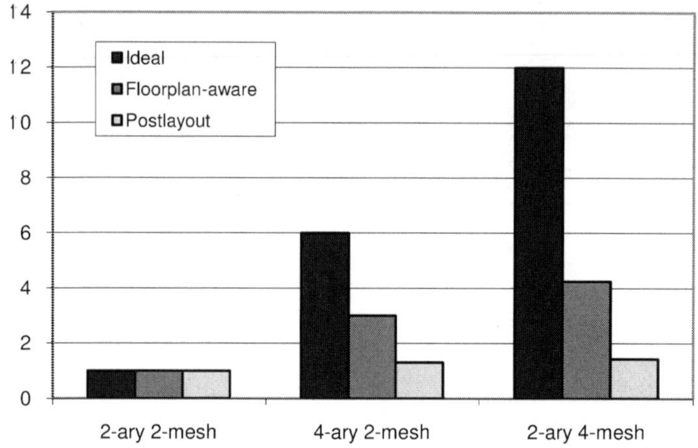

FIGURE 4.17: Total wire length (from [187] ©2009 ACM).

also features horizontal and vertical links of unequal length, indicating that the layout regularity often assumed does not materialize in practice. The assumed guiding principle for the floorplan definition consists of *shortening the longest links in each topology* and of *coming up with a scalable floorplanning style.* For the 2-ary 2-mesh, the network end nodes have been placed around the switch they are attached to, which is an ideal scenario for pipelining interswitch links. In all cases, network interfaces were placed close to their end node but also to the connected switch, so to move the critical path away from these links.

As a result of topology synthesis and place-and-route, Figure 4.17 shows the total wire length for the three topologies (*postlayout* bar), normalized to the least wire-hungry topology. It is compared with the results of traditional pencil-and-paper floorplanning considerations. Bar *Ideal* computes wire length based on the theorethical formula given above, which only considers the number of hops crossed by a link. Bar *Floorplan-aware* updates the previous formula with the knowledge of the asymmetric end node size and of switch placement. The ideal analysis largely overestimates the amount of wiring needed for the 4-hypercube. Floorplan awareness accounts for specific floorplanning techniques that optimize wiring of a given topology, and therefore leads to more conservative estimations of the wiring overhead. However, this is still far away from real-life, where the postlayout report of total wire length gives only a 10% overhead of the 4-hypercube wiring with respect to 2-D mesh and a 43% with respect to the 2-ary 2-mesh. This is because switch-to-switch and switch-to-network interface wiring only accounts for a relatively small percentage of total wiring, ranging from 7% for the 2-ary 2-mesh to 26% for the 4-hypercube. This explains the relatively small total wire length difference between the topologies. This scenario plays in favor of engineering performance-optimized interswitch links with a possibly minor impact on topology cost metrics.

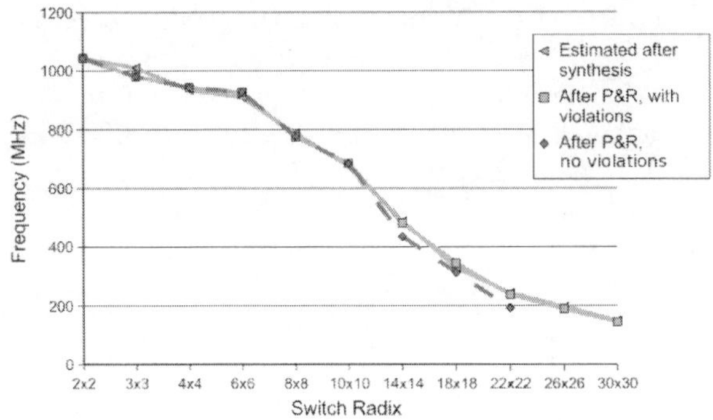

FIGURE 4.18: Switch frequency scaling with the switch degree (from [258] ©2007 IEEE).

4.4.2 Switch Degree

As mentioned many times throughout this chapter, there is a direct correlation between the switch radix and final operating frequency of the switch itself. Essentially, an increment of the switch radix has a direct consequence on the delay of the arbitration logic and of the crossbar selection logic, which adds up to the critical path. This trend is unmistakable when looking at Figure 4.18. Further considerations can be drawn with respect to area and power figures. In fact, as expected, they increase as the switch radix increases, while frequency goes down dramatically.

Another observation is that by leveraging tools performing placement-aware synthesis, the timing prediction after synthesis is rather accurate with respect to the final timing closure after placement and routing as pointed out in Figure 4.18.

The most interesting result concerns feasibility of switches with a large degree [258] with a standard cell design flow. In fact, logic synthesis tools are typically aware of placement but not yet of routing. As a consequence, for a 14×14 switch, the wire density in the switch crossbar becomes unmanageable to simultaneously cope with timing objectives, guarantee crosstalk freedom, and resolve design rule violations. Due to the goal priorities adopted (timing above all), the former two objectives are achieved while unfortunately causing from hundreds (14×14) to tens of thousands (30×30) of DRC (design rule check) violations. Such a large number of DRC violations is clearly unacceptable for manual fixing, and must be tackled automatically. Two possible solutions are available: increasing the switch area or decreasing the switch frequency. The former option proves only partially effective. Typical industrial rates are somewhere close to 85% area utilization. The 85% goal can be reproduced without issues until 10×10 cardinality; at this point, some widening of the target fences is necessary [258]. For example, the 14×14 switch can only be properly routed once its row utilization is tweaked to be close to 70%; the remaining "unused" space is in fact required to route resources. However, in the 30 × 30 case, the violations are not fixed even with a final row utilization of 50% (i.e., by leaving half of the switch floorplan unfilled). This result is clearly unacceptable due to its cost overhead.

The alternative solution to fix DRC issues by decreasing the target switch frequency returns somewhat better results, making 14×14 and 18×18 switches routable at a 25–30% frequency cost, but still fails on larger switches. Similarly to the above results, even after more than halving the frequency targets, 30×30 switches remain unroutable.

Even in cases where DRC violations can be fixed by some means, achieved results suggest that avoiding too large switches may be the best option. This is also due to system-level effects that would result from using large centralized blocks, which are not immediately apparent from the plots reported here. For example, the many cores connected to such a switch would ideally need to be physically placed just around it, causing obvious congestion in the floorplan. Alternatively, they could be spread around, but then several long links would be needed to connect remote cores to the switch. These links would require pipelining, bringing further latency, area, and power costs.

4.4.3 Link Length

An interesting physical design issue that is now addressed concerns the location of the critical path, which has been traditionally assumed to be inside switches or network interfaces. However, the reverse scaling of the delay of on-chip interconnects is giving rise to a different scenario.

When considering the switch in isolation, its critical path is obviously inside the switch block. Specifically, the critical path of the switch architecture assumed in this chapter goes from the input- to the output-buffer traversing arbitration and crossbar logic. Let us now consider a simple switch-to-switch system where initially, the blocks are placed close to each other at the distance of 1 mm. After placement and routing at such a distance, the critical path of the whole architecture is still inside the switch building block. The reason is that the delay of the interswitch channel is not dominating yet the overall timing of the system. By increasing the link length, results differ, as depicted in Figure 4.19.

Please note that each switch block has been synthesized with a very tight timing constraint in order to have a high performance switching fabric. This is the main reason that brings the critical path on the link channel already at the short distance of 2 mm. In fact, from this measure on, the critical path is determined by the switch-to-switch delay that is a typical phenomenon in interconnect dominated designs. For longer distances, the link delay dominates the whole design, therefore the critical path shifts to the communication channel from the output of the first switch to the input of the second one.

In general, this is a considerable result that points out the critical role

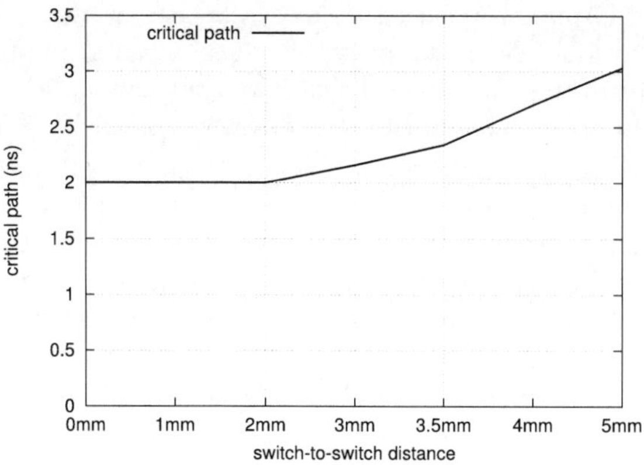

FIGURE 4.19: Critical path trend when increasing switch-to-switch link length.

played by the interconnects in the on-chip domain. In fact, it affects the performance of the network thoroughly. The key take-away here is that, if the main goal is performance, a topology with long links cannot be a good candidate for such high-speed systems as it will not be able to operate at high frequencies due to a large link delay. Another take-away is that as technology keeps scaling to the nanoscale regime, the critical path tends to move to the switch-to-switch links at progressively shorter link lengths than illustrated in Figure 4.19, although the break-even point depends on the actual constraints guiding switch synthesis.

In this context, link performance boosting techniques are beneficial for the performance of the entire network, and will be investigated in the next section.

4.4.4 Link Performance Boosting

This part of the chapter evaluates three fundamental link inference techniques applied when laying out a set of representative topologies. When moving from one technique to another, the objective is to speed up a topology by boosting its links. Therefore, the primary design objective is high performance.

4.4.4.1 Performance Boosting Techniques

The topologies under test are again those of Figure 4.16. In the first round of topology implementations, the inference of unrepeated links has been forced. In essence, the backend tools have been prevented from instantiating buffers or inverters along the link. Therefore, the only degree of freedom for such tools is to prevent timing violations on the links by inferring a driver of suitable driving strength in the output port of the upstream switch.

Second, a performance-boosting strategy consists of allowing repeater insertion along the link. While an increase of instantiated buffers can be expected, the lower driving strength of switch output gates partially offsets the added cost. In any case, a significant cut down on link delay is expected.

Finally, link pipelining has been applied to break long timing paths across interconnects. Synthesis tools do not provide a native support for this, thus requiring additional design effort. The ideal approach is not to manually place pipeline stages in the floorplan, but to let the tool handle this based on design constraints and optimization directives. The tool has better visibility of floorplan constraints and design rules. When implementing link pipelining, flow control issues need to be reconsidered. One solution is to implement pipeline stages not as simple retiming stages, but as retiming and flow control stages. In practice, data progresses from switch output buffers to downstream links by using the inherent storage of pipeline stages along those links. In the assumed architecture (see the Appendix), pipeline stages consist of 2 flit buffers plus a control logic and handle the stall/go flow control protocol.

4.4.4.2 Timing Closure

Since the final goal is high performance, the objective is to materialize, after topology place-and-route, the maximum speed achievable by the slowest NoC module (characterized in isolation with postlayout timing analysis). In the assumed architecture, the critical module turns out to be always the switch with the highest radix in the topology. The resulting speed upper bound, which ignores the effect of interswitch links, is reported in the first two rows in Table 4.5. For the slowest switch of each topology in isolation, the gap between postsynthesis and post-P&R speed ranges between 12% to 21%. The 2-ary 2-mesh exhibits a higher postsynthesis value due to the larger radix and a more severe post-P&R degradation due to the larger switch area compared with nonconcentrated topologies.

When considering timing closure for the entire topologies, the degradation associated with interswitch link routing and with their synthesis techniques becomes apparent. Let us consider repeaterless links first (Table 4.5, third row). While 2-D mesh and 4-hypercube had the same performance upper-bound (since they have switches with the same maximum radix), the more challenging routing of the 4-hypercube gave rise to a 39% degradation of the critical path, while routing of the 2-D mesh turns out to be less critical. The 2-ary 2-mesh has some links longer than 4 mm, therefore interswitch routing has an even more relevant impact. This also hides the effect of the higher switch radix, which is not the ultimate responsibility for the slower operating frequency. For all topologies, the critical path goes through the network links.

Although postlayout effects of interswitch wires paint a dismal picture of topologies using long wires (because of the use of more dimensions or of the switch separation in the layout), designers can optimize wires to overcome these large delays. Activating repeater-insertion during topology synthesis en-

Topology	4-ary 2-mesh	2-ary 4-mesh	2-ary 2-mesh
Postsynthesis (WC Switch))	1 ns	1 ns	1.15 ns
Post-P&R (WC Switch)	1.12 ns	1.12 ns	1.4 ns
Post-P&R repeaterless (Topology)	1.27 ns	1.56 ns	1.67 ns
Post-P&R with repeaters (Topology)	1.19 ns	1.56 ns	1.5 ns
Post-P&R with pipeline stages (Topology)	–	1.19 ns	1.42 ns

TABLE 4.5: Timing results.

ables the speedups illustrated by the fourth row of Table 4.5, denoting the best critical delay at which timing closure was achieved. Results are quite heterogeneous. The 2-D mesh further benefits from repeaters and achieves a 6% speedup of its critical path, which results in 1.19 ns. This value is quite close to the performance upper-bound (1.12 ns), thus indicating that while links are still critical in this topology, their length is such that their degradation of topology performance can be made irrelevant.

The opposite holds for the 4-hypercube, where repeater insertion did not surprisingly provide any performance improvement. The reason for this lies in the fact that horizontal routing channels (see Figure 4.16(b)) were sized conservatively small, approximately 2.5x the switch side. For this reason, switch-to-switch links sometimes end up finding another switch on their way. This is a placement constraint for the buffers that prevents their insertion with ideal spacing. The result is that link performance does not improve, indicating that buffer insertion should not be taken for granted in NoC design, but should be carefully engineered to materialize its expected advantages. In practice, widening the routing channel ideally to 4x the switch side would solve the problem but would also lead to a large floorplan area overhead. Finally, no such constraints exist for the 2-ary 2-mesh (its connectivity pattern is trivial), which improves its baseline performance by 10%.

Further critical path improvements were expected from link pipelining. When applying this technique, the objective was to materialize the same critical delay of the 2-D mesh (with buffers) even for the 4-hypercube and the 2-ary 2-mesh. Interestingly, link pipelining turns out to be effective even for the 4-hypercube, regardless of its undersized routing channels. In fact, the target critical delay of 1.19 ns was achieved. This result clearly indicates the lower sensitivity of link performance to nonideal pipeline stage placement compared with nonideal repeater spacing. Hence, link pipelining proves more robust for area-optimized floorplans with more challenging placements.

Pipelining was effective also for the 2-ary 2-mesh, however the upper bound for its performance is the postlayout critical delay of its high-radix switches (1.4 ns), far worse than the 2-D mesh link delay of 1.19 ns.

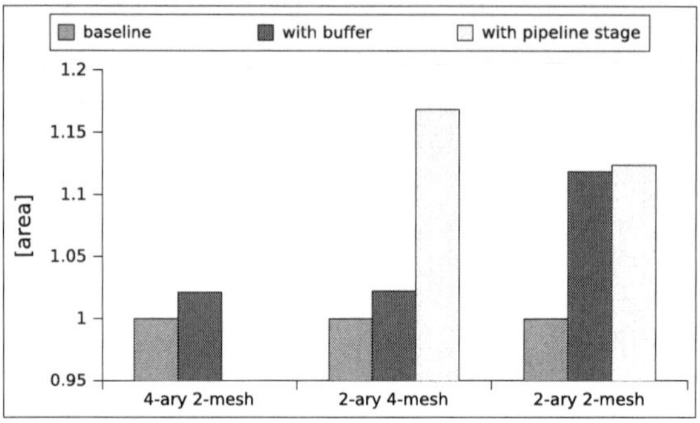

FIGURE 4.20: Normalized area (from [187] ©2009 ACM).

4.4.4.3 Area Overhead for Link Performance Boosting

In order to assess the implementation cost for link performance boosting techniques, we illustrate area reports in Figure 4.20. Results are grouped by topology and normalized to the baseline implementation of each topology.

Buffer insertion is quite cheap for the 2-D mesh and the 4-hypercube, while generates a significant area overhead for the 2-ary 2-mesh. This is due to the backend tools, that have dealt with the performance maximization of long links by dramatically increasing the number of buffering gates. As explained later on, this implies also a relevant power cost. However, when pipeline stages are inferred, the additional area overhead of the 2-ary 2-mesh is marginal. This is due to the fact that by inserting pipeline stages the tool was able to remove an equal amount of buffering area, so that the two contributions offset each other.

This does not hold for the 4-hypercube, which raises its area by 14% when pipelining is used. Not only the area overhead of the pipeline stages is incurred, but many buffers are kept in the links since the frequency boost is significant when moving from buffering to pipelining. The trend of leakage power fully reflects that of area overhead.

4.4.4.4 Energy Efficiency

Although there is a price to pay to boost link performance in terms of area and power, the speedup in job completion can be exploited to cut down on total energy of the on-chip network. However, this energy saving materializes only if the gain in execution time outweighs the power overhead. This section sheds light on this aspect. Moreover, it is also investigated whether the different link synthesis techniques can change the relative energy ratios between the different topologies.

Experiments were carried out with a parallel synthetic benchmark consisting of 1 producer task, 14 worker tasks, and 1 consumer task. Every task is assumed to be mapped on a different computation end node. The producer

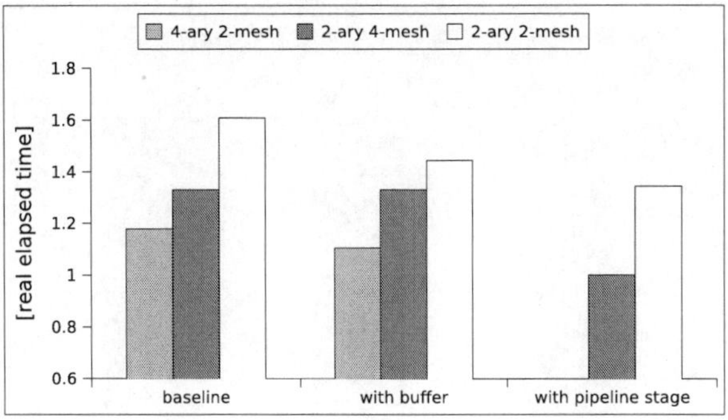

FIGURE 4.21: Normalized execution time (from [187] ©2009 ACM).

task reads in data units from the I/O interface of the chip and distributes it to the worker tasks. Output data from each worker end node is then collected by a consumer end node, which writes them back to the I/O interface. During computation, all-to-all accesses are generated to account for a cooperative computation process. Transactions are generated in compliance with the Open Core Protocol (OCP) protocol (burst accesses) by programmable OCP traffic generators [197]. Loose synchronization between the cores is assumed. In fact, whenever a producer has data available for a consumer, it sends them across the network without any previous check for consumer availability to accept the message. Similarly, different producers are enabled to send messages to the same consumer at the same time, thus generating more conflicts in the network. Traffic generation rates were set to have a balanced system operation, i.e., to avoid system bottlenecks (like the I/O), which would stress other design issues other than network bandwidth and link performance.

Real elapsed time results for the baseline repeaterless topologies along with their respective buffered and pipelined variants are reported in Figure 4.21. For each variant, the maximum achievable post-P&R speed is applied (see Table 4.5). Since the utilized network interfaces use a lightweight frequency-ratioed synchronization mechanism, a clock divider of 2 is applied at all network interfaces, instructing the end nodes to run at half the speed of the network and slowing down the end nodes as well when the network speed is low. The following indications come from Figure 4.21. Although the 4-hypercube reports a lower clock cycle count to complete the benchmark, its execution time is always higher in the baseline topologies and in the repeated ones since its running speed is much lower than the 2-D mesh. A 16-end node system is too small for a hypercube to take full advantage of its better properties (bisection bandwidth, diameter) and thus to offset the speed degradation associated with its relatively longer links. The situation is even worse for the 2-ary 2-mesh, which is worse than the other two topologies in the baseline and buffered variants both in execution cycles (because of poor bisection bandwidth) and in running speed (overly long links).

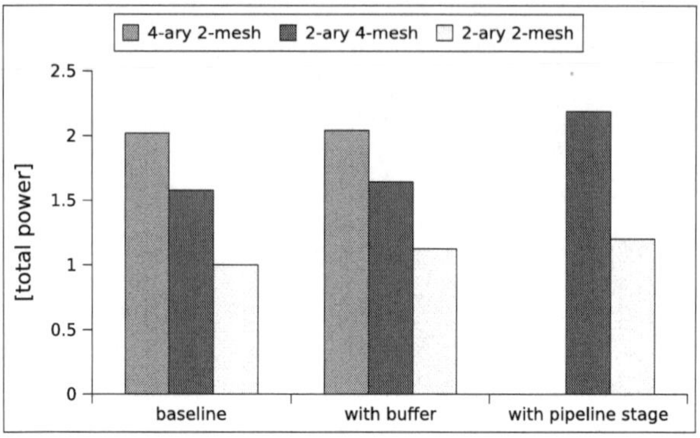

FIGURE 4.22: Total power (from [187] ©2009 ACM).

However, when the 4-hypercube can operate at the same speed of the 2-D mesh thanks to link pipelining, then it becomes the most performance-efficient solution. It is worth observing that link pipelining is not necessarily adverse to total performance. In this experiment, it enables a significant topology speedup and provides additional buffering to the topology itself, which is effective to handle a bandwidth-sensitive traffic pattern. The high switch degree of the 2-ary 2-mesh causes a saturation effect to the performance gain that this topology can achieve by means of link pipelining.

The power cost incurred to boost link and hence topology performance is illustrated in Figure 4.22. Power is affected by the operating frequency of each topology. By comparing topologies with boosted links with their baseline variants, it results that repeaters do not cost a lot in terms of power. This holds for the 2D mesh and for the 4-hypercube, which is in line with the hardly relevant share of interswitch links over total wire length. Power overhead of the 2-ary 2-mesh is an exception to this, due to the large power cost of driving a long link effectively. When it comes to link pipelining, the power cost abruptly increases, especially for the 4-hypercube, while the overhead for the 2-ary 2-mesh increases more smoothly. Fortunately, this is also the scenario where the 4-hypercube gains more in terms of performance. When we compare topologies with each other, we observe that the topology of choice when low power is the primary design goal is the 2-ary 2-mesh. The lower power of the 4-hypercube with respect to the 2-D mesh mostly derives from its lower speed, although this is not the only explanation. This is confirmed also by the marginal power overhead of the 4-hypercube when it can run at the same speed of the 2-D mesh (see buffered 2-D mesh vs. pipelined 4-hypercube in Figure 4.22). This is counterintuitive, since the hypercube has many more buffering resources than the 2-D mesh. The answer lies in the power breakdown. This is reported in Figure 4.24 for the baseline variants of the two topologies. The 2-D mesh was resynthesized and analyzed both at full speed (rightmost column) and at the same speed of the 4-hypercube for a fair comparison. In all cases, it is evident

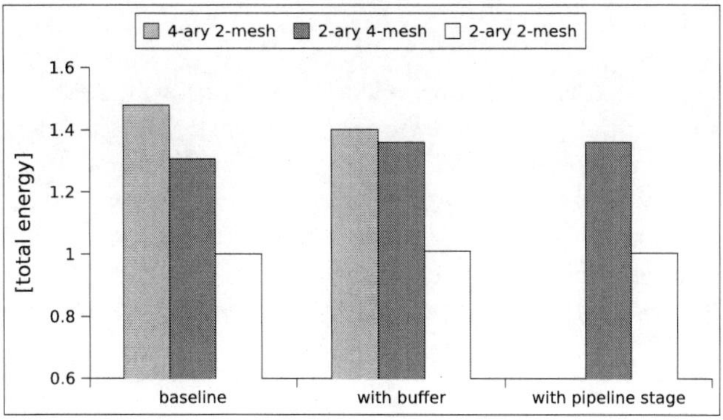

FIGURE 4.23: Normalized execution time, power, and energy with link synthesis techniques (from [187] ©2009 ACM).

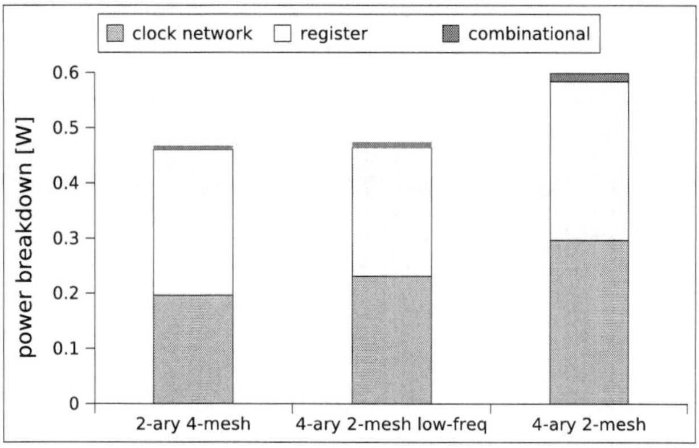

FIGURE 4.24: Clock tree power impact (from [187] ©2009 ACM).

that the clock tree has a relatively lower impact than the register power in the 4-hypercube, while in the 2-D mesh it weights more. As a consequence, when the same speed is inferred, the two topologies have the same power surprisingly. While register power obviously increases for the 4-hypercube, the clock tree is cheaper, and this explains the result. While the 2-D mesh has heterogeneous switches, the 4-hypercube has all switches with exactly the same radix. Hence, the clock tree is inherently better balanced and thus easier to synthesize while meeting skew constraints.

By combining the performance results of Figure 4.21 with the power reports of Figure 4.22, the energy results of Figure 4.23 are achieved. Overall, buffering the links of a 2-D mesh is always an energy-efficient strategy. On the contrary, 4-hypercube and 2-ary 2-mesh show minor energy variations when moving from one scenario to the other. This indicates that performance improvements have been achieved at the cost of a proportional power overhead.

4.4.4.5 Wrap-up

Summarizing, the contribution of Section 4.4.4 proved that the layout of a multidimensional topology hides a wide set of intricate physical design pitfalls. They span from a pure timing problem, were not all the topologies converging with the same timing closure (due to routing channel size as well as the intrinsic switch radix) up to a more complex trade-off between the deployed link boosting techniques and the area cost they come along with. The next section will extend the concepts of the above analysis to a wider range of topologies.

4.5 64-tile Systems

The goal of this section is to expand the physical design concepts presented so far to a wider set of topologies. Specifically, this analysis concerns some of the large topologies for 64-end nodes presented in Section 4.2.3. This section aims at assessing the feasibility of those connectivity patterns for largely integrated digital systems in the context of nanoscale silicon technologies.

Like for 16-tile topologies, the set of considered topologies for 64-node networks has been restricted for the need to keep the section focused. The criterion to select topologies for physical implementation was based on their suitability for a virtual channelless switch architecture. Introducing virtual channels is yet another degree of freedom that we prefer not to consider for the sake of clarification. So, the exclusion of a topology from the following analysis does not mean that the topology is not competitive. In this direction, we noticed that k-ary n-cube and k-ary n-rec topologies have issues associated with the implementation of routing algorithms, as both feature abundant cycles that must be broken in order to provide deadlock-free communication. In the case of torus topologies with wormhole switching, the most efficient ways of providing deadlock-free routing require the use of several virtual channels [85], while the WK-Recursive topology requires the use of virtual channels plus some additional hardware in order to be deadlock free [264].

Anyway, there are still several generic routing strategies that allow to provide deadlock-free deterministic routing in both cases without the need to modify the NoC architecture [85], but they are quite inefficient. In the case of the torus topology, the turn model has been proposed to break cycles by prohibiting some turns in the network. Unfortunately, this design choice transforms the torus into an equivalent mesh, by breaking the cycles introduced by wraparound links. Also, the Up/Down model, which transforms the topology into a spanning tree, has been proposed for its use in both topologies, but it incurs congestion issues in the root switches, greatly reducing the performance of the topology. For these reasons, k-ary n-cube and k-ary n-rec topologies have not been considered for the physical implementation analysis that follows.

Also multistage interconnection networks have not been considered since an assessment of their implementation quality reported in [188] does not support their use for large scale systems in light of their layout intricacy (unidirectional topologies) or of their large area and power cost (fat-trees). Obviously, full custom solutions to take advantage of the unique properties (especially latency scalability) of these topologies is always possible.

4.5.1 Characterization Methodology

The xpipesLite switch illustrated in the Appendix was used as the basic building block to construct the 64-end nodes topologies under test. However, exploring the design space of topologies with such a large number of cores by means of their actual physical synthesis proved impractical due to synthesis time and memory capacity requirements. Therefore, this section presents a way to cut down on the number of physical synthesis tests while still characterizing quality metrics of the full topology with high accuracy.

As already reported in [103, 102, 188], the performance bottleneck of a topology lies in its longest switch-to-switch communication channel. Aware of this, for each topology under investigation, a subsystem composed of two communicating switches at the maximum possible distance in the topology is built, thus capturing the largest link delay in that topology. Logic synthesis for *maximum performance* and place and route are then performed to capture the real minimized switch-to-switch delay. Such delay (which is the critical path delay of the network too) is then used as the *target delay* to resynthesize, place and route all the possible switch-to-switch system combinations at all the possible link distances in the topology under test. The reason for this is that the goal is to accurately capture the switch cell area at a certain distance and at a certain target speed. It is well known from logic synthesis theory that as the target speed is decreased, a large area can be saved. In this direction, it would make no sense to synthesize switches for maximum performance when a long link limits overall network speed (unless decoupling techniques like link pipelining are used, as we will see later on).

With this methodology, only a few selected synthesis runs for each topology need to be performed to characterize its delay and area as a whole. The approximation lies in the availability of enough routing channel space for regular routing of NoC links and in the accuracy of wire delays for links that undergo bending in the actual layout.

Moreover, with this method it is also possible to capture the link buffering cost, in fact, by leveraging the report of the utilized physical synthesis tool, the inferred buffers along switch-to-switch links can be traced. In order to be as accurate as possible, two communicating ports of both switches in the test subsystem are left unconnected. They are the ports connecting to the network interfaces of processing cores, which are typically placed close to their switch and therefore feature minimum capacitive load. This way, the tool is prevented from using large driving strengths for the gates on those ports.

Finally, for those cases where switches feature ports connected to links of different lengths, the area of each switch output port is corrected by a coefficient that reflects the driving strength needed to drive the corresponding link. Such coefficients are experimentally derived beforehand by means of a dedicated set of tests.

A further step of this section is the estimation of the number of required pipeline stages for each link to speed up a topology. For this purpose, such retiming and flow control stages are instantiated along the communication links thus breaking the switch-to-switch critical path. By incrementing the number of pipeline stages, the critical path can be brought back to the switch. However, in order to limit the area overhead, in the experiment that follows it was decided to bring back the critical path to the second topology dimension, thus having a milder impact on the link area.

The next subsection starts by commenting on physical synthesis results achieved for 64-end node topologies without link pipelining. Later, the analysis is shifted to pipelined networks. Section 4.6 will back-annotate physical synthesis results into a system-level exploration framework, thus gaining layout awareness during the topology selection process.

4.5.2 Physical Implementation

Table 4.6 shows the 64-node topologies considered for physical implementation. The maximum switch radix for the topologies under test ranges from a reasonable value of 6 to a large value of 12, which makes it already difficult to place and route switches without DRC violations (see Section 4.4.2), although still within the feasibility range according to [258]. Please note that the architectural assumption behind the switch radix column in Table 4.6 is that each tile attached to a switch occupies two ports (one for initiating transactions and one to serve transactions initiated by other tiles). This way, the switch radix parameter matches the content of Table 4.3, which is instead unaware of this physical implementation detail.

Postsynthesis frequency results are referred to the switches in isolation, hence reflect the critical path delay of the switch with the highest degree in the topology, thus confirming the principle illustrated in Section 4.4.2.

After placement and routing, the effect of the link delays comes noticeably into play. Most of the topologies suffer from long switch-to-switch channels that need to be routed across the chip. For the sake of the analysis, only the longest link length per topology is reported in the 5th column of Table 4.6. By comparing such column with the 4th one, it is possible to recognize a clear correlation between the increasing link length and the decreasing operating speed of the topology under test. Only topologies with short links (e.g., 8-ary 2-mesh and 4-ary 2-mesh) can work at a reasonable frequency for realistic application scenarios.

TOPOLOGY	Radix	postsynthesis (ns)	post-P&R (ns)	longest link
8-ary 2-mesh	6	1.08 GHz	890 MHz	1.5 mm
8-C-mesh	6	1.08 GHz	250 MHz	6.75 mm
4-ary 3-mesh	8	950 Mhz	220 MHz	6.9 mm
2-ary 6-mesh	8	950 Mhz	220 MHz	6.9 mm
2-ary 5-mesh	9	810 MHz	230 MHz	6.96 mm
4-ary 2-mesh	12	720 MHz	530 MHz	3.0 mm
2-ary 4-mesh	12	720 MHz	260 MHz	6.4 mm
4-C-mesh	12	720 MHz	260 MHz	6.4 mm
4-ary 3-flat	14	610 MHz	120 MHz	13.29 mm

TABLE 4.6: Post-place&route results of the 64-end nodes topologies under test.

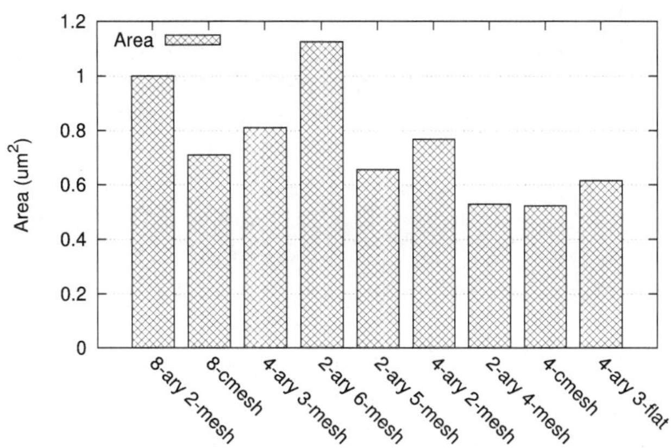

FIGURE 4.25: Normalized area for 64-end nodes topologies.

Besides stressing the role of interconnect delay in determining global network speed, it should be observed that also some logic gates end up in series to the critical links close to their far-ends. They are associated with flow control management and further contribute to the critical path delay.

From the area viewpoint (see Figure 4.25), it is interesting to note that this result is influenced by the combination of many parameters such as: number of switches in the topology, their radix and consequently their final working frequency, and link buffering. In fact, as explained above, in order to be accurate, all representative switches in every topology have been resynthesized at the final working speed of the whole network.

As an example, let us consider a very slow topology like the 2-ary 6-mesh that features a larger area footprint with respect to the 8-ary 2-mesh. Such a network is operating at a frequency much slower than the 8-ary 2-mesh, but since it has an equal number of switches (64) with a higher radix (8 vs. 4, 5 or 6), the overall area figures plays in favor of the 8-ary 2-mesh with a 10% saving.

Another interesting result concerns the 4-ary 2-mesh. This topology has a relatively short worst-case link (3 mm), thus it does not suffer from a large speed degradation after place-and-route. As reported in Table 4.6, this topology is the only one (along with 8-ary 2-mesh) to have a final working speed above 500 MHz. Interestingly, the area footprint of such topology has a 20% saving with respect to the 8-ary 2-mesh as it only has 16 switches. Although their radix is 10, 11, and 12, their final working speed along with the number of their instances results in area savings with respect to the 8-ary 2-mesh counterpart.

The overall conclusion is that most of the topologies are not competitive with the 8-ary 2-mesh because of their long links that influence the final working speed. A natural way to tackle this problem is to implement link pipelining on such long links but the insertion policy has to be carefully engineered. In fact, the studied 64-end nodes topologies feature a high number of long links that could rapidly bring the area overhead to an unaffordable budget.

4.5.3 Link Pipelining

In order to cope with the high-speed degradation of most topologies analyzed in the previous section, pipeline stages need to be inserted especially in very long links. As already explained in Section 4.4.4 by adding pipeline stages, it is possible to partially (if not completely) recover the initial operating frequency of the basic switch block. The criteria that has been adopted for the insertion of pipeline stages is to use them only from the third link dimension onwards. Therefore, topologies such as 8-ary 2-mesh and 4-ary 2-mesh have not been modified. Table 4.7 collects the results of this experiment. Please note that the 4-ary 3-flat topology (with 4 end nodes per switch) has not been considered for the following analysis because its switch radix (14) would be a major limiting factor for the speed of the whole topology and for its layout feasibility. A straightforward way to overcome such a limitation would be to consider a pipelined switch architecture or/and even a full custom design, that fall outside the scope of this chapter.

As clearly reported in the 3rd and 4th column (up), the insertion of pipeline stages is a very effective way to reduce postplace and route frequency degradation. Column 2 (down) reports the number of pipeline stages inferred in each link dimension, whereas the 3rd column (down) points out the number of links of each topology. The area weight comes from the combination of these two factors and it is reported in the 5th column (down). Total cell area of the topologies along with the contribution of such pipeline stages is reported in Figure 4.26.

Please note that the number of pipeline stages per link depends on the maximum achievable frequency (dictated by the maximum switch radix) along with the link length, which is an intrinsic characteristic of each topology. As

topology	radix	postsynthesis frequency	post-P&R frequency	
8-ary 2-mesh	6	1.08 GHz	893 MHz	
8-C-mesh	6	1.08 GHz	893 MHz	
4-ary 3-mesh	8	950 MHz	855 MHz	
2-ary 6-mesh	8	950 MHz	855 MHz	
2-ary 5-mesh	9	810 MHz	562 MHz	
4-ary 2-mesh	12	720 MHz	532 MHz	
2-ary 4-mesh	12	720 MHz	532 MHz	
4-C-mesh	12	720 MHz	532 MHz	
topology	**# of pipe-stage per dimension**	**num. links**	**total pipe-stage area (um2)**	**total switch area (um2)**
8-ary 2-mesh	0	112	0	2327712.8
8-C-mesh	express link⇒4	128	193425.9	2752108.8
4-ary 3-mesh	dim.3⇒4	144	660216.3	3182953.2
2-ary 6-mesh	dim.3,4⇒1, dim.5,6⇒5	192	1087918.1	4362092.8
2-ary 5-mesh	dim.3⇒1, dim.4.5⇒3	80	293081.6	2758480.4
4-ary 2-mesh	0	24	0	1860718.3
2-ary 4-mesh	dim.3,4⇒3	32	125574.7	2328426.4
4-C-mesh	express link⇒3	32	62787.4	2328426.4

TABLE 4.7: Post-place&route results of 64-end nodes topologies with pipeline stage insertion.

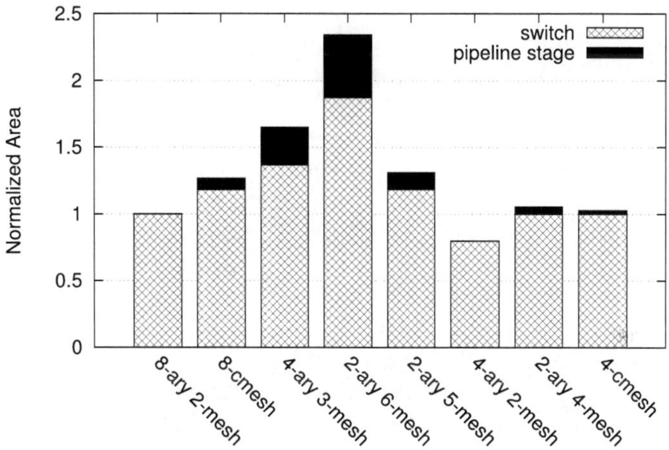

FIGURE 4.26: Normalized area for 64-end nodes topologies with pipeline stages.

reported in Figure 4.26, the 2-ary 6-mesh is the most area-greedy topology because it has the highest number of switches (64) that are placed and routed at the high frequency of 855 MHz. Moreover, this topology features 192 links with up to 5 pipeline stages on the longest interconnection channel. The key take-away is that, for each topology, there is a different price to pay to restore the possible working frequency allowed by the elementary switching fabric. For this reason, Section 4.6 will introduce the throughput/area metric (or

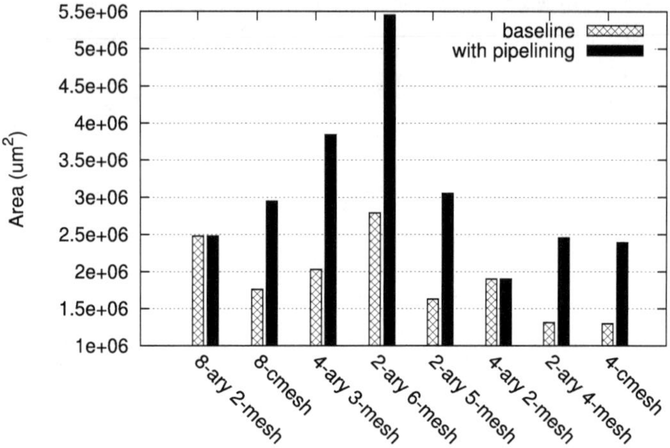

FIGURE 4.27: 64-end nodes topologies area overhead for pipeline stage insertion.

area efficiency) that provides a fair assessment of the cost of the achievable bandwidth in each topology.

To conclude the physical implementation part, it is interesting to observe the result depicted in Figure 4.27. For each topology, this plot reports area results before and after the insertion of pipeline stages. Interestingly, the substantial cell area increment in all cases (except topologies where pipeline stages were not inserted) comes from a twofold contribution: pipeline stage insertion (as discussed so far) and the restored higher frequency allowed by such insertion. In fact, the largest contribution in terms of area comes from the switch cell area devoted to achieve the higher working frequency. This relevant effect is typically overlooked by the vast majority of topology exploration frameworks in the open literature for the early NoC design stage.

4.6 The Performance Prediction Gap

Previous sections of this chapter focused at first on the theoretical evaluation of several topologies for NoCs and later on the properties of their physical implementations. It is now time to put everything together. This section in fact re-evaluates topology abstract quality metrics for 64-node networks in light of the efficiency of their physical implementation, thus pointing out how misleading conclusions can be when the physical synthesis effects are ignored or underevaluated.

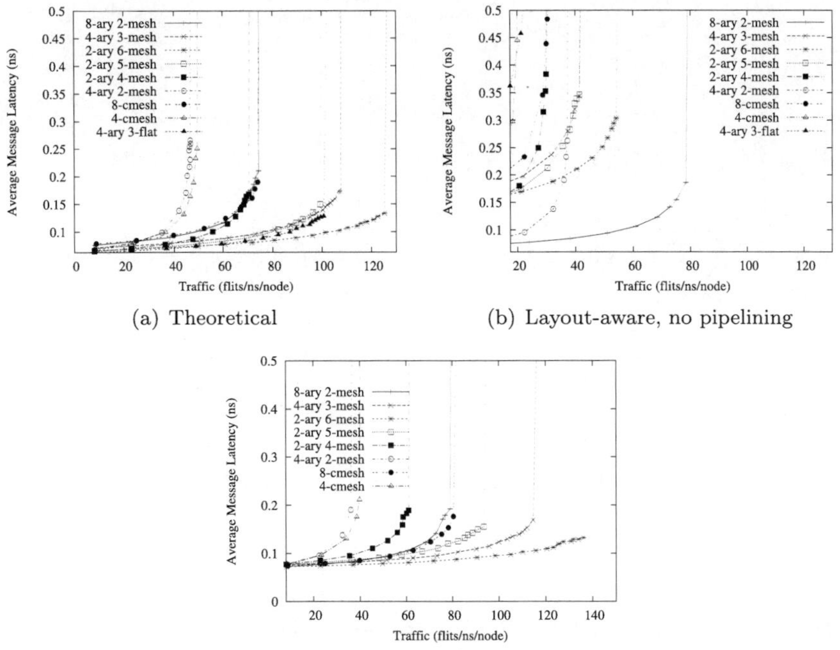

(a) Theoretical

(b) Layout-aware, no pipelining

(c) Layout-aware, with pipelining

FIGURE 4.28: Performance of 64-end node systems with uniform traffic.

For the layout-aware system-level evaluation of topologies, a restrictive assumption is made to keep the analysis focused: we consider source-based deterministic routing, and the dimension-order routing (DOR) in particular [85]. In order to remove the cycles introduced by the express links of k-ary n-C-mesh topologies, express links will be used only to advance messages that travel along the perimeter of the topology a distance equal or higher than the switches that this express link overcomes, like in [17].

4.6.1 Performance Comparison

Figure 4.28(a) depicts accepted traffic vs. average message latency for a uniform distribution of message destination for different topologies when considering high-level estimations. Obtained results reflect the conclusions drawn in Section 4.2.3 and do not take the actual operating speed of topologies into account. Figure 4.28(b) shows the same analysis where each topology works at the operating frequency reported in the previous section (Table 4.6, without pipeline stages). By comparing Figure 4.28(a) against Figure 4.28(b), when link pipelining is not considered, there is a misleading gap between the performance predictions of the high-level analysis and the layout-aware one.

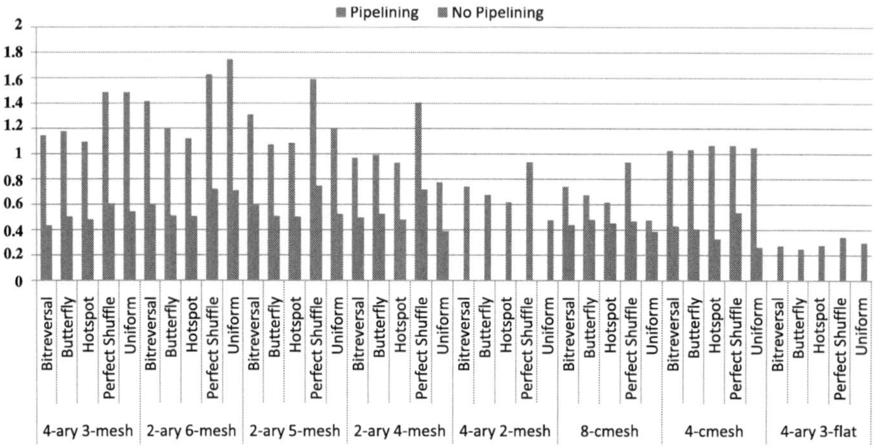

FIGURE 4.29: Normalized performance of 64-end node systems.

In fact, while the theoretical results reported in Figure 4.28(a) claim that several topologies outperform the 8-ary 2-mesh, this latter topology is proved to be the best solution in the layout-aware results of Figure 4.28(b). In fact, there is a direct correlation between the operating frequency and the achieved system-level performance: the lower the operating frequency, the higher the average latency and the lower the maximum achievable throughput, regardless of the results obtained in the high-level analysis. In practice, poor matching with silicon technology completely offsets the better theoretical properties of the topologies. However, when the impact of wiring complexity over the critical path is alleviated by using link pipelining techniques, different conclusions can be drawn.

Figure 4.28(c) reports the same analysis results when each topology works at the operating frequency (see Table 4.7) enabled by the usage of link pipelining, and also accounts for the higher latency of pipelined links. In this case, there are three network topologies that clearly outperform the 8-ary 2-mesh: 2-ary 6-mesh, 2-ary 5-mesh and 4-ary 3-mesh.

Similar curves have been drawn for several traffic patterns for each topology. Those results are summarized in Figure 4.29. This figure shows the normalized maximum throughput of each topology with respect to the 8-ary 2-mesh solution. In this plot, a bar higher than 1 implies an improvement of the maximum throughput over the 8-ary 2-mesh solution. Interestingly, those results follow the same trend as discussed for the uniform traffic pattern. All nonpipelined solutions are clearly worse than the 8-ary 2-mesh, while pipelined solutions follow the same trend than reported in the high-level analysis: most of the solutions outperforms the 8-ary 2-mesh, with the 2-ary 6-mesh being the best solutions for all the traffic patterns. Although in this case the obtained performance is closer to the high-level estimations, link pipelining techniques

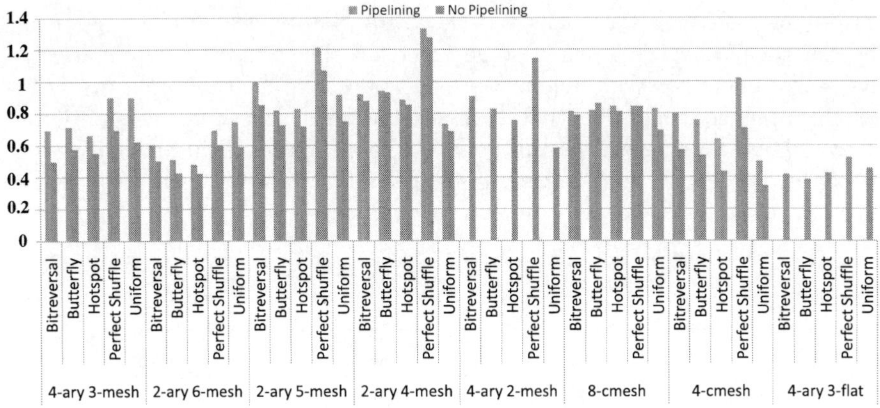

FIGURE 4.30: Normalized area efficiency of 64-end node systems.

may have a great impact over the implementation cost, thus requiring a new metric to assess the real effectiveness of link pipelining techniques.

In particular, we have considered the *area efficiency metric*, defined as *throughput/area*, which correlates the throughput improvement with the area cost that has been paid to achieve that. Results are shown in Figure 4.30, which depicts the area efficiency of each topology normalized with respect to that of the 8-ary 2-mesh. Results are reported with and without pipelining for several traffic patterns. In most of the cases, the area efficiency of both pipelined and nonpipelined solutions is clearly lower than the 8-ary 2-mesh solution. The key take-away is that the performance improvements achieved by complex topologies with pipelined links are not cost-effective.

The only exception is when the traffic pattern favors topologies with a low hop count, as in the case of the perfect shuffle traffic. This characteristic, along with the fact that some topologies feature a low area cost, leads to a higher area efficiency with respect to the 8-ary 2-mesh.

4.7 Conclusions

This chapter shows that the conclusions drawn by a pure high-level analysis of topology performance based on abstract quality metrics may be highly misleading if not enriched by the information provided by the physical synthesis. A basic implementation of investigated topologies (without link pipelining) leads to very poor performance due to their long links and a consequent unacceptable critical path delay. Nevertheless, it is possible to improve the performance of those topologies by alleviating the impact of long wires over the critical path with pipeline stage insertion. In spite of a very competitive performance for such augmented topologies, the incurred area overhead (area of pipeline stages plus that of the switches resynthesized to support the increased operating speed) is still too severe to make these solutions fully cost-effective with respect to a 2-D mesh. Therefore, they are a trade-off solution between communication performance and implementation cost.

Acknowledgment

This work was supported by the Spanish MEC and MICINN, as well as European Commission FEDER funds, under Grants CSD2006-00046 and TIN2009-14475-C04. It was also partly supported by the project NaNoC (project label 248972), which is funded by the European Commission within the Research Programme FP7.

Chapter 5

Routing Algorithms and Mechanisms[1]

José Flich

Technical University of Valencia, Spain

Samuel Rodrigo

Technical University of Valencia, Spain

Antoni Roca

Technical University of Valencia, Spain

Simone Medardoni

University of Ferrara, Italy

[1]Parts of this chapter are based on "Efficient unicast and multicast support for CMPs," by S. Rodrigo, J. Flich, J. Duato, and M. Hummel, which appeared in Proceedings of the 41st Annual IEEE/ACM International Symposium on Microarchitecture, 2008, [269], and "Addressing Manufacturing Challenges with Cost-Efficient Fault Tolerant Routing," by S. Rodrigo, J. Flich, A. Roca, S. Medardoni, D. Bertozzi, J. Camacho, F. Silla, and J. Duato, which appeared in Proceedings of the 4th ACM/IEEE International Symposium on Networks-on-Chip, 2010, [270].

Once the topology of the network is designed, messages need to be forwarded from their sources to their destinations, using the physical connections established by the topology. As the topology will offer many possible paths for every message, we need to select which path to use on a per message basis. In addition, the selection of specific paths may lead to situations where the performance of the system degrades dramatically or even the entire system collapses. Therefore, the way messages are routed is one of the key design decisions for an operative and efficient network-on-chip (NoC). Such decisions are taken based on a routing algorithm run on every switch in the network and potentially also at the network interface of the node that injects the message.

The new challenges that NoCs will face in the future (some of them are described in Chapter 1) will also influence the way routing algorithms are designed and implemented. In particular, heterogeneity induced by those challenges requires routing algorithms able to work on irregular topologies. Thus, compact and efficient algorithms designed exclusively for 2D meshes are no longer valid. Although we always may design a flexible routing algorithm through large memories at switches, such flexibility comes at the cost of overhead and delay. Implementation and access time to memories do not scale and thus seem not suitable for large systems.

In addition to unicast messages, in chip multiprocessor (CMP) systems the coherence protocol requires other means of communication where one end node sends a message to several or even all end nodes in the chip (see a detailed evaluation in Chapter 12). This type of communication is referred to as *multicast* (or *broadcast* when all end nodes are addressed) and is a variant of collective communication (one or more source end nodes send a message to one or more destination end nodes). Thus, we require some support at the network level for efficient collective communication.

In this chapter we address the issue of routing algorithms and their efficient implementation in the face of the new challenges. First, we describe the basic concepts for routing algorithms and their implementation, covering the issues of deadlock, livelock, unicast, and collective communication. Then, we revisit the literature for routing algorithms in the recent years for on-chip networks (and the literature for off-chip networks), focusing our criticism on the way the proposals deal with heterogeneity. Later, we describe some of the recent proposals that tackle the heterogeneity issue from a different perspective and use the concept of routing restrictions. Finally, we provide future directions in the research for routing algorithms and their implementation for the coming years.

5.1 Basic Routing Concepts

The routing algorithm is in charge of computing the path every message needs to take (through the topology) to reach its destination. There are many proposals in the literature for routing algorithms and their implementations. Covering all of them is daunting and at the same time worthless. For the sake of brevity we describe the major types of routing algorithms and their implementations in the following paragraphs.

Routing algorithms can be classified as deterministic or adaptive, based on the possibility to select among alternative paths at every hop a message takes. In a *deterministic* routing algorithm the path is known beforehand and is the same for a given source-destination pair regardless of the network status (e.g., traffic load). One of the most well-known deterministic routing algorithms is Dimension Order Routing (DOR) [308] for k-ary n-cube topologies. In a 2-D mesh network the message reaches first the X coordinate of the destination and then travels along the Y dimension. In contrast, an *adaptive* routing algorithm may take different paths for messages going from the same source to the same destination end node. Such decisions can be based on current traffic situations (e.g., a congested output port in a switch). Upon reception of a message, the switch analyses the current status of every possible output port. Ports already busy are filtered and thus the message takes one free output port. Notice that adaptive routing may introduce out-of-order delivery of messages for the same source-destination flow. However, in deterministic routing (and when no virtual channel multiplexing for the same flow is allowed) all messages for the same flow arrive in the same order they were injected into the network.

When considering the number of hops a message takes, we can further classify algorithms as minimal or nonminimal. Messages may always take output ports that get them closer to their destination. In this case the algorithm is *minimal*. In other algorithms, however, messages may take *nonminimal* paths, thus ending up in a nonminimal routing algorithm. Nonminimal path support is usually provided to enhance fault-tolerance, thus avoiding a faulty link along a minimal path for a source-destination pair.

As briefly introduced above, routing can be further classified based on unicast or collective communication. Unicast relates to a single message being forwarded from a source end node to a single destination end node. Collective communication refers to the other types of communication where the number of senders and/or receivers differs. Two particular cases, and the two most frequently used, are multicast and broadcast communication. In multicast, a sender sends the same message to a set of end nodes, whereas in broadcast, all the end nodes are the target of the message. Different ways exist to implement either multicast or broadcast routing in a network. The first solution, the simplest one with no network modifications required, is to send the same message sequentially as a series of unicast messages, each one addressed to one of the destinations. We refer to this method as unicast-based multicast (or unicast-based broadcast). The second method to implement multicast or broadcast communication is by creating a single path that reaches all the possible destinations. The message then, sequentially visits all the target end nodes. The message is ejected at intermediate destinations in the network until it reaches the final destination. We refer to this method as *path-based multicast*

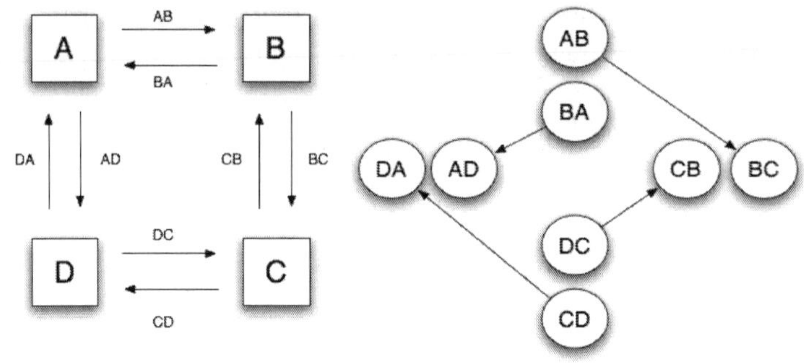

FIGURE 5.1: CDG for DOR routing in a 2×2 mesh network.

(or *path-based broadcast*). Both previous methods (unicast-based and path-based) add large latencies to messages, as messages either wait at the end node (unicast-based) or travel visiting sequentially all the nodes. The third method, addressing such inefficiency, is the so-called *tree-based multicast* (or *tree-based broadcast*). In such an approach a single message is injected into the network. On every switch the message is inspected and if different destinations can be reached through different ports, then the message is replicated through each port and being addressed to a disjoint set of destinations. Thus, a tree is formed within the network. This method takes less time to reach all the end nodes and usually requires less network resources.

One of the key issues routing algorithms face is the deadlock problem. A deadlock occurs when a set of messages occupy some network resources, typically buffers, and they request other resources (buffers) held by other messages in the same set. Thus, a cyclic dependency is created among resources. As no message can advance unless a buffer is guaranteed for the message, then all the messages block forever. These messages induce other messages to block (the situation is rapidly spread) and the network and system collapses.

A careful design of the routing algorithm prevents deadlocks from occurring. Basically, two sets of solutions exist. In the first one the algorithm is designed in such a way that cycles are prevented from occurring. For instance, dimension order routing sets rules for routing messages. A message can take X-Y transitions but cannot take Y-X transitions. Because of such restrictions no cycles can be formed. All this theory can be generalized by building the channel dependency graph (CDG) where nodes are resources[2] (buffers) and edges are dependencies between those resources. There exists a dependency if a message can use both resources sequentially. The resulting CDG for DOR routing is acyclic as shown in Figure 5.1.

The previous method is quite restrictive in some scenarios. Fortunately, there are routing algorithms that although allow the occurrence of cycles in

[2]If a channel is associated to a single buffer then the CDG can also be built from channels rather than buffers.

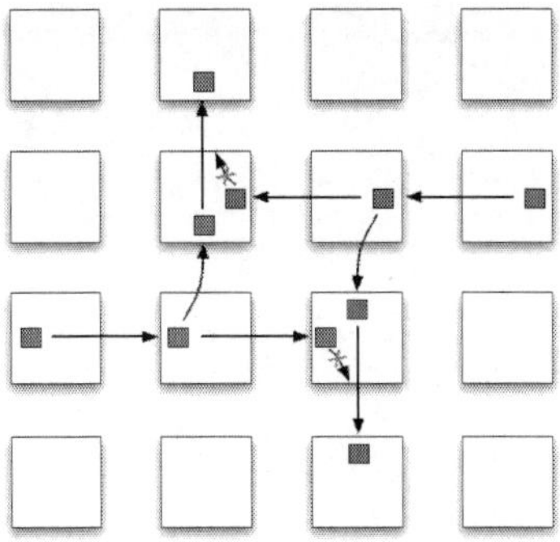

FIGURE 5.2: Two multicast messages induce a deadlock situation.

the CDG, the final outcome is a deadlock-free algorithm. This kind of algorithm allows fully adaptive routing and rely on the use of separate sets of resources for different messages. Virtual channels are used and classified as adaptive and escape path virtual channels. An adaptive virtual channel can be used without restriction, and usually the message takes any port that approaches the message to its destination. Such flexibility may introduce cycles in the CDG. In order to break such cycles, a message can take the escape virtual channel if all the adaptive virtual channels are busy (due to congestion or even due to a possible raising deadlock situation). Messages routed through the escape path are forwarded acyclically, for instance, using dimension order routing. Proving that this kind of routing algorithm is deadlock-free is complex. A design methodology and formal proof can be found in [82].

When addressing deadlock in a network with support for tree-based multicast or broadcast operations, the problem becomes more challenging. Indeed, although the underlying routing algorithm is deadlock-free for unicast messages, it could end up forming cycles for multicast/broadcast operations. Figure 5.2 shows an example where two multicast messages are deadlocked. Although both of them follow the dimension-order routing algorithm, different branches of the tree induce new dependencies. Indeed, there are new dependencies between channels used and requested by different branches of the trees. Notice that this problem is related to the use of wormhole switching, where the forwarding of the message depends on the advance of the message through other branches. If one of the branches blocks, then the other branch will also block (although resources are available). If virtual cut-through is used, then the problem simply vanishes. The blocked branch gets an entire copy of the message and, thus, allows the other branch to move forward. The branch will reach its destinations and then, will allow other branches to make further progress.

One key aspect of routing is the differentiation between algorithm and implementation. Indeed, few authors disassociate both issues and, when talking about NoCs the implementation becomes a big issue. Taking as an example

```
Inputs: Coordinates of current router (Xcurrent, Ycurrent) and destination router (Xdest, Ydest)
Output: Selected output port (Port)

Procedure:
Xoffset := Xdest - Xcurrent;
Yoffset := Ydest - Ycurrent;

if Xoffset=0 and Yoffset=0 then
  Port:=local        //Arrived to destination
endif

if Xoffset < 0 then
  Port:= X-
endif

if Xoffset > 0 then
  Port:= X+
endif

if Xoffset=0 and Yoffset<0 then
  Port:= Y-
endif

if Xoffset=0 and Yoffset>0 then
  Port:= Y+
endif

end procedure
```

FIGURE 5.3: DOR algorithm pseudocode.

the DOR routing algorithm, we can distinguish between the routing algorithm itself (the rules to apply to every incoming message) and the way such rules are implemented. Figure 5.3 shows the algorithm written in pseudocode and Figure 5.4 shows a possible implementation with some logic gates. This distinction will be key to properly identify and implement algorithms in NoCs addressing future challenges.

To conclude this overall description of the routing problem in NoCs, we need to identify the challenges that are directly impacted by the choice of routing algorithm. Indeed, as the routing algorithm instructs the paths messages need to take, congestion will be highly impacted by the algorithm. We need to design algorithms that are aware of potential congestion spots in the network. Please, refer to Chapter 12 for a further description of the congestion problem. Other challenges related to routing algorithms are fault-tolerance and heterogeneity by itself. Indeed, failures in the network will end up in irregular topologies. If we plan to make such chips still operative, then the routing algorithm needs to cope with such topologies. A routing algorithm able to work on any irregular topology will become of major importance, and an efficient implementation of such algorithm will be key for its success in a NoC environment.

5.2 Current Proposals for Routing in NoCs

In this section the most recent contributions to routing in NoCs are described, both for unicast and collective communication. Starting from unicast approaches and ending with techniques that deal with collective communi-

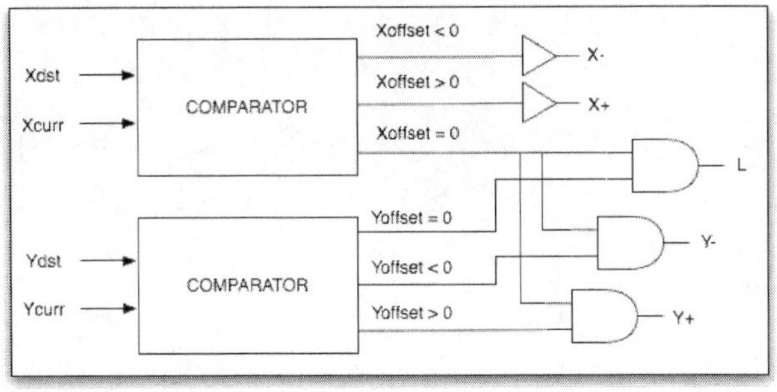

FIGURE 5.4: DOR routing implementation.

cation, the key is to identify their features and how they address the new challenges identified for NoCs. Designers must take into account that they have to build effective scalable mechanisms while looking for area, latency, and power consumption savings. Good designs require some effort at the on-chip network level.

Routing algorithms can be implemented in two straightforward ways: source-based or distributed-based. In the first case, *source-based routing*, the responsible agent for computing the path is the source end node. The computed path is stored in the message header and switches simply read the routing information. As for the implementation, it is by means of a routing table (frequently referred to as *look-up table*) stored at the end node. This solution allows to route messages in any irregular topology configuration (derived from any of the challenges mentioned in Chapter 1) as the header includes the entire path. However, it consumes network bandwidth as it is transmitted through the network. In addition, the size of look-up tables at every end node grows linearly with the system size, thus growing quadratically in the entire NoC. Examples of real NoCs using source-based routing implementations are xpipes [305] (see Appendix) and Intel Polaris Chip [126].

In *distributed-based routing*, however, each switch is in charge of computing the next hop that the message will take while it travels across the network. The header contains a reference to the destination node (usually an identifier or destination coordinate). There are two different methods to implement distributed-based routing. The first one is *algorithmic routing*, relying on combinational logic blocks. These kind of implementations are very efficient in terms of area, latency, and power consumption, but the routing algorithm is specific to the topology which is, usually, regular. To deal with nonregular topologies, and to support topology-agnostic routing algorithms, switches with *forwarding tables* were proposed. Each communication device stores a table that contains, for each destination end node, the output port that must be used. The main disadvantage is that look-up tables are implemented with memories, which do not scale in terms of latency, power consumption, and

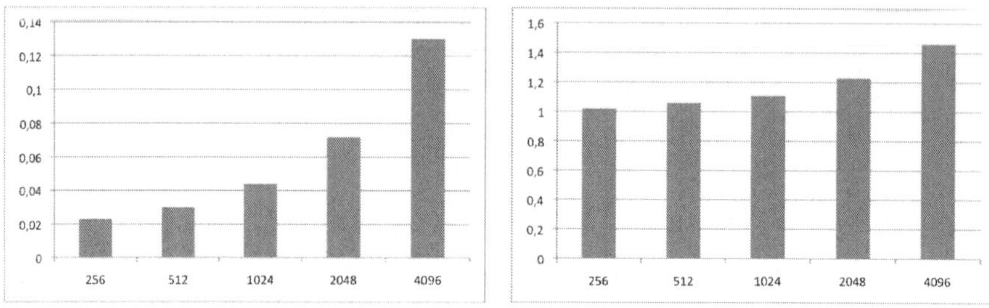

(a) Area (mm^2) (b) Latency (ns)

FIGURE 5.5: Area and latency for memory macros/blocks as a function of the number of entries for a 90 nm technology node.

area. An example can be seen in Figure 5.5 which shows the synthesis of memory macros with 90 nm technology obtained with Memaker[3] [56]. As can be seen, area and delay required for accessing memory blocks grow with the number of entries (table size).

5.2.1 Routing Algorithms

DOR [308] has been widely used for meshes and tori networks. This routing algorithm forwards every packet through one dimension at a time, following an established order of dimensions. Sometimes it is alternatively called XY, as when applied to meshes, packets first go through X dimension and then through Y dimension. Its implementation is very cost-effective, but it is not able to route packets in presence of any irregularity. Different alternative combinations have been proposed, like XY-YX routing [286], where separate virtual channels are used for packets routed in XY or in YX. Notice that the combination of both algorithms could end up in a deadlock situation (thus virtual channels guarantee traffic isolation and thus deadlock avoidance). Other derived solutions from DOR are pseudo-adaptive XY [74] and surrounding XY [34].

DOR is efficient in its implementation but rather inflexible (one single path for every source-destination pair). To support irregular topologies (e.g., due to failures), one popular routing algorithm is up*/down* (UD) [283]. It performs a breadth-first spanning tree search (BFS) from a root node, assigning one direction to each unidirectional channel as *up* (towards the root node) or *down*. The net result is a tree made of *up* and *down* channels. Computed paths are composed first of a sequence of *up* channels, followed by a sequence of *down* channels, and no packet is allowed to traverse an *up* channel after a

[3]Memaker (from Faraday Technology Corporation) produces memory macro models on UMC Logic LL-RVT (LowK) Process technology.

down channel has been used. Further refinements and derivations of this routing algorithm comprehend DFS (depth-first spanning tree) [276], the Flexible Routing scheme (FX) [275] and the left-up-first routing algorithm (LTURN) [166]. All these algorithms need to be implemented in look-up tables (either at the end node if source-based routing is implemented or at the switch if distributed-based routing is implemented). Indeed, most of these algorithms have been proposed for off-chip networks, where typically the use of a large look-up table on every switch is not a critical issue.

One of the main limitations of the UD algorithm is the concentration of the traffic (50% when assuming uniform distribution of message destinations) around the root node. Another limitation is the use of nonminimal paths that tends to increase average message latency. To overcome (or at least attenuate) both issues, there exists a collection of routing algorithms and solutions: In-Transit Buffers (ITB) [92], Layered Shortest Path (LASH) [295], Multiple UD (MUD) [192], Transition-Oriented Routing (TOR) [277], and Descending Layers (DL) [167]. Most of these algorithms use virtual channels to allocate different paths that, on the contrary, would create cycles in the CDG, thus inducing deadlock situations. As virtual channels are a precious and expensive resource in NoCs, their use to guarantee deadlock-free routing should be minimized if not used at all. Indeed, these algorithms were also initially proposed for off-chip networks.

One significant routing algorithm, able to deal with any topology, is the Segment-based Routing algorithm (SR) [206]. It renders in a up*/down* tree, however, the way is computed allows for much larger flexibility, ending up in many routing instances. SR uses a divide-and-conquer approach, partitioning a topology into subnets, and subnets into disjoint segments, and placing bidirectional turn restrictions (turn restrictions, or alternatively, routing restrictions, are introduced later in the chapter) locally within each segment. SR benefits from a larger degree of freedom when compared to previous routing strategies. Indeed, SR has been proposed to tackle the irregularities found in initial chip designs where a 2-D mesh has been broken in one or two locations, thus ending in an almost regular 2-D mesh structure.

Other routing algorithms, like adaptive-trail [261], minimal adaptive [292], fully adaptive [82], and smart-routing (SMART) [51] achieve performance improvements, but are specific to some (off-chip) network technologies and require extra functionality at the switches or have high computational costs.

All the routing algorithms can be classified as deterministic or adaptive. Most of the routing algorithms, however, are *partially adaptive* (e.g., UD, DFS, MUD, SR) in the sense that some alternative options are available by the routing algorithm and none of the options lead to a deadlock situation. However, the use of alternative paths may introduce out-of-order issues, and this could conflict with the application or the cache coherence protocol run on top of the network. Anyway, adaptive routing techniques like turn-model routing and planar adaptive routing [85, 69] offer better throughput and fault tolerance by providing alternative paths, depending on the network congestion and the presence of faults, and are subject of careful research for NoCs.

In the research of new routing algorithms that follow simplicity but effectiveness for NoCs, we must also point to deflective routing [232]. This technique routes messages through one of the profitable output channels (those getting the message closer to its destination). If such channels are busy, then, misrouting is applied. These algorithms can be implemented in buffer-less NoCs.

As opposed to the off-chip domain, power, thermal and reliability issues are important design restrictions in a NoC architecture. In [287], authors propose ThermalHerd, a distributed, collaborative runtime thermal management scheme for on-chip networks to tackle thermal emergencies ensuring thermal safety with little performance impact. In [198], authors describe the problem of transient failures of on-chip network and the impact on applications that require a guaranteed message arrival probability and response time. Their approach combines temporal and spatial redundancy taking into account energy consumption.

As previously commented, one important issue in routing algorithms for on-chip networks is the implementation cost of the algorithm. In the next sections we describe some of the recent proposals that deal also with the implementation cost of the algorithm. We focus our attention first on unicast-based routing algorithms and then on multicast/broadcast solutions. We describe the solutions that are conceived to support fault-tolerance in the network, and the assumed network topology is the 2-D mesh. Indeed, different causes may break the initial homogeneous and regular structure of the 2-D mesh. In such scenario, efficient routing of messages under irregularly shaped topologies becomes a challenge under the assumption of nonexpensive routing solutions. To reinforce the need of extending research in the field of fault-tolerance for NoCs, there are some studies [254, 256] that explore the possibilities of how NoC routing algorithms could be developed to provide routing around faults while maintaining the network operational. Specifically, in [254], authors make a comparative analysis between two methodologies of fault-tolerant routing algorithms, flood-based algorithms, and random-walk algorithms, showing the implementation costs and energy consumption. In [256], authors propose Immunet, a mechanism that tolerates failures for interconnection networks by doing a hardware reconfiguration of all the network resources that have not been affected by the failures. To effectively manage this objective, each switch incorporates two system-size tables for routing messages through different virtual channels (to avoid deadlocks).

5.2.2 Unicast-based Implementations

There are solutions from the off-chip network domain that could be applied to the NoC field. All these mechanisms do not fit properly in NoCs unless they are thoroughly redesigned. As an example, proposals for TCP/IP protocols

[313] are not suitable for NoCs as they rely on message dropping and later retransmissions, and would severely affect network performance. There are also techniques used in large parallel systems like the Blue Gene/L system [98] where entire sets of healthy nodes (lamb nodes) are switched off to keep topology and routing algorithm unchanged. Other mechanisms, focused on routing optimization [107], require the use of virtual channels (up to five in some cases) but they do not achieve 100% coverage[4] practically. Also, these mechanisms rely on adaptive routing, and the network must deal with out-of-order delivery issues, a feature that could be difficult to implement in NoCs. Other examples are Interval Routing [329] and extensions [106], which group sets of destinations requesting the same output port, and are an initial attempt to compress the routing table in switches. However, these techniques are not easily applicable to irregular networks.

Street-Sign Routing [39], a source-based routing implementation, compresses the message header so to minimize the impact on network bandwidth. Street-Sign Routing includes only the *switch identifier* (*switch ID*) of the next turn and the direction of the turn in the message header. Although message header is reduced it still consumes bandwidth. In addition, a table including the paths for every destination is required at every end node.

Regarding distributed routing solutions, first we focus on Region-Based Routing (RBR) [93] (and a similar proposal [240]). It tries to achieve fault-tolerance in regular and irregular on-chip networks while minimizing resources. At each switch RBR groups into a region different destinations that can be reached through a given output port. A set of registers and few logic gates are used to code all the destinations going through the same output port. The main drawback of such a mechanism is that, even with 16 regions defined, it still does not achieve 100% coverage [271]. Also, when the mechanism is implemented on a switch, it induces a long critical path [271].

Default-Backup Path (DBP) [168] tries to keep healthy processing elements (PEs) when the attached switch fails. It consists of adding redundant wiring and buffers that connect output and input ports directly. However, it does not address routing in irregular topologies and requires redundant hardware.

Adaptive-Stochastic Routing (ASR) [297] is a recently proposed routing algorithm (an improvement from the COSR algorithm [233]) that relies on a self-learning method to handle failures by assigning confidence fields to output ports for different tasks (or applications) running in the system. Thus, it requires a routing table at each switch, having n entries for n tasks and suffering from the same scalability problems and routing costs related to routing tables.

In [165], authors propose an architecture based on deflection routing that attempts to detect fault errors by adding CRC modules at input and output ports for crossbar faults, and SEC codes for link faults with the support of

[4] We define the term "coverage" as the percentage of irregular topologies derived from a 2-D mesh topology that are supported by the routing algorithm and its implementation.

routing matrices, one for each type of fault. The link fault matrix is $n \times n$, n being the switch radix (i.e., the number of inputs/outputs). Routing matrices are also $n \times n$, n being the switch radix, that represent the routing decisions on a message level, using a variant of deflection routing called delta XY with weighted priority. Deflection (or reflection) can lead to potential starvation solved with message dropping, thus impacting performance.

Another recent proposal, whose objective is to minimize the size of routing tables, either at end nodes or at switches, is described in [35]. In this article, three techniques are proposed: Turn-table (TT), XY Deviation Table (XYDT), and Source Routing Deviation Points (SRDP). All of these techniques consist of a routing table created by different routing algorithms to handle irregular cases in combination with routing strategies like XY (combined with YX), source routing or the *don't turn* technique (meaning that a message must not change direction when traversing the switch unless indicated). Deadlock freedom, however, is not assured in all these strategies (e.g., when changing from XY to YX) unless virtual channels are used, and if the technique supports all the possible failure cases. Anyway, in the worst case, the supporting routing table will need n entries for n destinations.

In [184], a compendium of state-of-the-art look-up tables (LUTs) implementations for routing purposes is proposed. These designs shift from being fully hardwired to being partially or fully configurable, depending on the degree of flexibility.

Novel implementations based on DOR, like Flexible Dimension-Ordered Routing (FDOR) [296], arise to provide coverage on irregular topologies. This routing methodology is based on the idea of dividing the dimensional mesh irregular topology into regular submeshes, a core mesh, and one or more flank meshes. Depending on the division, at the core mesh, messages are routed with XY routing and on the flank meshes with YX, or vice versa. One bit per switch is needed to configure XY or YX routing. FDOR provides a cheap and efficient routing solution to offer coverage on a set of irregular topologies that abide by certain conditions, but it does not offer full coverage.

As an overall view of the related work on fault-tolerant unicast routing for NoCs, the proposals use routing tables (either at sources or at destinations) and/or rely on an excessive number of resources (virtual channels) to avoid the deadlock problem. Also, none of the solutions (except when using tables) is able to provide full coverage (all the possible failure cases) for a 2-D mesh. Thus, existing solutions are very expensive in terms of routing delay and required silicon area.

5.2.3 Collective Communication Implementations

In this section we review the most recent contributions related to collective communication support in NoCs. Most of these are motivated by the existence of a cache coherence protocol with collective communication needs (see Chapter 12). Also, as will be seen, none of these solutions tackle the issue of fault tolerance as a fault-free 2-D mesh structure is typically assumed.

One of the most recent proposals for tree-based collective communication is Virtual Circuit Tree Multicasting (VCTM) [88]. VCTM builds a dynamic tree mapped on top of a mesh providing support for cache coherence protocols like Virtual Tree Coherence (VTC) [89]. For its implementation, VCTM uses two sets of tables. First, it uses a content-addressable memory at every end node, storing references for the current active trees. Second, it uses a VCT (Virtual Circuit Tree) table at every switch. The number of entries of the VCT table is the product of the number of end nodes times the number of trees used (e.g., 1024 entries for a 16-node system with a maximum of 64 trees per end node). This leads to significant (and nonscalable) area, delay, and power overheads. In addition, VCTM requires virtual channels to keep the ordering of the tree for different regions. Most important, VCTM is designed around the DOR routing algorithm, thus not being conceived for irregular topologies (2-D meshes with potential faulty links or switches).

XHiNoC (eXtendable Hierarchical NoC) [273] offers a very detailed multicast switch architecture with a different approach, not only focused on coherence protocols. XHiNoC provides a parallel pipelined multicast wormhole switching technique. While XHiNoC is flexible and extendible, it requires (on every switch) look-up tables (LUTs) and additional logic allocated at each port. In particular, it uses one table for routing (as many entries as destinations) and a table for message identity management (IDM) to break cyclic dependencies between multicast branches. XHiNoC uses a hardware logic called LCFS (Link Controller and Flow Supervisor) in order to control the links in the crossbar switch to prevent congestion states.

MRR (Multicast Rotary Router) [3] is a switch with multicast support based on two concepts. First, a topology-agnostic table-based solution, and second, two internal ring buffer structures at each switch, cyclically routing messages in opposite directions. All destinations of the collective operation are encoded in the message header (using one bit per end node). Instead of a centralized table, every output port has its own table with the reachable destinations from that output port as a bit-encoded (N destinations, N bits) solution. As said before, memory requirements increase quadratically with the number of end nodes.

RPM (Recursive Partitioning Multicast) [336] is a recent table-less solution for multicast. It uses a minimal logic to partition the entire network (from the perspective of the source end node) into eight partitions or quadrants. All the destinations of the collective operation are encoded in the message header. RPM generates message replicas at certain output ports based on the assigned partitions and on priorities. The proposal, however, is confined to a limited set of minimal routing paths and does not face the challenges mentioned in Chapter 1 that may end up in irregular topologies. In addition, it requires a number of virtual channels on every input port to solve the deadlock problem with multicast traffic. The number of virtual channels depends on the blocking possibilities due to wormhole switching, thus increasing with network and message size.

Following the summary for unicast-related work, multicast/broadcast proposals either rely on the use of tables for routing purposes (and for building multicast trees) and/or rely on simple routing algorithms not adapted to faulty networks.

In the following section we describe a new methodology to tackle irregularities in a initial 2-D mesh structure. The method will enable a compact implementation of a topology-agnostic[5] routing algorithm, either for unicast solutions and for multicast/broadcast solutions. Although this is not the only way to proceed in the implementation of routing algorithms, and most important, although it is particular for 2-D mesh-based designs, the method drives the efforts towards an efficient and compact implementation while providing flexibility. This is the suggested way to proceed when dealing with routing algorithms for future NoCs.

5.3 Routing Restrictions as a Way to Effective Routing Implementation

In this section we provide an alternative view of representing a routing algorithm. This will enable the efficient implementation of routing algorithms for NoCs. Although the view can be applied on many topologies, we restrict the applicability to 2-D meshes and derived topologies from the 2-D mesh due to link or switch failures. Also, we apply the new method to deterministic routing algorithms, although they can provide alternative deadlock-free routing paths for the same source-destination pair.

Figure 5.6(a) shows a 4×4 mesh topology. In the figure, the routing restrictions of the XY routing algorithm are shown. A routing restriction is defined between two consecutive links and simply indicate that a message is not allowed to use both links sequentially. Notice that routing restrictions are unidirectional, thus messages can cross the routing restriction in the opposite direction of the arrow shown in the figure. The use of routing restrictions will help in the search for an efficient implementation of the algorithm.

Although routing restrictions can be used to represent any routing algorithm, they are better suited for deterministic (although partially adaptive) routing algorithms. Indeed, algorithms that lead to acyclic CDGs are best suited. Also, the routing restriction representation can be applied on top of different topologies. As an example, Figure 5.6(b) shows the resulting routing restrictions when applying the SR routing algorithm on top of a 2-D mesh

[5] A routing algorithm is said to be agnostic to the topology if is able to be applied on any topology.

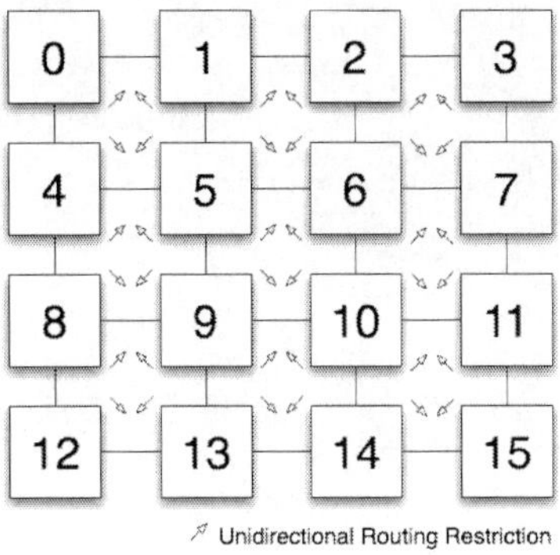

(a) 4×4 mesh with XY routing restrictions

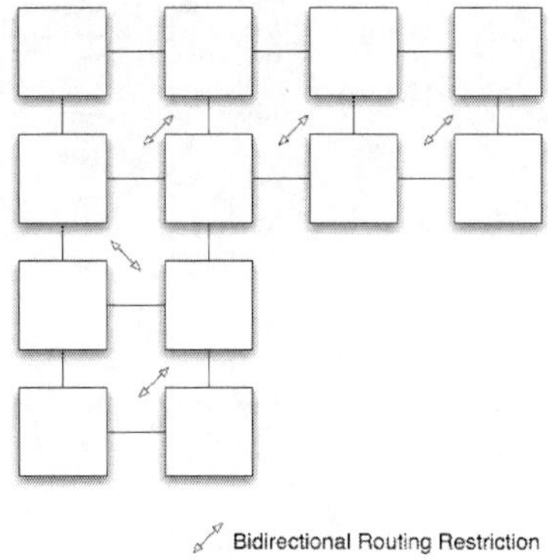

(b) p mesh with SR routing restrictions

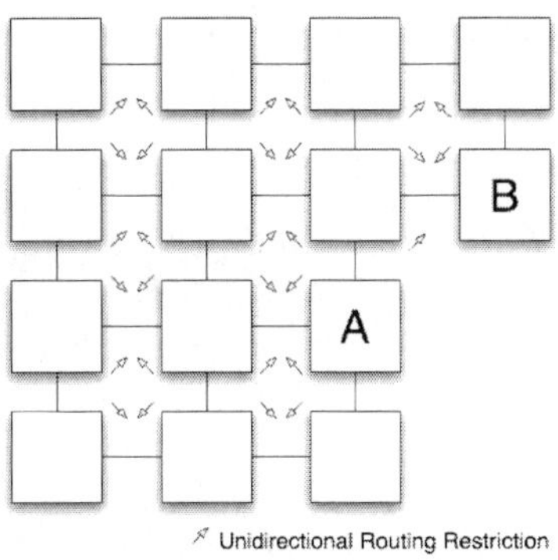

(c) p mesh with XY routing restrictions

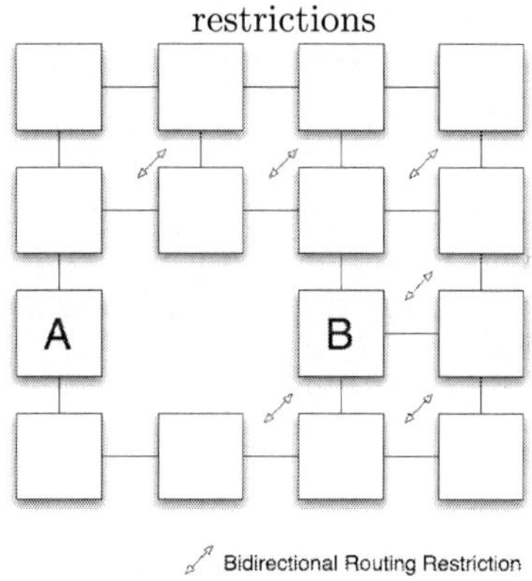

(d) 4×4 mesh with UD routing restrictions with a failed switch

FIGURE 5.6: Several topologies with routing restrictions.

topology with some failed (or powered down) switches and links. As can be observed in the figure, routing restrictions in the SR case are bidirectional, meaning both directions are forbidden. This is the usual case for up*/down*-based routing algorithms.

Notice that the drawing of routing restrictions for the XY routing algorithm in an irregular topology can be done. However, in this case, the network is not logically connected. This is the case shown in Figure 5.6(c). No message can reach switch B from switch A without crossing a routing restriction. Therefore, the XY routing algorithm in that topology is not applicable.

In addition, routing restrictions can be used to represent routing algorithms with nonminimal path needs. This is the case when the topology has, for instance, a failed switch inside the 2-D mesh structure, and in that situation nonminimal paths (from the perspective of the original fault-free 2-D mesh) are needed to communicate at least the pair of switches in the same

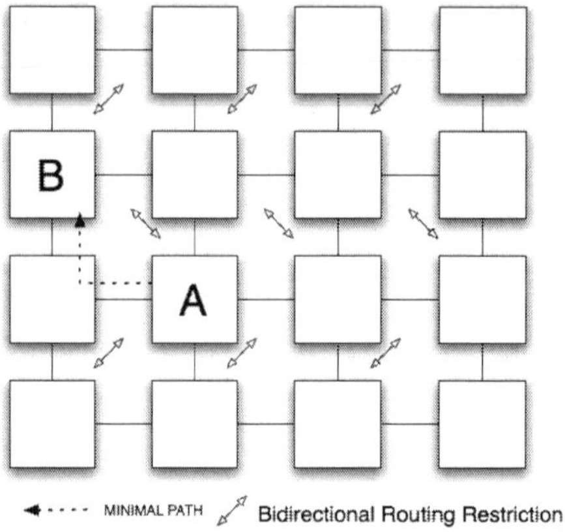

FIGURE 5.7: A minimal path between switches.

row and column of the failed switch. Figure 5.6(d) shows the case where the UD algorithm is applied. Notice that nonminimal path support is needed to reach switch B from switch A (and vice versa).

Once we are aware of the routing restrictions, imposed by the applied routing algorithm, we are in a position to implement the routing algorithm in a compact and efficient way. Indeed, the key to achieve such success lies on the way of storing the routing restrictions at switches and using them locally at each switch. Figure 5.7 shows a path from switch A to switch B. As can be seen, the path does not cross any routing restriction. Indeed, it is a valid deadlock-free path. Note that on every visited switch by the message, local actions can be performed if every switch is aware of the routing restrictions surrounding the switch (at the neighbor switches). In that sense, switch A forwards the message through the *west* port since it knows the message should be forwarded towards the north-west quadrant (by just comparing current switch coordinates and destination coordinates of the message) and that there is a routing restriction at the *north* port that forbids messages to later take the *west* port. From that perspective, the *west* option is safe as there is no routing restriction that prevents messages from taking the *north* port at the next switch. The same reasoning can be applied on every visited switch until the message reaches its final destination.

Notice that only routing restrictions local to the switch (at the neighboring switches) and a comparison of coordinates (similar to how DOR works) is needed. Therefore, implementation should be compact and similar in complexity to the DOR routing algorithm, but flexibility will be increased (when compared to DOR) as routing restrictions can be computed for different algorithms and on top of different topologies. Also, notice that a minimal path is expected when comparing current switch coordinates and a destination's coordinates. Indeed, assuming such conditions simplifies the implementation

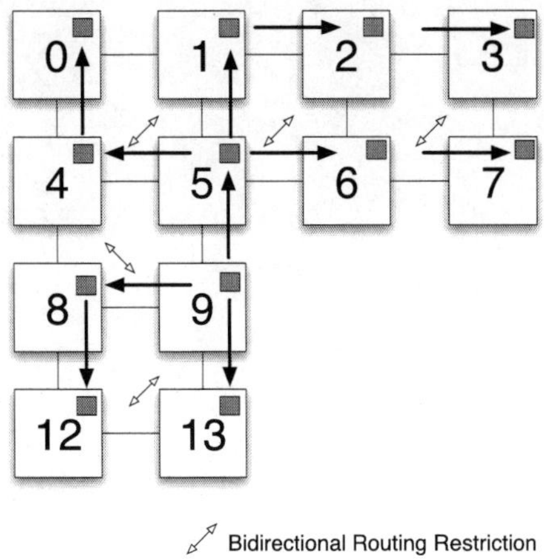

FIGURE 5.8: Broadcast routing in a p mesh with routing restrictions.

but reduces applicability. For instance, the topology shown in Figure 5.6(d) should rely on nonminimal path support. As we will see in the next section this support will require more logic.

In addition, tree-based broadcast (or tree-based multicast) solutions can be engineered also on top of routing restrictions. Figure 5.8 shows a tree-based broadcast that takes into account the routing restrictions locally. Indeed, control messaging can be performed between the switches in order to spread a broadcast tree over the entire network without crossing routing restrictions. In the next section we will develop more on this issue.

To conclude the discussion for routing restrictions, an efficient way to store the routing restrictions is by the use of a single configuration bit per routing restriction. Indeed, we can define up to eight configuration bits, that we will refer to as routing bits, on every switch. In that sense, bit R_{xy} is defined as the routing bit that indicates if there exists a routing restriction through the x port with the y port at the next switch (reached through the x port). Thus, we can define R_{ne}, R_{nw}, R_{en}, R_{es}, R_{wn}, R_{ws}, R_{se}, and R_{sw} bits on every switch. Figure 5.9 shows the routing bits on every switch for the provided topology and SR routing algorithm (represented by the drawn routing restrictions).

Note that for nonminimal path detection and support, four routing bits (R_{nn}, R_{ee}, R_{ww} and R_{ss}) are added to the first set of routing bits. These bits indicate whether a message can advance along the same direction in X or Y dimensions or, on the contrary, either should take a nonminimal path, or has reached the mesh border. Following the description in Figure 5.9, in the case of switch 5, bit R_{nn} is reset due to the fact that there is no path to the north after traversing switch 1.

In addition, and to provide connectivity information, we can also enrich the information a switch has with the connectivity through its output ports.

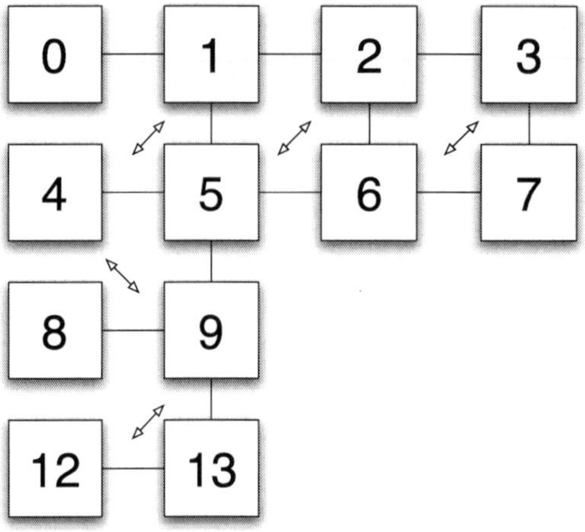

Router	Cn	Ce	Cw	Cs
0	0	1	0	1
1	0	1	1	1
2	0	1	1	1
3	0	0	1	1
4	1	1	0	1
5	1	1	1	1
6	1	1	1	0
7	1	0	1	0
8	1	1	0	1
9	1	0	1	1
10	-	-	-	-
11	-	-	-	-
12	1	1	0	0
13	1	0	1	0
14	-	-	-	-
15	-	-	-	-

Router	Rne	Rnw	Rnn	Ren	Res	Ree	Rwn	Rws	Rww	Rse	Rsw	Rss
0	1	1	0	1	1	1	1	1	0	1	1	1
1	1	1	0	1	1	1	1	1	0	1	0	1
2	1	1	0	1	0	1	1	1	1	1	0	1
3	1	1	0	1	0	1	1	1	1	1	0	1
4	1	1	0	0	1	1	1	1	0	0	1	1
5	1	1	0	0	1	1	1	1	0	1	1	1
6	1	1	0	0	0	1	1	1	1	1	1	1
7	1	1	0	1	0	1	1	1	1	1	1	1
8	1	1	1	1	0	1	1	1	0	1	1	1
9	1	1	1	1	0	1	0	1	0	1	0	1
10	-	-	-	-	-	-	-	-	-	-	-	-
11	-	-	-	-	-	-	-	-	-	-	-	-
12	1	1	1	0	0	1	1	1	0	1	1	1
13	1	1	1	1	0	1	1	1	0	1	1	1
14	-	-	-	-	-	-	-	-	-	-	-	-
15	-	-	-	-	-	-	-	-	-	-	-	-

FIGURE 5.9: *p* mesh with routing and connectivity configuration bits.

In that sense, we can define four connectivity bits, one per output port. These bits indicate if there is a switch connected through that output port or not. So, we have the C_n, C_e, C_w, C_s bits. Figure 5.9 shows the connectivity bits as well.

Although a single bit needs to be used per output port to provide information on topology connectivity, C_x bits may be extended to 8-bit registers, thus providing flexibility when defining regions (which enable effective virtualization and aggressive power saving mechanisms) and for broadcast/multicast purposes. This means, for example, that the *north* port of a switch has its connectivity defined from $C_n[0]$ to $C_n[7]$ for regions 0 to 7. Note that while defining regions or domains, by using connectivity bits, those regions may even overlap. We can find an example in Figure 5.10. Switch A is shared by two different regions. For region 0 the C_s bit is reset and for region 1 the bit is set. With this configuration, messages (labelled with the appropriate region iden-

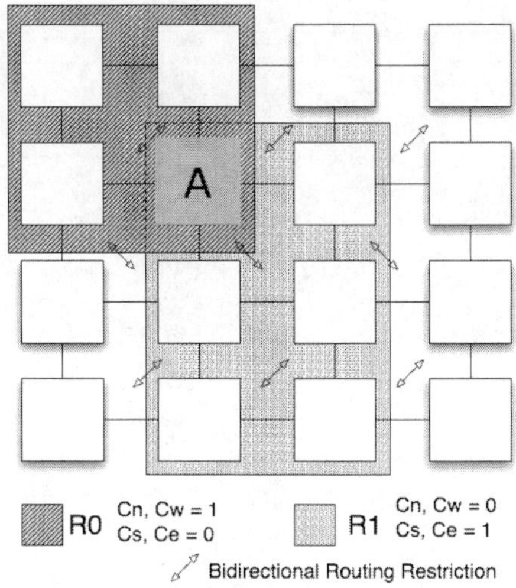

FIGURE 5.10: Different regions defined, different connectivity bits.

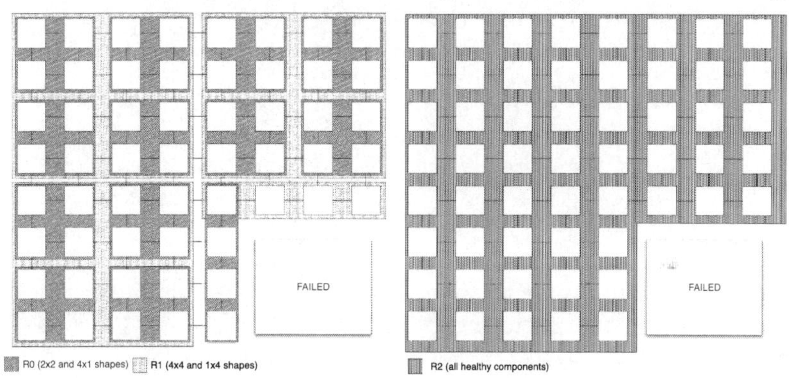

FIGURE 5.11: Flexibility when defining regions with connectivity layers.

tifier) can be managed appropriately. Also, failed links and boundary switches can be configured accordingly. Figure 5.11 shows an example of the different regions that can be defined in a chip with 16 nodes.

In the next section we derive reduced logic implementations of the routing algorithms represented by those bits. Also, extensions to cover nonminimal path support and broadcast deadlock-free implementations will be covered.

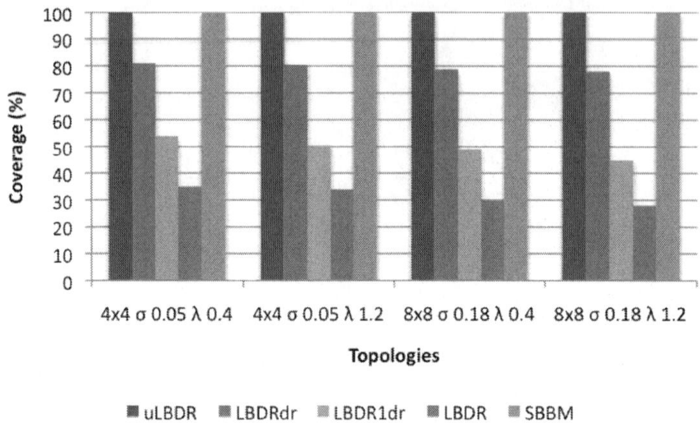

FIGURE 5.12: Coverage of several routing implementations.

5.4 Unicast Logic-Based Distributed Routing

The overall goal in this section is to achieve reduced (small logic blocks) implementations of the routing algorithms while providing support for most of the topologies that will be derived from an initial 2-D mesh. The more the cases are supported the more efficient the mechanism becomes to deal with the challenges addressed in Chapter 1. To better address this issue, we define the *coverage* term as the percentage of topologies the routing mechanism is able to successfully route (a topology is correctly routed if the algorithm and the implementation are able to provide at least a valid path for every source-destination pair without incurring into deadlocks).

To assess the coverage of a particular mechanism and algorithm we use a set of 1423 topologies, derived from a process variation analysis shown in [271] (see also Chapter 12). Chips are modeled on a real 65 nm implementation NoC layout where all cores are identical, and their size is 1 mm^2. In particular, different NoC operating frequency thresholds were set and links not reaching those thresholds (due to variability effects) were labelled as faulty. Two different configurations are used, 4×4 and 8×8 NoCs with different values of spatial correlation ($\lambda = 0.4$ or 1.2) and variance ($\sigma = 0.05$ or 0.18).

The mechanisms will be designed from the simplest one (minimum logic requirements) with low coverage percentages, to the more complex one with full coverage. All the mechanisms rely on the use of the routing bits and connectivity bits introduced in the previous section. However, the most advanced methods will incorporate some additional hint bits. The coverage results for all of the mechanisms described here are shown in Figure 5.12.

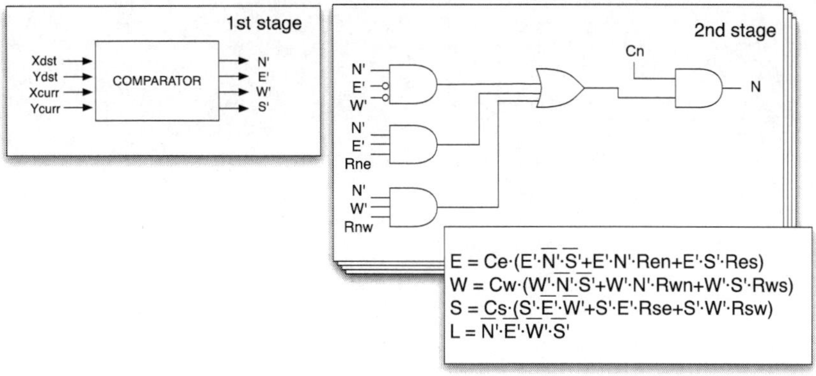

FIGURE 5.13: LBDR implementation for an input port, details for the logic associated to the north output port case.

5.4.1 The Basic Mechanism

Logic-Based Distributed Routing (LBDR) [272] is the basic mechanism, shown in Figure 5.13. LBDR is based on the relative position in the mesh of the destination switch and the current switch. In order to compare both, a COMPARATOR module is used, which generates four control signals. Signals N', E', W', and S' indicate the relative position of the final switch. For example, in Figure 5.14, if the current switch is 8 and message's destination is switch 5, signals N' and E' will be activated. With these control signals, and using the R_{xy} and C_x bits, the LBDR module computes an initial set of routing decisions. Figure 5.13 shows the required logic at every input port. Notice, however that the routing bits and connectivity bits do not need, in principle, to be replicated on every input port.

This logic provides support for minimal paths in the network, and generates a signal per output port. When signal N is set it means that the north output port can be used for the message. As shown in Figure 5.13, the N port will be selected if any of the following three cases is met:

- First (first AND gate), the destination is on the same column (to the north).
- Second (second AND gate), the destination is on the NE quadrant from the switch viewpoint and the message can take the south-east turn at the next switch (bit R_{ne} is set).
- Third, (third AND gate), the destination is located on the NW quadrant from the switch viewpoint and the message is allowed to take the south-west turn at the next switch (bit R_{nw} is set).

Note that the connectivity bit filters the final decision. Also, if the current

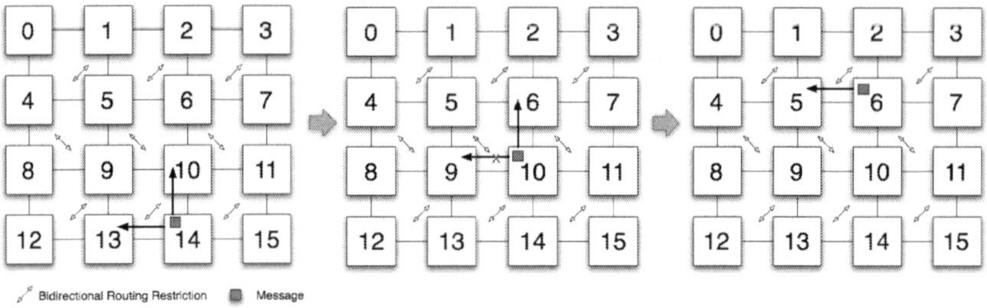

FIGURE 5.14: Example of routing decisions in LBDR.

switch ID is the same of the destination switch ID (when signals N', E', W', and S' are reset), then the message has reached the final destination and is forwarded through the local port (signal L in the figure). Also, consider the fact that two possible output ports may be activated by the logic at the same time. This can later be filtered (if adaptive routing is to be avoided) by the arbiter.

In Figure 5.14 there is an example of routing with LBDR. Source node attached to switch 14 wants to send a message to the destination node attached to switch 5. At switch 14, signals N' and W' are activated at the first stage as the destination is on the NW quadrant. At the second stage, for N signal, as N' and W' are active, R_{nw} is set at switch 14, and there is connectivity to the north (C_n is set too at that switch), north arises as a valid routing choice. Also, W output port is valid to route the message, as R_{wn} is set, and there is connectivity to the west. Let us assume that the arbitration process chooses the N output port between both options. At the next hop, at switch 10, a routing operation is started again. N' and W' are active, again. After the second stage, LBDR promotes the N direction as a valid choice, but unlike in the previous switch, W is discarded, as the R_{wn} bit is not set, which represents the routing restriction at switch 9. Therefore, the message is routed north to switch 6. At this switch, the process is repeated again, as the message is still not at its destination switch. In this case, only the W' signal is active, and after the routing process, W is the only routing option available. The message is routed to switch 5, where it will be forwarded to the node connected to the switch, as the message has arrived to its destination.

Notice that the LBDR logic does not define the routing algorithm (the actual paths messages must follow). Indeed, this is enforced by the configured routing bits. Thus, a routing algorithm needs to be used to compute the actual values of the routing bits. For our discussion, and the coverage analysis we perform, we use the SR algorithm, as it allows many instances of paths for the same topology and indeed can be applied on any topology (is topology-agnostic).

However, take into account that the just-described LBDR mechanism is not able to support all the possible topologies we may face. Indeed, the LBDR mechanism relies on the fact of using always minimal paths for every source-destination pair (minimal paths computed in the initial 2-D mesh structure). This is because the output signals of the COMPARATOR is a set of possible paths and all these options lead to minimal paths. In the case of having a single hole (a failed switch) in the center of the mesh prevents LBDR from obtaining a valid output port for a combination of nonminimal paths around the failed switch. This is the case shown in Figure 5.6(d).

5.4.2 Deroute Operations

With the LBDR mechanism, 30% of the failure cases are supported, as shown in Figure 5.12. This means, in 30% of the topologies LBDR succeeds in routing all the messages from every source to every destination. Indeed, the topology is simply supported if minimal paths are guaranteed to exist for each pair of communicating devices.

Let us assume now a case where a switch at the center of the mesh presents a manufacturing defect that renders it inoperative. In this case, a message coming from one neighbor trying to reach the adjacent switch requires a nonminimal path. Figure 5.15 shows such a case, which is not supported by the LBDR mechanism presented above. The reason is that the path from A to B cannot be supported. At switch A, the possible minimal directions to reach B are N and E, however, both links are missing, and therefore there is no possible way (with LBDR) to reach B. This motivates for the first extension to the LBDR mechanism: the deroute logic. The deroute logic, when properly configured, acts as a filter to the output choices provided by LBDR by introducing the possibility of using nonminimal paths. In Figure 5.15, at switch A messages coming from the south port are derouted towards the west direction, whereas packets coming from the west port are derouted towards the south direction.

We need to provide nonminimal support in an efficient way, that is, with as minimum logic as possible. Figure 5.16 shows the DEROUTE logic. In particular, at every input port of the switch a deroute option (dr_0, dr_1 bits) is provided. This set of two bits encodes the deroute option. Whenever the basic mechanism is unable to provide a valid output port for a switch (NOR gate with five inputs) the deroute option is selected. The logic and the dr_x bits are replicated for every input port. Therefore, the deroute option for a packet is the one configured at the input port the message arrives through. Alternatively, a single deroute option could be used for the entire switch. However, flexibility is reduced as shown in Figure 5.12, where this solution, named LBDR1dr, provides a coverage of around 50%. With different deroute bits per input port, coverage is increased to 80%.

It is worth mentioning that the deroute bits need to be computed in accordance with the routing algorithm. In fact, the deroute option must not introduce potential cycles that could lead to deadlocks. In Figure 5.15 the

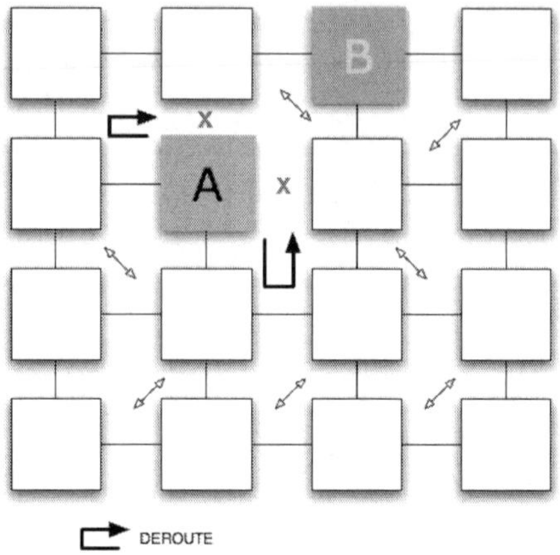

FIGURE 5.15: Some links are inoperative forcing nonminimal paths.

deroute option at input port *west* in switch *B* cannot be set to *south* since this would lead to packets crossing a routing restriction.

5.4.3 Fork Operations

The DEROUTE logic enhances greatly the percentage of irregular topologies supported (see Figure 5.12). Indeed, 80% of the topologies are now covered with a deroute per input port. However, there are subtle cases that are still not covered (the remaining 20%). Figure 5.17 shows an example. The problem in this example comes by the fact that for some destinations located at the same quadrant, at switch *B* the routing engine should provide one port (*north*) for some destinations (destination *C*) and another port (*west*) for other destinations (destination *A*). As the mechanism works in quadrants for routing, there is no way to indicate the switch which port should be granted for a particular message.

To solve the previous problem, the mechanism is enhanced with an additional and final feature. At switch *B*, the message is simply forked through *north* and *west* output ports. This leads, however, to important changes in the switch design. First, the arbiter must allow one message to compete for more than one output port at the same time. Two design alternatives are possible. In the first one, the arbiter may consider a request from a message to two output ports as an indivisible request, therefore, granting or denying access to both outputs at the same time. This leads to a simpler design of the buffering at the input port, since only one read pointer is needed (as in the normal case of messages competing for a single output port). In the second one, however, the arbiter may grant or deny access to one output port regardless of the action performed for the other output port. This leads to a more complex input buffering, since forwarding of the message is shifted for both

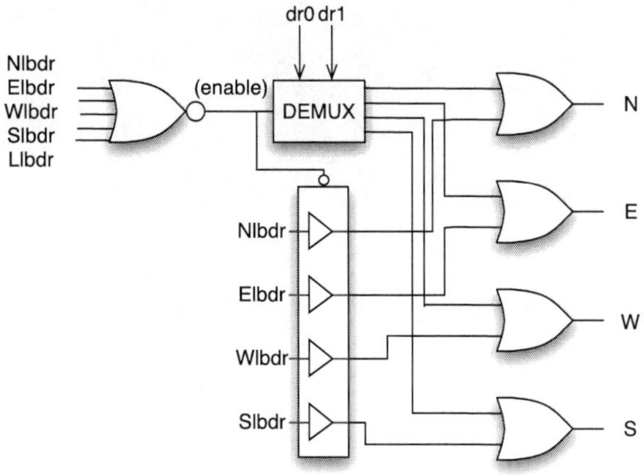

FIGURE 5.16: Deroute logic.

output ports, thus each requiring a read pointer. We assume in this discussion and implementation the first option because of its simplicity and the fact that fork operations are used in some rare cases.

The second change in the switch (to allow fork operations) is related with deadlock. Indeed, deadlock may occur in wormhole switching as two forked messages may compete for the same set of resources. Although the routing algorithm used is deadlock-free (thus, no message crosses a routing restriction), performing fork actions (like collective communication) may lead to deadlock. Imagine that a broadcast packet m_1 gets access to output port p_1 at switch r_1 and requests access to output port p_2 at switch r_2. However, broadcast packet m_2 gets access to output port p_2 at switch r_2 and requests access to output port p_1 at switch r_1. If packets are long enough they will block the current resources being used while requesting the new ones. Indeed, none of the packets will advance since the input buffers will fill and the output ports will never be released. There are two solutions to this problem. The easiest way is the use of virtual cut-through (VCT) switching, thus ensuring a packet will fit always in a buffer. Thus, output ports in the previous example will be released (the packet has been forwarded entirely) and the requests for the output ports will be granted. Other options rely on allowing internal switch flit-level multiplexing and wormhole switching among switches. Basically, flits are labelled with identifiers, thus flits from different packets can be mixed in the same buffer. The problem with this kind of solutions is that internal tables are required to keep flit identifiers.

As pointed out previously, one solution is the use of VCT switching, where a packet[6] fits in a buffer, thus breaking the dependencies between branches.

[6]Here we use the term "packet" as a message may be needed to be packetized to comply with the rule of packets being shorter than buffers.

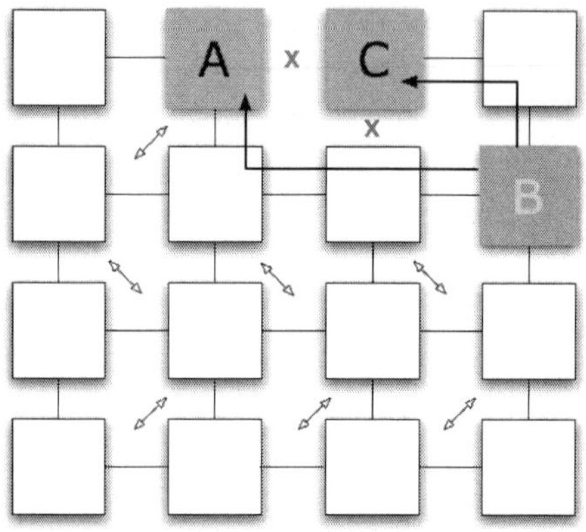

FIGURE 5.17: Case not handled by the deroute logic.

Although VCT is seen as demanding much buffer space at switches, a careful design of the switch can minimize this effect (an implementation example can be seen in the Appendix for the gNoC switch design). Note that fork operations resemble tree-based broadcast/multicast operations as a packet is replicated through different output ports.

Finally, packets being forked will reach the final destination. However, one of the forked replicas will not reach the destination. In this situation, the packet needs to be removed from the network. This will be easily achieved by silently destroying the packet at a switch (in the Appendix we will show the implementation details for two switch designs).

The FORK logic is shown in Figure 5.18. As shown, it relies on four additional configuration bits (fork bits): F_n, F_e, F_w, and F_s. These bits are set to reflect the output ports that must be used to fork a packet. Whenever a packet comes and its destination is in the same quadrant defined by the fork bits, then the packet needs to be forked. The FORK signal is set and forwarded to the arbiter. Also F_n, F_e, F_w, and F_s are set appropriately. Fork bits are checked in parallel with LBDR, just after computing the N', E', W', and S' signals. With the FORK logic, we obtain full coverage (100%) as seen in Figure 5.12.

5.4.4 uLBDR: A Complete Unicast Solution

In order to offer a complete unicast solution, LBDR and the extensions, DEROUTES and FORKS are gathered in one mechanism, known as uLBDR (Universal Logic-Based Distributed Routing) [270]. LBDR is integrated as the core module and extended to include the additional set of R_{xx} routing bits, to reflect routing restrictions that prevent horizontal or vertical traversals of a switch (EW, WE, NS, and SN) and the extension of the connectivity bits layer to support overlapped regions or domains as described in Section

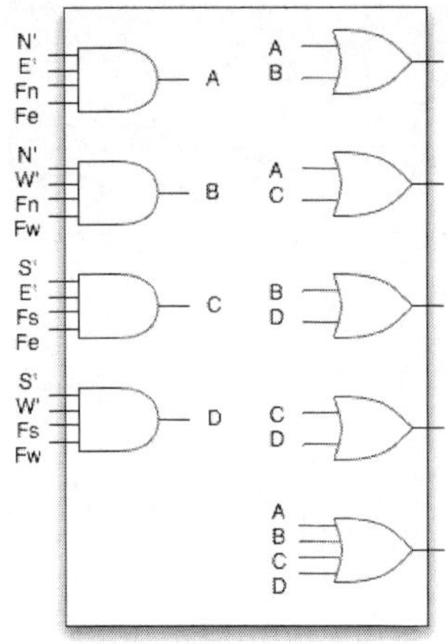

FIGURE 5.18: Fork logic.

5.3. This led to an additional (but small) overhead of the LBDR module as pictured in Figure 5.19.

Now, on the first stage, at the *COMPARATOR* module, eight control signals are generated. Apart from the regular ones, N', E', W', and S', signals $N1$, $E1$, $W1$, and $S1$ are computed. They indicate whether the final node is one hop away in each direction. As shown in Figure 5.19, the N port will be selected (UN' signal activated) if any of the following four conditions is met. First (first AND gate), the destination is on the same column (to the north) and one hop away. Second (second AND gate), the destination is on the same column (to the north) but more than one hop away and the packet is allowed to cross the next switch from south to north (bit R_{nn} is set). Third (third AND gate), the destination is on the NE quadrant from the switch viewpoint and the packet can take the south-east turn at the next switch (bit R_{ne} is set). And finally (fourth AND gate), the destination is located on the NW quadrant from the switch viewpoint and the packet is allowed to take the south-west turn at the next switch (bit R_{nw} is set). Note that the connectivity bit filters the final decision. The UL signal is set when the packet has reached the final switch destination, which translates when N', E', W', and S' signals are reset.

An overview of the schematic of the complete mechanism can be seen in Figure 5.20. The inputs of the mechanism are the coordinates decoded from the header, as in the original LBDR, and the region identifier, primarily for selecting the correct connectivity bit layer. The first signals are computed on the COMPARATOR module, and then are forwarded to the LBDR and FORK modules. Fork bits are checked in parallel with LBDR, just after computing the N', E', W', and S' signals. The outputs of the FORK logic are combined with the outputs of the LBDR mechanism to produce a possible set of valid signals.

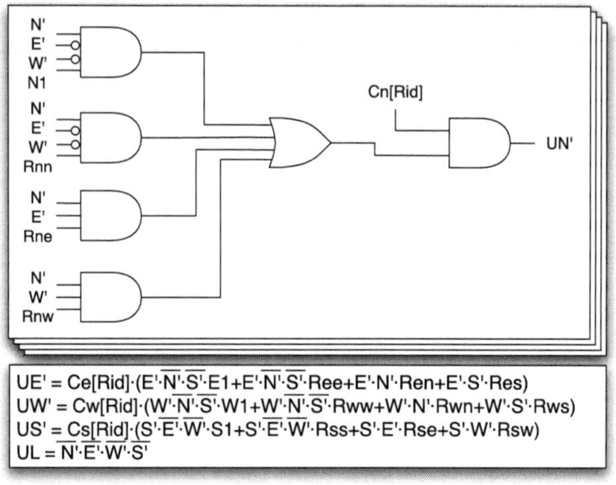

FIGURE 5.19: Improving the LBDR mechanism (support for R_{xx} bits), details for the logic associated to the north output port case.

After crossing the DEROUTE module, they are exposed to the arbiter. Note that fork operations have priority over deroute operations, as the DEROUTE logic is only enabled if none of the combined signals are set. If not enabled, the DEROUTE module just acts as a wire.

5.4.5 Algorithm for the Computation of Routing Bits

As described, the *uLBDR* mechanism relies on some configuration bits. The way these bits are computed is critical for the success of the mechanism. Indeed, not all the combinations of bits guarantee connectivity and/or deadlock freedom. In this section we describe the method followed to compute the bits. It is worth mentioning this process is performed offline and before starting any normal operation of the chip.

The first step is computing the routing bits (R_{xy}) and connectivity bits (C_x). This is done by analyzing the topology (including the failed/powered down switches and links) and applying the routing algorithm. The choice of the routing algorithm is critical since it must guarantee deadlock freedom and connectivity.[7] The routing algorithm is represented by the routing restrictions it enforces. Routing bits (R_{xy}) are then computed by taking into account the location of the routing restrictions. Note that when the location of routing restrictions is known, the computation of R_{xy} bits is straightforward and thus,

[7]As previously commented, SR is chosen as it is topology-agnostic and does not require virtual channels. Nevertheless, any other topology-agnostic routing algorithm may be used. Alternatively, for a healthy chip (no failures) the *XY* routing algorithm can be used by computing the corresponding bits.

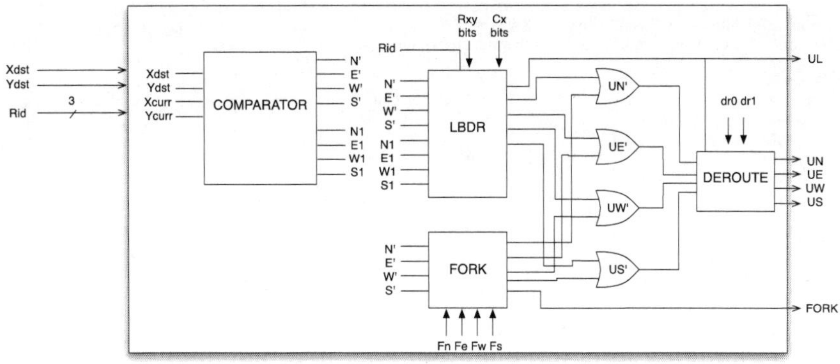

FIGURE 5.20: uLBDR logic with LBDR, DEROUTE, and FORK modules.

its computation complexity is low (linear with the number of routing restrictions). Connectivity bits (C_x) are computed based on the existence of links in the router.

Once the routing and connectivity bits are computed, the deroute options are searched. To do this, the algorithm checks the existence of valid paths for every source-destination pair. The algorithm starts with the first deroute option and keeps following the path, thus taking the deroute, checking if the path (and all their possible alternative paths) will reach the destination. In case of success, the deroute option is set and the deroute bit is configured. In case of failure (destination is not reached), then another deroute option is tried. Note that several deroute options may be required for a single path. A similar task is done to compute fork operations. If the algorithm fails in placing deroutes, fork operations are considered. To do so, the algorithm tests any possible fork operation at switches where no deroute succeeded. Preventing deadlock scenarios when applying deroutes or fork operations is guaranteed by an additional restriction on the computation algorithm. Any nonminimal path that uses deroute or fork options is checked before validating the deroute and fork bits. If it crosses any routing restriction, by encountering any routing bit reset, then the path is discarded. Figure 5.21 shows the algorithm in pseudocode for computing the deroute and fork bits.

5.5 SBBM: Signal Bit-Based Multicast

Besides the support for unicast traffic, we should also provide support for multicast/broadcast traffic, specially in CMP systems where coherence actions (e.g., invalidations of multiple memory block copies) might be required (an

```
Inputs: Routing and connectivity bits of all switches
Outputs: Deroute and fork bits set for each switch (if needed)

Procedure:
for each switch
 for each other_switch
  testPath(switch, other_switch);
 end for
end for
end procedure

function testPath(current, destination)
variables ok, neighbour, outport

ok := FALSE
outport := doLBDRrouting(current, destination);
if outport != NO outport //LBDR has found a valid outport
then
 if outport == LOCAL PORT //Destination
 then
 ok := TRUE;
 else
  neighbour := getNeighbour(current, outport);
  ok := testPath(neighbour, destination);
 endif
else
if outport == NO outport //LBDR can not provide a valid outport (presence of non-minimal routing)
then
 outport_array := testDeroute(current); //this function returns an array of valid outports for derouting,
                                         //i.e., checking with routing bits and destination quadrant signals
 for each outport in outport_array
  neighbour := getNeighbour(current, outport);
  ok := testPath(neighbour, destination);
  if ok := TRUE;
  then
   setDeroute(current, outport); //we set the deroute bits to the outport that allows misrouting
  endif
 end for
endif

if ok :=  FALSE; //Deroute and LBDR operations have not succeeded, try again with fork operations
then
  outport_array := testFork(current); //this function gives an array of valid outport with forks, again checking
                                       //with routing bits and destination quadrant signals (N', E', W' or S')
  for each outport in outport_array
   neighbour := getNeighbour(current, outport);
   ok := testPath(neighbour, destination);
 end for
 if ok := TRUE;
 then
   setFork(current, outport_array); //we set fork bits to all outports provided by fork operations
 endif
endif

return ok
end function
```

FIGURE 5.21: Algorithm in pseudocode for the computation of the LBDR bits.

impact on performance when supporting broadcast communication can be seen in Chapter 12). As with the previous implementations and extensions, the goal is to provide a compact and simple mechanism offering full coverage on topologies derived from a 2-D mesh. In particular, the mechanism will rely on the concept of routing and connectivity bits exposed before.

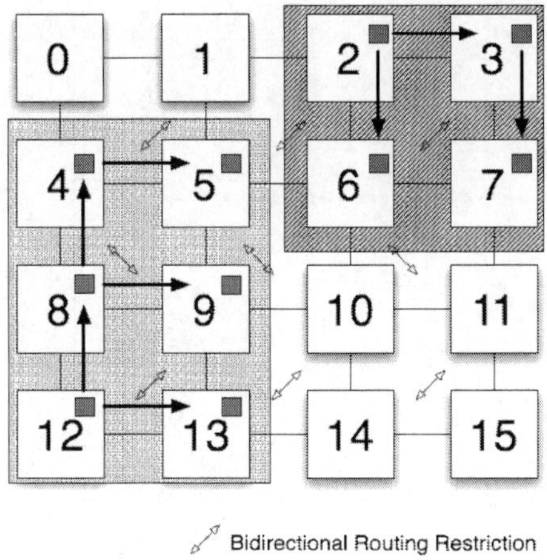

FIGURE 5.22: Broadcast operations in different regions.

SBBM (Signal Bit-Based Multicast) offers a tree-based broadcast operation at the network/chip level. The broadcast operation can be initiated by any of the nodes contained within a region defined by the connectivity C_x bits, and the broadcast packet reaches only the end nodes defined within the region. Only one packet is received by every destination of the collective communication, thus avoiding any possible duplicate. Figure 5.22 shows an example with regions defined and broadcast operations within the regions.

Before normal operation and after configuration bits are computed (see Section 5.4.5), an acyclic spanning tree for broadcast operations is computed within each defined region. SBBM assumes the existence of four connectivity regions per switch, using two connectivity bits per layer (e.g., connectivity bits 0 and 1 for region 0, connectivity bits 2 and 3 for region 1, and so on). Therefore, the eight connectivity layers defined previously are now packed into four layers, each one now with broadcast support at the region level.

The spanning tree will allow the dissemination of a broadcast message throughout the entire region. To do this, at each switch the tree is coded in one of the connectivity bits layer. Imagine we want to define an entire region covering all the chip and we want to provide support for broadcast operations. At every switch, $C_x[0]$ (connectivity layer 0) will be used for unicast actions (with the uLBDR mechanism) and $C_x[1]$ (connectivity layer 1) will be used for broadcast actions. Initially, $C_x[0]$ bits are computed based on the connectivity pattern derived from the final topology (as described in Section 5.4.5). Those $C_x[0]$ bits are, then, copied to $C_x[1]$ bits. In order to represent the acyclic spanning tree, some $C_x[1]$ bits are reset by applying the following rule: If one switch has a routing restriction between two links, one of the links is virtually disconnected ($C_x[1]$ bit is reset) in both directions. If, when processing a routing restriction, one link belongs to two routing restrictions, then the link

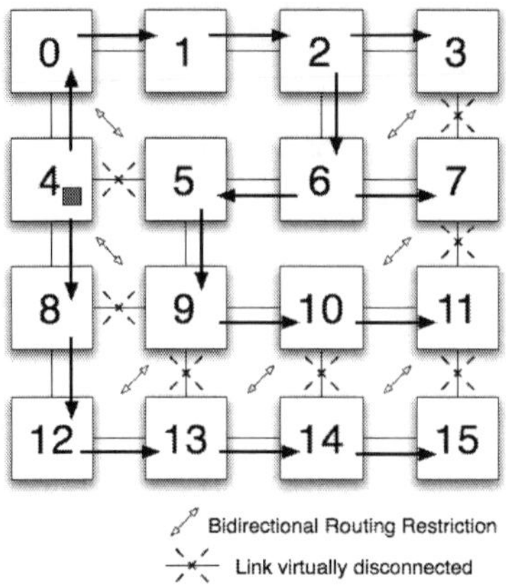

Router	Cn[0]	Ce[0]	Cw[0]	Cs[0]	Cn[1]	Ce[1]	Cw[1]	Cs[1]
0	0	1	0	1	0	1	0	1
1	0	1	1	0	0	1	1	0
2	0	1	1	1	0	1	1	1
3	0	0	1	1	0	0	1	0
4	1	1	0	1	1	0	0	1
5	0	1	1	1	0	1	0	1
6	1	1	1	0	1	1	1	0
7	1	0	1	1	0	0	1	0
8	1	1	0	1	1	0	0	1
9	1	1	1	1	1	1	0	0
10	0	1	1	1	0	1	1	0
11	1	0	1	1	0	0	1	0
12	1	1	0	0	1	1	0	0
13	1	1	1	0	0	1	1	0
14	1	1	1	0	0	1	1	0
15	1	0	1	0	0	0	1	0

FIGURE 5.23: Broadcast operation from switch 4.

is virtually disconnected for both routing restrictions, that is, two neighbor routing restrictions end up in only one link being virtually disconnected.

Let us explain this operation with an example (Figure 5.23). In this topology two links, one between switches 1 and 5 and the other between switches 6 and 10, have failed. According to this, routing restrictions (bidirectional arrows in the figure) have been computed and so routing bits and connectivity bits (e.g., $C_s[0]$ on switch 1 will be set to zero meaning it has no connectivity to the south). Connectivity bits from $C_x[0]$ are then copied to $C_x[1]$ and the appropriate links are virtually disconnected (applying the previous rule). As it is shown, seven links are virtually disconnected (fourteen connectivity bits are reset, those shadowed in Figure 5.23). Notice that one link being virtually disconnected only affects the collective operations in the network, as the link is still being used by the unicast mechanism.

Once the bits are computed the tree is ready for broadcast operations at the region defined (if the region covers all the mesh then is a global tree for the entire chip). Figure 5.23 shows an example where a broadcast operation is initiated at switch 4. Notice that the example shows a broadcast operation where only three fork operations are performed (at switches 4, 2, and 6). If the network has no failed links the number of fork operations in the broadcast increases, thus reducing the broadcast operation latency. A different example is provided in Figure 5.24.

With SBBM, when an end node initiates a broadcast operation a packet is created. A control signal (B_h) is used to differentiate broadcast packets from unicast ones. Additionally, four control signals (B_n, B_e, B_w, and B_s) are used. These signals travel along with the packet through control lines. In SBBM control signals are computed on each switch but they are used by the next switch to decide which output ports may be used for the packet. As the first switch has to inject the packet through all possible directions

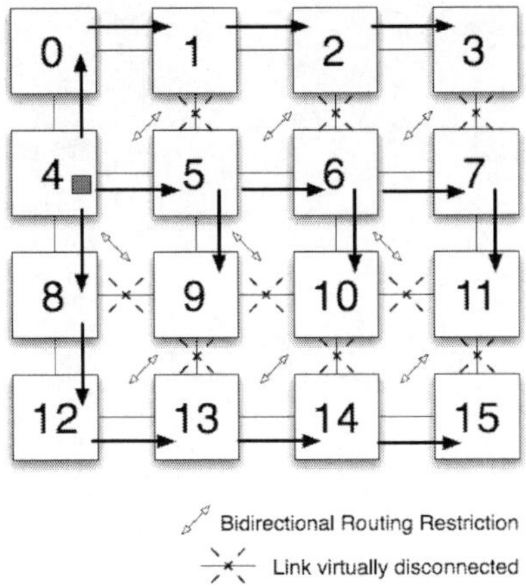

FIGURE 5.24: Broadcast operation from switch 4, no failed links.

(regardless of the connectivity), the initiating end node sets all the signals to one, regardless of the routing and connectivity bits. Thus, the network interface does not need to have any routing/topology bit definition. Only, the region identifier is set within the packet header.

A switch may receive a packet either from its local port (from the end node) or through the incoming ports. Signals for a replica packet generated to be sent through port X are labelled as XB_h, XB_n, XB_e, XB_w, and XB_s. Signals for the incoming packet are labelled as IB_h, IB_n, IB_e, IB_w, and IB_s regardless of the port identifier the packet comes from.

Upon reception of a packet, the switch may forward it through up to four ports (local port is included and U turns are forbidden). The switch generates a replica through an output port (e.g., N port) if the corresponding control signal received (e.g., IB_n signal) is set, and the connectivity bit through the output port (e.g., $C_n[1]$ bit) is set. For each replica control signals are computed again (to be used at the next switch) and attached to the packet replica.

Figure 5.25 illustrates the SBBM mechanism. Both the packet and control signals arrive through an input port. In particular, 8 control signals are received: three signals encoding the region the packet belongs to (Rid), and five signals with the broadcast information. The SBBM logic computes all the output ports that must be used to inject all possible broadcast packets. To do this, the unit computes all the signals (20 signals, five signals per output port). The logic is replicated at each input port. As it can be seen the logic is small. The reason for this simplicity is the fact deadlock freedom is currently guaranteed by the routing bits, and replication of packets is prevented with the connectivity bits building the acyclic tree ($C_x[1]$). Therefore, SBBM just checks the available paths at the next switch, which matches with the routing bits present on the current switch. As an example of a computed signal, the

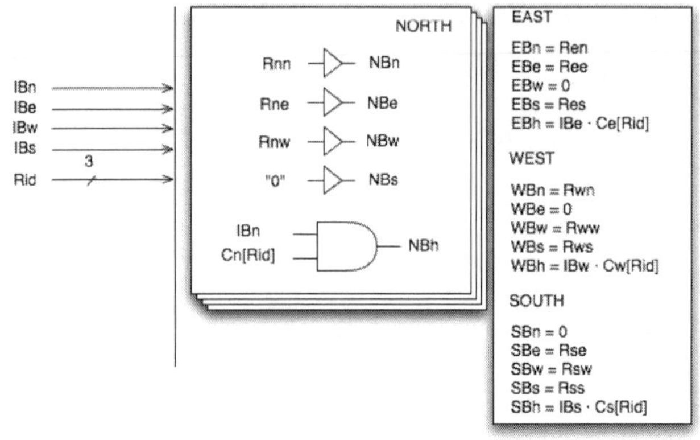

FIGURE 5.25: SBBM logic.

NB_n signal is set if there is no routing restriction between channels S and N at the next switch through the north output port, i.e., the R_{nn} routing bit at the current switch. This signal tells the next switch that the broadcast packet is allowed to go north. Similar conclusions are obtained for the remaining signals and output ports of the switch.

With this description in mind, let's describe the example provided in Figure 5.23. A broadcast action in a 2-D mesh with two disconnected links is initiated at switch 4. The switch receives the packet from its local node and immediately sends two replicas through N and S output ports. The packet is not replicated through the E port as this link is virtually disconnected ($C_e[1]$ bit is reset). The incoming packet received at switch 0 will have its IBh, IBn, IBe, and IBw signals set, but as there is only connectivity to the east, just one packet will be forwarded from switch 0. Broadcast actions follow until the last node is reached with this mechanism. Note that the packet is delivered to the local node on every visited switch.

Figure 5.12 shows the coverage results for SBBM. As can be seen, coverage in SBBM is optimal since all the topologies analyzed were supported. In all the cases (1423 topologies) SBBM successfully implemented a broadcast operation through a valid acyclic spanning tree.

5.6 eLBDR: Gathering Unicast and Broadcast Implementations

In the previous sections, we have seen an evolution of a basic mechanism, LBDR, to mechanisms able to obtain full coverage in unicast and collective

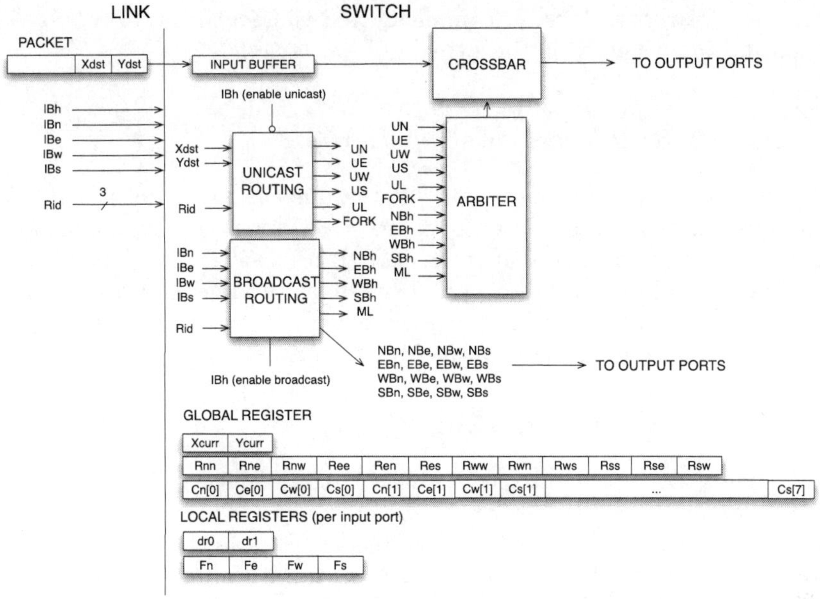

FIGURE 5.26: A general overview of the eLBDR architecture inside a switch.

communication based on the foundations of routing restrictions. An ideal routing architecture integrated at a switch should be capable of giving support to both types of routing. eLBDR (effective Logic-Based Distributed Routing) is the addition of both unicast and broadcast mechanisms, uLBDR, and SBBM. Figure 5.26 shows a general schematic of the eLBDR architecture.

As soon as the packet header is decoded in the input buffer of a switch, the first operation is to distinguish whether the packet is meant for unicast or for broadcast routing. A control bit, IB_h, is associated with the packet and enables the appropriate module (unicast or broadcast module). SBBM additional control signals (IB_n, IB_e, IB_w, and IB_s) are used for broadcast operations together with the R_{id} bits that indicate the region the packet belongs to. Each switch has a global register storing all the eLBDR bits (routing and connectivity bits). Additionally, each switch has a register with an ID (i.e., the coordinates in the 2-D mesh). For each input port a local 6-bit register, including the dr_x and F_x bits, for deroute and fork operations is included. The global register is used by both modules, unicast and broadcast, whereas the local register is only used by the unicast module.

Both modules generate the corresponding signals to the arbiter. There is a signal for each output port (N, E, W, S, or a local port) from each module. The arbiter configures the crossbar to route the packet to the corresponding output port (or replicates it to several output ports when needed). However, notice that the way the arbiter is implemented is independent of the routing mechanisms implemented in the switch.

5.6.1 eLBDR: Implementation Details

In this section we provide details for the implementation of the eLBDR mechanism. To do so, both switches (described in the Appendix) are modified and enhanced with the inclusion of the eLBDR mechanism. The first switch is a pipelined switch design targeting CMP systems (we refer to it as the CMP switch) while the other switch is a nonpipelined switch targeting high-end MPSoC systems (we refer to it as the MPSoC switch).

5.6.1.1 CMP Switch with eLBDR

We analyze the area and latency penalty introduced in a switch when using the new routing techniques described before. The switch used throughout this section is the gNoC switch, being a baseline switch design targeted for CMP systems. The gNoC switch is a academical pipelined wormhole-based switch. In order to evolve the XY-based switch design to new routing techniques as $LBDR_{fork}$, $LBDR_{fork+dr}$, or SBBM broadcast routing, virtual cut-through (VCT) switching must be supported. For that purpose, the basic wormhole switch has been modified to allow VCT switching. The main changes are related to the way flow control is performed. For a more detailed description of the wormhole switch design and the changes applied to support VCT switching, see the Appendix.

In order to implement LBDR, $LBDR_{dr}$, $LBDR_{fork+dr}$, and broadcast SBBM, the routing (RT) stage has been modified with respect to the basic XY routing implementation in the gNoC switch design. The configuration bits necessary for these new routing mechanisms (described in Sections 5.3, 5.4.2, and 5.4.3) have been implemented using registers. Furthermore, the associated logic (Figures 5.13, 5.16, 5.18, 5.19, 5.20, 5.25, and 5.26) has been implemented.

Figure 5.27(a) summarizes frequency and area results of the gNoC switch for different switching techniques and routing mechanisms. The first thing to highlight is the unexpected improvements of both area and frequency of the VCT switch. As mentioned in the Appendix, the reason for this improvement is due to the more efficient input buffer (IB) stage implemented in the VCT version, as the flow control has been simplified.

There is no difference in the operating frequency when using either XY, LBDR or $LBDR_{dr}$, and only a marginal increase in area (differences fall within the margin of the synthesis optimization process). This is due to the RT stage not setting the maximum frequency of the switch. eLBDR experiences, however, a small impact in performance and area. This performance degradation is due to the overhead added to the crossbar switch stage (SW). The support

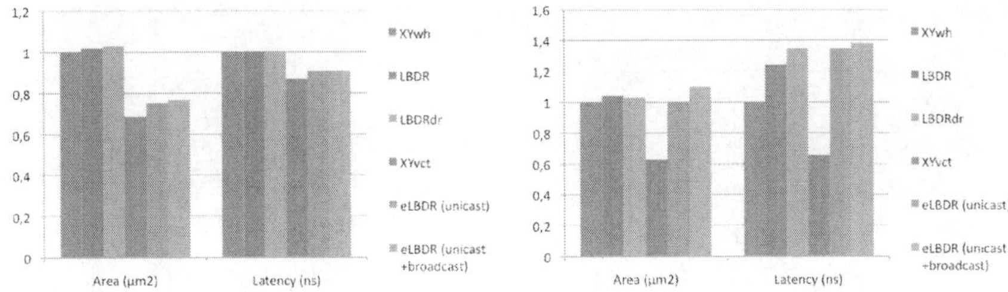

(a) Complete switch (b) RT module

FIGURE 5.27: CMP complete switch (gNoC) and RT stage overhead, normalized results.

for broadcast operations comes with no further penalty in performance and with a marginal increase of area.

Figure 5.27(b) shows also the area overhead and frequencies of the different routing modules (RT stage), thus not considering the entire switch. Now, there are significant differences between the XY module for wormhole switching (WH) and for VCT switching. These differences are due to the different flow control mechanisms used in both versions. Also, the different input buffer design affects the routing module. On the other hand, LBDR and $LBDR_{dr}$ mechanisms have a small impact on area but a large one in frequency. Also, the complexity of eLBDR and the support for broadcast has a large impact on both area and frequency. However, remember that this module is not the one setting the switch frequency in the gNoC switch design.

5.6.1.2 MPSoC Switch with eLBDR

Postsynthesis area and performance results for the Multiprocessor System-on-Chip (MPSoC) switch are shown in Figure 5.28. Switches with just the unicast module and full eLBDR (with unicast and broadcast modules) were synthesized. All switches implement the same amount of buffering (4 slots per input port) and were synthesized for maximum performance.

The choice of a specific flow control, switching and routing mechanism configuration affects the maximum achievable speed by each switch. By looking at the resulting clock cycle time T it is possible to split the results into two parts: the switches with the wormhole switching and the switches with VCT. In fact, changing only the routing mechanism affects maximum performance in a marginal way (around 5% in the worst case). Instead, when we also use different switching/flow control mechanisms the performance drops by 20%. If we only focus on VCT switches it should be noticed that eLBDR (unicast+broadcast) is only less than 1% slower than the unicast implementation. This is due to the fact that the broadcast module works in parallel with the unicast module, which is more complex from a combinational logic viewpoint. The actual difference between both critical paths is the combinational logic

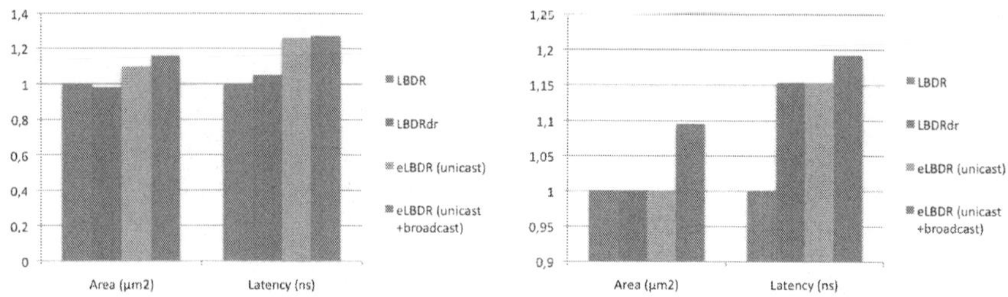

(a) Complete switch (b) RT module

FIGURE 5.28: MPSoC complete switch and RT stage overhead, normalized results.

added to couple the unicast and broadcast routing modules together. However, this logic is efficiently handled by the synthesis tool and its impact is minimized.

By looking at the area results, eLBDR (unicast) is about 10% larger than the baseline LBDR, clearly due to the more complex port arbiters and to the need to support full credit-based flow control (see the Appendix). To make this relatively more complex circuit faster, the synthesis tool tried to speed up the crossbar at the cost of further increased area. We also observe that LBDR and $LBDR_{dr}$ feature approximately the same maximum area, except for hardly controllable specific optimizations that the synthesis tool applies to the two netlists. The take-away message here is that the logic complexity of these two routing mechanisms is pretty much equivalent.

The breakdown of the critical timing path in the VCT switch is also of interest. In both cases the routing function takes around 24% of the total critical path, while the arbitration, input buffer control logic, and crossbar take 35%, 22%, and 19%, respectively. Moreover, by looking at area results, it can be noticed that the routing module of eLBDR (unicast+broadcast) increases the total area only by a negligible 5% respect to the eLBDR (unicast).

5.7 Future Challenges for Routing

In the previous section we have outlined a possible method to implement flexible routing algorithms in NoCs. The main challenge was to achieve efficient implementation while addressing manufacturing defects in chips relying on 2-D mesh structures. Both unicast and broadcast (multicast at the chip-level when applied to regions within the chip) have been addressed. Although this is not the only way to address routing implementation, it highlights the need for compact and flexible implementations. Future directions should address other topologies suitable for on-chip networks (see Chapter 4 for a description of possible topologies). The concept of routing bits and connectivity bits can be extended to support new topologies, thus obtaining efficient implementations that rely on such bits. For collective communication support, new mechanisms need to be covered as well. As an example, in many cache coherence protocols several end nodes send an acknowledgment packet to the same destination (e.g., because of invalidations of several copies of a memory block), thus demanding for an efficient implementation of many-to-one collective traffic.

The routing algorithm affects in many ways the performance of the system. The paths taken by messages may lead to the building of congestion within the network (see Chapter 12), thus routing algorithms able to balance the traffic over the network need to be researched. The location of the main components in the system (how they are attached to the network) also influences the traffic shape, thus routing algorithms must be co-designed jointly with the topology and the location of resources within the network. Quality of service guarantees are also linked with routing. Bandwidth guarantees, maximum transfer latencies, and minimization of jitter must be co-designed with the routing algorithm. In this sense, the knowledge of the traffic shape, although not known before the CMP chip design, can be estimated. Indeed, if a cache coherence protocol is run within the chip, the traffic shape is mostly determined by the coherence protocol. Percentages of short and large messages, and the location of key components (like memory controllers or L3 cache locations) can be taken into account in order to design the routing algorithm. Although many research proposals exist for quality of service (QoS) and congestion management strategies, the routing algorithm suitable for NoCs needs to be finally implemented in the chip, thus an efficient implementation will be required.

Fully adaptive routing algorithms open the possibility for higher performance levels. Indeed, as topologies with higher number of dimensions are expected (e.g., 3-D meshes due to 3-D stacking), higher adaptivity levels within the network can be achieved. The future research directions are linked with the problem of out-of-order delivery issues. Indeed, how to efficiently implement strict in-order guarantees (for those applications or cache coherence protocols requiring in-order delivery) while allowing adaptive algorithms is an important research topic. A possibility is to guarantee in-order delivery to those traffic classes demanding such strict ordering, while allowing more efficient handling of traffic classes that allow out-of-order delivery of messages.

Virtualization is another emerging concept in on-chip networks. Efficient distribution of chip resources over different applications, the existence of regions with different voltage and frequency domains, and the effective use of power-saving techniques, demand for a mechanism able to virtualize the chip. The previous concept of routing bits and connectivity bits can also be applied for that purpose. However, the way those bits are modified to adapt the chip to the changing system conditions needs to be explored. Indeed, focusing the attention on the routing implementation, and closely linked to routing and

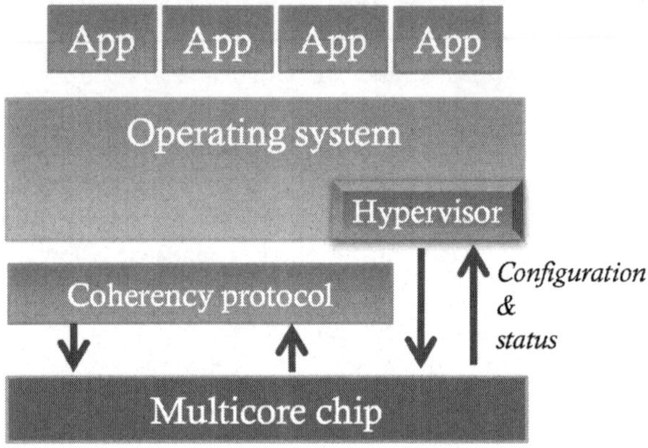

FIGURE 5.29: A possible CMP system with the OS having access to the on-chip network resources.

connectivity bits, the system must allow for new capabilities and functionalities. A clear example is the reconfiguration of the entire chip based on the current user and/or application demands. We can think of an environment where a data server is accepting requests, and those requests need to be processed in certain locations of the chip, using a subset of chip resources. The chip must be reconfigured (even at the network level) to make room for the new request. Figure 5.29 shows a possible scenario where the operating system running on top of the CMP system is involved. The operating system (through a hypervisor) may have access to the current network configuration, and based on the changing demands by the running applications, the hypervisor may reconfigure the network to adapt to the changing requirements. Indeed, the operating system may need to change the routing algorithm run inside the network to avoid conflicting traffic patterns from different applications, thus avoiding deadlocks and guaranteeing traffic isolation.

Acknowledgment

This work was supported by the Spanish MEC and MICINN, as well as European Commission FEDER funds, under Grants CSD2006-00046 and TIN2009-14475-C04. It was also partly supported by the project NaNoC (project label 248972), which is funded by the European Commission within the Research Programme FP7.

Chapter 6

The Synchronization Challenge

Davide Bertozzi

University of Ferrara, Italy

Alessandro Strano

University of Ferrara, Italy

Daniele Ludovici

TUDelft, The Netherlands

Vasileios Pavlidis

Integrated Systems Lab (LSI), EPFL, Lausanne, Switzerland

Federico Angiolini

iNoCs, Switzerland

Milos Krstic

Innovations for High-Performance Microelectronics (IHP), Germany

6.1 Introduction

Network-on-chip (NoC) communication architectures are being widely adopted in multicore chip design to ensure scalability and facilitate a component-based approach to large-scale system integration. As technology advances into aggressive nanometer-level scaling, a number of design challenges emerge from technology constraints that require a continuous evolution of NoC implementation strategies at the circuit and architectural level.

Synchronization is today definitely among the most critical challenges in the design of a global on-chip communication infrastructure, as emerging variability, signal integrity, power dissipation limits are contributing to a severe break-down of the global synchronicity assumption when a logical structure spans more than a couple of millimeters on die [91]. NoCs typically span the entire chip area and there is today little doubt on the fact that a high-performance and cost-effective NoC can only be designed in 45 nm and beyond under a relaxed synchronization assumption [38].

One effective method to address this issue is through the use of globally asynchronous locally synchronous (GALS) architectures where the chip is partitioned into multiple independent frequency domains. Each domain is clocked synchronously while interdomain communication is achieved through specific interconnect techniques and circuits. Due to its flexible portability and transparent features regardless of the differences among computational cores, GALS interconnect architecture becomes a top candidate for multi- and many-core chips that wish to do away with complex global clock distribution networks.

In addition, GALS allows the possibility of fine-grained power reduction through frequency and voltage scaling [37]. Since each core or cluster of cores operates in its own frequency domain, it is possible to reduce the power dissipation, increase energy efficiency, and compensate for some circuit variations on a fine-grained level.

Among the advantages of a GALS clocking style, it is worth mentioning [322]:

- GALS clocking design with a simple local ring oscillator for each core eliminates the need for complex and power hungry global clock trees.
- Unused cores can be effectively disconnected by power gating, thus reducing leakage.

- When workloads distributed to cores are not identical or feature different performance requirements, we can allocate different clock frequencies and supply voltages for these cores either statically or dynamically. This allows the total system to consume a lower power than if all active cores had been operated at a single frequency and supply voltage [119].

- We can reduce more power by architecture-driven methods such as parallelizing or pipelining a serial algorithm over multiple cores [47].

- We can also spread computationally intensive workloads around the chip to eliminate hot spots and balance temperature

The methodology of interdomain communication is a crucial design point for GALS architectures. One approach is the purely asynchronous clockless handshaking, that uses multiple phases (normally two or four phases) of exchanging control signals (request and ack) for transferring data words across clock domains [227, 299]. Unfortunately, these asynchronous handshaking techniques are complex and use unconventional circuits (such as the Muller C-element [263]) typically unavailable in generic standard cell libraries. Besides that, because the arrival times of events are arbitrary without a reference timing signal, their activities are difficult to verify in traditional digital Computer-Aided Design (CAD) design flows.

The so-called delay-insensitive interconnection method extends clockless handshaking techniques by using coding techniques such as dual-rail or 1-of-4 to avoid the requirement of delay matching between data bits and control signals [46]. These circuits also require specific cells that do not exist in common Application Specific Integrated Circuit (ASIC) design libraries. Quinton et al. implemented a delay-insensitive asynchronous interconnect network using only digital standard cells; however, the final circuit has large area and energy costs [262].

Another asynchronous interconnect technique uses a pausible or stretchable clock where the rising edge of the receiving clock is paused following the requirements of the control signals from the sender. This makes the synchronizer at the receiver wait until the data signals stabilize before sampling [349, 224]. The receiving clock is artificial meaning its period can vary cycle by cycle; so it is not particularly suitable for processing elements with synchronous clocking that need a stable signal clock in a long-enough time. Besides that, this technique is difficult to manage when applied to a multiport design due to the arbitrary and unpredictable arrival times of multiple input signals.

An alternative for transferring data across clock domains is the source-synchronous communication technique that was originally proposed for off-chip interconnects. In this approach, the source clock signal is sent along with the data to the destination. At the destination, the source clock is used to sample and write the input data into a First-In First-Out (FIFO) queue while the destination clock is used to read the data from the queue for processing.

This method achieves high efficiency by obtaining an ideal throughput of one data word per source clock cycle with a very simple design that is also similar to the synchronous design methodology; hence, it is easily compatible with common standard cell design flows [348, 25, 126, 303]. Unfortunately, correct operation of source-synchronous links is very sensitive to routing delay mismatches between data and the strobe signals, and hence to delay variability. It is therefore very challenging in the context of nanoscale silicon technologies.

In general, each GALS paradigm has its own pros and cons. Fully asynchronous design techniques are a more disruptive yet appealing solution, although their widespread industrial adoption might be slowed down by the relevant verification and design automation concerns they raise. Moreover, they tend to be quite area greedy, at least when timing robustness is enforced. In this context, a full custom design style is still the safer strategy for their successful utilization, hence asynchronous NoC building blocks are often instantiated as hard macros.

For this reason, this chapter will not review fully asynchronous NoC architectures, but will rather focus on synchronizer-based GALS architectures, and on source synchronous communication in particular. Using synchronizers for GALS NoC design implies an incremental evolution of mainstream Electronic Design Automation (EDA) design tools and paves the way for compatible architectures (with careful synchronizer engineering) with a standard cell design flow. This chapter is fully devoted to this kind of GALS architecture. First, it will be showed that GALS technology has fundamental system-level implications, and pushes the development of novel architecture-level design techniques. Second, the basic building blocks for GALS architecture design will be illustrated, aiming at an overview of promising design methods for synchronization interfaces rather than at their comparative evaluation (which is left for the actual designer).

Finally, this chapter addresses the synchronization concerns and promising clock tree synthesis solutions for the fast evolving domain of vertical integration. 3-D stacking is being today introduced as an effective means of relieving the memory bottleneck and scalability limitations in the bidimensional silicon surface. Synchronization is a critical challenge there too, and a mixture of both synchronous and asynchronous approaches will be most likely utilized to deliver satisfying performance with a reasonable power dissipation and low design, testing, and cooling cost.

6.2 The System Architecture

A GALS-based design style fits nicely with the concept of voltage and frequency islands (VFIs), which has been introduced to achieve fine-grain system-level power management. The use of VFIs in the NoC context is likely

Benchmark	Network size	Total Energy Consumption (mJ) 1-VFI —— 2-VFI —— 3-VFI
Consumer	3×3	18.9 —— **12.1*** —— **12.2**
Network	3×3	12.9 —— **6.6*** —— **6.7**
Auto-Industry	4×4	1.67—— **0.34*** —— **0.40**
telecom	5×5	6.9 —— 2.6 —— **1.5***

TABLE 6.1: Total energy consumption of some benchmarks as a function of the number of voltage and frequency islands. Results from [236].

to provide better power-performance trade-offs than its single voltage, single clock frequency counterpart, while taking advantage of the natural partitioning and mapping of applications onto the NoC platform.

While an NoC architecture where each processing/storage element constitutes a separate VFI exhibits the largest potential savings for energy consumption, this solution is very costly. Indeed, the associated design complexity increases due to the overhead in implementing synchronizers and voltage converters required for communication across different VFIs, as well the power distribution network needed to cover multiple VFIs. Additionally, the VFI partitioning needs to be performed together with the assignment of the supply and threshold voltages and the corresponding clock speeds to each VFI.

Clearly, a design methodology is required to partition a given NoC architecture into multiple voltage-frequency domains, and to assign the supply and threshold voltages (hence the corresponding clock frequencies) to each domain such that the total energy consumption is minimized under given performance constraints.

A unified approach for solving the VFI partitioning and voltage/speed assignment problems is reported in [236]. The basic idea is to start with an NoC configuration where each core belongs to a separate VFI characterized by given supply and threshold voltages and local clock speed (i.e., having initially N VFIs, for N cores). This solution may achieve the minimum application energy consumption, but not necessarily the minimum total energy consumption due to the additional overhead incurred by implementing a large number of VFIs. Therefore, the proposed approach finds two candidate VFIs to merge such that the decrease in the energy consumption is the largest among all possible merges, while performance constraints are still met. The process is repeated until a single VFI implementation is obtained. Consequently, for all possible VFIs (i.e., 1,2,...,N), the outcome consists of partitioning and corresponding voltage level assignments such that the total energy is minimized, subject to given performance constraints. Finally, among all VFI partitionings determined by the iterative process, the one providing the minimum energy consumption is selected as being the solution of the VFI partitioning problem.

As an example, a set of benchmarks is chosen from a public benchmark suite. *Consumer*, *networking*, *auto-industry* and *telecom benchmark* applications are collected from E3S [77]. These benchmarks are scheduled onto 3×3,

3×3, 4×4, and 5×5 mesh networks, respectively, using the Energy Aware Scheduling (EAS) scheme presented in [130]. Then, the approach in [236] is used for VFI partitioning and static voltage assignment. The third column of Table 6.1 (1-VFI) shows the minimum energy consumption when the entire network consists of a single island. The remaining columns show the energy consumption values obtained for the best partitioning with two and three islands, respectively. The energy consumption of the partitioning picked by the algorithm is marked with an asterisk.

Apart from the specific methodology and the energy model parameters used in this case-study, the key take-away is that using VFIs in the NoC context provides better power-performance trade-offs than its single voltage, single clock frequency counterpart. However, since such partitioning comes with an implementation cost, the best partioning might not be the one that associates one VFI to each core.

Let us now address the challenge of implementing the GALS synchronization in a system architecture structured into multiple VFIs. Two main design paradigms do exist and are hereafter illustrated.

6.2.1 Network Components Distributed across VFIs

The former one consists of subdividing the system into disjoint physical domains, where each domain can be operated at its own independent voltage and frequency. Network components are interspersed into the different domains as well, therefore the network results from the combination of baseline switching elements, each belonging to a distinct VFI. There is no specific physical domain for the network as such, therefore a synchronization element needs to be inserted when connecting two network components with each other. This is the case of Figure 6.8(a), where a dual-clock FIFO is placed between any two connected switches. While the construction methodology of this architecture template is very modular, the resulting architecture has a bottleneck in the switching elements belonging to domains that at a certaint point in time are running at a slow speed. Since many communication flows will be crossing such switching element at runtime, they will be all slowed down. More in general, each flow will cross switching elements running at different speeds, and the slowest one ends up limiting performance of the entire flow. To the limit, when power management techniques are implemented, a physical domain can be shutdown or placed in sleep mode, while cores in the other domains can be operational. For instance, power gating using sleep transistors is a popular way to shutdown cores. To achieve power gating, the sleep transistors are added between the actual ground lines and the circuit ground (also called the *virtual ground*), which are turned off in the sleep mode to cut-off the leakage path. The interconnect architecture is a bottleneck in allowing the shutdown of voltage and frequency islands. If the switches are spread across the different VFIs and if a VFI needs to be shutdown, then packets between cores on the

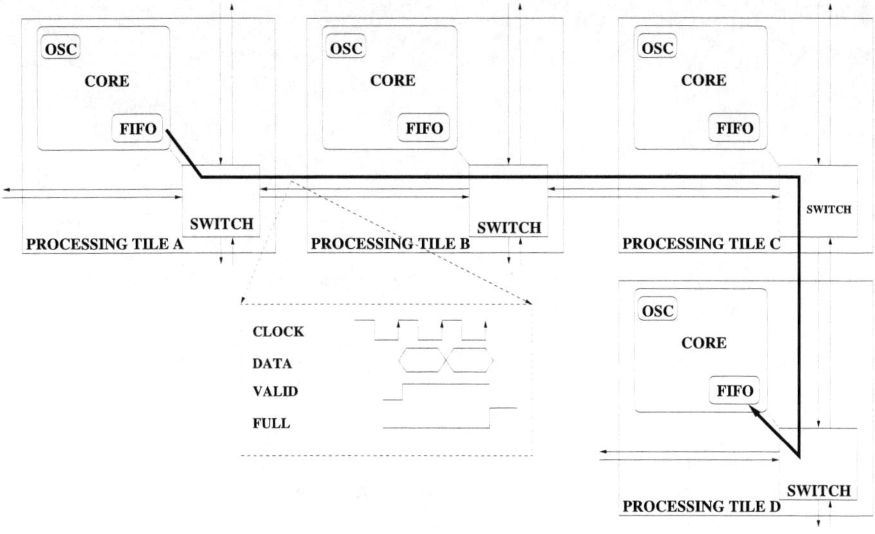

FIGURE 6.1: Long distance source synchronous interconnect path through intermediate switches (from [322] ©2010 IEEE).

other VFIs that use switches in this VFI cannot be transmitted. This will prevent the shutdown of the entire island.

Architecture-level techniques are required to cope with these effects. Use of *virtual channels* can help relieve the HoL blocking effect that arises when packets are destined to a physical domain running at low speed. An example virtual channel allocation policy that can be used in this case was first proposed in [228] and reused in [101] in the network on chip domain, named destination-based buffer management (DBBM). DBBM maps packets addressed to different destination nodes to different queues, thus eliminating the Head-of-Line (HoL) blocking among particular flows heading to different destinations. Since however the number of queues might be unbounded, DBBM enables different flows to share the same queue. Although HoL blocking will be introduced among flows sharing the same queue, this approach uses a reduced set of queues. The simplest mapping method of the DBBM family is based on the following coding of destinations. The lowest bits of the destination address indicate the target queue (*modulo mapping*). For instance, if we have 16 queues and the network has 256 destinations, the 4 Least Significant Bits (LSB) of the destination address indicate the queue to map the flow into. With 2 Virtual Channels (VC), all even nodes are mapped to one VC, while all odd nodes to the other one. However, all possible assignments of destinations to virtual channels are feasible within the DBBM approach. In all cases, these mappings guarantee that at every point in the network, all the packets with the same destination will use the same VC and will be delivered in order.

Another solution consists of using *long distance interconnect paths* and

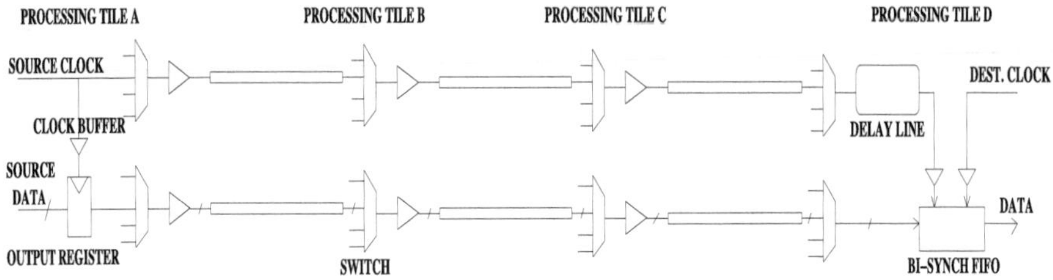

FIGURE 6.2: Simplified view of the interconnect path with a delay line in the clock signal path (from [322] ©2010 IEEE).

clockless switches [322], and is illustrated in Figure 6.1. Switches are reconfigurable and consists only of five 4 input multiplexers. Switches have 5 ports: the core port that is connected to its local core, and the North, South, West, and East ports that are connected to the four nearest neighbor switches. Long distance interconnections can be established from a source to a destination core by configuring the multiplexers in the switches belonging to intermediate cores. The configuration is done pre-runtime and fixes the communication paths; thus, these static circuit-switched interconnections are guaranteed to be independent and never shared. So long as the destination core's FIFO is not full, a very high throughput of one data word per cycle can be sustained. On these interconnect paths, the source core sends its data along with its clock to the destination. The destination core uses a dual-clock FIFO to buffer the received data before processing. Its FIFO's write port is clocked by the source clock of the initiator, while its read port is clocked by its own oscillator, and thus supports GALS communication. Intermediate switches along the interconnect paths are not only responsible for routing data on the links but also act as repeaters along the long path when combined with wire drivers. Although care is taken to match the propagation delay of data and clock, a timing uncertainty concerning their transitions at the destination FIFO does exist, which is associated with the different logic gates generating them or fed by them at the source and destination cores, respectively. Moreover, random process variations instill a probabilistic flavor to their exact transition times. Given this, the authors of [322] suggest to insert a delay line on the clock signal before driving the destination FIFO, effectively moving the rising clock edge into the stable window between two edges of the data bits, as illustrated in Figure 6.2.

A similar idea inspires the work in [135], which however still adheres to a packet switching technique at the cost of increased switch complexity. The underlying intuition is that synchronizing a packet at each intermediate hop does come with a significant latency overhead. Such synchronization latency is then avoided by means of *asynchronous bypass channels* (ABCs) around intermediate nodes. The switches should independently and dynamically determine

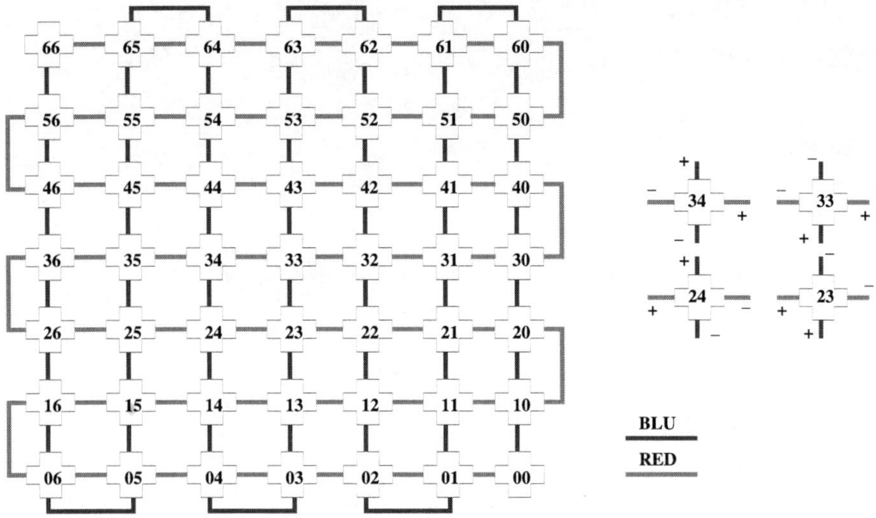

FIGURE 6.3: A double chain topology relying on asynchronous bypass channels (from [135] ©2010 IEEE).

the availability of the bypass channels without the overheads of allocation or setup and tear-down.

In the envisioned network, packets traversing a given intermediate node switch are most likely to travel straight, that is from the input port on one side of the switch to the output port on the opposide side. In a typical 2-D mesh network, a minimum length path between any two nodes may be found that has at most one turn. This observation is leveraged in [135] by biasing the switch towards the straight path over the turn path. In 2-D mesh networks, only a limited set of source-destination pairs require no turn during packet traversal, namely those pairs that are in the same column or row. It should be possible to gain a further benefit from the switch's straight path bias if the network topology provides a greater number of source-destination pairs that does not require a turn. To this end, a new class of network topologies (the *double-chain topologies*) was proposed by [135] (see Figure 6.3). A double-chain topology is comprised of two disjoint but overlapping chains, each of which connects all network nodes. As the traditional 2-D mesh terminology of North, South, East, and West is less useful in a double-chain topology, these chains are labelled as the red chain and the blue chain, respectively. As showed in Figure 6.3, nodes within each chain are perceived to be ordered such that traversing a given chain in one direction is considered to be increasing (e.g., the red+ direction) and traversing that chain in the opposite direction is considered to be decreasing (e.g., the red– direction). Jumping from one chain to another is referred to as a *turn*. To avoid deadlock, the chains themselves

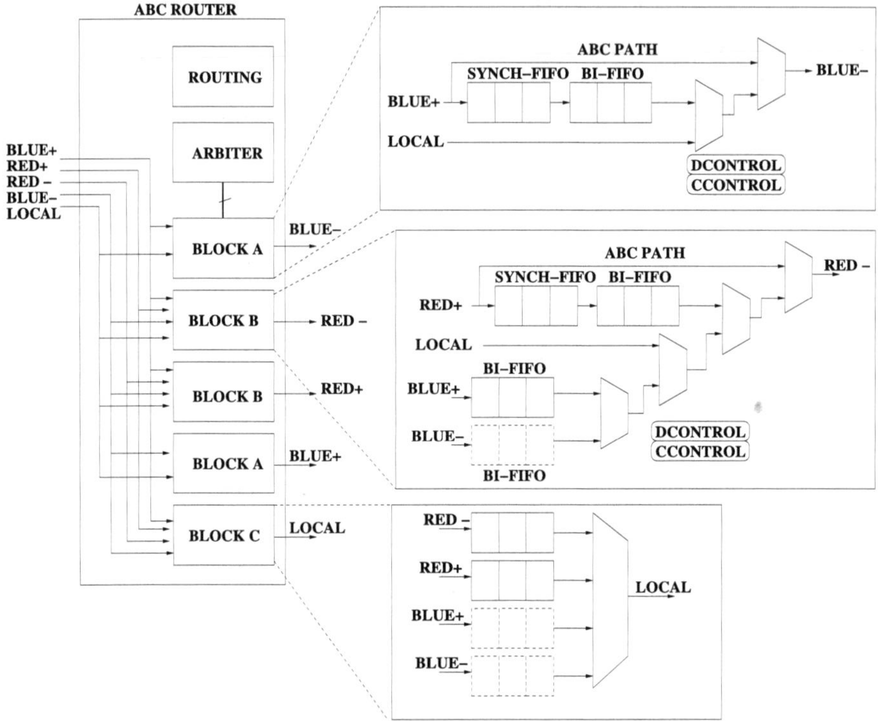

FIGURE 6.4: High-level block diagram of the switch with asynchronous bypass channels (from [135] ©2010 IEEE).

are ordered, and a jump is allowed only from the blue chain to the red chain and not vice versa.

Figure 6.4 shows a high-level block diagram of the switch. The switch is source-synchronous, hence each port's link is comprised of a clock bit along with data bits and an on/off flow control bit. The switch contains five output unit blocks corresponding to each direction. These output unit blocks are broken into three types. Output block A connects the blue (+ or –) and local inputs to the opposite blue chain output, while output block B connects all the possible inputs to the corresponding red output. Block C connects all possible inputs to the switch's local core. Block A has a smaller set of inputs because it is illegal to route from a red input to a blue output as previously discussed. At the intermediate nodes, if a packet travels straight through the switch, it is desirable that the packet be forwarded without latching and synchronization. To this end, output blocks A and B both contain an "ABC path," which the switch is predisposed to use in the absence of congestion. When an incoming flit arrives, it is simultaneously latched into the synchronous FIFO (Synch-FIFO) and propagated along the ABC path towards the output multiplexer

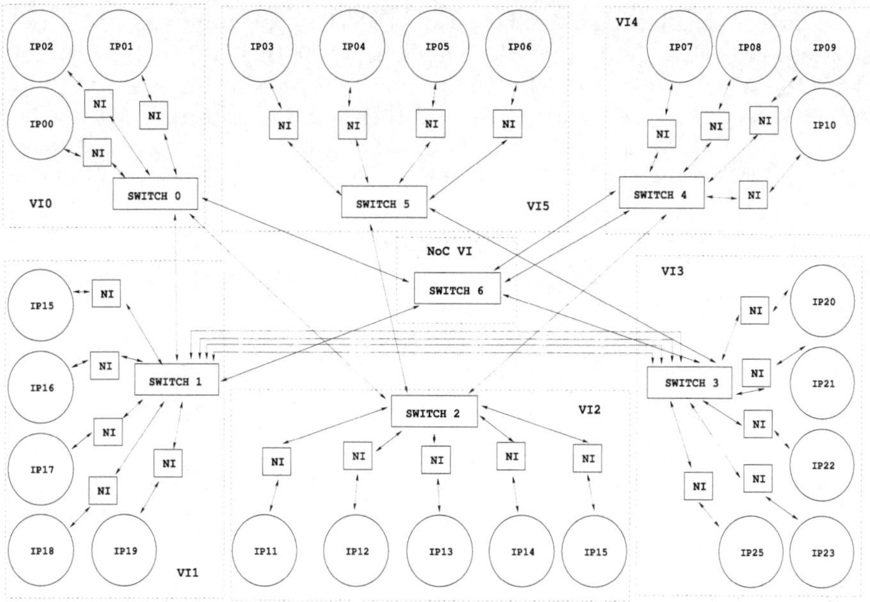

FIGURE 6.5: NoC topology where the shutdown of voltage and frequency islands is enabled by an always-on island with a subset of NoC components (from [285] ©2009 IEEE).

(MUX). In the absence of congestion, indicated by a lack of packets currently waiting for the output port and the presence of favorable downstream flow control, the flit is propagated through the output port multiplexer via the ABC and the contents of the synch-FIFO are flushed away. Alternately, if the buffer FIFOs are not empty, indicating the presence of older packets, the contents of the synch-FIFO are forwarded to the straight path bisynchronous FIFOs (bi-FIFOs) and the output port is fed by one of the bi-FIFOs. Thus, the synch-FIFO acts as a backup in case the flit was not able to successfully utilize ABC. Finally, if the packet needs to turn at that particular node, the flits are latched into one of the turn path bi-FIFOs. Although there are no buffers along the ABC path, bi-FIFOs are employed to store packets in the event of congestion. Bi-FIFOs allow writes and reads to be done in different clock domains, although they require an overhead in clock cycles to avoid metastability. In the switch, bi-FIFO writes are done in the incoming clock domain and reads are done in the local clock domain. The path bridging a given input port with the same color output port (e.g., red− to red+) is referred to as a *straight path.* All other paths are referred to as *turn paths*, as showed in Figure 6.4. The ABC path is only a bypass around the straight path FIFOs. Providing ABC paths for the turns as well is in principle feasible, but would require a significant amount of FIFOs.

The output channel is shared between ABC, the straight path bi-FIFO and turn path bi-FIFOs. To avoid livelock and ensure routing fairness, the arbiter selects the ABC path in the ABC mode, and arbitrates among the bi-FIFOs according to the round robin algorithm. However, on transition from ABC mode to FIFO mode, the straight path bi-FIFO is given a priority. As showed in Figure 6.4, the switch arbiter is divided into two types of control modules for each output unit block, the clock control module (Ccontrol) and the data control module (Dcontrol). The Ccontrol module arbitrates the output clock while the Dcontrol module controls the output data. As the Ccontrol and Dcontrol modules receive input from two different clock domains, metastability must be addressed. Control logic synchronization is used to avoid all metastable conditions. Further details on this logic, and on possible double-chain topologies and their routing algorithms, are provided in [135].

Other architecture-level techniques have been conceived to support the shutdown of one or more voltage and frequency islands while preserving correct operation of the network. An example comes from [285], where the problem is addressed in the context of topology synthesis for Systems-on-Chip. When a switch in one VFI is to be connected to a switch in another VFI, a bisynchronous FIFO is used to connect them together. The FIFO takes care of the voltage and frequency conversion across the islands. The frequency conversion is needed because the switches in the different VFIs could be operating at different frequencies. Even if they are operating at the same frequency, the clock tree is usually built separately for each VFI. Thus, there may be a clock skew between the two synchronous islands. [285] uses over the cell routing with unpipelined links to connect switches across different VFIs, as the wires could be routed on top of other VFIs. In the topology synthesis procedure, switches are generated in each VFI to connect the cores in the VFI itself. Optionally, the method can explore solutions where a separate NoC VFI can be created. The availability of power and ground lines for the intermediate VFI is taken as an input, and the method will use the intermediate island, only if the resources are available. The method produces topologies such that a traffic flow across two different VFIs can be routed in two ways: (i) the flow can go either directly from a switch in the VFI containing the source core to another switch in the VFI containing the destination core, or (ii) it can go through a switch that is placed in the intermediate NoC VFI, if the VFI is available. The switches in the intermediate VFI *are never shutdown*. The method will automatically explore both alternatives and choose the best one for meeting the application constraints.

The objective of the synthesis method is to determine the number of switches needed in each VFI, the size of the switches, their operating frequency and routing paths across the switches, such that application constraints are satisfied and VFIs can be shutdown, if needed. The method determines if an intermediate NoC VFI needs to be used to connect the switches in the different VFIs and if so, the number of switches in the intermediate island, their sizes, frequency of operation, connectivity, and paths. For the details, the interested reader should refer to [285].

An example topology synthesized with this methodology is illustrated in Figure 6.5 and reported from [285]. The mobile communication and multimedia benchmark has 26 cores, consisting of several processors, Digital Signal Processors (DSP), caches, Direct Memory Access (DMA) controller, integrated memory, video decoder engines, and a multitude of peripheral Input/Output (I/O) ports. Cores have been assigned to islands based on a logical partitioning, i.e., based on the functionality of the cores. For example, shared memories are placed in the same VFI, as they have the same functionality and therefore are expected to operate at the same frequency and voltage. The island with the shared memories is also expected not to be shutdown, since memories are shared and should be accessible at any time and this is another reason to cluster them in the same VFI. Similar reasoning was used to partition all the cores. The ad-hoc voltage island for switch 6 is clearly visibile.

Finally, support for VFI shutdown while preserving overall network operation can be provided by means of approaches routing packets even when parts of the NoC are disconnected. This is the case of components that have been shutdown or that have failed due to manufacturing defects.

An example comes from [87], advocating for on-chip stochastic communication. The cores communicate using a probabilistic broadcast scheme, similar to the randomized gossip protocols [76]. If a core has a packet to send, it will forward it to a randomly chosen subset of the tiles in its neighborhood. This way, the packets are diffused from tile to tile to the entire NoC. Every core then selects from the set of received messages only the ones that have its own IDentifier (ID) as the destination. Since the message might reach its destination before the broadcast is completed, the spreading could be terminated even earlier. For this purpose, a time-to-leave (TTL) is assigned to every message upon creation. The key observation behind this stategy is that at chip-level bandwidth is less expensive than in traditional networks, therefore having more packet transmissions to deal with topology irregularities is affordable. Although randomized gossip algorithms make a more conscious use of bandwidth than network flooding techniques (which send messages to all neighboring tiles, hence resulting into deterministic broadcast), we find they are not the first choice when low power and/or hard real-time constraints need to be enforced.

For those cases, more conservative solutions with a milder impact on network traffic are being currently developed. Dynamic routing has often been suggested as the answer to best handle these scenarios. Dynamic routing involves forwarding packets along different paths depending on a choice of decision variables, which are evaluated cycle-by-cycle at runtime [10, 24]. For example, with dynamic routing, packets would automatically find an alternate way around a shutdown NoC node. Unfortunately, dynamic routing introduces two major problems: deadlocks and packet ordering. Since every packet may

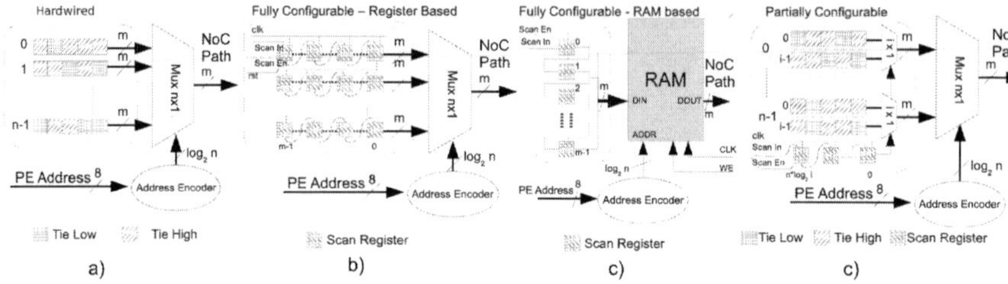

FIGURE 6.6: Different logic implementation of the Source Routing LUT: (a) Hardwired: each route is encoded as hardwired sequence of 0s and 1s; (b) Fully configurable RAM-based: each route is stored in a memory cell and can be fully reprogrammed at any time with a scan chain; (c) Fully configurable register-based: each route is stored in a register and can be fully reprogrammed at any time with a scan chain; (d) Partially configurable: alternate fixed routes are available for each destination and can be selected by programming a few control bits via a scan chain (from [184] ©2009 IEEE).

follow a different route, it becomes hard to avoid routing loops, which induce deadlocks. Sequential packets travelling among the same end points on different routes may also encounter different congestion, and reach their common destination in swapped order—a condition that is forbidden in several implementations, requiring reordering queues. Therefore, dynamic routing can become impractical in practice, at least under tight resource constraints.

An intermediate approach between deterministic and dynamic routing consists of *configurable routing*. The basic idea is to add reconfiguration capabilities into the NoC routing mechanism. The runtime reconfiguration of the NoC routes then occurs upon one of the rare events that really demand it (e.g., a power down message), and not cycle-by-cycle.

The main issue with configurable routing is that programmable NoC routing tables can have a prohibitive hardware cost. Depending on how reconfigurable they are, we classify routing look-up tables (LUTs) in three categories [184]:

- Hardwired
- Fully Configurable
- Partially Configurable

As LUT configurability increases, the system becomes more flexible. A highly configurable LUT could for example be reprogrammed to route packets around a large number of NoC faults, while a less configurable LUT may not be able to work around more than one fault, and a hard wired LUT may not tolerate any single faulty link in the NoC. Unfortunately, in general, more configurable LUTs are also more expensive in area. Next, a review from [184] of LUT

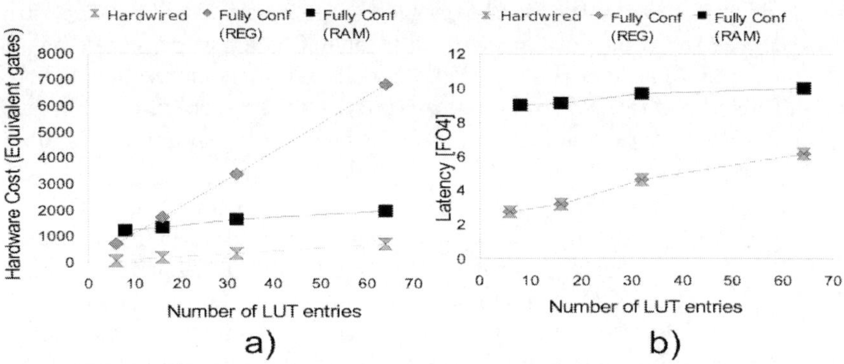

FIGURE 6.7: Cost of hardwired and fully configurable LUTs for a 3×3 mesh: (a) area cost (equivalent NAND2 gates); (b) propagation delay between input and output ports (FO4 delays) (from [184] ©2009 IEEE).

implementation styles follows. Without lack of generality, source-based routing is assumed, where routing LUTs are implemented at network interfaces.

Logically, a *hardwired LUT* is simply a Read Only Memory (ROM); circuit synthesis tools will actually render it with a netlist of combinational gates. The routing information is permanently stored: No changes are possible at runtime. The address decoder translates the addresses issued by the attached core to properly drive the multiplexer selector. Figure 6.6(a) illustrates in detail the logic implementation.

In a *fully configurable LUT*, the routing information is stored on either registers or memory banks. Unlimited remapping of NoC routes is allowed at runtime, with maximum routing flexibility. Very different reconfiguration circuits are needed depending on whether the memory elements are rendered as plain flip-flops or whether the designer instantiates a RAM macro instead. The register programming can take a place by injecting a setup vector through a scan chain, depicted in Figure 6.6(b). On the contrary, the RAM-based LUT uses a dedicated data structure, and as depicted in Figure 6.6(c), a serial-to parallel converter is needed to properly initialize the memory. The number of memory elements is in both cases $n * m$, n being the number of destinations (LUT entries) and m being the number of bits required to encode the longest path. Both solutions use a large amount of scan registers—namely $n * m$ for the former and m for the latter—for the programming operations. This also means that several clock cycles are needed to change the route map, during which the Network Interface (NI) is forced to stay idle. This high programming latency is, however, consistent with the fact that the LUT reprogramming is expected to happen only upon rare events, such as upon failures or power downs.

Partially configurable LUTs represent a hybrid solution between the previous schemes. Figure 6.6(d) gives an overview of this solution. Up to i NoC paths are allowed for each possible destination. These are hard coded, and through a setting logic block, the desired path is enabled. No memory element is required to store the LUT content itself; however, some flip-flops are still needed as the choice of which route to enable is again performed via a scan chain. Since the scan chain injects only the ID of the desired configuration, and not the entire NoC map connectivity, sequential resources decrease drastically and the total amount of clock cycles needed to reprogram the table is much lower compared to the fully configurable scheme.

To quantify the cost for LUT reconfigurability, Figure 6.7(a) compares the silicon cost, expressed in equivalent gates, of three possible LUT implementations [184]. 14 bits of routing information per LUT entry are assumed, and the UMC-Faraday 130 nm CMOS Technology library is used. The hard wired LUT shows a very efficient area utilization, respectively 18 and 30 times smaller compared to the fully configurable register- and Random Access Memory-based (RAM-based) LUTs for 6 to 8 LUT entries (the lower bound). This can be explained by the fact that in completely static LUTs the synthesis tool can tap into a huge optimization potential, as the address decoder and the multiplexer can be merged as glue logic. As the number of LUT entries increases, the silicon cost increases in a linear fashion, mainly due to the multiplexer. Concerning the RAM-based solution, four memories have been generated with the UMC-FARADAY 130 nm memory compiler. This tool allows the designer to specify some parameters (word length, number of columns, number of words, etc.). Four values were selected for the number of words in the LUT—8, 16, 32, and 64. As expected, RAMs of few words experience a high hardware overhead due to memory handling logic, but this solution is quite efficient when the number of entries grows, since its cost slope is comparable to the hardwired solution. Finally, the register-based fully configurable LUT is very efficient for small LUTs, but its cost slope is steep. Beyond about 10 LUT entries, it shows poor area efficiency as compared to the RAM-based solution.

In many designs, the LUT may fall in the critical path of the Network Interface block, therefore the timing performance of the whole NoC may degrade if the LUT is too slow. Figure 6.7(b) summarizes the timing cost: hardwired and register-based solutions show the same timing properties, since the latency is dominated by the multiplexer, while in case of a RAM-based LUT the latency is dominated by the memory access time. This cost however grows slowly as compared to the other solutions, confirming the finding that the RAM usage becomes suitable for very complex designs with large LUTs.

An alternative idea is to rely on the same implementation principle of hard-wired LUTs but trying to extend their flexibility. This is the intuition behind the logic-based distributed routing approach in [94]. The routing path is computed on-the-fly at each switch by means of combinational logic augmented with a few configuration bits. Such bits enable the routing mechanism to implement a specific routing algorithm and to deal with missing network nodes. When the actual topology changes as an effect of power management decisions or of faulty nodes, configuration bits need to be updated at runtime in a similar way to routing tables (although largely minimizing memory requirements). Unlike routing tables, Logic-Based Distributed Routing (LBDR) logic and bits stay the same and feature the same area and delay cost regardless of the number of network nodes, hence resulting in excellent scalability. Vice versa, more complex routing logic and configuration bits are required to tolerate an increasing amount of faulty or shutdown links and switches. Chapter 5 of this book covers logic-based distributed routing more in detail, given its potential impact on next-generation NoC architectures.

6.2.2 Network as an Independent VFI

As illustrated in the previous section, one safe solution for the shutdown of selected VFIs is to have an always-on island containing critical NoC components to preserve connectivity under all usage scenarios. By pushing this approach to the limit, we could think of having the entire NoC implemented as an independent voltage and frequency island. This way, the network would be loosely coupled with the cores' VFIs, and each core/cluster of cores could be shutdown without any impact on global network connectivity. From a synchronization viewpoint, the network would then be implemented as a large synchronous clock domain distributed throughout the entire chip.

Clearly, this solution inherently avoids many of the architecture-level concerns raised by the paradigm in Section 6.2.1, but its feasibility completely depends on physical implementation cost considerations. In fact, if the entire NoC lies in the same island, it is difficult to route the VDD (supply voltage) and ground lines for the NoC across the entire chip. Next, we assume that the resources for routing the additional voltage and ground lines are available, and each designer can verify this assumption for the design at hand.

Apart from the power distribution network concerns, the main issue with an independent NoC VFI consists of the feasibility of its clock tree. There is today little doubt of the fact that successful NoC design on nanoscale silicon technologies will only be feasible under relaxed synchronization assumptions. Even though inferring a global clock tree for the entire network will be still feasible for some time, it will come at a significant power cost. Moreover, there is a clear roadblock ahead, associated with the intricacy to tightly and globally enforce the clock skew constraint. The reverse scaling of interconnect delays and the growing role of process variations are some of the root causes for this.

However, a workaround for this problem does exist, as illustrated in Figure 6.8(b). The network could be inferred as a collection of mesochronous domains, instead of a global synchronous domain, yet retaining a globally synchronous perspective of the network itself. There are several methods to do this. One simple way is to go through a hierarchical clock tree synthesis process, pictorially illustrated in Figure 6.9. In practice, a local clock tree is synthesized for each mesochronous domain, where the skew constraint is enforced to be as tight as in traditional synchronous designs. Then, a top-level clock tree is synthesized, connecting the leaf trees with the centralized clock source, with a very loose clock skew constraint. This way, many repeaters and buffers, which are used to keep signals in phase, can be removed, reducing power and ther-

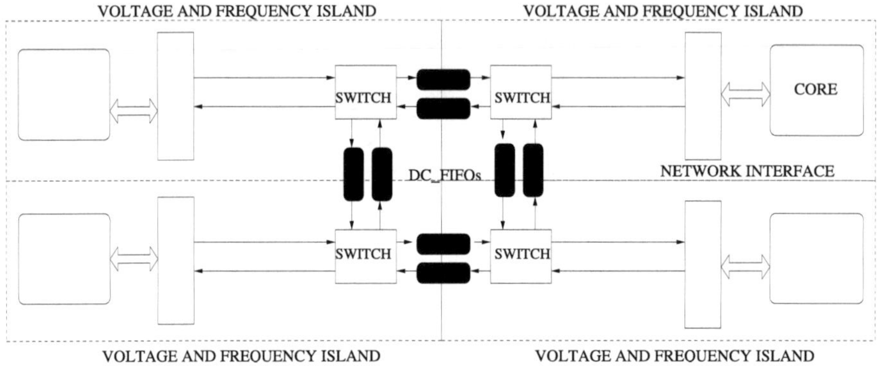

(a) System architecture split into voltage and frequency islands.

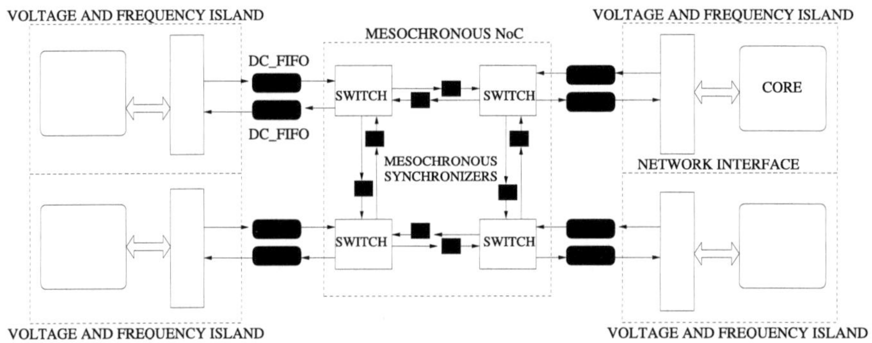

(b) The network-on-chip is a disjoint physical domain, possibly implemented with mesochronous clocking.

FIGURE 6.8: Paradigms for GALS-system architecture.

mal dissipation of the top-level clock tree. The granularity of a mesochronous domain can be as fine as a NoC switch block.

The communication between neighboring switches is then mesochronous as the clock tree is not equilibrated, while the communications between switch and Intellectual Property (IP) cores are fully asynchronous because they belong to different clock domains. Bisynchronous FIFOs are therefore used to connect the network switches to the network interfaces of the cores, as showed in Figure 6.8(b).

This synchronization paradigm comes with additional advantages. First, it makes a conscious use of area/power-hungry dual-clock FIFOs, which end up being instantiated only at network boundaries. Instead, more compact mesochronous synchronizers are used inside the network, thus minimizing the cost for GALS technology.

Finally, unlike fully asynchronous interconnect fabrics, both the synchronizer-based source-synchronous GALS architectures illustrated in Figure 6.8(b) and Figure 6.8(a) are within reach of current mainstream de-

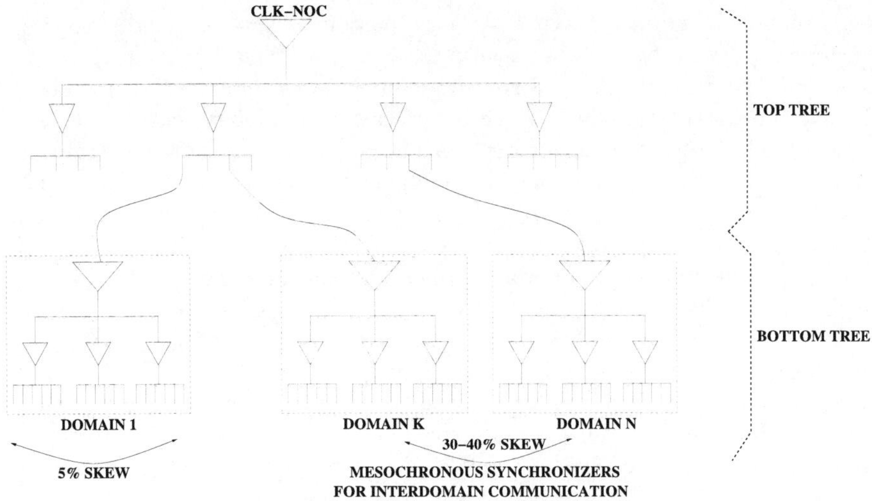

FIGURE 6.9: Output of a hierarchical clock tree synthesis run.

sign toolflows with just an incremental effort. Typically, some scripting effort within commercial synthesis frameworks enables these latter to meet the physical requirements of source-synchronous designs. Two relevant examples follow.

First, a methodology taken from [214] is reported, showing how to implement a hierarchical clock tree while inferring a mesochronous NoC. See reference picture in Figure 6.9. A buffer or an inverter is added on the clock input of each switch for the mesochronous communications. These buffers/inverters are used to build the clock tree of the switches while supporting the GALS approach. The clock tree implementation follows four steps:

1. The buffer/inverter on the clock input pin of each switch is manually placed in the middle of the area occupied by the cluster. Thus, the switch clock tree wires are as short as possible.

2. A clock tree is synthesized for each switch. The starting point of the clock-tree is the buffer/inverter on the clock input pin of the switch. Each clock tree is synthesized with 5% skew target.

3. Each clock tree is characterized with its input delay, skew, and input capacitance.

4. A top clock tree is synthesized to balance the clock trees of all the switches. Following the GALS approach, the top clock tree is balanced with (say) a 30% skew while the leaves have a 5% skew.

This methodology can be implemented with state-of-the-art synthesis and place&route tools.

A second example of synthesis tool customization to handle source-synchronous links concerns the routing step of the physical design process. At a first glance, link physical routing can be carried out with a standard physical implementation flow, which routes each net independently, while trying to match design constraints on delay, area, and power. However, routing algorithms are extremely complex and influenced by a number of factors (such as wire congestion). Hence, in practice it is extremely difficult to ensure that wires that take widely different routes maintain closely-matched delay and power consumption. In other words, when the logical atomicity of a link is broken at the physical level, significant length, resistance and load deviations are experienced among different wires of a link. Reducing delay and power variation between wires of a link is a key requirement for advanced NoC implementation styles. First and foremost, in source-synchronous signaling, data lines and the strobe signal have to arrive at the destination within a narrow timing window to achieve correct and high-performance operation.

The work in [139] proposes a link routing approach that considers all nets of a link together and efficiently routes NoCs links as an atomic entity (a bundle). This routing step is performed early in the design implementation flow, at the global routing step, and it integrates seamlessly with the design implementation tools to ensure that link quality will be fully preserved during detailed routing. As a result, the variability in link net attributes such as length, resistance, load, and delay is much reduced. Moreover, enforcing the same trajectory on all nets of a link facilitates wire spacing and shielding. Finally, bundled routing improves the routability of the NoCs link, as sometimes even a powerful place and route tool struggles to perform detailed routing of a large number of links. This can significantly reduce routing run time as well as the time for parasitics extraction.

The routing methodology is applied step-by-step in Figure 6.10. Assume a part of an NoC layout shown in Figure 6.10(a) in which there are some existing routing blockages. Pin placement and ordering have already been performed on this layout, and link L12 is going to be routed from switch S1 to switch S2. Note that a link is a group of logically related interconnects. First, the existing routing blockages are extended by W/2, where W is the width of Link L12, which leads to the layout showed in Figure 6.10(b).

After the preparation step, the middle net of the current link is considered as the link agent, and the routing engine is required to route that net. The middle net trajectory is showed in Figure 6.10(c). These virtual routing blockages force the routing engine to route the link agent at a distance of W/2 from the existing actual routing blockages.

After routing the middle net, the virtual routing blockages are removed, and then the trajectory of the middle net is applied to other nets of the link. To do this, a two-step algorithm is used. In the first step the vias of the middle net are replicated for all other nets at the appropriate position determined by the pin placement and ordering of the related link. In the second step, for each net the source pin is connected to the first via and then the vias are connected

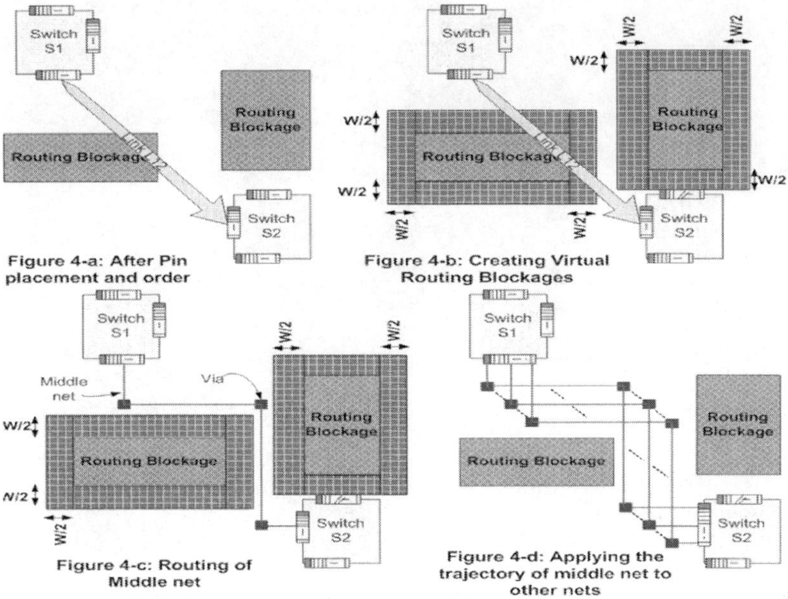

FIGURE 6.10: Steps for link-bundled routing.

together sequentially, and finally the last via is connected to the destination pin. For each net of the link, via positions are computed in such a way that the via-to-via length of all nets of a link is equal. The length difference between different nets of a link from source pin to the first via will be compensated by length difference from the last via to the destination pin. Figure 6.10(d) shows the routed link. Further details on robust bundled routing signaling in NoC links can be found in [139].

After viewing the main architecture templates embodying the GALS concept, we now illustrate the basic building blocks the designer can use to build such architectures, essentially mesochronous synchronizers and dual-clock FIFOs.

6.3 Building Blocks for GALS NoC Architectures

6.3.1 Mesochronous Synchronizers

When mesochronous clocking is used, two architecture blocks end up operating at the same clock frequency but with unknown phase offset. Hence,

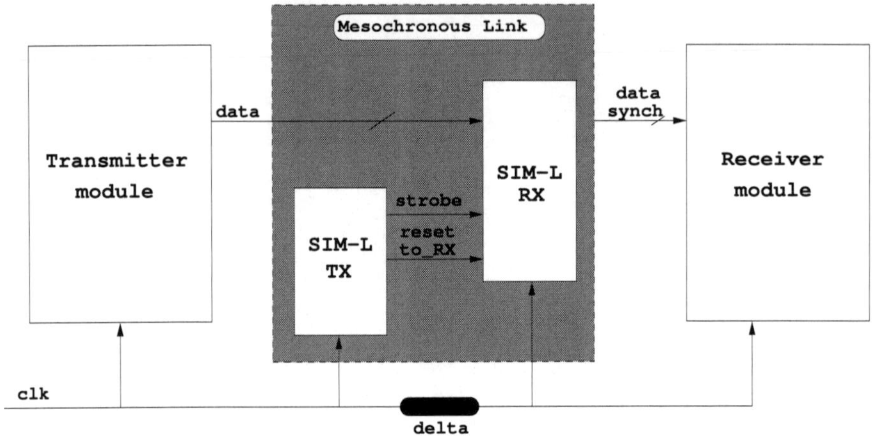

FIGURE 6.11: SIM-L architecture for mesochronous synchronization (from [331] ©2008 IEEE).

a mesochronous synchronizer needs to be inserted in between for interblock communication.

Unfortunately, mesochronous synchronizers come with their own set of problems. A traditional approach to their design consists of delaying either data or the clock signal so to sample data only when it is stable. This solution requires components often not available in a standard-cell design flow (e.g., configurable digital delay lines) or explicitly targeting full-custom design (e.g., Muller C-elements).

Another issue concerns the synchronization latency, which impacts not just communication performance but has architecture-level implications as well, thus resulting in a more costly system overall. As an example, in the network-on-chip domain, increasing the latency for across-link communication implies a larger amount of buffering resources at the end nodes to properly handle flow control and thus not to loose packets.

In the direction of providing a synthesizable, low-footprint and low-latency realization of a mesochronous synchronizer, some recent works suggest a source-synchronous design style combined with some form of ping-pong buffering to counter timing and metastability concerns [331, 100, 190]. While [331] avoids a phase detector but requires the link delay to be less than one clock period, the scheme in [100] can handle slow and long links but requires a phase detector.

Next, this chapter focuses on mesochronous synchronizers for NoCs avoiding the use of a phase detector, since this latter is a tedious component often incurring a high area footprint, timing uncertainty, or noncompliance with a standard cell design flow.

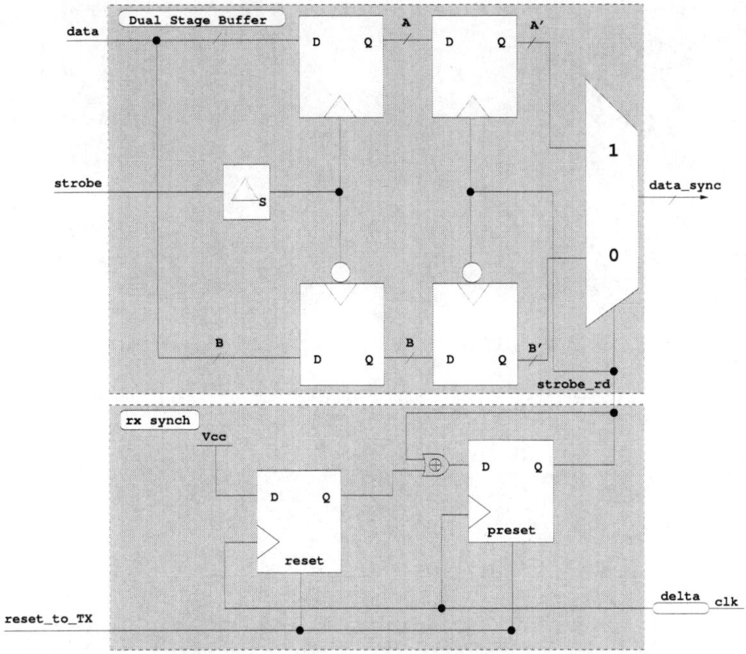

FIGURE 6.12: SIM-L architecture: receiver side (from [331] ©2008 IEEE).

6.3.1.1 The SIM-L Architecture

The top-level view of a Skew-Insensitive Mesochronous Link (SIM-L) is illustrated in Figure 6.11, based on the architecture proposed in [331]. The following details are also taken from [331] to show that with GALS technology architecture design needs to be aware of physical design issues.

δ indicates the clock skew. SIM-L is composed of two units: SIM-L Transmitter (TX) and SIM-L Receiver (RX). Link output (*data_sync*) is a copy of the incoming data synchronized with the receiver local clock, thus absorbing the skew between the transmitter and the receiver modules. A TX/RX SIM-L pair supports communication in only one direction. If a full-duplex link is needed, two SIM-L TX/RX pairs must be instantiated, one for each direction.

Figures 6.12 and 6.13 show the SIM-L RX and SIM-L TX architectures, respectively.

During the steady-state operation, the SIM-L TX module generates a signal (strobe) toggling at half the clock frequency. The strobe signal is routed together with transmitted data and is used in the SIM-L RX module to synchronize them in the skewed clock domain.

SIM-L RX module is composed of an *rx_synch* block and a dual stage buffer. Like the corresponding *tx_synch* block in SIM-L TX, the *rx_synch* block produces a *strobe_rd* signal synchronous with the skewed clock on the

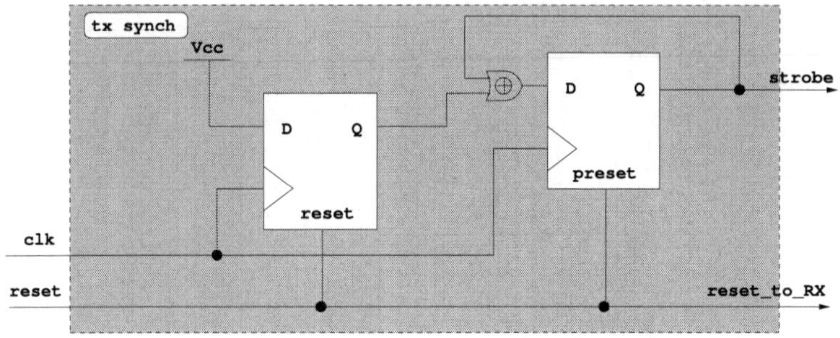

FIGURE 6.13: SIM-L architecture: transmitter side (from [331] ©2008 IEEE).

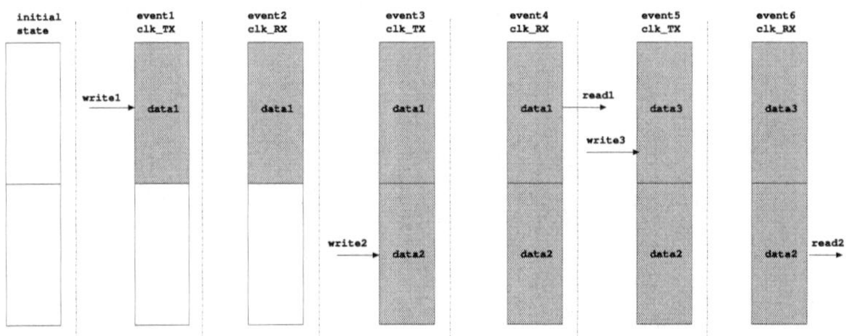

FIGURE 6.14: SIM-L architecture: behavior of the 2-stage register. Registers A/A' and B/B' are collapsed into a single abstract register (from [331] ©2008 IEEE).

receiver side and toggling at half the clock frequency. SIM-L guarantees that data are sampled when they are stable, thanks to the dual stage buffer structure in Figure 6.12, which is written by the transmitter (*strobe*) and is read by the receiver (*strobe_rd*). Incoming data, which are synchronous with the transmitter clock, are alternatively stored in registers A and B. Data stored in A and B are then read from registers A' and B' and are alternatively multiplexed toward the output. Figure 6.14 shows the "ping-pong" operating principle of the two-stage buffer. The key to correct data transfer during the steady-state operation is in reading or writing both the A/A' and B/B' registers without causing metastability. This is achieved by ensuring that the sampling edges of *strobe_rd* occur enough time later than the corresponding sampling edges of *strobe*. For instance, this happens when *strobe* and *strobe_rd* have opposite phases. In general, the valid phase relationships between *strobe_rd* and *strobe* are defined by a skewing window whose width is equal to the master clock period T_{clock}, as showed in Figure 6.15. Correct SIM-L operation for any skew δ is ensured when the skewing window is placed across the strobe falling edge.

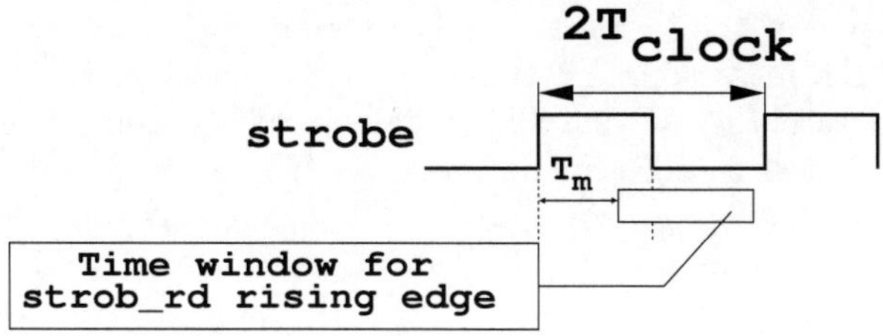

FIGURE 6.15: SIM-L architecture: phase relationship between strobe and strobe_rd signals. The rising edge of strobe_rd may fall in any point within the temporal interval denoted by the rectangle. It depends on the amount of skew between the transmitter clock and the receiver clock (from [331] ©2008 IEEE).

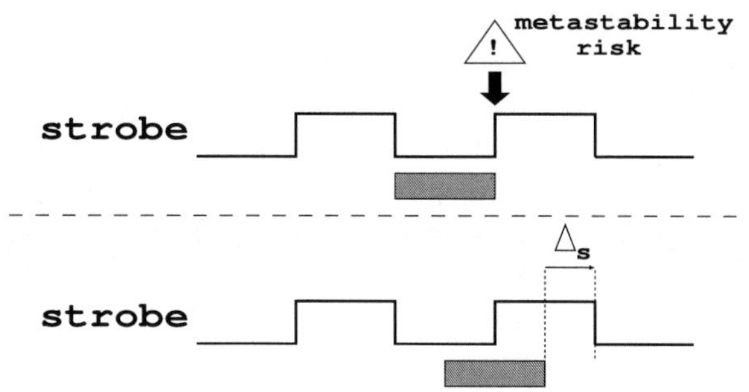

FIGURE 6.16: SIM-L architecture: relative position of the strobe_rd skewing window (a) without delay Δs and (b) with delay. In these scenarios, FFs in the tx_synch and rx_synch blocks are initialized with opposite values (from [331] ©2008 IEEE).

The condition on T_m is expressed as follows, where Tcq is the propagation delay of the A/B registers:

$$T_{cq} + T_{setup} < T_m < T_{clock} \tag{6.1}$$

The sizing of the delay Δs on the strobe line in Figure 6.12 is fundamental to a proper SIM-L operation. Such a delay guarantees that the $Tsetup$ rule for registers A/B is not violated. In fact, both data and strobe signals are generated synchronously with the transmitter clock and they are routed together toward the receiver. Since strobe is used to sample data in the A/B registers, it must be delayed of $\Delta s \geq Tsetup$ to give the data signal enough time to become stable. To this aim, a cascade of a few standard-cell buffers or inverters is enough since a fine-tuning of Δs is not required.

Bootstrap conditions are crucial to ensuring a correct steady-state operation. When the reset condition is removed in the *tx_synch* block (see Figure

6.13), the output of the first FF is set, thus enabling the toggle of strobe signal at half the master clock frequency. *Strobe* and *strobe_rd* should toggle with opposite phases so that the time interval between two consecutives *strobe/strobe_rd* rising (falling) edges is enough to allow FFs data sampling outside the metastability window ($T_{setup} + T_{hold}$). To this aim, the second FF in the *rx_synch* block is initialized to an opposite value (1) with respect to the second FF in *tx_synch* (initialized to 0). However, a successful operation strongly depends on the delay Δs. Figure 6.16 shows two different scenarios for the skewing window when the FFs in the *tx_synch* and *rx_synch* blocks are initialized with opposite values. Consider the scenario in Figure 6.16(a), where no delay on the strobe line is present. The skew amount between the two copies of the clock signal is not known a priori, so it may happen that the *strobe_rd* rising edge falls on the right boundary of the skewing window; this situation may cause metastability. On the contrary, in the scenario in Figure 6.16(b), the risk is avoided thanks to the delay Δs that puts the skewing window in a safe position.

The latency introduced by SIM-L amounts to two clock cycles due to the cascade of A/A' and B/B' registers. Since the SIM-L data path is mainly a cascade of registers, no bottlenecks are introduced on the maximum achievable clock frequency. The line delay (i.e., the signal propagation delay from the transmitter to the receiver) that SIM-L tolerates is around 1 clock cycle; in fact, if $\Delta s = T_{setup}$, then T_m in Figure 6.15 is $T_{clock} - T_{setup}$. Note that the line delay adds up to Δs, thus shifting ahead the strobe signal and pushing its rising edge near the one of *strobe_rd*. If strobe gets delayed by T_{clock}, metastability risks are possible (a scenario similar to the one in Figure 6.16(a)). In order to keep the SIM-L operation safe, the line length may add delay only up to $T_{clock} - 2T_{setup}$; the factor 2 keeps a safe margin around the left boundary of the skewing window.

6.3.1.2 The Rotating Synchronizer

Another kind of synchronizer is now presented, which is based on a similar ping-pong buffering scheme but features a different front-end architecture at the receiver side. The scheme was first presented in [100], but its adaptation to the network-on-chip domain was illustrated in [126] and [183]. We will move from the architecture variant presented in [183], which follows the concept scheme illustrated in Figure 6.17, and present its evolution toward a phase detector-free scheme [190].

The circuit receives as its inputs a bundle of NoC wires representing a regular NoC link, carrying data and/or flow control commands, and a copy of the clock signal of the sender. Since the latter wire experiences the same propagation delay as the data and flow control wires, it can be used as a strobe signal for them. The circuit is composed by a front-end and a back-end. The front-end is driven by the incoming clock signal, and strobes the incoming data and flow control wires onto a set of parallel latches in a rotating fashion,

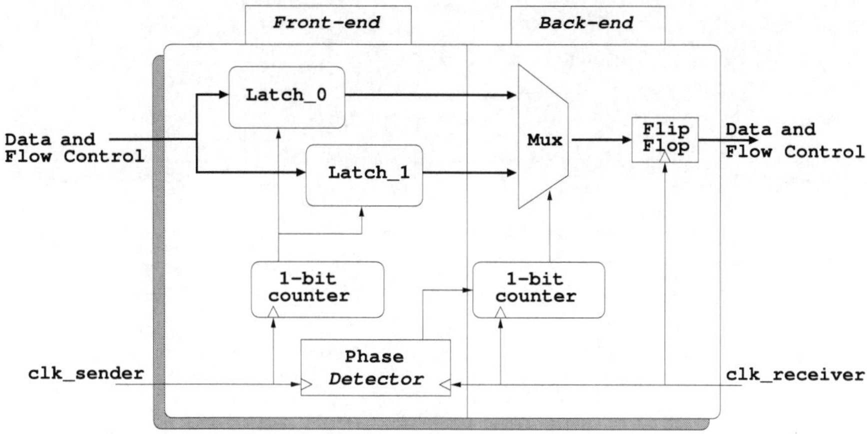

FIGURE 6.17: Baseline rotating synchronizer architecture with phase detector.

based on a counter. The back-end of the circuit leverages the local clock, and samples data from one of the latches in the front-end thanks to multiplexing logic, which is also based on a counter. The rationale is to temporarily store incoming information in one of the front-end latches, using the incoming clock wire to avoid any timing problem related to the clock phase offset. Once the information stored in the latch is stable, it can be read by the target clock domain and sampled by a regular flip-flop.

Counters in the front-end and back-end are initialized upon reset, after observing the actual clock skew among the sender and receiver with a phase detector, so as to establish a proper offset. The phase detector only operates upon the system reset, but given the mesochronous nature of the link, its findings hold equally well during normal operation. We will see shortly hereafter how to remove the phase detector.

[183] claims that it is always possible to choose a counter setup so that the sampling clock edge in the back-end captures the output of the latches in a stable condition, even accounting for timing margin to neutralize jitter. Therefore, no more than 2 latches in parallel are needed in the front-end for short-range (i.e., single cycle) mesochronous communication with this scheme.

However, other considerations may suggest a different choice. In particular, by increasing the number of input latches by one more stage, it becomes possible to remove the phase detector (see the new architecture in Figure 6.18). A third latch bank in fact keeps latched data stable for a longer time window and to find a unique and safe bootstrap configuration (i.e., counters initialization) that turns out to be robust in any phase skew scenario.

Postsynthesis simulations confirmed that the circuit works properly when sweeping the clock skew from –360° to +360°. At regime, the output multiplexer always selects the output of the latch bank preceding the bank that is

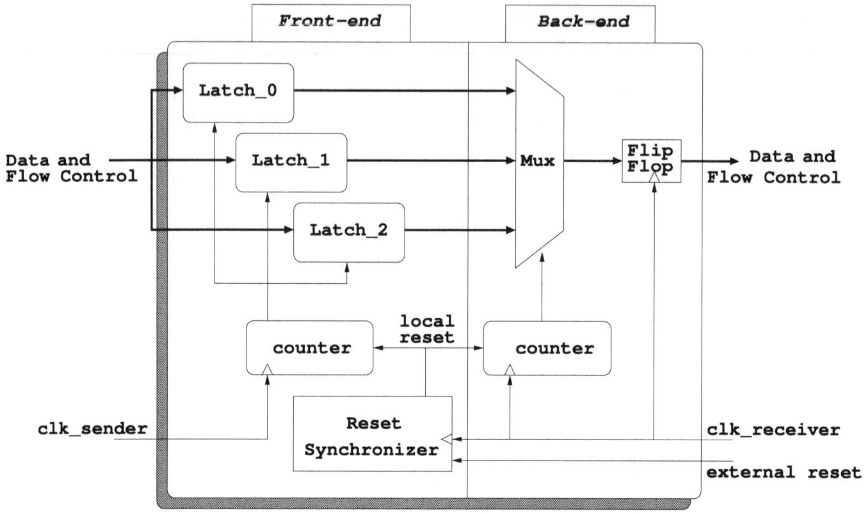

FIGURE 6.18: The loosely coupled rotating synchronizer.

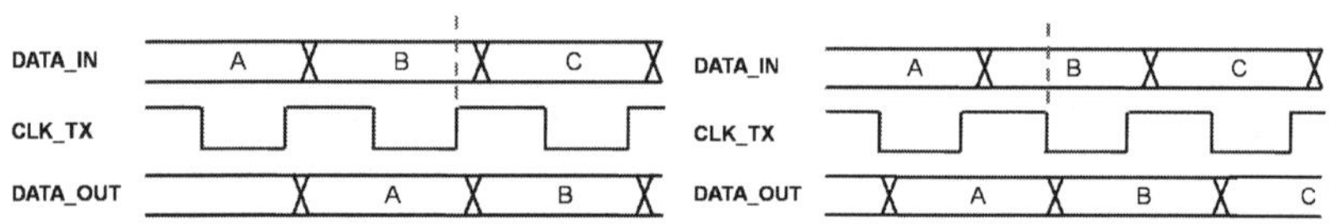

(a) Sampling on clock rising edge. (b) Sampling on clock falling edge

FIGURE 6.19: Need for timing margins at the synchronizer front-end.

being enabled by the front-end counter. Rotating operation of both front- and back-end counters preserves this order.

In most systems, the reset signal coming into the chip is an asynchronous input. Therefore, reset de-assertion should be synchronized in the receiver clock domain. In fact, if a reset removal to a flip-flop occurs close to the active edge of its clock, flip-flops can enter a metastable state. A brute-force synchronizer (available in some new technology libraries as a standard cell) is used for reset synchronization with the receiver clock. Now the problem arises about how to reset the front-end. Typically, a reset signal can be sent by the upstream switch together with data and transmitter clock. In the architecture of Figure 6.18, instead, metastability in the front-end is prevented by delaying the strobe generation in the upstream switch by one clock cycle after reset de-assertion. This way, on the first strobe edge, the receiver synchronizer is already reset. This strobe generation delay is compliant with network packet injection delay after reset.

The transmitter clock signal is used as the strobe signal in this rotating synchronizer architecture (see Figure 6.19(a)), although an ad-hoc strobe signal at half the clock frequency is feasible too. Some design tricks can be implemented to enforce larger timing margins for safe input data sampling at the synchronizer front-end. In fact, the transmitter clock signal has to be processed at the receiver in order to drive the latch enable signals. In actual layouts, this processing time adds up to the routing skew between data and strobe and to the delay for driving the latch enable high-fanout nets. As a result, the latch enable signal might be activated too late, and the input data signal might have already changed.

In order to make the synchronizer more robust to these events, it can be ensured that input data sampling occurs in the middle of the clock period and is triggered by the falling edge of the transmitter clock (see Figure 6.19(b)). In fact, a switching latch enable signal opens the sampling window of the next latch during the rising edge of the transmitter clock, and closes it during the falling edge. As a result, closing of the latch transparency window has a margin of half clock cycle to occur.

6.3.1.3 Loose Coupling with the NoC Switch

The typical approach for synchronizer insertion into a NoC link consists of placing it in front of the downstream switch, as done in [126]. This approach will be hereafter denoted as *loose coupling* of the synchronizer with the NoC switch. However, this intuitive choice comes with its own disadvantages. First, it increases the link latency due to the timing overhead for synchronization. In turn, the increased link latency forces the designer to increase the size of the downstream input buffer because of flow control concerns. In fact, assume the simple stall/go flow control protocol is used. The forward-propagating "valid" signal flags data availability, while the backward-propagating "stall" signal instructs the upstream switch to stop sending because, for instance, the destination buffer is full. If the stall command is not received within 1 clock cycle, multiple flits would be in flight that cannot be stored at the destination buffer. Therefore, it is important that the size of this buffer reflects the round-trip latency of the link. When the mesochronous synchronizer in Section 6.3.1.2 was applied to the xpipesLite NoC architecture, the input buffer of the switches was enlarged from 2 slots to 4 slots [183], which is a significant synchronizer-induced overhead.

In addition to this, the synchronizer itself comes with its own buffering needs. Let us recall that buffers are the most power-hungry components in NoCs, therefore synchronizer buffers further contribute to drain the tight power budgets.

When flow control is considered, signaling in a NoC link becomes bi-directional, as in the case of the stall/go protocol. For this reason, synchronizers need to be inserted not only in the forward path (synchronizing data and forward propagating flow control signals), but also in the backward path (synchronizing backward flow control signals). The stall signal is an example of such a backward path, and an additional 1-bit synchronizer needs to be

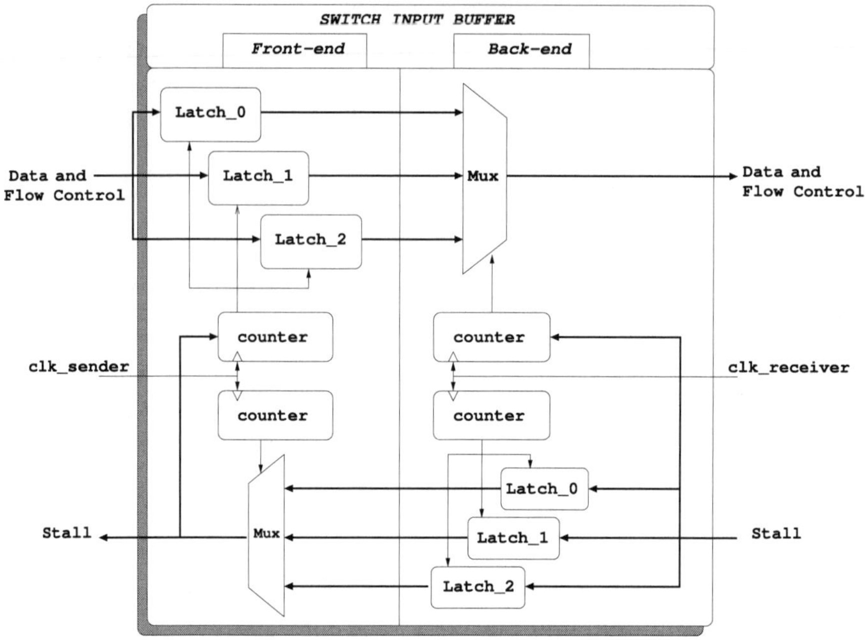

FIGURE 6.20: Tightly coupled mesochronous synchronizer (from [191] ©2010 IEEE).

instantiated in front of the upstream switch, before the stall signal is sampled by the flow control logic.

Finally, adding one more NoC building block (the synchronizer) in a NoC link breaks the modularity of NoC design and gives rise to placement constraints (of the synchronizer close to its reference switch) that need to be enforced during physical synthesis.

6.3.1.4 Tight Coupling with the NoC Switch

Given the large cost of a loosely coupled design style, we might think of bringing the synchronizer deeper into the downstream switch. As a case study of this architecture optimization, let us consider the rotating synchonizer illustrated in Section 6.3.1.2. A first optimization consists of omitting the FF in the synchronizer back-end, since sampling takes place in the switch input buffer as well.

However, a tighter integration between the synchronizer and the switch architecture is possible. In particular, we notice that the latch enable signals in the synchronizer front-end could be conditioned with backward-propagating flow control signals (the stall/go in this case), so to exploit input latches as useful buffer stages and not just as an overhead for synchronization. This means that in case of a stall command driven by the downstream switch, data could be frozen in the synchronizer front-end, while in the implementation of Figure 6.18 they are propagated all the way to the switch input stage and only once there they are stored. Should we implement this modification, we would realize that the synchronizer and the switch input buffer perform exactly the same overlapped task: data storage when further downstream propagation is forbidden by the flow control. In addition to this, the synchronizer performs also mesochronous synchronization. Hence, the straightforward consequence is that the switch input buffer could be omitted and it could be completely replaced by the synchronizer. This is what has been done in the upper part of Figure 6.20.

The synchronizer output now directly feeds the switch arbitration logic and its internal crossbar, thus materializing the tight coupling concept of the mesochronous synchronizer with the switch architecture. The ultimate consequence is that the mesochronous synchronizer becomes the actual switch input stage, with its latching banks serving both for performance-oriented buffering and synchronization. A side benefit is that the latency of the synchronization stage in front of the switch is removed, since now the synchronizer and the switch input buffer coincide. The buffering overhead in the switch input buffer because of flow control is also removed accordingly.

The main change required for the correct operation of the new architecture is to bring the stall/go flow control signal to the front-end and back-end counters of the synchronizer, in order to freeze their operation in case of a stall. While this signal is already in synch with the back-end counter, it should be synchronized with the transmitter clock before feeding the front-end counter. The backward-propagating stall/go is then directly synchronized with the transmitter clock available in the front-end by means of a similar but smaller (1-bit) synchronizer.

This paves the way for a relevant optimization. The output of the 1-bit synchronizer could be brought not only to the front-end counter, but also directly to the upstream switch without the need for any further synchronizer since synchronization has already been performed in the downstream switch. The ultimate result is the architecture illustrated in Figure 6.20. For this architecture solution, only 3 latching banks are needed in the synchronizer front-end, since link latency has been reduced to a minimum. In practice, only 1 slot more than the fully synchronous input buffer, when the xpipesLite architecture in [183] is considered. With the same architecture, we recall from Section 6.3.1.3 that when a loosely coupled synchronizer is inserted, the switch input buffer consisted of 4 slots in addition to the 3-slot synchronizer latches. Also, the tightly coupled synchronizer makes the mesochronous NoC design fully modular like the synchronous one, since no external blocks to the switches have to be instantiated for switch-to-switch communication.

The link latency improvements of a tightly coupled design style are relevant. In the xpipesLite NoC architecture [305], a fully synchronous switch takes 1 cycle in the upstream link and 1 cycle in the switch. The tightly coupled switch takes from 1 to 3 clock cycles to cross the same path, depending

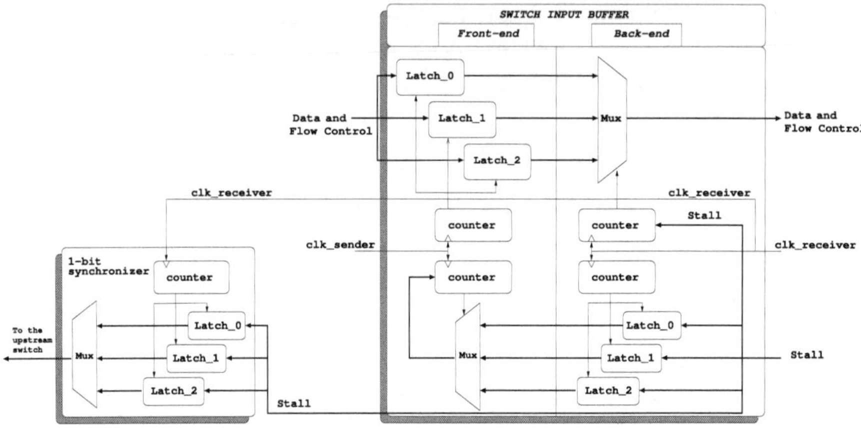

FIGURE 6.21: Hybrid mesochronous synchronizer (from [191] ©2010 IEEE).

on the negative or positive skew between transmitter and receiver clocks. Interestingly, the link is always crossed in 1 cycle, therefore there is no need of increasing the buffer size of the switch input port since the round-trip latency is unaffected.

Unfortunately, the way the tightly coupled architecture synchronizes the stall signal gives rise to a severe constraint on the link round trip time. In fact, the transmitter clock used for stall synchronization at the downstream switch has already undergone a link delay T_{link}, and the stall signal itself takes another link delay to go back to the upstream switch:

$$2 \cdot T_{link} + T_{counter} + T_{mux} + T_{setup} \leq T_{clock} \tag{6.2}$$

Whenever this constraint cannot be met, then the tightly coupled architecture proves flexible enough to provide an alternative solution that meets the constraints at a small cost in area and modularity. The solution, denoted as *Hybrid*, is illustrated in Figure 6.21. In practice, the simple 1-bit synchronizer can be replicated in front of the upstream switch, thereby breaking the round-trip dependency of the timing. The hybrid architecture now exposes a new timing path going from the switch arbiter to the upstream switch through the stall signal, which is a one-way path, not a round-trip one.

Next, the implications of a tightly vs. loosely coupled design style on physical design parameters will be evaluated. The xpipesLite NoC switch and the rotating synchronizer are chosen as an experimental platform, although the considerations that follow are not limited to them.

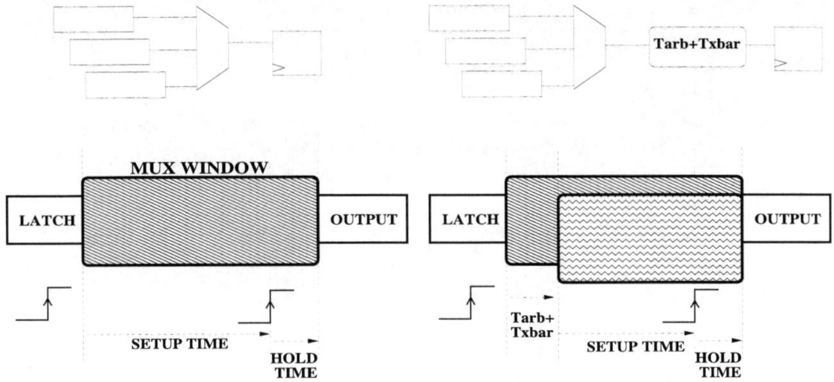

(a) Loosely coupled synchronizer. (b) Tightly coupled synchronizer.

FIGURE 6.22: Basic mechanisms affecting skew tolerance (from [191] ©2010 IEEE).

6.3.1.5 Skew Tolerance

Skew tolerance depends on the relative alignment of data arrival time at latch outputs, multiplexer (MUX) selection window, and sampling edge in the receiver clock domain. A few basic definitions help to assess the interaction among these parameters in determining skew tolerance.

For the loosely coupled rotating synchronizer, such definitions are pictorially illustrated in Figure 6.22(a). During the *MUX window*, data at latch outputs is selected for forwarding to the sampling flip-flop in the switch input port. Its duration closely follows that of the clock period. Sampling occurs on the next rising edge of the receiver clock inside the MUX window. We denote the time between the starting point of the MUX window and such sampling instant as the *Setup time*. Conversely, after an *Hold time* since the rising edge of the clock the MUX window terminates. This is the time required by the back-end counter to switch the multiplexer selection signals.

When we consider the tightly coupled architecture of the rotating synchronizer, then the same metrics are taken at the switch output port rather than at the multiplexer output of the synchronizer (Figure 6.22(b)). Therefore, the starting time of the *MUX window* is delayed due to the worst-case timing path between the synchronizer output and the switch output port, which includes the arbitration time, crossbar selection time, and some more combinational logic delay for header processing. At the same time, the sampling rising edge of the receiver clock remains unaltered, therefore the ultimate effect is a shortening of the *Setup time* for the tightly integrated mesochronous switch architecture.

Figure 6.23(a) quantifies these timing margins for the loosely coupled switch architecture on a 65 nm industrial technology library. Results are referred to a 2×2 switch working at 660 MHz after place&route. The x-axis

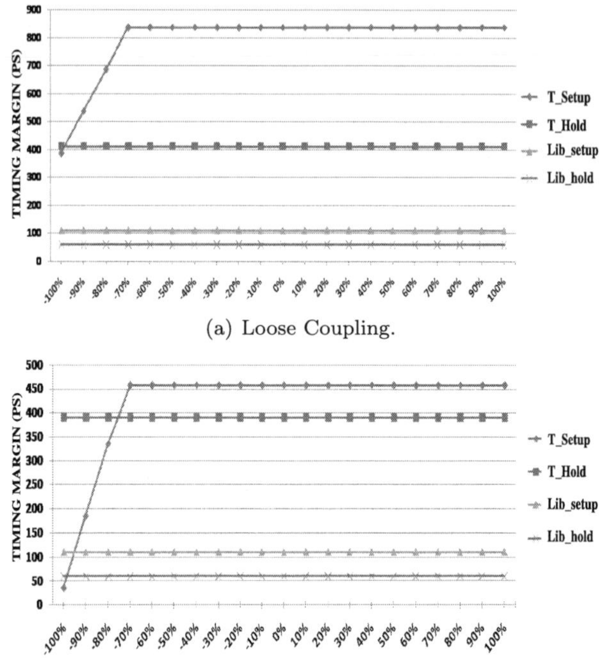

(a) Loose Coupling.

(b) Tight Coupling

FIGURE 6.23: Setup and hold times for the loosely vs. tightly coupled architectures as a function of clock phase offset (from [191] ©2010 IEEE).

reports negative and positive values of the skew, expressed as percentage of the clock period. Setup and hold times have been experimentally measured by driving the switch under test with a clocked testbench, by inducing phase offset with the switch clock and by monitoring waveforms at the switch. The connecting link between the testbench and the switch is assumed to have zero delay. A positive skew means that the clock at the switch is delayed with respect to the one at the testbench. The figure also compares setup and hold times with the minimum values required by the technology library for correct sampling (denoted *Lib_setup* and *Lib_hold*).

First of all, we observe that both times are well above the library constraints, thus creating some margin against variability. For the whole range of the skew, the hold time stays the same. The result is relevant for positive skew, since its effect is to shift the *MUX window* to the right, close to the region where latch output data switches. However, the stability window of the latch output data is long enough to always enable correct sampling of stable data before the point in time where it switches.

In contrast, a negative skew causes the *MUX window* to shift to the left, therefore as the negative skew grows (in absolute values) the latch output data

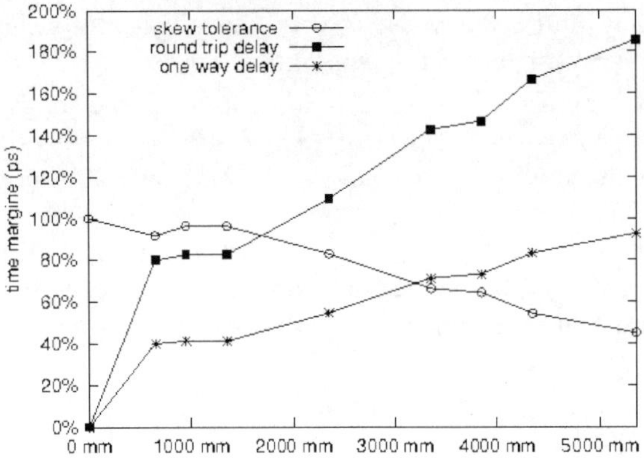

FIGURE 6.24: Effect of link length on skew tolerance and feasibility of the tightly coupled rotating synchronizer (from [191] ©2010 IEEE).

ends up switching inside the *MUX window*, which corresponds to the knee of the setup time in Figure 6.23(a). From there on, the switching transient of data becomes closer to the sampling edge of the receiver clock and correct sampling can be guaranteed until the setup time curve equals the *Lib_setup* one. However, even for –100% skew, synchronizer operation is correct.

Figure 6.23(b) illustrates the same results for the tightly coupled mesochronous switch. As anticipated above, the setup time is decreased by 370 ps, corresponding to the time for arbitration, crossbar selection, and shifting of routing bits. Hence, the margins against process variations are reduced and could be restored by synthesizing the switch at a higher target frequency than the one it will be actually working at (power-reliability trade-off). Interestingly, the knee of the setup time occurs for the same value of the negative skew, in that the switching instant of the latch output data enters the *MUX window* at exactly the same point in time. The ultimate implication is that the tightly coupled synchronizer cannot work properly with –100% skew, since the crossing point with the *Lib_setup* occurs at around –95%. This is, however, a marginal degradation of the skew tolerance.

6.3.1.6 Interconnect Delay

The impact of the interconnect delay on the setup time curves in Figures 6.23(a) and 6.23(b) is to shift them to the right. In fact, such a delay implies that new data appears later at the output of synchronizer latches. As a consequence, the knee of the setup time will occur earlier, when moving from a condition of null skew towards more negative skews, hence the maximum tolerated negative skew will be reduced as well. How fast the curve moves to the right and negative skew tolerance is degraded depends on the link length, on the target frequency and on how effectively the physical synthesis tool enforces the timing constraint on that specific link.

Figure 6.24 shows how skew tolerance is impacted by the length of the upstream link in the *tightly coupled rotating synchronizer architecture.* It is derived from a postlayout analysis of two connected 2×2 switches synthesized at a target speed of 660 MHz. Skew tolerance of the downstream switch is analyzed. It should be observed that by playing with the input/output delay constraints for switches during the synthesis process, we were able to demand 20% more performance to the switch-to-switch wires to conservatively meet the constraint in Equation 6.2.

Each point represents the percentage of negative skew at which the setup time curve intersects the *Lib_setup* one, i.e., it shows how the maximum tolerable negative skew degrades with link length. Not only interconnect delay, but also clock propagation time and a small combinational logic delay at the output of the upstream switch contribute to degrade negative skew robustness of the downstream switch. The irregular decay of the skew tolerance curve depends on the interconnect delay enforced by the physical synthesis tool for each value of the link length. Skew tolerance remains always above 50% for links of less than 5 mm.

However, correct operation of the tightly coupled synchronizer depends not only on skew tolerance, but also on the satisfaction of the constraint in Equation 6.2. The timing path indicated by that equation is displayed in Figure 6.24 as *Round-Trip Delay.* The intersection of this curve with a y value of 100% indicates that from there on Equation 6.2 is violated, and the stall signal cannot reach the upstream switch within one clock cycle. In practice, a link length of 2 mm is the *feasibility range* for the tightly coupled synchronization architecture in this experimental setting. For longer links, the hybrid architecture might be the right choice, and its feasibility is also illustrated in Figure 6.24 by reporting the *One Way Delay* constraint. This solution leaves the skew tolerance of the tightly coupled synchronizer unaffected but extends the feasibility range to around 5.5 mm at the small area cost of an additional 1-bit synchronizer.

6.3.1.7 Mesochronous NoCs

It has been demonstrated in [102] that as technology keeps scaling to sub-65 nm technologies it is more and more likely that the critical path moves from the switch internal microarchitecture to the switch-to-switch links due to the reverse scaling of on-chip interconnects. Therefore, it is essential to assess the pressure that each synchronization interface puts on the propagation delay of such links.

The following experiment (from [191]) assesses this property. In practice, two 5×5 xpipesLite switches (ideally extracted from the center of a 2-D mesh) are considered, connected with each other in three architectural variants: tight coupling, loose coupling, and hybrid coupling with the rotating synchroniz-

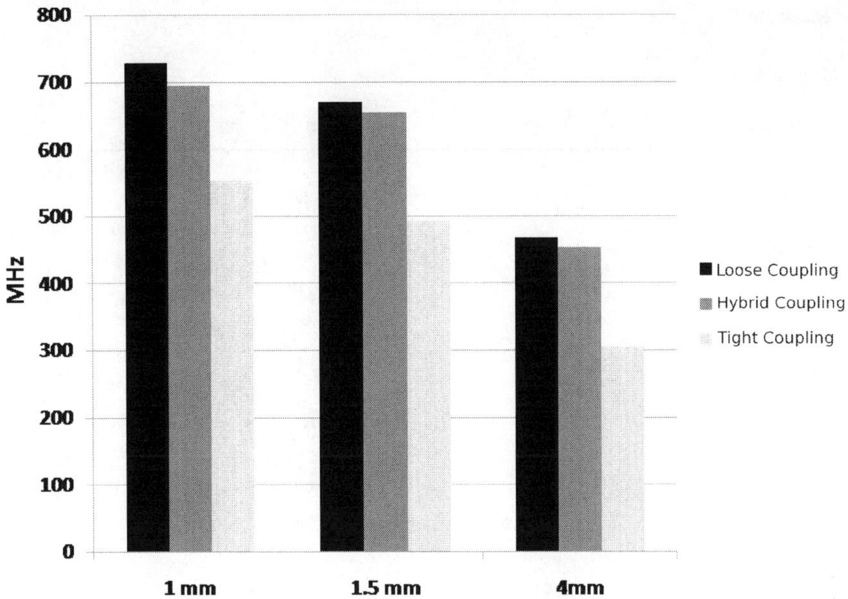

FIGURE 6.25: Post-P&R critical path for the different synchronization schemes.

ers. These switches are synthesized with a very tight timing constraint (1.25 GHz), so that after place&route the critical path for all architectures will be certainly in the switch-to-switch link. The comparative maximum post-P&R speed denotes a low criticality of link delay, while the lowest performance denotes the need to place ad-hoc timing constraints on the links during physical synthesis or the suitability of this architecture for lower operating speeds.

Critical path results as a function of the length of interswitch links are illustrated in Figure 6.25. The round-trip dependency of the critical path in the tightly coupled architecture determines a maximum speed that is on average 28% lower than the one of the loosely coupled architecture. In contrast, the figure proves that the hybrid architecture can easily approach the same maximum performance of the loosely coupled one but at a much lower area. In practice, different circuits in the two architecture variants end up contributing the same delay. Stall generation delay, link propagation delay, and driving delay of the control logic of a 1-bit synchronizer in the hybrid scheme almost equal the delays in the ouput buffer, in the link, and in the control logic of the external synchronizer in the loosely coupled scheme. The hybrid solution is an effective way of relieving link timing constraints of the tightly coupled architecture.

The next experiment (again from [191]) proves the ability of tightly coupled

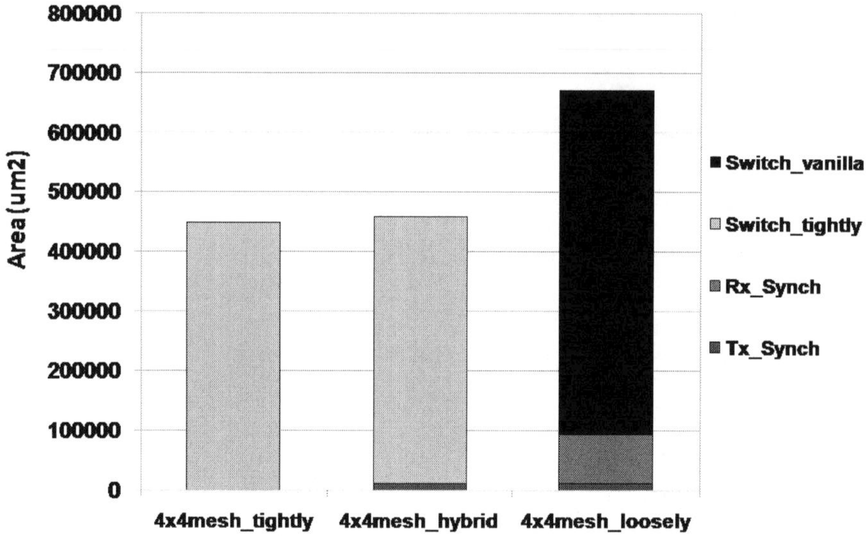

FIGURE 6.26: Area of 4 × 4 2-D meshes (from [191] ©2010 IEEE).

and hybrid design platforms to build a cost-effective GALS NoC. A 4 × 4 2-D mesh topology is designed in the tightly coupled, loosely coupled and hybrid architecture variants. The target speed during physical synthesis for the three topologies is the same, in order to assess system-level area savings with the tightly coupled approach. In light of the study on the critical paths, switches for the loosely coupled and hybrid schemes have been synthesized at a 25% lower target speed with respect to their maximum performance, thus largely saving area, since they have to match the speed of the slowest architecture. Switches are placed 1 mm apart from each other.

Timing closure of the three platforms was achieved at 600 MHz. Post-P&R area figures are compared in Figure 6.26. Clearly, the hybrid and the tightly coupled variants provide a 40% area saving over the loosely coupled one. This latter has to implement 2 more slots in the input buffer, which makes the area of the vanilla switches (i.e., switches without any support for mesochronous synchronization) larger than that of the tightly coupled ones. The hybrid architecture limits its area overhead with respect to the tightly coupled one because of the low footprint of 1-bit external synchronizers on the *stall* wires. In contrast, external synchronizers at receiver switches in the loosely coupled architecture consume about 12% of total mesh area.

Although mesochronous synchronizers can be used inside a network to make its CTS feasible and/or low power across a wide chip area, whenever a link is established between two separate voltage and frequency islands potentially operating at different frequencies, dual-clock FIFOs need to be used. The next sections are devoted to them.

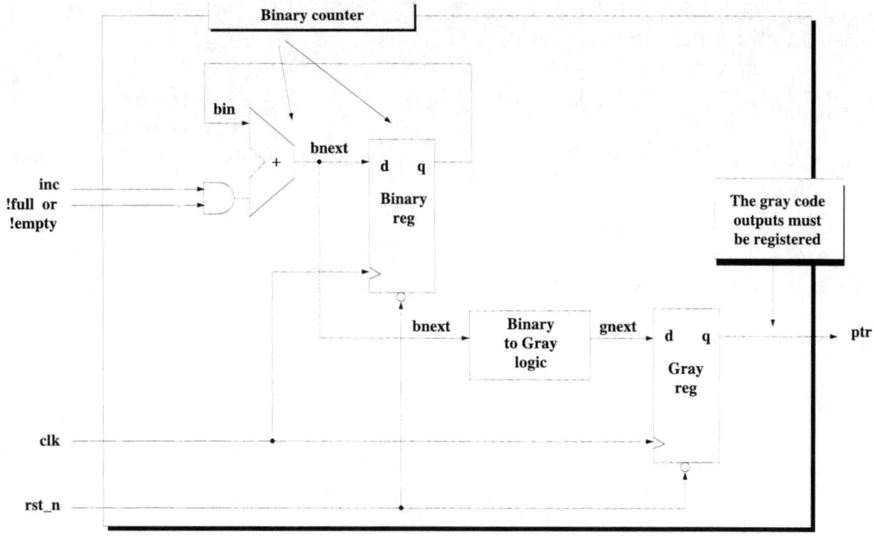

FIGURE 6.27: Dual n-bit Gray code counter.

6.3.2 Dual-clock FIFOs

Next, a review of architecture design techniques for Dual-Clock (DC) FIFOs will be illustrated. Since a wide range of design methods are embodied into two relevant DC-FIFO architectures reported in the open literature [60, 215], such architectures will be analyzed next.

6.3.2.1 Asynchronous Pointer Comparison

An asynchronous FIFO refers to a FIFO design where data values are written sequentially into a FIFO buffer using one clock domain, and the data values are sequentially read from the same FIFO buffer using another clock domain, where the two clock domains are asynchronous to each other. One common technique for designing an asynchronous FIFO is to use Gray [112] code pointers that are synchronized into the opposite clock domain before generating synchronous FIFO full or empty status signals [59]. An interesting and different approach to FIFO full and empty generation is to do an asynchronous comparison of the pointers and then asynchronously set the full or empty status bits.

One Gray code counter style uses a single set of flip-flops as the Gray code register with accompanying Gray-to-binary conversion, binary increment, and binary-to-Gray conversion [59]. A second Gray code counter style, the one described in this chapter and mutuated from [60], uses two sets of registers, one is a binary counter and the second is used to capture a binary-to-Gray converted value. The intent of this latter Gray code counter is to utilize the binary carry structure, simplify the Gray-to-binary conversion, reduce combinational

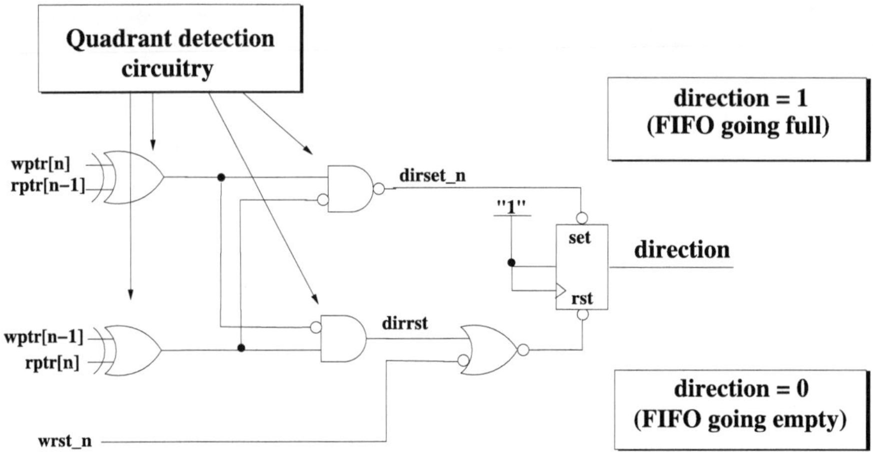

FIGURE 6.28: FIFO direction quadrant detection circuitry.

logic, and increase the upper frequency limit of the Gray code counter. The binary counter conditionally increments the binary value, which is passed to both the inputs of the binary counter as the next-binary-count value, and is also passed to the simple binary-to-Gray conversion logic, consisting of one 2-input XOR gate per bit position. The converted binary value is the next Gray-count value and drives the Gray code register inputs. Figure 6.27 shows the block diagram for an n-bit Gray code counter. This implementation requires twice the number of flip-flops, but reduces the combinational logic and can operate at a higher frequency.

As with any FIFO design, correct implementation of full and empty is the most difficult part of the design. There are two problems with the generation of full and empty. First, both full and empty are indicated by the fact that the read and write pointers are identical. Therefore, something else has to distinguish between full and empty. One known solution to this problem appends an extra bit to both pointers and then compares the extra bit for equality (for FIFO empty) or inequality (for FIFO full), along with equality of the other read and write pointer bits [59]. Another solution, reported in [60], divides the address space into four quadrants and decodes the two MSBs of the two counters to determine whether the FIFO was going full or going empty at the time the two pointers became equal. If the write pointer is one quadrant behind the read pointer, this indicates a "possibly going full" situation. When this condition occurs, the direction latch of Figure 6.28 is set. If the write pointer is one quadrant ahead of the read pointer, this indicates a "possibly going empty" situation. When this condition occurs, the direction latch of Figure 6.28 is cleared. This is the solution considered next.

The second, and more difficult, problem stems from the asynchronous nature of the write and read clocks. Comparing two counters that are clocked

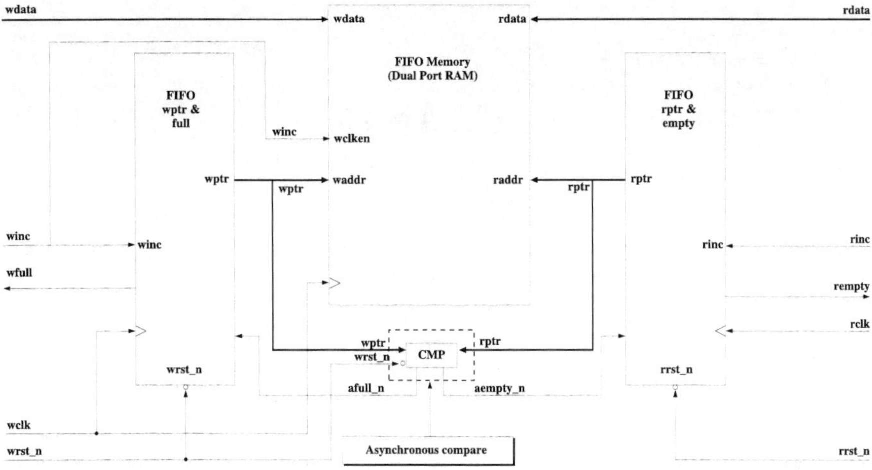

FIGURE 6.29: DC-FIFO architecture based on asynchronous pointer comparison.

asynchronously can lead to unreliable decoding spikes when either or both counters change multiple bits more or less simultaneously. The solution described in [60] and considered here uses a Gray count sequence, where only one bit changes from any count to the next. Any decoder or comparator will then switch only from one valid output to the next one, with no danger of spurious decoding glitches.

The block diagram for the resulting asynchronous FIFO is showed in Figure 6.29. The logic used to determine the full or empty status on the FIFO is the most distinctive feature of this architecture. The asynchronous comparison module is detailed in Figure 6.30.

aempty_n and *afull_n* are the asynchronously decoded signals. The *aempty_n* signal is asserted on the rising edge of an *rclk*, but is de-asserted on the rising edge of a *wclk*. Similarly, the *afull_n* signal is asserted on a *wclk* and removed on an *rclk*. The empty signal will be used to stop the next read operation, and the leading edge of *aempty_n* is properly synchronous with the read clock, but the trailing edge needs to be synchronized to the read clock. This is done in a two-stage synchronizer that generates *rempty*. The *wfull* signal is generated in the symmetrically equivalent way. It is not difficult to notice that the direction bit is set long before the FIFO is full and is therefore not timing-critical to the assertion of the *afull_n* signal, and vice versa.

When the write pointer is incremented starting from an empty condition (e.g., upon reset), then the FIFO pointers are no longer equal so the *aempty_n* signal is de-asserted, releasing the preset control of the *rempty* flip-flops. After two rising edges on *rclk*, the FIFO will de-assert the *rempty* signal. Because the de-assertion of *aempty_n* happens on a rising *wclk* and because the *rempty* signal is clocked by the *rclk*, the two-flip-flop synchronizer is required to re-

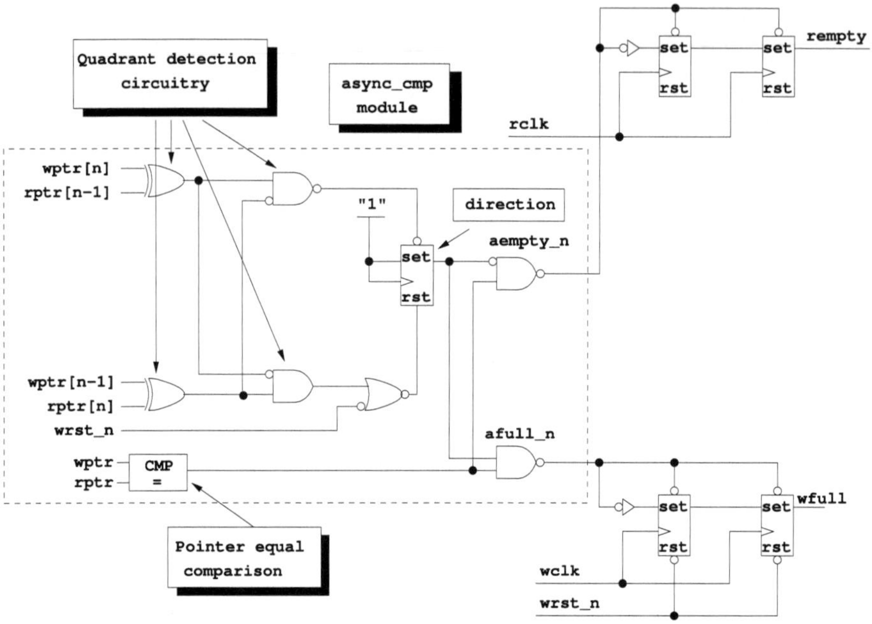

FIGURE 6.30: Asynchronous pointer comparison module.

move the metastability that could be generated by the first *rempty* flip-flop (see Figure 6.30).

When the *wptr* catches up to the *rptr* (and the *direction* bit is set), the $afull\_n$ signal presets the $wfull$ flip-flops. The $afull\_n$ signal is asserted on a FIFO-write operation and is synchronous to the rising edge of the *wclk*; therefore, asserting full is synchronous to the *wclk*. Then, a FIFO-read operation takes place and the *rptr* is incremented. At this point, the FIFO pointers are no longer equal so the $afull\_n$ signal is de-asserted, releasing the preset control of the $wfull$ flip-flops. After two rising edges on *wclk*, the FIFO will de-assert the $wfull$ signal. Because the de-assertion of $afull\_n$ happens on a rising *rclk* and because the $wfull$ signal is clocked by the *wclk*, the two-flip-flop synchronizer is required to remove metastability that could be generated by the first $wfull$ flip-flop capturing the inverted and asynchronously generated $afull\_n$ data input.

Using the asynchronous comparison technique described so far, there are critical timing paths associated with the generation of both the *rempty* and $wfull$ signals. The *rempty* critical timing path, showed in Figure 6.31, consists of (i) the $rclk-to-q$ incrementing of the *rptr*, (ii) comparison logic of the *rptr* to the *wptr*, (iii) combining the comparator output with the direction latch output to generate the *aempty_n* signal, (iv) presetting the *rempty* signal, (v) any logic that is driven by the *rempty* signal, and (vi) resultant signals meeting the setup time of any downstream flip-flops clocked within the *rclk* domain.

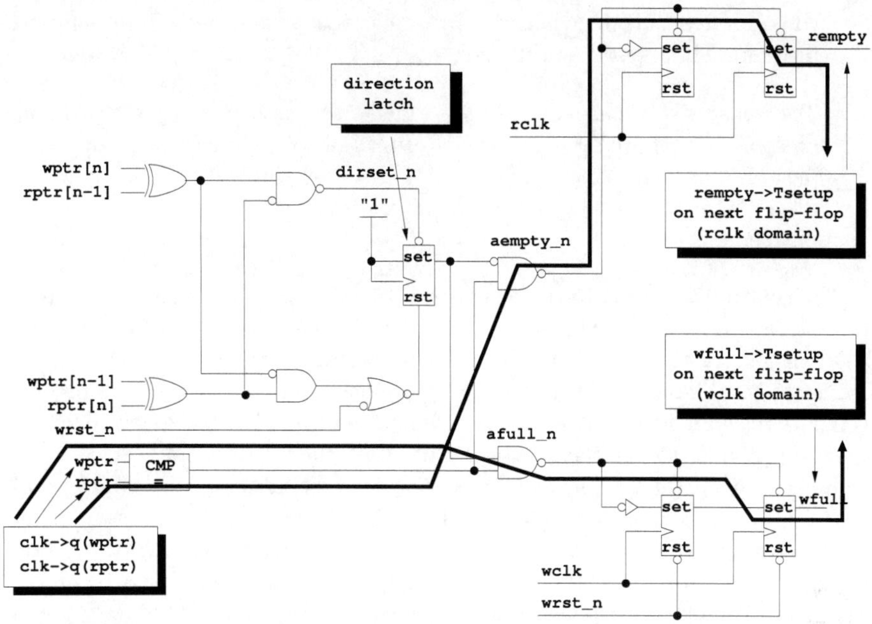

FIGURE 6.31: Critical timing paths for asserting *rempty* and *wfull*.

This critical timing path has a symmetrically equivalent critical timing path for the generation of the *wfull* signal, also showed in Figure 6.31.

Asynchronous comparison between pointers and full/empty synchronization in the opposite clock domain come with their own inconveniences. Since *aempty_n* is started by one clock and terminated by the other, it has an undefined duration, and might even be a runt pulse. A runt pulse is a Low-High-Low signal transition where the transition to High may or may not pass through the logic-1 threshold level of the logic family being used.

If the *aempty_n* control signal is a runt pulse, there are four possible scenarios that should be addressed:

- the runt signal is not recognized by the *rempty* flip-flops and *empty* is not asserted. This is not a problem.

- The runt pulse might preset the first synchronizer flip-flop, but not the second flip-flop. This is highly unlikely, but would result in an unnecessary, but properly synchronized *rempty* output, that will show up on the output of the second flip-flop one read clock later. This is not a problem.

- The runt pulse might preset the second synchronizer flip-flop, but not the first flip-flop. This is highly unlikely, but would result in an unnecessary, but properly synchronized *rempty* output (as long as the empty critical timing is met), that will be set on the output of the second flip-flop until the next read clock, when it will be cleared by the zero from the first flip-flop. This is not a problem.

- The most likely case is that the runt pulse sets both flip-flops, thus creating a properly synchronized *rempty* output that is two read-clock periods long. The longer duration is caused by the two-flip-flop synchronizer (to avoid metastable problems as described below). This is not a problem.

The runt pulse cannot have any effect on the synchronizer data-input, since an *aempty_n* runt pulse can only occur immediately after a read clock edge, thus long before the next read clock edge (as long as critical timing is met). The *aempty_n* signal might also stay high longer and go low at any moment, even perhaps coincident with the next read clock edge. If it goes low well before the set-up time of the first synchronizer flip-flop, the result is like the last item discussed above. If it goes low well after the set-up time, the synchronizer will stretch *rempty* by one more read clock period. If *aempty_n* goes low within the metstability-catching set-up time window, the first synchronizer flip-flop output will be indeterminate for a few nanoseconds, but will then be either high or low. In either case, the output of the second synchronizer flip-flop will create the appropriate synchronized *rempty* output.

Now, what happens if the write clock de-asserts the *aempty_n* signal coincident with the rising *rclk* on the dual synchronizer? The first flip-flop could go metastable, which is why there is a second flip-flop in the dual synchronizer. But the removal of the setting signal on the second flip-flop will violate the recovery time of the second flip-flop. Will this cause the second flip-flop to go metastable? The authors of [60] do not believe this can happen because the preset to the flip-flop forced the output high and the input to the same flip-flop is already high, which they believe is not subject to a recovery time instability on the flip-flop.

Finally, can a runt-preset pulse, where the trailing edge of the runt pulse is caused by the *wclk*, preset the second synchronizer flip-flop in close proximity to a rising *rclk*, violate the preset recovery time and cause metastability on the output of the second flip-flop? The answer is no as long as the *aempty_n* critical timing path is met. Assuming that critical timing is met, the *aempty_n* signal going low should occur shortly after a rising *rclk* and well before the rising edge of the second flip-flop, so runt pulses can only occur well before the rising edge of an *rclk*.

Again, symmetrically equivalent scenarios and arguments can be made about the generation of the $wfull$ flag.

6.3.2.2 Token Ring-based Pointers

The second DC-FIFO architecture that this chapter presents is taken from [215] and features two fundamental differences with respect to the one presented above: pointers are synchronized from one clock domain to another

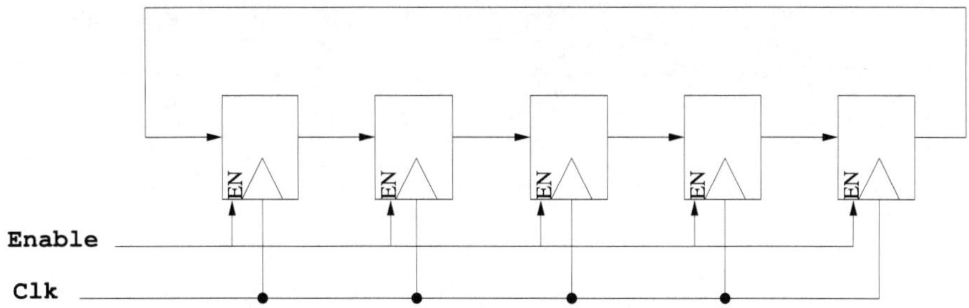

FIGURE 6.32: Token ring (from [215] ©2007 IEEE).

before generating full and empty flags, and a novel token ring-based encoding method is used to implement finite state machines for read and write pointers, hence replacing Gray coding.

A token ring is a chain of nodes interconnected in a circular manner that contains tokens. It can be described with N cascaded registers (with enable signal), like a cyclic shift-register. Figure 6.32 shows an example of a token ring with 5 registers. If the *enable* signal is true, the content of the register is shifted (register i is shifted to register $i + 1$, and register $N - 1$ to register 0) at the rising edge of the clock, otherwise the register maintains the data. A token is represented by the logic state 1 of the register. The number of tokens in a token ring can be from one to N, where N is the number of registers. A token ring with one token can be used as a state machine. The position of the token defines the state of the state machine. The *enable* signal of the token ring is equivalent to the next-state of the state machine. If this condition is true, the token ring shifts and the state machine changes. It is also possible to define a state machine when the token ring contains two consecutives tokens; the state of the state machine can be defined, for example, as the position of the first one.

Since the position of the tokens defines the state of the state machine, the synchronization of the position can be exploited to interface two clock domains. To synchronize the state of the state machine, a parallel synchronizer (two registers per bit) can be used, as showed in Figure 6.33.

However, as described by [104], the parallel synchronizer does not guarantee the correctness of the result. Figure 6.34 shows an example of synchronization. Solutions A, B, C, and D are all the possible solutions when the metastability of the register changes its content. Solutions A and B are correct since the token is well determined. Solution C is exploitable using some logic but Solution D is useless due to the absence of information. Because the token was moving, the metastability of the register can be resolved to a useless result. In that case, one clock cycle should be waited on to attempt to obtain a useful data.

To solve this issue, [215] proposes to use two consecutive tokens (*bubble encoding*) in the token ring. As the metastability affects the changing regis-

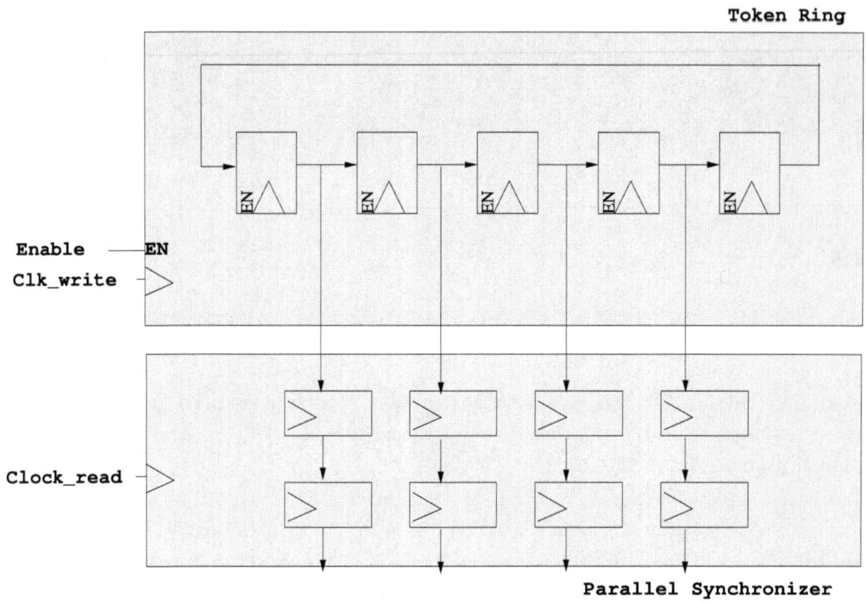

FIGURE 6.33: Synchronization of a token ring (from [215] ©2007 IEEE).

ters, the use of two consecutive tokens prevents some registers from changing. Assuming that registers i and $i+1$ have the tokens, if the token ring shifts, register $i+2$ gets a token, register i loses its token, and register $i+1$ does not change (it shifts its token and gets a token). In terms of logic value, register i and $i+2$ change state but register $i+1$ remains unchanged. Because there always exists a register that does not change state, it is always possible to detect a token. Figure 6.35 shows an example of synchronization. For example, we can define the position of the detected token by the position of the first logic 1 after a logic 0 (starting from the left). In this case, all solutions A, B, C, and D are correct because the token can be well defined; it is always possible to detect a transition between 0 and 1. This encoding algorithm does not avoid the metastability on the synchronizer. It just guarantees that the position of the token will be detected and a useless situation will never occur.

The described token rings with the bubble-encoding algorithm can be used to implement the write and read pointers in the FIFO architecture in Figure 6.37 [215]. The position of the tokens determines the position of the pointer. The position of the *Write_pointer* is defined by the position of the register containing the first token (starting from the left) as showed in Figure 6.36. Likewise, the position of the *Read_pointer* is defined by the position of the register after the second token (starting from the left). The full and empty detectors exploit this particular definition of the pointers and will be explained hereafter.

The *Write_pointer* shifts right when the FIFO is not full and the Write

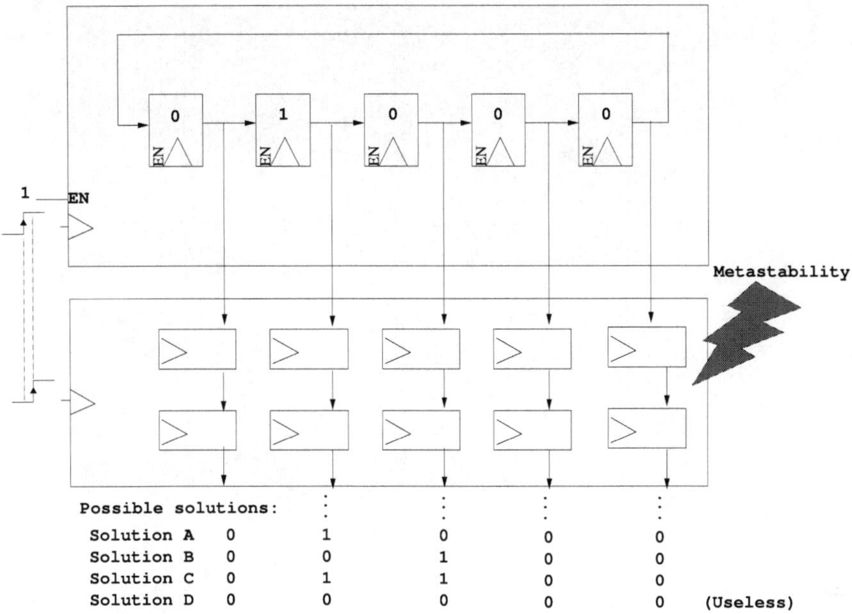

FIGURE 6.34: Possible solutions in the synchronization of a single token ring containing one token (from [215] ©2007 IEEE).

signal is true. Likewise, the *Read_pointer* shifts right when the FIFO is not empty and the Read signal is true. As the write and read interfaces belong to different clock domains, the token rings are clocked by their clock signal, *Clk_write* and *Clk_read*, respectively.

The Data buffer module is the storage unit of the FIFO. Its interfaces are: *Data_write*, *Data_read*, *Write_pointer*, *Read_pointer*, and *Clk_write*. It is composed of a collection of data-registers, logical AND gates, and tristate buffers as showed in Figure 6.38. The input data, *Data_write*, is stored into the data register pointed by the *Write_pointer* at the rising edge of *Clk_write*. AND gates recode the *Write_pointer* into a one-hot encoding that controls the enable signals of the data-registers. Likewise, the *Read_pointer* is recoded into one-hot encoding that controls the tristate buffer on each data-register. Finally, the *Data_read* signal collects the outputs of the tristate buffers. It is also possible to replace the tristate buffers with multiplexers to simplify the Design for Test (DfT) of the FIFO. The width and number of data-registers determine the width and the depth of the FIFO. The depth also determines the range of the Write and Read pointers.

The full detector computes the Full signal using the *Write_pointer* and *Read_pointer* contents. No status register is used as in Chelcea-Nowick [48] solution. The full detector requires N two input AND gates, one N-input OR gate, and one synchronizer, where N is the FIFO depth (Figure 6.39). The de-

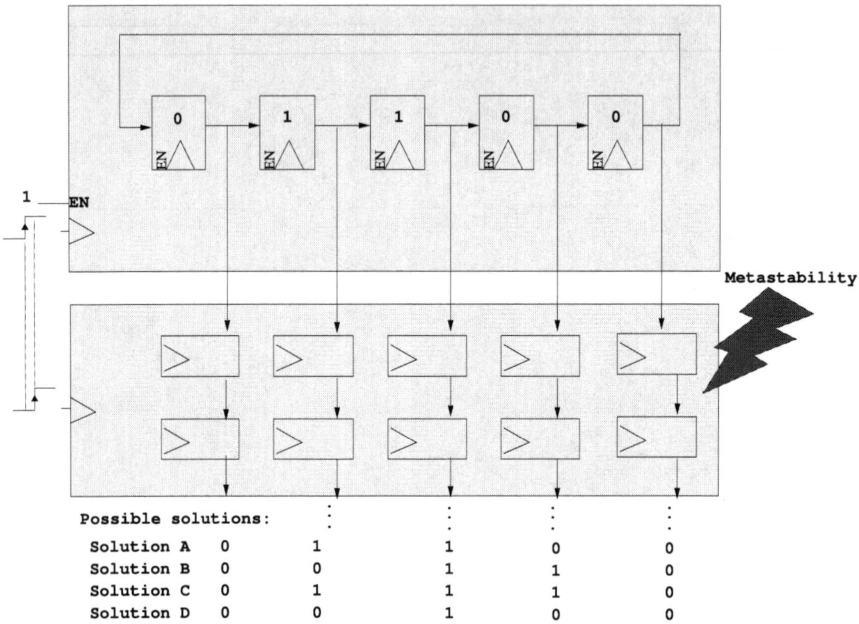

FIGURE 6.35: Possible solutions in the synchronization of a token ring containing two successive tokens (from [215] ©2007 IEEE).

tector computes the logic AND operation between the Write and Read pointer and then collects it with an OR gate, obtaining logic value 1 if the FIFO is full (Figure 6.36(e)) or quasi-full (Figures 6.36(c) and 6.36(d)) otherwise is 0. This value is finally synchronized to the *Clk_write* clock domain into *Full_s*. Since the synchronization has a latency of one clock cycle and the situation on Figure 6.36(c) can potentially be metastable, the detector has to anticipate the detection of the full condition. It is for this reason that the output of the OR gate detects the full and the quasi-full conditions. The full detector in Figure 6.39 can be optimized since the synchronization latency inhibits, in some cases, the FIFO from being completely filled. For example, if the FIFO is in the situation of Figure 6.36(c), and the sender does not write any other data, the *Full_s* will be asserted even if the FIFO is not filled completely.

However, a nonoptimal full detector does not penalize the throughput of the FIFO as much as a nonoptimal empty detector, since in the former case the receiver limits the throughput of the FIFO anyway. Therefore, design effort and chip area should be devoted to improving the performance of the empty detector.

The implementation of the empty detector is similar to the full detector because both employ the Write and Read pointer contents. As seen in the previous paragraph, the full detector has to anticipate the detection of the full condition to avoid FIFO overflow. As the empty detector is correlated to

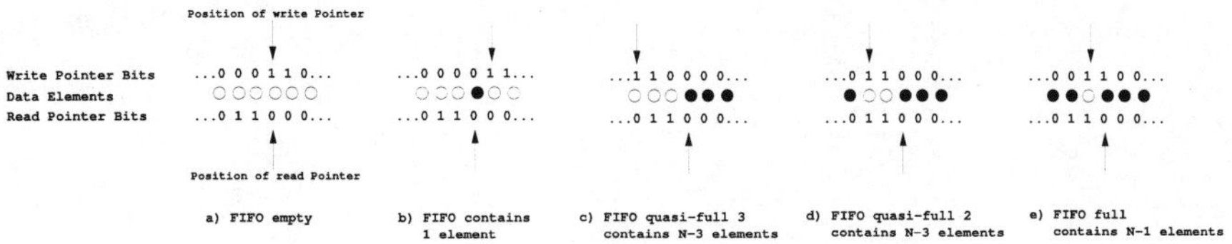

FIGURE 6.36: Write and Read pointer position definition and Full and Empty conditions in terms of token position (from [215] ©2007 IEEE).

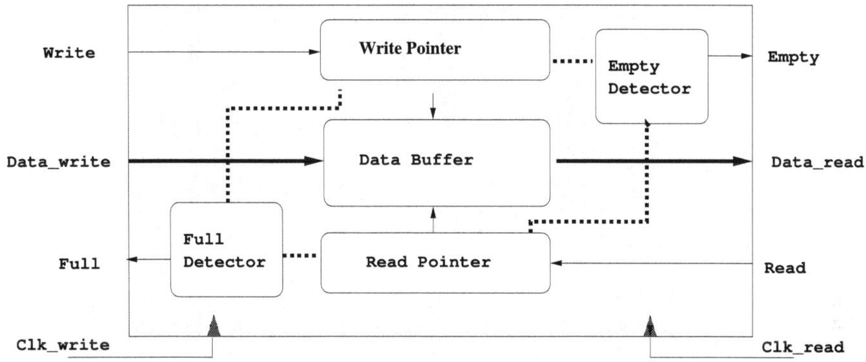

FIGURE 6.37: Block-level schematic of the DC-FIFO architecture (from [215] ©2007 IEEE).

the FIFO throughput, its detection has to be optimized, and no anticipation detector should be used. Figure 6.40 shows the empty detector for a five-word FIFO.

First, the $Write\_pointer$ is synchronized with the read clock into the $Synchronized\_Write\_pointer(SW)$ using a parallel synchronizer. Next, the $Read\_pointer$ is recoded into the $AND\_Read\_pointer(AR)$ using two input AND gates, operation also done in the $Data\_buffer$ module. The output of AR is a one-hot encoded version of the $Read\_pointer$. Finally, the empty condition is detected comparing the SW and AR values using three-input AND gates. As the metastability can perturb some bits of the SW (as seen on Figure 6.35), each pair of consecutive bits is compared to find a transition between 0 and 1. Their analysis is as follows, if the values of $SW_i = 0$ and $SW_{i+1} = 1$ that means that the SW pointer is on position $i + 1$. Furthermore, when $AR_i = 1$ that means that the AR pointer is on position $i + 1$ (Figure 6.40). The FIFO is considered empty (see Figure 6.36(a)) when the $Write\_pointer$ points the same position of the $Read\_pointer$. This can be detected when $SW_i = 0$, $SW_{i+1} = 1$, and $AR_i = 1$ for any i. These comparisons are computed by means of the three-input AND gates. Finally, a N-input OR gate collects all the values of the three-input AND gates to generate the Empty signal. This N-input OR gate and the one on the full detector can be decom-

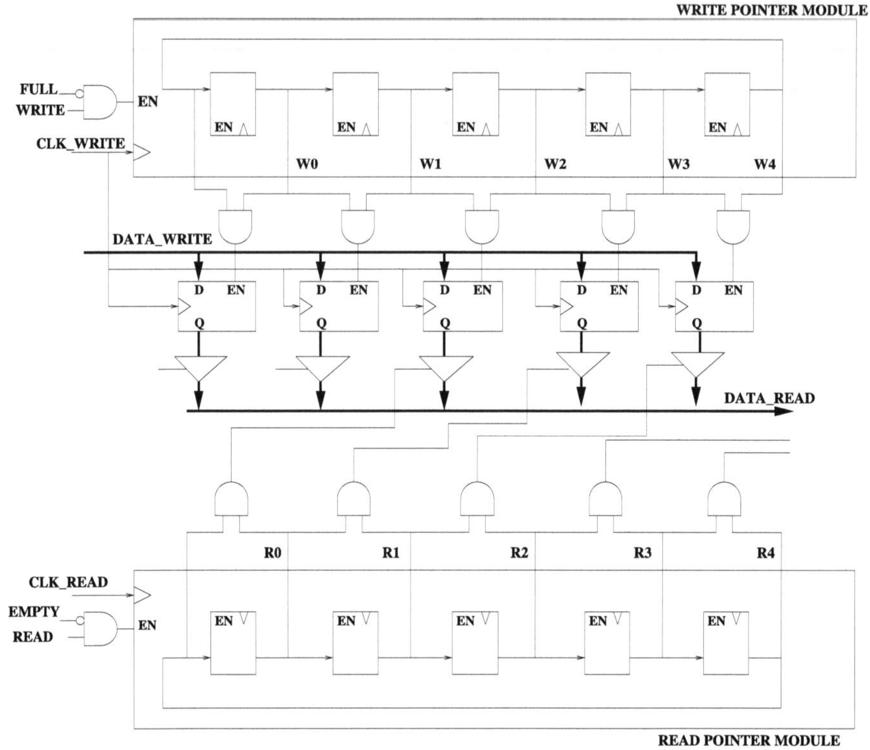

FIGURE 6.38: Write pointer, Read pointer, and Data buffer detail (from [215] ©2007 IEEE).

posed with $log_2 N$ levels of two-input OR gates. The latency introduced by the synchronization of the $Write\_pointer$ cannot corrupt the FIFO, because a change in this pointer cannot underflow/overflow the FIFO, it just introduces latency into the detector. The advantage of the bubble-encoding algorithm in this detector relies on the continuous detection of the $Write\_pointer$ position. Otherwise, as seen in Figure 6.34 Solution D, its position cannot be detected and the empty condition should be asserted to avoid a possible underflow of the FIFO, thus, introducing one additional clock cycle latency to the FIFO.

6.3.2.3 Tight Coupling of the DC-FIFO with the NoC Switch

Using DC-FIFOs as stand-alone components fully decoupled from NoC switches and/or network interfaces incurs a large latency and buffering overhead for clock domain crossing. The same approach followed for bringing mesochronous synchronizers deeper into the NoC switch can be taken for a DC-FIFO. The basic idea is to share the buffering resources of the DC-FIFO with those of the switch input stage, thus resulting in a more compact and higher performance realization. The tight coupling process is even easier in

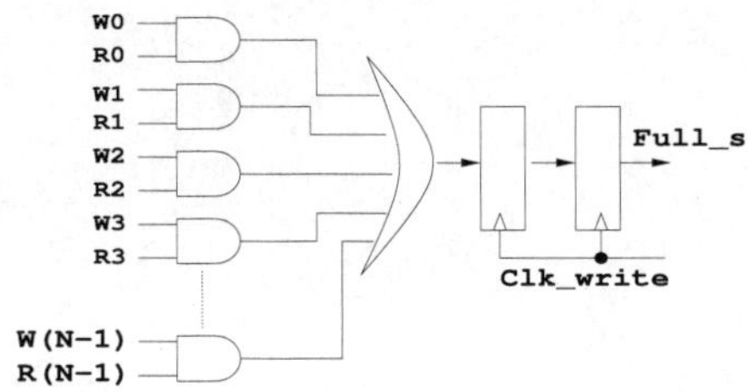

FIGURE 6.39: Full detector detail (from [215] ©2007 IEEE).

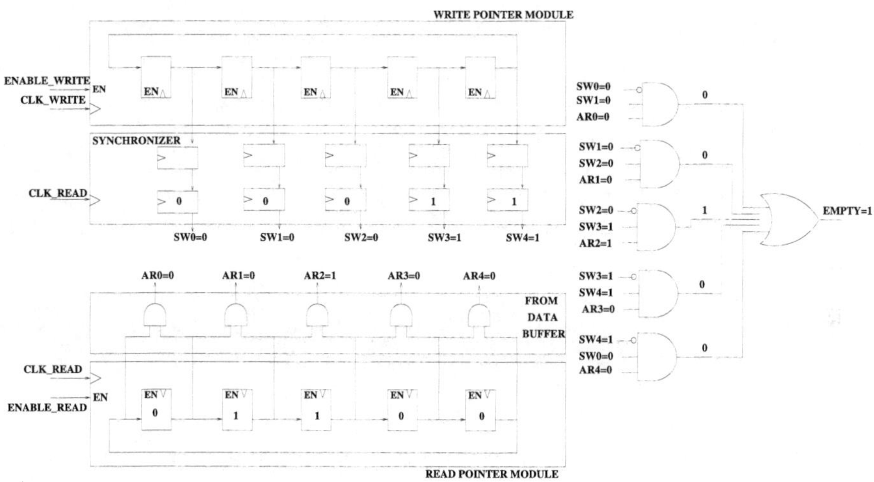

FIGURE 6.40: Empty detector details (from [215] ©2007 IEEE).

this case, since there is a perfect matching between the full/emtpy protocol of the FIFO and the simplest mechanism for NoC flow control (stall/go). A DC-FIFO simply notifies a stall signal upstream when it becomes full, and keeps the valid signal low for downstream communication when it is empty. Therefore, insertion of a DC-FIFO into a NoC path is straightforward. Similarly, replacing the switch input stage with a DC-FIFO is equally straightforward. Not only the flow control protocol matches, but also the architecture of the DC-FIFO is not significantly different from that of a switch input stage, apart from the control logic implementation. Figure 6.41 shows integration of a generic DC-FIFO into the xpipesLite NoC architecture [189]. Interestingly, the switch can be designed in such a way that the most suitable port-level synchronization option can be configured. In fact, two input stages are pointed out in the figure: a tightly coupled dual-clock FIFO and a traditional synchronous N-slot input buffer. Whether their area is the same actually depends on the size of the DC-FIFO, which is a function of the target throughput and hence of the frequency ratios between transmitter and receiver.

Three different designs are now evaluated and compared to materialize the

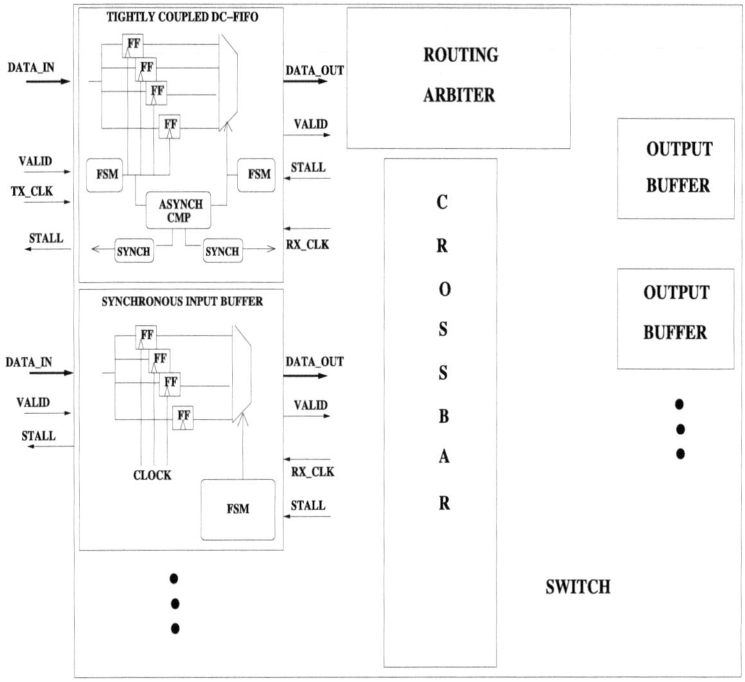

FIGURE 6.41: An asynchronous FIFO tightly coupled with a NoC switch that supports port-level configuration of the most suitable synchronization option.

benefits of a tightly coupled design style. The xpipesLite architecture is again used as experimental platform, while the DC-FIFO in [189] is considered, which is very similar to the one presented in Section 6.3.2.1. The first design is the conventional 5×5 vanilla (fully synchronous) switch with a 6-slot input buffer and a 2-slot output buffer per port. The second one is a switch where a dual-clock FIFO with 6 buffer slots has been merged with each input port. In order to carry out a fair comparison with the vanilla switch, total buffering resources have been kept equal, i.e., the output buffer size in the switch with the FIFO synchronizers has been reduced from 6 to 2 slots. The last configuration is a vanilla switch (6-slot inputs, 2-slot outputs) with external (i.e., loosely coupled) dual-clock FIFO (6 buffer slots) per input port.

To assess area occupancy, all the above switch configurations have been synthesized, placed, and routed at the same target frequency of 1 GHz. Total area of the tightly coupled system exhibits almost the same area footprint of the vanilla switch. This is a direct consequence of the fact that exactly the same buffering resources have been deployed in a specular fashion (between input and output). As showed in Figure 6.42(a), being the input buffer size of the three systems the same (6 slots), there is a similar amount of cell area devoted to either only buffering (vanilla switch) or buffering and synchronization (tightly and loosely coupled switch). Moreover, the loosely coupled system features the same area overhead (with the same distribution of input buffer and other cell area) of the other switches plus a further synchronization area due to the external blocks implementing the dual-clock FIFOs.

These results point out that the merging approach applied to the dual-clock FIFO design achieves up to 24% of area saving with respect to the loosely coupled design methodology. To assess the power consumption of a switch integrating dual-clock FIFOs on the input ports, the vanilla, tightly and loosely coupled designs have been tested with different traffic patterns: idle (to measure standby power), random (the target output port of input packets is randomly selected), and parallel (no switch internal conflicts). Postlayout simulations have been carried out at 800 MHz. The switch with the external dual-clock FIFO is the most power greedy under all possible traffic patterns, as showed in Figure 6.42(b). This is due to a larger amount of buffering resources. From the power viewpoint, there is a substantial benefit when integrating the dual-clock FIFO in the switch architecture. In fact, the tightly coupled design is the most power efficient among those under test and achieves up to 51% power saving (under random traffic). The motivation lies in the inherent architecture-level clock gating that is implemented by the dual-clock FIFO under test, which clocks only one bank of flip-flops at a time out of the total input buffer. If the incoming data is not valid, then the write pointer update circuit does not even switch thus gating the entire input buffer. Obviously, a similar clock gating technique (either at architecture or logic level) can be applied to the vanilla switch as well, and in fact the key take-away here is that integration of the dual-clock FIFO into the switch does not imply any major power overhead, as long as total buffering resources used in all switch variants are kept the same. Above all, this test shows that a fully synchronous switch can be evolved to a more advanced one supporting GALS communication with a marginal overhead.

6.4 NoC Clocking and Synchronization in 3-D Stacked ICs

The increasing emphasis on three-dimensional (3-D) integration, as an alternative to enhance circuit performance without facing many of the deep submicron scaling issues [246], has recently led to the proposal of several 3-D NoC architectures [247, 90]. These architectures exploit a salient feature of this design paradigm—the decrease in interconnect length—to considerably enhance the performance of the on-chip network. In addition to these complex NoC-based systems, the performance characteristics of the ASIC [137] and

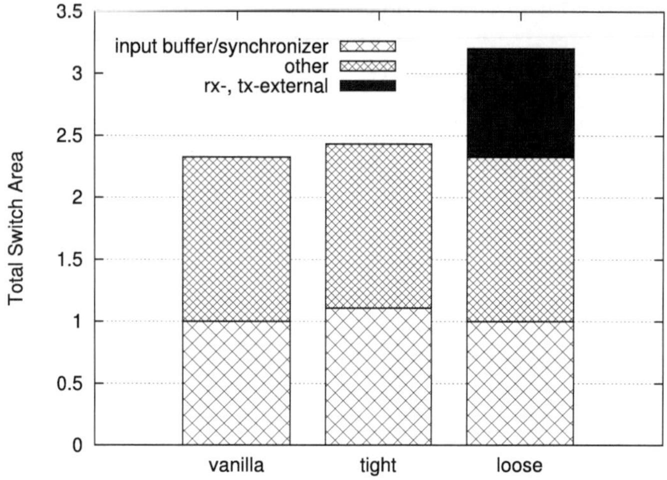

(a) Area breakdown.

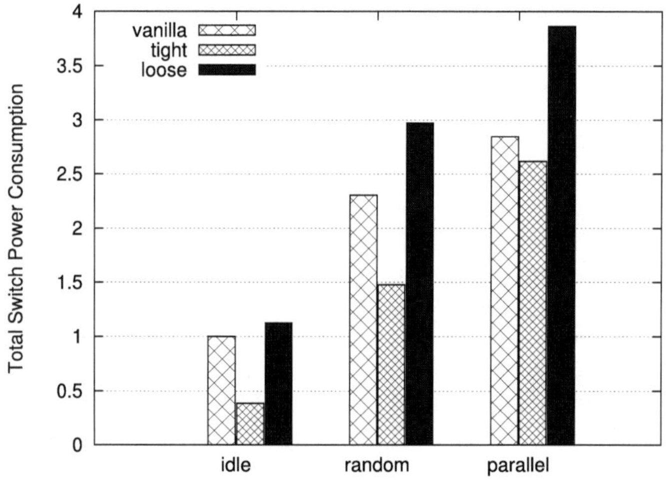

(b) Power consumption.

FIGURE 6.42: Postlayout normalized results of area (a) and power (b) for a switch with and without dual-clock FIFOs. These latter are considered to be either loosely or tightly coupled with the switch.

Field Programmable Gate Array (FPGA) [293, 9] design paradigms are also expected to benefit from the exploitation of the third dimension. Since an increase in clock frequency can constitute a primary objective for 3-D systems, providing a robust synchronization mechanism is an important issue [248, 318].

The different approaches that can be adapted for synchronizing 3-D systems and the related difficulties in the implementation of these approaches are concisely discussed in this section.

Several mechanisms have been developed for synchronizing planar on-chip networks each with specific advantages [171, 224]. These methods include fully synchronous, combinations of synchronous and asynchronous approaches with a different degree of periodicity, and fully asynchronous designs. Extending each of these approaches into three dimensions is not a straightforward task. Although the different synchronization strategies can be applied to any 3-D circuit, the discussion herein emphasizes NoC-based 3-D systems. Further, although several physical approaches to 3-D integration have been proposed, it is assumed that the vertical interconnects connecting circuits located on different planes are implemented with Through-Silicon Vias (TSVs) [301], as TSVs represent a high-performance and low-overhead vertical connectivity solution.

6.4.1 Synchronous Clocking

One of the advantages of a fully synchronous clocking scheme for 3-D circuits is easier NoC architectural design and better performance predictability as compared to nonsynchronous approaches. To implement a 3-D clock distribution network, however, several challenges need to be addressed, since 3-D circuits include two types of interconnects with substantially different characteristics. Compared to regular planar wires, vertical TSVs exhibit a different electrical behavior [248], occupy an active silicon area [281], and may pose manufacturing yield concerns [185]. Any 3-D clock distribution network requires a specific number of vias to connect the clock sinks throughout the multiplane stack.

Several clock distribution networks to globally distribute the clock signal to specific locations within each plane have been proposed [177, 249]. These topologies can be an immediate extension of H- and X-trees where either the root or the leaves of the tree connect to TSVs propagating the clock signal from the plane that contains the main clock driver to the other planes. Alternatively, clock synthesis algorithms can be used to generate a 3-D clock network [355, 162]. The main objectives of the synthesis process are to minimize (or bound) the clock skew for the entire system while using the least wire resources also including the TSVs. The application of these techniques on benchmark circuits reveals that there is a trade-off between the number of the employed TSVs and the resulting wire length of the network. A basic, single-TSV topology requires significantly greater wire length [210, 218], with a direct impact on the power dissipated by the interconnects, which may negatively affect the thermal properties of 3-D stacks. On the other hand, in a multi-TSV network, multiple local interplane clock trees are connected with TSVs, and each local tree spans a small area; the local clock trees allow for a greater reduction in wire length. However, this comes at a small increase in silicon area consumed

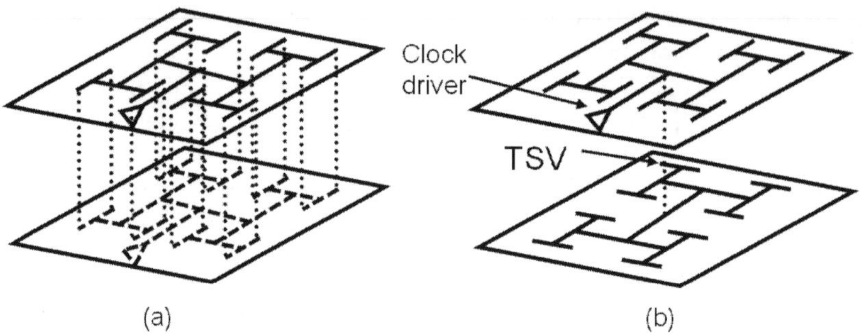

FIGURE 6.43: Candidate H-tree based 3-D clock distribution networks: (a) multi-TSV topology with redundant interconnect resources for improved testability, (b) single-TSV topology with higher testability and larger total wire length.

by the TSVs and a potential yield degradation depending on the quality of the TSV manufacturing process. The work in [218], based on the compensation of the driving strength of the 3-D clock tree buffers, shows an architectural solution to the minimization of clock skew while in the presence of nonuniform heating patterns in 3-D chips. Other works, such as [185], propose methods to tolerate TSV faults, but of course at an area penalty.

Another important criterion for choosing a 3-D clock distribution network is how many additional resources become required to test each of the planes of a 3-D circuit prior to bonding. Some topologies are more amicable to testing as compared to other clock networks, which may require some interconnect overhead to produce the same level of testability [177]. These additional interconnects are utilized only during the individual testing of each plane ensuring correct functionality prior to stacking the planes. For example, in Figure 6.43(a), each plane can contain an H-tree. The H-tree in the lower plane is activated only during the testing phase (before the planes are bonded) while for normal operation the TSVs provide the clock signal at the lower plane. During normal operation the clock tree showed with dashed lines is not used to lower the total power consumption of circuit.

A technique that provides a clock tree connecting all the intraplane clock sinks of each plane during testing has been added in a clock synthesis algorithm [354]. Traditional and properly-sized transmission gates are used to disconnect these intraplane clock networks from the overall clock distribution network of the 3-D stack used during the normal operation of the circuit. Inserting these gates increases the capacitive load of the clock network. This increased capacitance of the network, which is used only during testing, can be decoupled, however, by switching off the transmission gates. In this way, this capacitance is not considered during the synthesis procedure of the clock tree that connects the clock sinks in normal operation.

Note that although the existing clock synthesis techniques can manage the nominal skew of a 3-D clock network, the effect of process variation on the generated networks has yet to be sufficiently explored. The additional effect of die-to-die variations that are present in 3-D circuits can increase the overall variability of the clock signal, thereby degrading the performance of a 3-D clock distribution network [344]. Incorporating a process variation model in the methods used to construct these clock networks becomes, consequently, necessary to produce high-performance 3-D systems with a low skew variation.

Targeting 3-D NoCs where each processing element (PE) is assumed to be contained within a single physical plane, a moderate number of sinks needs to be connected since each PE is considered to include an individually synthesized intra-PE clock distribution network. Consequently, a global clock network similar to the topologies showed in Figure 6.43 could potentially be adopted. Alternatively, a custom clock tree based on a technique as described in [162] could be a better solution for those 3-D NoCs where the PEs are split into more than one plane. A synchronous approach can support such multiplane PEs since only a clock domain (or synchronous multiples of a primary clock frequency) is exploited.

6.4.2 Nonsynchronous Clocking

3-D NoCs are expected to improve performance over 2-D NoCs due to the smaller number of hops for data transmission featured by 3-D topologies. However, another potential performance gain derives from the greater bandwidth or shorter delay provided by the short vertical interconnects [182]. In fully synchronous 3-D NoCs, this latter advantage is not exploited since the horizontal links will mainly determine the speed that can be supported by the network. In other words, data among planes is propagated with the same speed as in the slower horizontal links, and the vertical links (i.e., the TSVs), which exhibit significantly shorter delays than the intraplane buses, are not fully exploited.

To exploit the higher bandwidth offered by the TSVs and facilitate back-end design, other nonsynchronous clocking schemes can be adapted. As pointed out in [244], serialization can be adopted for communication across TSVs. By adopting shift registers at both ends of TSVs, it is possible to carry out vertical communication over a smaller number of TSVs at virtually unchanged NoC performance and power levels. However, while drastically optimizing the area overhead of TSVs, this approach does not solve interplane synchronization challenges, actually requiring an additional fast clock at sending and receiving ends of TSVs, nor does it in itself actually improve performance.

Another potential candidate, which has been utilized in several 2-D circuits [224], is the GALS approach. An immediate extension of the GALS approach into three physical dimensions would include a number of clock domains equal to the number of planes comprising a 3-D circuit. This physical and, simultaneously, logical separation of clock domains can greatly facilitate the prior-to-bonding testing process for a 3-D system, since an independent clock network is included in each plane. In addition, the low latency of the vertical interconnects can be better exploited, since higher frequencies can be used for interplane communication [182]. Alternatively, clock generation circuitry in each plane (or for a small group of planes) is required to support a GALS approach, which can increase the overall cost of a 3-D system. This increase in design cost, however, can be counterbalanced by a decrease in the cost of testing since each plane can be independently tested without the need for additional interconnect resources as mentioned above.

Another implication of choosing a GALS mechanism for a 3-D system is related to the employed level of integration, which, in general, is considered to be determined by the size and density of the TSVs. Typically, the fewer the TSVs are, the coarser the integration level will be (i.e., core or circuit module level) and vice versa. In case a clock domain is considered not to be limited within a physical plane, a finer integration can be supported (i.e., at the register or logic gate level). Since the application of a GALS clocking scheme at a small number of large circuit blocks (where each block is synchronous) has diminishing benefits [224], a greater number of clock domains can be a better choice from a performance perspective. The existence of a clock domain on multiple physical planes faces, nonetheless, similar functional verification challenges to a fully synchronous approach.

The synchronizer architectures discussed in the previous sections of this chapter can be used within such clocking schemes. In particular, mesochronous approaches [183] allow for a "quasi-synchronous" design style while removing the need for interplane clock skew minimization. In general, however, the use of synchronizers at the boundaries of the clock domains, which are electrically connected with short vertical buses implemented with TSVs, poses different time constraints. Due to the short length and low latency of the TSVs, certain parts of these circuits should be properly adapted to avoid data latching failures due to hold time violations or metastability occurrences in the vertical direction. In addition, for data-driven clock generation circuits [224], proper data sequences should be generated as input patterns during testing to enable prior-to-bonding verification.

Finally, replacing the clock signal with handshaking-based communication for 3-D circuits can have some benefits related to the power consumption and the thermal behavior of these circuits. An asynchronous 3-D circuit can significantly benefit from the absence of the clock signal and the capacitance of the clock network that comprises a considerable portion of the dissipated power. This power decrease can, in turn, result in sufficiently low temperatures across the entire stack of planes. Lower temperatures, or equivalently, power densities relax the cooling as well as packaging requirements, which are expected to be a critical component determining the viability and the application domains within which 3-D circuits can be utilized. A 3-D test circuit including asynchronous circuits has recently been fabricated, demonstrating close to gigahertz data throughput [7]. Although the demonstrated circuit does not contain the complex functionality envisioned for 3-D systems, asynchronous design could potentially be a useful approach for those heterogeneous 3-D systems with low processing requirements but with a strict power budget due to thermal concerns.

6.5 Conclusions

Clock distributions are becoming more challenging with technology scaling because of several factors. The size of a typical synchronous domain is not decreasing as more function is integrated onto chips with the availability of additional transistors. The relative nonscaling of wire delay and the increasing amount of capacitance per unit area exacerbate clock latency and increase the required gain of the clock network. Variability in process, temperature, and voltage, both temporal and spatial, make skew and jitter management increasingly difficult.

These challenges have led not only to a search for circuit alternatives for clock distribution networks, but also to the definition of new system-level architectural templates relying on relaxed synchronization assumptions. In this direction, there is today consensus on the fact that future architectures will reflect a globally asynchronous and locally synchronous design style, with major implications on network-on-chip design. This chapter has focused on synchronizer-based GALS NoC architectures, which are within reach of current mainstream design tools and which will be probably more flexible in the short-to-medium term. The chapter has illustrated the main system-level GALS paradigms and design techniques to make them viable in real-life systems. A lot of space has then been devoted to the basic building blocks of GALS NoC architectures: mesochronous synchronizers and dual-clock FIFOs. The reader has certainly noticed that architecture design techniques for such synchronization interfaces need to be developed with the awareness of physical synthesis implications. This is a key requirement for timing closure.

Finally, the chapter has considered timing issues in vertical integrated circuits, which have only recently started to be explored. The new type of interconnects included in these circuits (i.e., the TSVs), the elevated power densities, the intra- and interplane sources of process variations, and the increased testing requirements advocate that synchronization is a critical challenge for vertical integration. While there is plenty of space for the different clocking approaches that have been discussed, it is not certain which of these schemes would best exploit the advantages of this technology. Since 3-D integration is expected to be applicable to a large number and types of electronic systems, a mixture of both synchronous and asynchronous approaches can be utilized to deliver satisfying performance with a reasonable power dissipation and low design, testing, and cooling cost.

Acknowledgment

This work was supported by the project Galaxy (project label 214364) that is funded by the European Commission within the Research Programme FP7.

Part II

The Industrial Perspective

Chapter 7

Networks of the Tilera Multicore Processor[1]

David Wentzlaff
Patrick Griffin
Henry Hoffmann
Liewei Bao
Bruce Edwards
Carl Ramey
Matthew Mattina
Chyi-Chang Miao
John F. Brown III
Anant Agarwal
Tilera, USA

---

[1] Based on [339], "On-Chip Interconnection Architecture of the Tile Processor," by David Wentzlaff, Patrick Griffin, Henry Hoffmann, Liewei Bao, Bruce Edwards, Carl Ramey, Matthew Mattina, Chyi-Chang Miao, John F. Brown III, and Anant Agarwal that appeared in *IEEE Micro*, September/October 2007 (vol. 27 no. 5). ©2007 IEEE.

7.1 Introduction

As a greater number of processor cores is integrated onto a single die, the design space for interconnecting these cores becomes more fertile. One manner to interconnect those cores is simply to mimic multichip multiprocessor computers of the past. Following the past, simple bus-based shared memory multiprocessors can be integrated onto a single piece of silicon. But, by following the past, we do not take advantage of the unique opportunities afforded by single-chip integration. Specifically, buses require global broadcast and do not scale to more than about 8 or 16 cores. Some multicore processors have used one-dimensional rings, but rings do not scale either because their bisection bandwidth does not increase as more cores as added.

In this work we describe the architecture of iMesh, the mesh-based on-chip interconnection network of the Tile Processor. The Tile Processor is a tiled multicore architecture developed by Tilera Corporation, and is inspired by MIT's Raw processor [333, 317], which was the first tiled multicore processor. Tiled multicore architectures are Multiple Instruction Multiple Data (MIMD) machines consisting of a two-dimensional grid of general-purpose compute elements. The compute elements are called *tiles* or *cores* and are homogeneous. They are coupled together by packet-routed mesh interconnects, which provide the transport medium for off-chip memory access, Input/Output (I/O), interrupts, and other communication activity.

Each tile in the Tile Processor is a powerful, full-featured computing system that can independently run an entire operating system, such as Linux. Likewise, multiple tiles can be used together to run a multiprocessor operating system such as Symmetric Multi-Processors (SMP) Linux. Each tile implements a three-way Very Long Instruction Word (VLIW) processor architecture with independent program counter (PC), a two-level cache hierarchy, 2-D Direct Memory Access (DMA) subsystem, and support for interrupts, protection, and virtual memory.

The first implementation of the Tile Processor architecture is the TILE64, a 64-core processor implemented in 90 nm, which contains five independent networks and is capable of executing 192 billion 32-bit operations per second at 1 GHz. The second implementation of the Tile Processor architecture is the TILE*PRO*64, a 64-core processor implemented in 90 nm, which contains six independent networks, improved cache sharing, and improved memory controllers. Because Tile processors are fully cache coherent, they run SMP Linux and off-the-shelf shared memory pthreads-based programs. Tile processors also support a light-weight multicore synchronization and messaging Application Programming Interface (API) called TMC (Tile Multicore Components), and a sockets-like streaming API called iLib.[2]

[2]iMesh^{TM}, iLib^{TM}, Multicore Hardwall^{TM}, TILE64^{TM}, TILE*PRO*64^{TM}, and Tile Processor^{TM} are trademarks of Tilera^{TM} Corporation.

The interconnect of the Tile Architecture is a departure from the traditional bus-based multicore processor. Instead of using buses or rings to connect many on-chip cores, the Tile Architecture utilizes six two-dimensional mesh networks to connect processors. Having six mesh networks leverages the on-chip wiring resources to provide massive on-chip communication bandwidth. The mesh networks on the TILE*PRO*64 afford 1.54 Tbps of bandwidth into and out of a single tile and 3.07 Tbps of bisection bandwidth for an 8×8 mesh. By utilizing mesh networks, the Tile Architecture can support a small to large number of processors without the need to modify the communication fabric. In fact, the amount of in-core (tile) communications infrastructure stays constant as the number of cores grows. While the in-core resources do not grow as tiles are added, the bandwidth connecting the cores grows as the number of cores increases.

It is not sufficient to have a massive amount of interconnect resources if they cannot be effectively utilized. In order to be effectively utilized, the interconnect must be flexible enough to efficiently support many different communication needs and programming models. The Tile Architecture's interconnect provides communication via shared memory and direct user accessible communication networks. The direct user accessible communication networks allow for scalar operands, streams of data, and messages to be passed between tiles without the need for system software intervention. The iMesh interconnect architecture also contains specialized hardware to disambiguate flows of dynamic network packets and sort them directly into distinct processor registers. Hardware disambiguation, register mapping, and direct pipeline integration of the dynamic networks provide register-like intertile communication latencies and enable scalar operand transport on dynamic networks. The interconnect architecture also includes a mechanism called Multicore Hardwall, which protects one program or operating system from another program or operating system when using directly connected networks.

The Tile Architecture benefits by having six separate networks each with a different use. By separating the usage of the networks and specializing the interface to their usage, the architecture allows programs with varied requirements to map in an efficient manner. As an example, the Tile Architecture has separate networks for communicating with main memory, communicating with I/O devices, and for user-level scalar operand and stream communication between tiles. Thus, many applications pull in their data over the I/O network, access memory over the memory networks and communicate among themselves all at the same time. This diversity allows a natural manner by which additional bandwidth can be utilized and separates traffic so that unrelated traffic does not interfere.

The desirability of separate physical networks is another instance where on-chip integration significantly alters the design trade-offs. Logical or virtual channel networks are useful in wire-dominated situations in which buffers are cheap because they allow logically separate networks to share a single physical channel. However, when integrated on-chip, near-neighbor wires are cheap, and buffer area becomes a significant proportion of network area. In this situation, separate physical networks become superior to logical channel networks because they offer proportionally more bandwidth without taking up significantly more area, due to their buffer requirements being the same. Furthermore, they are simpler to implement.

Being able to take advantage of the huge amount of bandwidth afforded by the on-chip integration of multiple mesh networks requires new programming APIs and a tuned software runtime system. In this work, we describe iLib, Tilera's C-based API for stream programming. iLib is a user-level API library, which provides primitives for streaming and messaging, much like a lightweight form of the familiar sockets API. This work describes how the library maps onto the user-level networks without the need for the overheads of system software.

The rest of this work is organized as follows. Section 7.2 describes the overall architecture of the Tile multicore processor, and Section 7.3 focuses on the interconnect hardware. Section 7.4 describes the interconnection software (iLib) and Section 7.5 presents results utilizing iLib.

7.2 Tile Processor Architecture Overview

The Tile Processor architecture consists of a two-dimensional grid of compute elements. The compute elements are called *tiles* and are identical. Each tile is a powerful, full-featured computing system that can independently run an entire operating system, such as Linux. Likewise, multiple tiles can be used together to run a multiprocessor operating system such as SMP Linux. Figure 7.1 is a block diagram of the 64-tile TILE64 processor with Figure 7.2 showing the major components inside a tile.

As illustrated in Figure 7.1, the perimeters of the mesh networks in a Tile Processor connect to I/O and memory controllers, which in turn connect to the respective off-chip I/O devices and Dynamic Random Access Memories (DRAM) through the pins of the chip. Each tile combines a processor and its associated cache hierarchy with a switch, which implements the various interconnection networks of the Tile Processor. Specifically, each tile implements a three-way Very Long Instruction Word (VLIW) processor architecture with an independent program counter (PC), a two-level cache hierarchy, 2D DMA subsystem, and support for interrupts, protection, and virtual memory.

7.2.1 TILE64 Implementation

The first implementation of the Tile Processor architecture is the TILE64, a 64-core processor implemented in 90 nm. TILE64 clocks at speeds of up to 1 GHz and is capable of 192 billion 32-bit operations per second. It supports

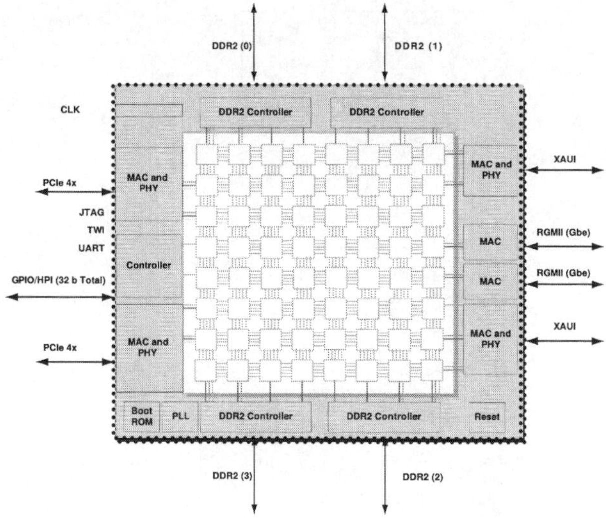

FIGURE 7.1: Block diagram of the TILE64 processor with on-chip I/Os (from [339] ©2007 IEEE).

subword arithmetic and can achieve 256 billion 16-bit operations, or a half TeraOp for 8-bit operations. The TILE64 Processor consists of an 8×8 grid of tiles. The chip includes 4.8 MB of on-chip cache distributed among the processors, per-tile Translation Lookaside Buffers (TLB) for instruction and data, and 2D DMA between the cores, and between main memory and the cores. The Tile Processor provides a coherent shared memory environment in which a core can directly access the cache of any other core using the on-chip interconnects. The cores provide support for virtual memory and run SMP Linux that implements a paged memory system. To meet the power requirements of embedded systems, the TILE64 employs extensive clock gating and processor napping modes. To support its target markets of intelligent networking and multimedia, TILE64 implements all the required memory and I/O interfaces on the System-on-Chip (SoC). Specifically, it provides off-chip memory bandwidth up to 200 Gbps using four DDR2 interfaces, and I/O bandwidth in excess of 40 Gbps through two full-duplex 4x PCI-e, two full-duplex XAUI (X Attachment Unit Interface), and a pair of Gigabit Ethernet interfaces. The high-speed I/O and memory interfaces are directly coupled to the on-chip mesh interconnect through an innovative, universal I/O shim mechanism.

7.2.2 TILE*PRO*64 Implementation

The second implementation of the Tile Processor architecture is the TILE*PRO*64. The TILE*PRO*64 contains 64 cores and has much similarity to the TILE64 processor. The TILE*PRO*64 processor contains several improve-

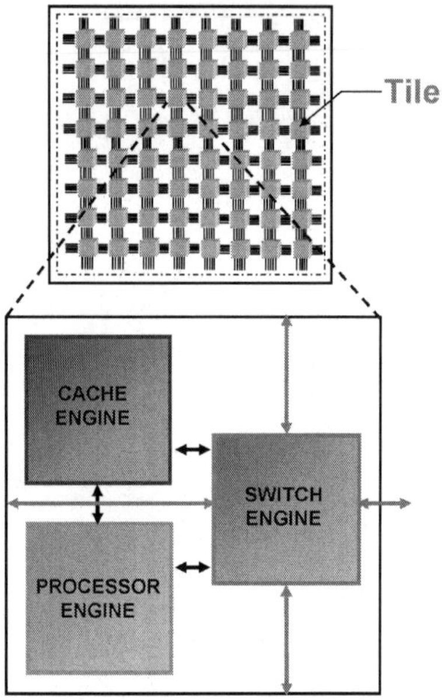

FIGURE 7.2: The view inside a tile (from [339] ©2007 IEEE).

ments, mainly the performance of intercore data sharing has been improved through the addition of a higher performance cache coherence system. To support this higher performance cache coherence system, an additional dynamic network was added that is used by the cache subsystem. Also, the TILE*PRO*64 gained a larger L1 Instruction cache and optimized DDR2 main memory controllers.

7.3 Interconnect Hardware

The Tile Architecture provides ample on-chip interconnect bandwidth through the use of six low-latency mesh networks. This section describes the six networks, their uses, performance characteristics, and connections to tiles and I/O devices.

The Tile Architecture is organized as a two-dimensional mesh topology. This topology is used because two-dimensional mesh topologies effectively map onto two-dimensional silicon substrates. The networks are not toroidal in nature, but rather simple meshes. While it is possible to map two-dimensional toroids onto a two-dimensional substrate, there is added cost in wire length and wiring congestion, approximately by a factor of two.

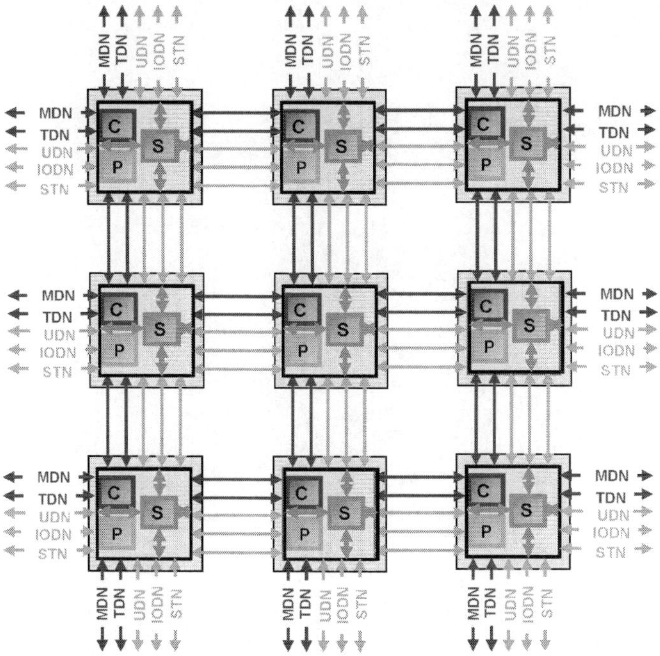

FIGURE 7.3: A 3×3 array of tiles connected by networks (from [339] ©2007 IEEE).

The six networks are the User Dynamic Network (UDN), I/O Dynamic Network (IDN), Static Network (STN), Memory Dynamic Network (MDN), Tile Dynamic Network (TDN), and inValidation Dynamic Network (VDN). The TILE64 contains the first five of these networks, while the TILE*PRO*64 added the VDN to accelerate cache traffic. Each network connects five directions, north, south, east, west, and to the processor. Each link consists of two 32-bit wide unidirectional links, thus traffic can flow in both directions on a link at one time. A fully connected crossbar is utilized in each tile that allows an all-to-all five-way communication. Figure 7.3 shows a grid of tiles connected by the five-networks as in the five network TILE64 and Figure 7.4 shows a single dynamic switch point in detail.

Five of the networks are dynamic networks. The Tile Architecture's dynamic networks provide a packetized fire-and-forget interface. Each packet contains a header word that denotes the "x" and "y" destination location for the packet along with the packet's length up to a maximum of 128 words per packet. The dynamic networks are dimension-ordered wormhole routed. The latency of each hop through the network is one cycle per hop when packets are going straight and one extra cycle for route calculation when a packet needs to make a turn at a switch. Because the networks are wormhole routed, a minimum of in-network buffering is utilized. In fact, the only buffering in

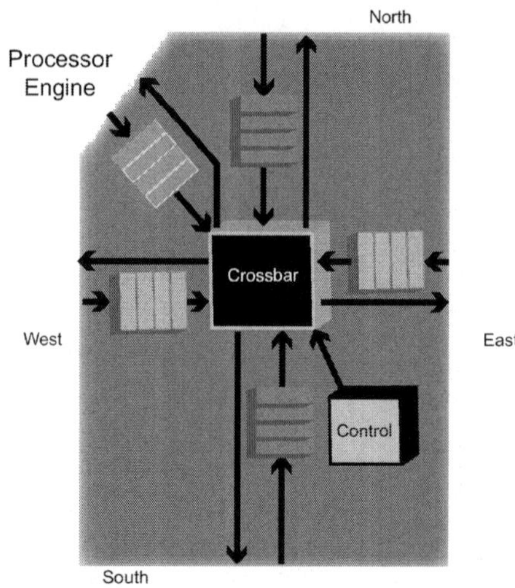

FIGURE 7.4: A single network crossbar (from [339] ©2007 IEEE).

the network are small three-entry First-In First-Out (FIFO) buffers utilized only to cover the link level flow control cost.

The dynamic networks preserve ordering of messages between any two nodes, but do not guarantee ordering between sets of nodes. A packet is considered to be atomic and is guaranteed not to be interrupted at the receiving node. The dynamic networks are flow controlled and guarantee reliable delivery. As discussed in the following sections, the dynamic networks support scalar and stream data transport.

The Static Network is a network that allows for static communications. The Static Network does not have a packetized format but rather allows for the static configuration of the routing decisions at each switch point. Thus, with the STN, streams of data can be sent from one tile to another by simply setting up a route through the network and injecting a stream of data. The data stream then traverses the already set-up routers until the data reaches the destination tile to be consumed. This allows for a circuit switched communications channel to be built that is ideal for streaming data. The Static Network also contains an auxiliary processor that can be utilized to reconfigure the network in a programmatic manner. This functionality is an improved form of the static network processor functionality of the Raw Processor [315].

7.3.1 Network Uses

The UDN is primarily used by userland processes or threads to communicate with low-latency in a flexible and dynamic manner. This is a departure from a typical computer architecture where the only userland communication between threads is via shared memory communication. By providing an extremely low-latency user accessible network, streams of data, scalar operands, or messages can be directly communicated between threads running in parallel on multiple tiles without involving the operating system. The Tile Architecture also supplies userland interrupts to provide a fast mechanism for userland programs to be notified of data arrival. Section 7.4 details how the Tilera software stack utilizes the low latency and flexibility provided by the UDN.

The Tile Architecture does not contain any unified bus for communication with I/O devices. I/O devices are connected to the networks just as the tiles are connected to the networks. In order to support direct communication with I/O devices and to allow system level communications, the I/O Dynamic Network (IDN) is used. The IDN connects to each tile processor and extends off of the fabric of processors into I/O devices. Both control information and streams of I/O data are sent over the IDN. The IDN is also used for operating system and hypervisor level communications. It is important to have a dedicated network for I/O and system level traffic to protect this traffic from userland code.

The MDN is utilized by the caches in each tile to communicate with off-chip DRAM. The TILE64 and TILE*PRO*64 have four 64-bit Double Data Rate-2/800 (DDR-2/800) DRAM controllers on-chip, which connect to the MDN at the edge of the tile arrays. These memory controllers allow for glueless interfacing to DRAM. The MDN provides a mechanism for every tile to communicate with every Random Access Memory (RAM) controller on the chip. When the cache needs to retrieve backing memory from off-chip DRAM, the in-tile cache controller constructs a message that it sends across the MDN to the DRAM controller. After servicing the transaction, the memory controller constructs up a reply message across the MDN. A buffer preallocation scheme is utilized to ensure that the MDN runs in a deadlock-free manner.

The TDN works in concert with the MDN as a portion of the memory system. The Tile Architecture allows for direct tile-to-tile cache transfers. The request portion of tile-to-tile cache transfers transit the TDN, while responses transit the MDN. Tile-to-tile requests are not sent over the MDN to prevent deadlock in the memory protocol. Thus, an independent communications channel was needed. The VDN, which was added with the TILE*PRO*64, carries memory invalidation traffic separate than the TDN and MDN. This traffic was required to be separate than the TDN and MDN traffic in order to avoid deadlock.

The STN is a userland network. Userland programs are free to map communications channels onto the STN thereby allowing an extremely low-latency high bandwidth channelized network great for streaming of data.

7.3.2 Logical versus Physical Networks

When designing a multicore processor with communications networks, it is often desirable to have multiple independent logical networks. Having multiple logical networks allows for the privileged isolation of traffic, independent flow control, and traffic prioritization. As described above, the Tile Architecture utilizes five different networks. When designing the TILE64 Processor, these logical networks could have been implemented as logical or virtual channels over one large network or as independent physical networks.

The TILE64 Processor implements the five logical networks with five physically independent networks. This choice is contrary to much previous work and is motivated by how the relative costs of network design change when implemented on a single die. The first surprising realization is that network wiring between tiles is effectively free. Modern day multilayer fabrication processes provide a wealth of wiring as long as that wiring is nearest neighbor and stays on-chip. If a tile shrinks small enough, the perimeter of the tile may get small enough that wiring may be a challenge, but for our small tile this issue was nowhere in sight.

The second trade-off in on-chip network design is the amount of buffer space compared to wire bandwidth. In traditional off-chip networks, the wire bandwidth is at a premium, while on-chip buffering is relatively inexpensive. With on-chip networks, the wiring bandwidth is high and the buffer space takes up silicon area, the critical resource. In the TILE64 Processor each network accounts for approximately 1.1% of a tile's die area of which greater than 60% of that area is dedicated to the buffering. Because the buffering is the dominating factor, the TILE64 minimizes the buffering to the absolute minimum needed to build efficient link-level flow control. If virtual channels were built with the same style of network, additional buffering is needed for each virtual channel equal to that of another physical network. Thus, the savings of building virtual channels is minimized as they do not save on the 60% of the silicon area dedicated to buffering. Also, by building multiple physical networks via replication, the design is simplified and more intertile communication bandwidth is provided. If buffers larger than the minimum needed to cover the flow control loop are used, then there would be opportunities to share buffering in a virtual channel network that do not exist in the multiple physical channel solution. Finally on-chip networks are different from interchip networks in that they can be considered reliable. This reliability mitigates the need for much buffering used to manage link-failure in less reliable systems.

7.3.3 Network to Tile Interface

In order to reduce latency for tile-to-tile communications and reduce instruction occupancy, the Tile Architecture provides access to the on-chip networks through register access tightly integrated into the processor pipeline. Any instruction executed in the processor within a tile can read or write to the UDN, IDN, and STN. The MDN, TDN, and VND are connected to each tile, but connect only to the cache logic and are only indirectly used by cache misses. There are no restrictions with the number of networks that can be written or read in a particular instruction bundle and reading and writing networks can cause the processor to stall. Stalling occurs if read data is not available or a network write is writing to a full network. By providing register mapped network access, the latency of network communication can be reduced and instruction occupancy can be reduced. In example, if an "add" must be done and the result value needs to be sent to another tile, the "add" can directly target the register mapped network without the additional occupancy of a network move.

7.3.4 Receive Side Hardware De-multiplexing

Experience has shown that the overhead associated with dynamic networks is rarely in the network itself, rather it is in the software logic at the receive side to "de-multiplex" data. On a dynamic network, each node can receive messages from a large number of other nodes (a node can be a process on a tile, or it may even be a specific channel port within a process on a given tile). For many applications, the receive node needs to be able to quickly determine for any data item that it receives which node sent the data. Unfortunately determining this in software by looking at a header is many times too costly, thus we look to hardware structures to accelerate de-multiplexing of packets at a receiving node.

One software-only solution to receive side de-multiplexing is to augment each message with a tag. When a new message arrives the receiving node takes an interrupt. The interrupt handler then inspects the tag and determines a queue in memory or cache into which the message should be enqueued. Then when the node wants to read from a particular sending node, it looks into the corresponding queue stored in memory and dequeues from a particular queue that the data was already sorted into. While this approach is flexible, the cost associated with taking an interrupt and implementing the sorting based on a tag in software can be quite expensive. Also, reading out of memory on the receive side is more costly than reading directly from a register provided by register mapped networks.

To address this problem, the Tile Architecture contains hardware de-multiplexing based on a tag word associated with each network packet. The tag word can signify the sending node, a stream number, a message type, or some combination of these characteristics. Receive side hardware de-multiplexing is implemented on the UDN and IDN. Figure 7.5 shows an example use of the de-multiplexing hardware with two neighboring tiles communicating via the UDN.

Hardware de-multiplexing (de-mux) is implemented by having several independent hardware queues with settable tags that data can be sorted into at the receive endpoint. The tags are set by the receiving node to be some

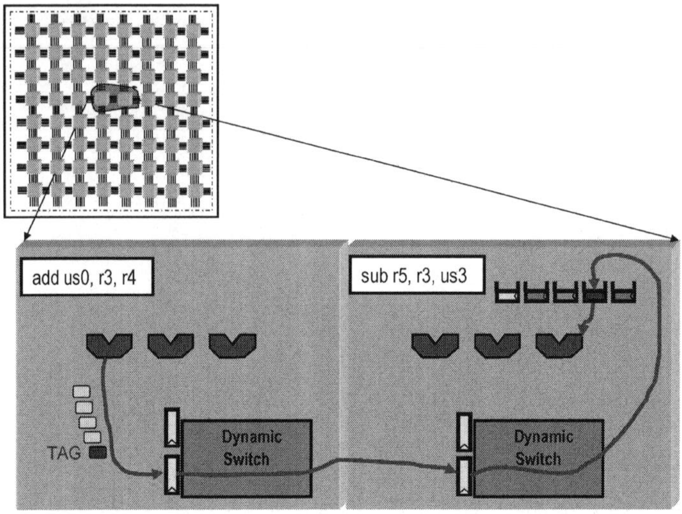

FIGURE 7.5: De-multiplexing overview (from [339] ©2007 IEEE).

value. When a message is received, a tag check occurs. If the tag matches one of the tags set by the receiving node as one of the tags that it is interested in, the hardware de-multiplexes the data into the appropriate queue. If an incoming tag does not match any of the receive side set tags, a tag-miss occurs; the data is shunt to a catch-all queue, and a configurable interrupt can be raised. By having a large tag space and a catch-all queue, the physical queues can be virtualized to implement a very large stream namespace. In effect the implemented queues serve as a cache of the most recent streams that a receive node has seen. Figure 7.6 shows the implementation of the hardware de-multiplexing logic.

The UDN contains four hardware de-multiplexing queues and a catch-all queue while the IDN contains two de-multiplexing queues and a catch-all queue. The TILE64 and TILE*PRO*64 implementations contain 128 words (512 bytes) of shared receive side buffering per tile that can be allocated between the different queues by system software.

Together the de-multiplexing logic and the streamlined network to tile interface allows the Tile Processor to support operations that require extremely low latency such as scalar operand transport, which facilitates streaming as discussed in Section 7.4.

7.3.5 Flow Control

It is desirable that on-chip networks be a reliable data transport mechanism. Therefore, all of the networks in the Tile Architecture contain link-level forward and reverses flow control. The TILE64 and TILE*PRO*64 implementa-

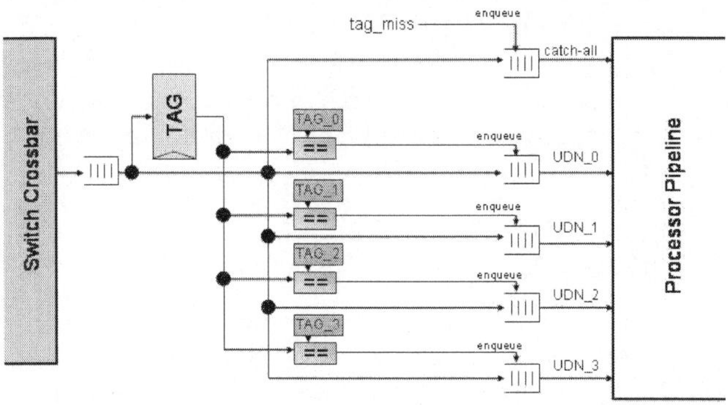

FIGURE 7.6: Receive side de-multiplexing hardware with tags (from [339] ©2007 IEEE).

tions utilize a three-entry credit-based flow control system on each tile-to-tile boundary. Three entries is the minimum buffering needed to maintain full bandwidth acknowledged communications in the design. This link-level flow control is also utilized for connecting the switch to the processor and memory system.

While link-level flow control provides a reliable base to work with, dynamic on-chip networks require higher level flow control to prevent deadlocks on a shared network and provide equitable sharing of network resources. All four dynamic networks on the TILE64 are replicas of each other, but surprisingly utilize quite varied solutions to end-to-end flow control. The strictest flow control is enforced on the MDN. On the MDN, a conservative deadlock avoidance protocol is utilized. Every node that can communicate with a DRAM memory controller is allocated message storage in the memory controller. Each node guarantees that it will never utilize more storage than it has been allocated. Acknowledgments are issued when the DRAM controller processes a request. The storage space is assigned such that multiple in-flight memory transactions are possible to cover the latency of acknowledgments. Because all in-flight traffic has a pre-allocated buffer at the endpoint, no traffic can ever congest the MDN. On the other memory network, the TDN, no end-to-end flow control is used. The TDN relies solely on link-level flow control. The TDN is guaranteed not to deadlock as it only requires forward progress of the MDN in order to make forward progress.

The two software accessible dynamic networks, IDN and UDN both implement mechanisms to drain and refill the networks. By doing so, in the case of an insufficient buffering deadlock, the networks can be drained and virtualized by utilizing the off-chip DRAM as extra in-network buffering. In addition to this deadlock recovery mechanism, the IDN utilizes pre-allocated buffering with explicit acknowledgments when communicating with IDN connected I/O devices. Communications on the UDN use different end-to-end flow control dependent on the programming model used. Buffered channels and message passing utilize software generated end-to-end acknowledgments to implement flow control. When using raw channels, only the de-multiplexing buffering is utilized and it is up to the programmer to orchestrate usage.

7.3.6 Protection

The Tile Architecture has novel features not typically found in conventional multicore processors such as user accessible networks and directly accessible networks. By having directly accessible and particularly user-accessible networks, usability and modularity require that programs be protected from one another. A network protection scheme is needed as it is not desirable for one program to be able to communicate with another program unrestrained. Likewise it is not desirable for a userland program to be able to message an I/O device or another operating system running on another set of tiles directly. To address these problems, the Tile Architecture implements a mechanism called Multicore Hardwall.

Multicore Hardwalling is a hardware protection scheme where individual links in the network can be blocked from having traffic flow across them. Every link on the UDN, IDN, and STN is protected by Multicore Hardwalling, while the traffic on the MDN and TDN are protected via standard memory protection mechanisms via a TLB. Figure 7.7 shows multiple protection domains between tiles for a 4×4 fabric of tiles. If traffic attempts to be sent over a hardwalled link, the traffic is blocked and an interrupt is signaled up to system software where appropriate action can be taken. Typically the system software will kill the process, but it is also possible that Multicore Hardwalling is being used to virtualize larger-sized fabrics of tiles, thus the system software can remove the offending message and play it back later on a different link. The network protection is implemented on outbound links, thus it is possible to have unidirectional links in the network where protection is set up in only one direction. An example of this is denoted as link L0 in Figure 7.7.

7.3.7 Shared Memory Communication and Ordering

When designing single-chip tiled processors, communications cost is low enough that alternative approaches to traditional problems such as shared memory can be utilized. The TILE64 processor uses neighborhood caching to provide an on-chip distributed shared cache. Neighborhood caching functions by homing data in a single tile's on-chip cache. This homing decision is made

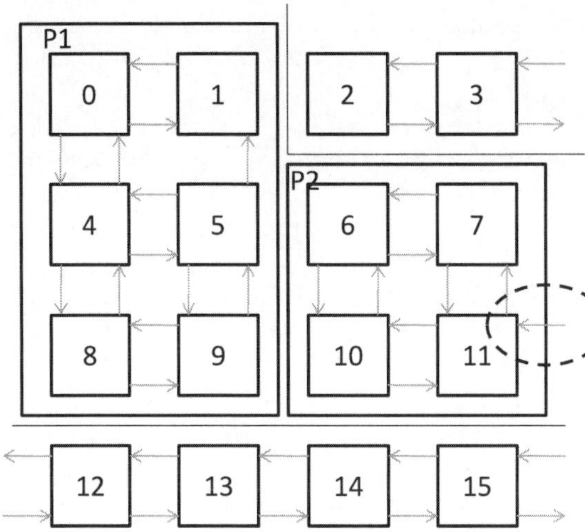

FIGURE 7.7: Protection domains on a dynamic network (from [339] ©2007 IEEE).

by system software and is implemented on a page basis via a tile's memory management unit. If a memory address is marked local, the data is simply retrieved from the local cache or on miss, from main memory. If an address is marked as remote, a message is constructed over the TDN to the home tile to retrieve the needed data. Coherency is maintained at the home tile for any given piece of data and data is not cached at nonhome tiles. The relative proximity of tiles and high bandwidth networks allow for neighborhood caching to achieve suitable performance. Read-only data can be cached throughout the system.

The TILE*PRO*64 implements a more sophisticated cache coherence system than the TILE64 processor. The TILE*PRO*64 processor allows data to be brought dirty into a cache that is not the home cache for the data. This can improve read performance for data that is read often, but not homed locally. This feature requires extra bookkeeping data to be kept in the home cache to track cores that have read-only copies. Also, the TILE*PRO*64 adds the ability to interleave the homing location based on address bits, thereby possibly reducing cache home hotspots.

When communicating with multiple networks or networks and shared memory, the question of ordering arises. The Tile Architecture guarantees that network injection and network removal occurs in programmatic order. But, there are no internetwork guarantees within the network, thus synchronization primitives are constructed out of the unordered networks. When utilizing both memory and network, a memory fence instruction is required in order to make memory visible if memory is being used to synchronize traffic that flows over the networks or vice versa.

7.4 Interconnect Software

The Tile Architecture's high-bandwidth, programmable networks allow software to implement many different communication interfaces at hardware-accelerated speeds. Tilera's iLib library provides programmers with a set of commonly used communication primitives, all implemented via on-chip communication on the UDN. For example, iLib provides lightweight socket-like "streaming channels" used for streaming algorithms and it provides a Message Passing Interface (MPI)-like interface for ad hoc messaging.

By providing several different communication primitives, iLib allows the programmer to use whichever communication interface is best for the problem at hand. The UDN network design, in particular the de-mux hardware, allows iLib to provide all of these communication interfaces simultaneously, using some de-mux queues for channel communication and others for message passing. As we will show later, the high-bandwidth, scalable network enables efficient burst communication for applications that require data reorganization between phases of computation.

The following subsections describe the "C" programming language communication APIs provided by iLib and the implementation of those APIs using the UDN and de-mux hardware.

7.4.1 Communication Interfaces

iLib provides two broad categories of UDN-based communication: socket-like channels and message passing.

The iLib channels interface provides long-lived connections between processes or threads. Semantically, each channel is a FIFO connection between two processes. A channel send operation always sends data to the same receiver. Generally, a channel provides a point-to-point connection between two processes (a sender and a receiver); this type of connection is often used to implement producer-consumer communication without worrying about shared memory race conditions. The all-to-all connectivity of the UDN also allows iLib to provide "sink" channel topologies, in which many senders send to one receiver. In this case, the receive operation implicitly selects the next available packet from any of the incoming FIFOs. Sink channels are often used to collect results from process-parallel programs, in which many workers must forward their results to a single process for aggregation.

iLib actually provides several channel APIs, each optimized for different communication needs. The next section will describe two possible implementations. "Raw channels" allow very low overhead communication but each FIFO can only use as much storage as is available in the hardware de-mux buffer in each tile. "Buffered channels" have slightly higher overhead but allow arbitrary amounts of storage in each FIFO by virtualizing the storage in raw channels:

```
int a = 2, b = 3;
ilib_rawchan_send_2(send_port, a, b);

int a_in, b_in;
ilib_rawchan_receive_2(receive_port, a_in, b_in);
```

buffered channels:

```
char packet[SIZE];
ilib_bufchan_send(send_port, packet, sizeof(packet));

size_t size_in;
char* packet_in = ilib_bufchan_receive(receive_port, &size_in);
```

messages:

```
char message[SIZE];
ilib_msg_send(ILIB_GROUP_SIBLINGS, receiver_rank, TAG,
              message, sizeof(message));

char message_in[SIZE];
ilibStatus status;
ilib_msg_receive(ILIB_GROUP_SIBLINGS, sender_rank, TAG,
                 message_in, sizeof(message_in), &status);
```

FIGURE 7.8: iLib code examples (from [339] ©2007 IEEE).

memory. Thus, buffered channels are appropriate for applications that require a large amount of buffering to decouple burstiness in a producer-consumer relationship, whereas raw channels are appropriate for finely-synchronized applications that require very low latency communication.

The message passing API is similar to MPI[207]. Message passing allows any process in the application to send a message to any other process in the application at any time. The message send operation specifies a destination tile and a "message key" to identify the message. The message receive operation allows the user to specify that the received message should come from a particular tile or have a particular message key. The ability to restrict which message is being received simplifies cases in which several messages are sent to the same tile simultaneously; the receiving tile can choose the order in which the messages are received and the iLib runtime will save off the other messages until the receiver is ready to process them. The ability to save off messages for later processing makes message passing the most flexible iLib communication mechanism: any message can be sent to any process at any time without any need to establish a connection or worry about the order in which the messages will be received. Figure 7.8 presents example code for the raw channels, buffered channels, and messaging iLib APIs.

Table 7.1 shows the relative performance and flexibility of the different iLib communication APIs. Raw channels achieve single-cycle occupancy by sending and receiving via register-mapped network ports. Buffered channels are more flexible and allow the user to create FIFO connections with unlimited amounts of in-channel buffering (the amount of buffering is determined at channel creation time), but they incur more overhead because they use an interrupt handler to drain data from the network. Message passing is the

Mechanism	Latency (cycles)	Occupancy (cycles)	BW (B/c)	Buffering	Ordering
Raw Channels	9	3 send 1 receive	3.93	hardware	FIFO
Buffered Channels	150	100	1.25	unlimited, static	FIFO
Message Passing	900	500	1.0	unlimited, dynamic	out-of-order FIFO by key

TABLE 7.1: Performance and ordering properties of different UDN communication APIs (from [339] ©2007 IEEE).

most flexible but most expensive interface. It provides unlimited, dynamically allocated buffering and out-of-order delivery of messages with different "message keys," but at the expense of more interrupt overhead. Note that the data transmission bandwidth for buffered channel messages is actually the same as raw channels (they all utilize the same UDN), but the interrupt and synchronization overhead has a significant performance impact.

7.4.2 Implementation

The Tile interconnect architecture allows the iLib communication library to implement many different forms of communication using the same network. This makes the Tile Architecture a more flexible, more programmable approach to parallel processing than SoC designs that only provide interconnect between neighboring cores. In particular, each tile's de-mux queue and de-mux buffer allow iLib to efficiently separate different flows of incoming traffic and handle each flow differently.

For example, the raw channels implementation allows the programmer to reserve a de-mux queue for a single channel's worth of incoming data. To connect a raw channel, iLib allocates one of the receiving tile's four UDN de-mux queues and assigns its tag value to a per-channel unique identifier. The sending process (or processes) then send data packets to that receiving tile, specifying the same UID that was mapped to the now-reserved de-mux queue. Thus, all of the packets for a particular raw channel are filtered into a particular de-mux queue. The receive operation then simply reads data directly from that queue. Because raw channel packets can be injected directly into the network, the send operation requires only a few cycles. Similarly, the receive operation requires only a single register move instruction because the incoming data is directed into a particular register-mapped de-mux queue. In fact, because the receive queue is register mapped, the incoming data can be read directly into a branch or computation instruction without any intermediate register copy.

Users of raw channels must implement flow control in order to manage the amount of available buffers without overflowing it. Often the necessary flow control is implemented as an "acked channel," in which a second FIFO is connected in order to transmit acks in the reverse direction (from receiver to sender). In such implementation, the channel receiver begins by sending several "credit packets" to the sender. When the sender needs to send, it first blocks on the receive of a credit and then sends the data packet. When the receiver dequeues a data packet, it will send a credit packet back to the sender.

The raw channels implementation demonstrates how iLib can reserve a de-mux queue to separate out traffic for a particular point-to-point channel. However, some algorithms require a large amount of buffering in the semantic FIFO between the sender and receiver. The required buffering may be significantly larger than the amount of storage in the UDN de-mux buffer. In such cases, a de-mux queue can be reserved for "buffered channel" traffic. When using buffered channels, the de-mux is configured to generate an interrupt when the de-mux buffer fills with data, allowing the interrupt handler to drain the incoming traffic into a memory buffer associated with each buffered channel. The receive operation then pulls data out of the memory buffer instead of from a register-mapped de-mux queue. Two key features of the Tile Architecture allow efficient implementation of these virtualized, buffered channels: configurable interrupt delivery on de-mux buffer overflow and low latency interrupts. Configurable interrupt delivery allows interrupts to be delivered for buffered channel data but not for raw channel data, and low-latency interrupts allows the data to be rapidly drained into memory. In fact, an optimized interrupt handler can interrupt the processor, save off enough registers to do work, and return to the interrupted code in under 30 cycles.

Finally, the message passing interface uses yet a third de-mux configuration option to implement immediate processing of incoming messages. When the message passing interface is enabled, the catch-all de-mux queue is configured to interrupt the processor immediately when a packet arrives. To send a message, the ilib_msg_send() routine first sends a packet containing the "message key" and size to the receiver. The receiver is interrupted and the messaging engine checks to see whether the receiver is currently trying to receive a message with that key. If so, a packet is sent back to the sender telling it to transfer the message data. If no receive operation matches the "message key," the messaging engine saves off the notification and returns from the interrupt handler. When the receiver eventually issues an ilib_msg_receive() with the same "message key," the messaging engine will send a packet back to the sender, interrupting it and telling it to transfer data. Thus, the ability to configure a particular de-mux queue to interrupt the processor when packets arrive allows iLib to implement zero-copy, MPI-style message passing.

We have shown how the Tile interconnect architecture allows iLib to separate out different traffic flows and handle each flow differently. De-mux queues can be used to separate out individual channels' traffic and map it to register mapped queues that will only drain when the receiver reads from the network registers. Alternatively, a de-mux queue can be configured to interrupt the tile when the de-mux buffer is full of data, so that incoming traffic can be drained in large bursts. And as a third option, a de-mux queue can be configured to generate an interrupt whenever traffic arrives, so that the incoming packet

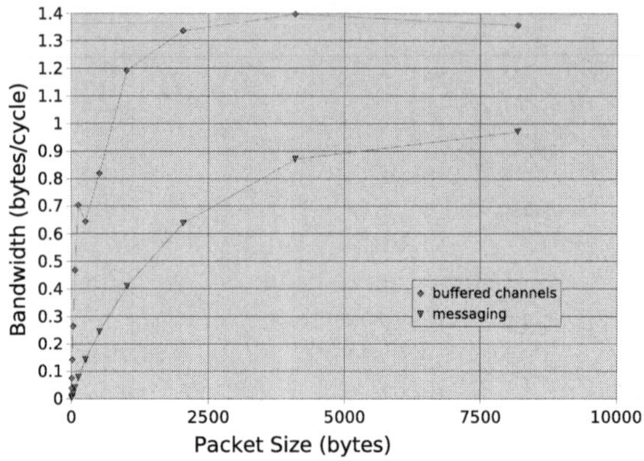

FIGURE 7.9: Bandwidth versus packet size for buffer channels and messaging (from [339] ©2007 IEEE).

can be processed promptly and a response generated. These different modes of operation can be used to implement raw channels, buffered channels, and message passing, respectively, all using the same hardware and running as needed by the application.

7.4.3 iLib Characterization

Software libraries introduce overheads on communication channels while providing ease of programming and flow control. iLib is no exception, thus this section characterizes the performance of the different forms of iLib channels.

iLib communications flow over the UDN. The hardware of the UDN provides a maximum of 4 bytes (one word) per cycle. UDN links consist of two unidirectional connections. The most primitive link type is raw channels. For raw channels, communicating data occurs at a maximum of 3.93 bytes/cycle. The overhead is due to header word injection and tag word injection cost. Figure 7.9 presents the performance of transferring differing sized packets utilizing buffered channels and the iLib messaging API. As can be seen, there is a decent amount of overhead related to reading and writing memory for the buffered channels case. The messaging interface incurs additional overhead related to interrupting the receive tile. Both buffered channels and messaging utilize the same packets for bulk data transfer, an 18-word packet consisting of one header word, a tag word, and 16 words of data. Because buffered channels and messaging utilize the same messaging primitives, asymptotically, they can reach the same maximum bandwidth for large packet sizes.

Figure 7.10 examines the latency of buffered channels and messaging as a function of packet size. This test was conducted by sending a packet of fixed size from one tile to a neighboring tile and then having that tile respond back

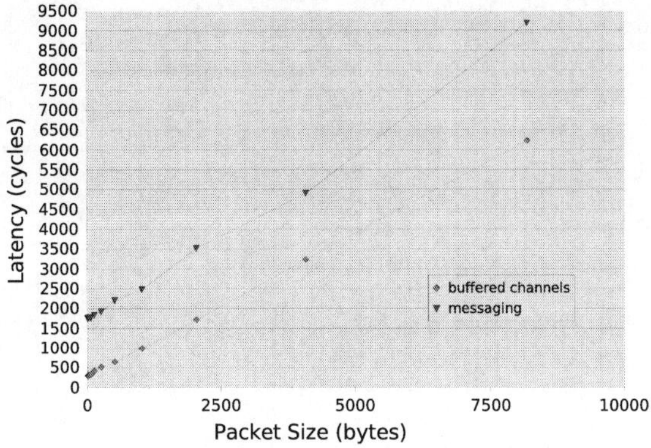

FIGURE 7.10: Latency versus packet size for buffered channels and messaging (from [339] ©2007 IEEE).

with a packet of the same size. The latency is the time taken to complete this operation divided by two. Messaging adds approximately a 1500 cycle overhead over the buffered channel overhead regardless of packet size.

7.5 Applications

This section presents several micro-benchmarks that are useful in characterizing a chip's interconnect. The same applications are implemented in multiple programming styles to demonstrate the relative communication overhead of a communication mechanism.

7.5.1 Corner Turn

In order to illustrate the benefits of Tile's flexible, software-accessible, all-to-all interconnect network, we will now examine a micro-benchmark commonly seen between phases of DSP applications.

Image processing applications operate on multidimensional data. The image itself is 2-D and often pixels from multiple images are required for computation. On a multicore processor, these multidimensional arrays are generally distributed across cores to exploit data parallelism.

It is often the case that the most efficient data distribution for one stage of the application is inefficient for another. For example, a 2-D frequency transform is implemented as a series of 1-D transforms on rows of data and then a series of 1-D transforms on columns of data. In this scenario, we would

Tile Config.	Matrix Size (words)	Shared Mem/ UDN Sync Achieved Bandwidth (Gbps)	Shared Mem/ STN Sync Achieved Bandwidth (Gbps)	Raw Channels/ UDN Sync Achieved Bandwidth (Gbps)	Raw Channels/ STN Sync Achieved Bandwidth (Gbps)
2×2	256x128	9.54	13.97	9.73	13.91
4×4	512x256	17.67	23.69	19.18	46.58
8×8	1024x512	37.99	42.92	11.24	96.85

TABLE 7.2: Performance comparison of shared memory vs. raw channels for corner turn on the TILE64 (from [339] ©2007 IEEE).

like each core to contain entire rows of the array during the first phase, and then entire columns during the second phase.

The process of reorganizing a distributed array from distribution in one dimension to distribution in another is known as a corner turn [176]. Implementing the corner turn requires each core to send a distinct message to every other core. Furthermore, these messages can be relatively small.

To perform well on the corner turn, a multicore processor needs a high bandwidth network with minimal contention and low message overhead.

To illustrate the performance and behavior of the various networks, the corner turn is implemented in four different ways. These implementations are distinguished by two factors: the network used to redistribute the data and the network used to synchronize the tiles. Data is transmitted on either the TDN, using shared memory, or the UDN, using iLib's raw channels. For each of the data transmission methods, synchronization is implemented using either the STN or the UDN. The combination of data transmission methods and synchronization methods yields four implementations. To measure the efficacy of an implementation, the achieved bandwidth of the data reorganization is measured.

The results for the four implementations of the corner turn are shown in Table 7.2. The best performing implementation is clearly the one that transmits data using raw channels and synchronizes using the STN. Raw channels provide direct access to the UDN and allow each word of data to be transmitted with minimal overhead. Furthermore, using the STN for synchronization keeps the synchronization messages from interfering with the data for maximum performance. The one drawback of this implementation is that it requires the programmer to carefully manage the UDN and it requires extra implementation time to fully optimize.

The corner turn implementation that uses raw channels for data transmission and the UDN for synchronization gets much worse performance than the one that uses raw channels and the STN. This is due to the overhead of virtualizing the UDN for multiple logical streams (the data and the synchronization). When two logically distinct types of messages must share the same network, the user must add extra data to distinguish between the types

Number Tiles	Shared Mem Cycles	Raw Channels Cycles
1	420360	418626
2	219644	209630
4	87114	54967
8	56113	28091
16	48840	14836
32	59306	6262
64	105318	4075

TABLE 7.3: Performance comparison of shared memory vs. raw channels for dot product on the TILE64. Lower cycle count is faster (from [339] ©2007 IEEE).

of messages. Deadlock is avoided by sharing the available buffering between the two types of messages. Both of these issues add overhead to the implementation and in the 64-tile case, this overhead becomes very destructive to performance.

The shared memory implementations of the corner turn are much simpler to program, but their performance is also lower than that of the raw channels and STN based corner turn. Both the ease of implementation and the performance difference are due to the extra overhead of sending shared data on the TDN. For each data word the user sends on the TDN, the hardware adds four extra header words. These extra words allow the hardware to manage the network and avoid deadlocks, so the program is much easier to write, but it is limited in performance. When data is sent over the TDN, we see that the synchronization method makes less difference than in the raw channels case. This is due to the fact that in the TDN implementation, data and synchronization are kept separate no matter the synchronization method.

7.5.2 Dot Product

The dot product is another widely used DSP computation. The dot product takes two vectors of equal length, pairwise multiplies the elements and sums the results of the multiplications returning a scalar value. The dot product has wide applications in signal filtering where a sequence of input samples are scaled by differing constants and is the basic building block for FIR filters.

In order to map a dot product across a multicore processor, the input vectors are evenly distributed across the array of processors. A processor then completes all of the needed multiplications and sums across the results. Finally a distributed gather and reduction add is performed across all of the individual processor's results. In order to measure performance, a 65536 ele-

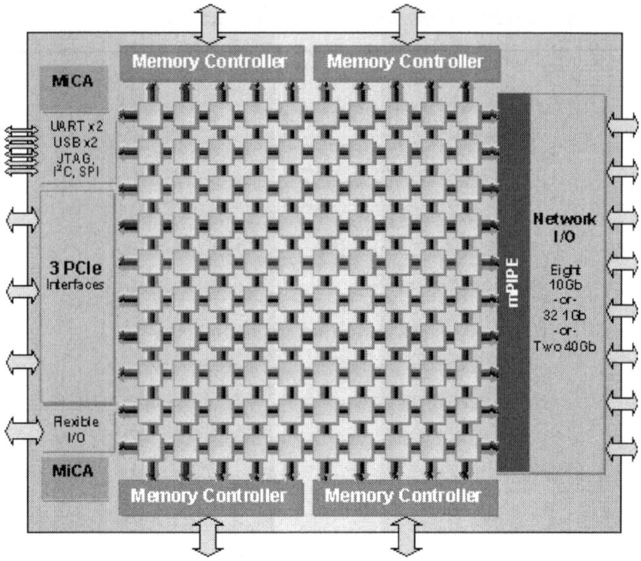

FIGURE 7.11: Block diagram of the TILE-Gx100 processor.

ment dot product was used. The input dataset was comprised of 16-bit values and computed a 32-bit result.

The dot product was mapped using two communications mechanisms as in the corner turn. Tilera shared memory is compared with iLib's raw channels. Both methods were optimized for the architecture. Table 7.3 presents results for the two communication mechanisms. As can be seen, the shared memory implementation contains higher communication overheads and hence does not scale as well as the raw channel implementation. Another result of note is the performance jump from 2 to 4 tiles. While the compute resources only double in this case, the application exhibits super-linear speedup because this is the point where the datasets completely fit in a tile's L2 cache.

7.6 Future Architectures

Tilera has announced its third generation product called the Tile-Gx line of processors. The Tile-Gx line of processors is a family of processors. One design point of the Tile-Gx lines is the Tile-Gx100 which contains 100 cores as shown in Figure 7.11. The Tile-Gx family extends the TILE*PRO* line of processors into a 64-bit architecture. The Tile-Gx will also sport larger caches, higher clock frequencies, and richer instruction set. Like the TilePro family of processors, the caches in the TileGx family of processors are fully cache coherent. The Tile-Gx line of processors will be implemented on a 40 nm process and will contain DDR3 memory controllers. In addition to higher performance cores and lower power used, the Tile-Gx processor will contain on-chip hardware encryption and compression engines.

7.7 Conclusion

In conclusion, this work describes the networks of the Tile Processor architecture embodied in its the TILE64 and TILE*PRO*64 processors. The work discussed the synergy between the hardware architectures of on-chip interconnect and the software APIs that use the interconnect. Specifically, we explored how numerous directly accessible networks are utilized by the iLib communications library to provide flexible communication mechanisms to meet the needs of varied parallel applications. We also discussed cases where multicore interconnect decisions fly in the face of conventional wisdom. For example, we showed that when networks are integrated on-chip, multiple physical networks can be superior to logical or virtual channel networks.

Acknowledgment

We thank the whole Tilera team for all their effort in designing the Tilera chip and accompanying software tools.

Chapter 8

On-chip Interconnect Trade-offs for Tera-scale Many-core Processors

Mani Azimi

Intel Labs, Intel Corporation, USA

Donglai Dai

Intel Labs, Intel Corporation, USA

Akhilesh Kumar

Intel Labs, Intel Corporation, USA

Aniruddha S. Vaidya

Intel Labs, Intel Corporation, USA

With the stringent requirements on power and silicon area, building a processor with many cores has been widely accepted in the industry as the primary approach to continue delivering ever-increasing performance. Processors with tens to hundreds of cores, with aggregate performance exceeding one trillion operations per second are expected soon. Such tera-scale many-core processors will be highly integrated system-on-chip designs containing a variety of on-die storage elements, memory controllers, and input/output (I/O) functional blocks. As a result, the on-chip interconnect becomes a crucial factor in designing such highly integrated heterogeneous systems. Demanding workloads create hot-spots, jitters, and congestion during communication among various blocks. An adaptive interconnect capable of responding gracefully to such transients can address such issues. Other challenges, such as manufacturing defects, on-chip variations, and dynamic power management can also be better served through flexible and adaptive interconnect solutions.

In this chapter we discuss on-chip interconnect solutions for enabling efficient many-core communication. We motivate the desired flexibility and adaptivity to entertain the requirements of various designs and their ramifications. We also present an on-chip interconnect design with aggressive latency, bandwidth, and energy characteristics. We discuss the related design choices and policies based on the constraints of the on-chip interconnect domain and demonstrate the effectiveness of these choices for different usage scenarios.

8.1 Introduction

Designing processors with many cores has been widely accepted in the industry as the primary approach for delivering ever-increasing performance under hard power and area constraints. The number of cores in a general purpose processor is expected to scale to several tens and possibly over a hundred cores by the end of the decade, leading towards the aggregated performance of trillions of operations per second per chip. Such a tera-scale processor provides a platform for use across a wide array of application domains, taking advantage of increasing device densities offered by Moore's law. A high degree of system integration, along with an architecture that can exploit different types of parallelism, characterizes this evolution [117].

A typical implementation of such a processor will include, in addition to tens or hundreds of general-purpose cores, multiple levels of cache memory hierarchy to mitigate memory latency and bandwidth bottlenecks, and interfaces to off-chip memory and I/O devices. Most many-core chip proposals use a modular approach to make implementation tractable, wherein the building blocks with well-defined physical and logical interfaces are connected through an on-chip interconnect to realize specific products. The definition of products targeting specific application domains may be done using a combination of a few building blocks that meet appropriate cost, power, and performance goals.

A flexible, capable, and optimized on-chip interconnect is central to realizing this vision.

On-chip interconnects can take advantage of abundant wires, smaller clock synchronization overhead, lower error rate, and lower power dissipation in links compared to off-chip networks, where chip pin counts and power dissipation in the transceivers and links dominate design considerations. However, efficient mapping to a planar substrate places restrictions on suitable topology choices [69]. Furthermore, a need to support performance isolation, aggressive power-performance management using dynamic voltage-frequency scaling (DVFS) techniques, and handling within-die process variation effectively may place additional requirements on topology selection at design time [78] (process variation impact is discussed in Chapter 12).

In addition to physical design considerations that affect topology choices as above, there are workload considerations. A general-purpose many-core processor must efficiently run diverse workloads spanning legacy and emerging applications from domains as varied as scientific computing, transaction processing, visual computing, and those under the nebulous umbrella of cloud computing. Such workloads may exhibit communication characteristics with transitory hot-spots, jitters, and congestion among various functional blocks. Therefore, it is an imperative design requirement for on-chip interconnects to respond to these conditions gracefully.

There are several plausible approaches for designing the on-chip interconnect for a many-core tera-scale processor chip. For example, one approach could be to partition the overall design into smaller subsystems, design separate interconnects suitable for each subsystem, and use an appropriate global interconnect to connect the subsystems together. Such a methodology requires the design, analysis, and composition of multiple interconnects. It takes a static design time view of interconnects, which may be acceptable in some specific contexts but may be not for general-purpose computing. An alternative approach is to use a parameterized single interconnect design and apply it across all the subsystems with specific parameters tuned for each subsystem. However, such a design would not only be challenging for tuning the parameters appropriately but also relatively inflexible. We propose taking a third approach, where we incorporate a novel set of flexible and adaptive features into the interconnect enabling it to meet the diverse requirements expected of a many-core processor interconnect in a more systematic and tractable manner. We will outline the design of one such interconnect in this chapter.

The rest of this chapter is organized as follows. In Section 8.2, we briefly review several generations of interconnect designs in Intel products and silicon prototypes. In Section 8.3, we discuss some typical usage models and the desired attributes, and then present the design of a particular on-chip interconnect, primarily focusing on the microarchitecture of the router[1] and highlighting the choices made to optimize for latency, energy efficiency, sup-

[1]In this chapter we use the term "router" instead of "switch".

port for flexibility, and adaptivity. Details of a field programmable gate array (FPGA) prototype and performance results are then presented for different usage scenarios, indicating the benefit derived from the various features in interconnect. Section 8.4 concludes with a summary of our work.

8.2 Evolution of Interconnects in Intel Processors: External to Internal

The requirements of performance scalability and cost reduction drove a major transition from shared bus interconnect to point-to-point interconnect in multisocket systems. This transition is apparent in off-chip socket-to-socket multiprocessor (MP) interconnect architectures of Intel platforms. The multiprocessor platforms evolved from multidrop buses such as the Intel® Pentium® 4 bus [122] and bridged buses as with the Twincastle 4-way Intel® Xeon® MP chipset [341] to point-to-point interconnects with the Intel® QuickPath Interconnect [341], all supporting cache-coherent shared-memory architectures.

Another significant change was driven by Moore's law and the power wall, which led to designs that moved away from the sole focus on improvements in instruction-level parallelism (ILP) towards more throughput-oriented energy-efficient architectures. In such designs, the additional performance was delivered through an increase in number of cores within a processor socket with successive process generation. The cores within a single die/socket operate at a more modest clock rate to keep the processor under the required power rating for the socket, while providing higher throughput performance. This transition, starting with dual-core processors, has accelerated the multicore era in the mainstream of computer architecture.

Multicore processors have been a great commercial success and have found applicability in running high-bandwidth, compute-intensive applications including high-performance throughput-oriented and scientific computing applications, high-performance graphics and 3-D immersive visual interfaces, as well as in decision support systems and data-mining applications. Systems using multicore processors are now the norm rather than the exception. Diverse demanding applications including emerging applications are driving processor performance towards the level that will require several tens to over a hundred cores per processor. This many-core processor architecture vision was outlined in the Intel Tera-scale computing initiative [117].

The trend shows that high-performance computer systems that used to be made of a huge number of single-core processor sockets can be made using a smaller number of sockets with many-core processors. The advantages of the higher level of integration are clear in aspects such as power, performance, cost, reliability, and system size. Such an acceleration in integration brings about other bottlenecks and roadblocks that require addressing over time such as in memory bandwidth and memory capacity.

The intersocket connectivity, including the associated protocols, has remained as the key complexity and cost contributor to single-core processor systems. The design and validation innovations have made this scalability more manageable but the intersocket bandwidth still remains as a major cost contributor. The era of many-core processors requires high scalability within a single socket with additional constraints such as on power, latency, and physical design flexibility. With a rather simplistic functional viewpoint, one could regard the evolution of intersocket interconnects and on-die interconnects as being on a continuum. However, in reality, very distinct physical design and power constraints differentiate the evolutionary trajectories of off-chip and on-chip interconnects. Below, we look at several generations of interconnect solutions in Intel products and silicon prototypes as samples on the evolutionary path of the overall system interconnects exemplifying the change in the solution space in response to the change in requirements over a decade.

8.2.1 Off-chip Interconnect: Intel® QuickPath Interconnect Example

Intel® QuickPath Interconnect [195] was first introduced in 2008 with the 45 nm Intel® Core® i7 processor (codenamed Nehalem) to support scalable shared memory multiprocessor architecture of Intel server processors. The Intel® QuickPath Interconnect is a high-speed, message-oriented, point-to-point interconnect. Narrow high-speed links connect the processors in a distributed shared memory-style platform architecture. The cache coherence protocol is optimized for low latency and high scalability, with message and lane structures enabling quick completions of transactions. Reliability, availability, and serviceability features (RAS) are built into the architecture to meet the needs of even the most mission-critical servers. The typical configuration for a 4-socket multiprocessor system introduced later is shown in Figure 8.1.

Intel® QuickPath Interconnect's key advantages can be summarized as:

- Scalability: Intel® QuickPath Interconnect uses a point-to-point link with high-speed unidirectional differential signaling resulting in much higher data rates per pin than possible with bus based designs. Each processor normally supports multiple Intel® QuickPath Interconnect interfaces and an on-die memory controller. Compared to the previous generation front-side bus (FSB), the Intel® QuickPath Interconnect eliminates the competition between processors for memory or I/O bandwidth. Coupled with integrated memory controllers on processor sockets, Intel® QuickPath Interconnect significantly improves the performance of multiple chip servers and workstations.

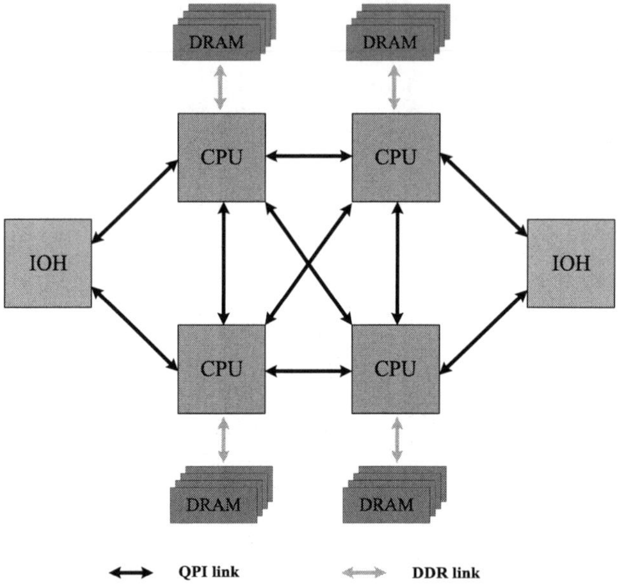

FIGURE 8.1: Example of server system connected using Intel® QuickPath Interconnect links.

- Cache coherence: Intel® QuickPath Interconnect supports a family of cache coherence protocols optimized for different classes of systems, built on top of precisely defined agents, message types and classes, and transactions, with many performance optimization options.

- Performance: An Intel® QuickPath Interconnect link can operate at 6.4 GT/s, delivering a total bandwidth of up to 25.6 GB/s.

- Efficiency: Intel® QuickPath Interconnect reduces the amount of communication required in the interface of multiprocessor systems to deliver faster payloads. A full-width link contains only 84 differential signals (20 lanes in each direction). Half-width (10 lanes) and quarter-width (5 lanes) are possible. The dense message and lane structure allow more data transfers in less time, improving overall system performance.

- Reliability: Implicit Cyclic Redundancy Check (CRC) with link-level retry ensures data quality and performance by providing CRC without the performance penalty of additional cycles. Link-level retry retransmits data to make certain the transmission is completed without loss of data integrity. Advanced servers may enable additional RAS features such as:

    1. self-healing links that avoid persistent errors by reconfiguring themselves to use the good parts of the link;

2. clock fail-over to automatically reroute clock function to a data lane in the event of clock-pin failure;
3. hot-plug capability to enable hot-plugging of nodes, such as processor cards.

Given the Intel® QuickPath Interconnect's design goals as a chip-to-chip interconnect, let's discuss some of the architectural choices made for the Intel® QuickPath Interconnect that may or may not be suitable for an on-chip interconnect. Since the Intel® QuickPath Interconnect is a system-level interconnect, the topology used by specific systems is not known at design time or possibly even at system shipment time since not all boards and sockets in a system may be populated at once. This necessitates that the Intel® QuickPath Interconnect allows a much larger degree of topological flexibility, which is manifested as a very flexible table-based routing mechanism that can be programmed through system software. A narrow message-oriented interface that must provide the same interface across multiple generations of products also limits flexibility on how routing information can be passed between routers. Both of these routing-related limitations are not as critical for an on-chip interconnect and can be relaxed to take advantage of simpler routing mechanisms specific to a topology of interest and enabling schemes such as route precomputation at the expense of additional wires or minor change in the interface between different implementations, thus reducing latency, complexity, area, and the power cost of routing.

Similarly, the error characteristics and frequency for an off-chip interconnect requires a robust detection and recovery mechanisms, such as CRC and link-level retry used in Intel® QuickPath Interconnect. An on-chip interconnect can meet the reliability requirements through much simpler schemes, resulting in latency, area, and power savings. Intel® QuickPath Interconnect uses the virtual cut-through [69] scheme for resource allocation and flow control, where buffers for the entire message are allocated at an input port. Since the buffer sizes at the input ports have to be large to allow for large round-trip delay in the credit loop, the use of virtual cut-through switching does not add to the overall buffering cost. However, in an on-chip interconnect the buffering needed to cover the credit loop is typically small, and other design choices such as the switching mechanism, flow control, and number of virtual channels used play a larger role in determining the buffer sizes at each router.

The above discussion illustrates that the design choices and trade-offs that are suitable for off-chip interconnects may not necessarily be the best choices for on-chip interconnects. In the next section, we will describe a few examples of on-chip interconnect in current designs.

8.2.2 On-chip Interconnect

In this subsection, we will describe some examples of on-chip interconnects in current products and research prototypes.

8.2.2.1 The Intel® Xeon® 7500 Series Processor

The Intel® Xeon® 7500 series processor, also known as the Nehalem-EX processor, is Intel's latest commercial product for enterprise compute servers [170]. Overall, the chip contains 2.3 billion transistors and is manufactured using 45 nm process technology. It supports Intel® QuickPath Interconnect cache coherence protocol enabling systems of 2, 4, or 8 sockets in glue-less configurations or even larger systems using external Node Controller (NC) chipsets. The Nehalem-EX processor chip has 8 cores and a 24 MB shared last level (an inclusive instruction-data unified L3) cache with each core paired with a 3 MB last-level cache (LLC) slice. In addition to the 8 core-LLC slice pairs there are 2 integrated memory controllers with Scalable Memory Interfaces (SMI), and 4 Intel® QuickPath Interconnect link interfaces. Each core can support 2 threads, resulting in 16 threads per chip. Each of the memory controllers can support 4 DDR3[2] channels. Each Intel® QuickPath Interconnect interface supports link operating at 6.4 GT/s transfer rate, resulting in 25.6 GB/s peak bandwidth per link.

Figure 8.2 shows a simplified version of the on-chip interconnect on the Nehalem-EX processor. The bidirectional ring in the middle of the chip is the main interconnection network interfacing the various agents together in an efficient manner. It consists of 4 physically distinct bidirectional rings, 3 of which are used for transferring different classes of control messages and 1 for carrying the data messages. The link width of data ring is 32 bytes in each direction, resulting in a peak bisection bandwidth of 250 GB/s or more depending on clock speed. The link width of each control ring is determined by message classes. Each ring is partitioned into many fixed segments by ring stops, each of which can carry a valid message at a given time (ring slot). Each ring slot advances one stop per clock enabling a pipelined operation. All rings use simple *rotary rules* for distributed arbitration, with the right of way for messages already on the ring. No new message can be injected into a ring if the ring slot passing by is occupied by a valid packet. Antideadlock and antistarvation mechanisms are provided to ensure that rings can operate correctly and fairly under extreme conditions.

Each pair of core and LLC slice shares a single ring stop on each of the rings. At a given clock, a core may receive/sink one message from one direction of a ring and in parallel the associated LLC slice may receive a different message from the opposite direction of the same ring. The next clock, the core and LLC slice will alternate their receiving directions. Such an arrangement is called the *odd/even polarity rule* for sinking messages. Each agent connecting to a ring has polarity rules assigned as part of the system configuration. A message can be injected into an available ring slot only if it matches the correct polarity at the destination.

In addition to the 8 pairs of core and LLC slices, 2 Intel® QuickPath

[2] DDR3 is the double data rate three synchronous dynamic random access memory interface technology.

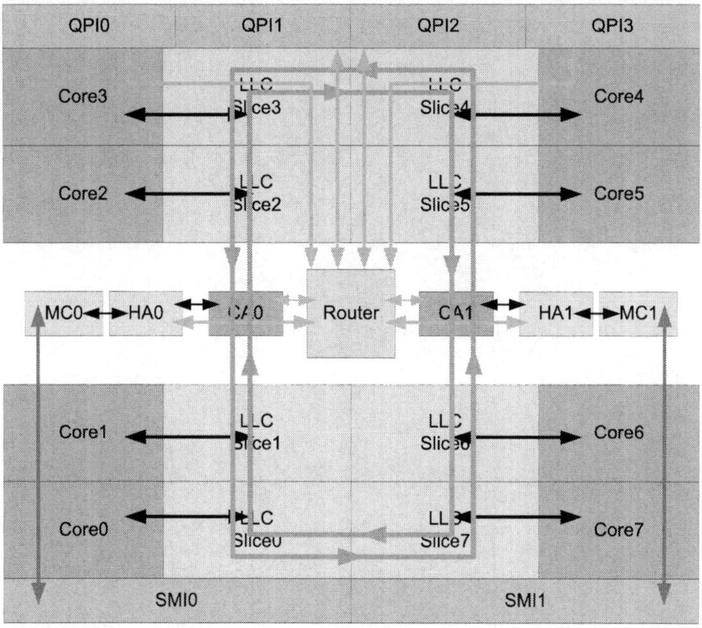

FIGURE 8.2: Block diagram of the Intel® Xeon® 7500 Series Processor (from [170] ©2009 IEEE).

Interconnect caching agents (CAs) are connected to the rings. The caching agents ensure coherence of all cache blocks held by the LLC slices. They interact with 2 Intel® QuickPath Interconnect home agents (HAs) and the associated memory controllers (MCs). All caching agents and home agents interact with 4 Intel® QuickPath Interconnect interfaces via an 8-port router.

The selection of the distinct physical channels and low-latency transfer of ring agents for a small number of agents in this design distinguishes this design for the critical functions in a server architecture system with distributed shared LLC. The large LLC size enables the placement of the ring physical channels over cache regions using the upper metal layers of the chip, rather than allocate additional area. It would not be possible to overlay the ring channels on cores or other logic regions on the tile.

8.2.2.2 The Teraflops Processor

The Teraflops processor chip is the first generation silicon prototype of the tera-scale computing research initiative at Intel [330] [117] [126]. As shown in Figure 8.3, the chip mainly consists of 80 homogeneous tiles arranged in an 8×10 2-D mesh topology operating at 5 GHz and 1.2 V. Each tile contains a simple processing engine (PE) connected to a 5-port router. Each port has two 39-bit unidirectional point-to-point phase-tolerant mesochronous links, one in

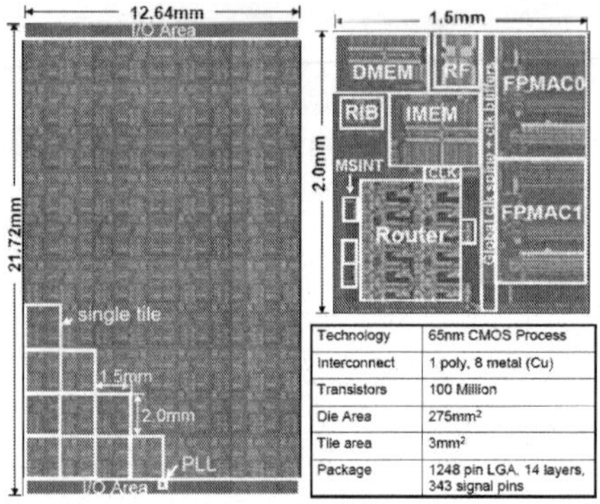

(a) micrograph

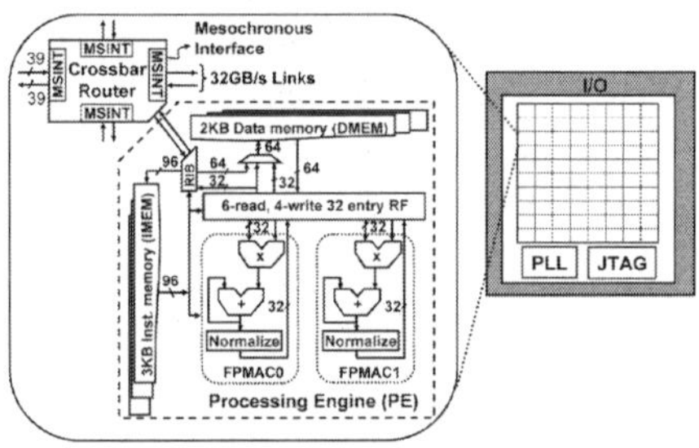

(b) tile architecture

FIGURE 8.3: Intel 80-core Teraflops processor chip (from [330] ©2007 IEEE).

each direction. The total data bandwidth of the router is 80 GB/s (4 bytes × 4 GHz × 5 ports), enabling a bisection bandwidth of 256 GB/s for the network. A router interface block (RIB) handles message encapsulation between the PE and router, transferring instruction and data between different tiles.

The router has a 5-port fully nonblocking crossbar and uses wormhole switching for messages. Each port or link supports two virtual lanes to avoid deadlock. Each lane has a 16-entry First-In First-Out (FIFO) flit buffer. The router uses a 5-stage pipeline with a 2-stage round-robin arbitration scheme. It first binds input port to an output port in each lane and then selects a pending flit from one of the two lanes. A shared data path allows crossbar switch reuse across both lanes on a per flit basis. To save the wire routing area, the crossbar operates in a double-pumped fashion by interleaving alternate data bits using dual edge triggered flip-flops. The entire router is implemented in 0.34 mm^2 using 65 nm technology process.

The on-chip interconnection focus in this prototype chip was the clock

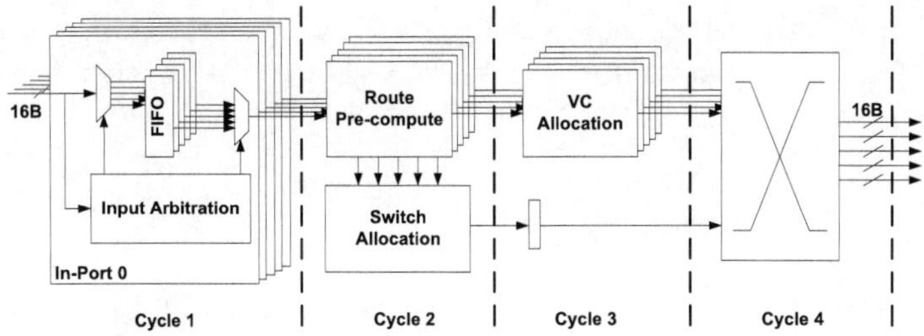

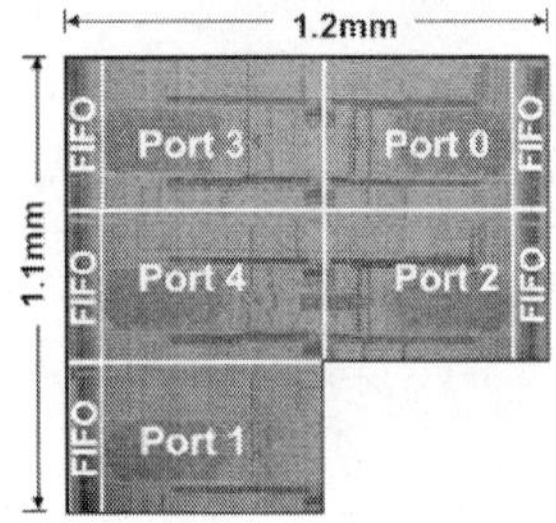

Frequency	2GHz @ 1.1V
Latency	4 cycles
Link Width	16 Bytes
Bandwidth	64GB/s
Architecture	8 VCs over 2 MCs
Power Consumption	500mW

FIGURE 8.4: Architecture of the router in Intel 48-Core Single-chip Cloud Computer (from [127] ©2010 IEEE).

frequency and associated bandwidth for a large number of agents but not latency. The physical channels are narrow to ease the channel width constraint.

8.2.2.3 48-core Single-chip Cloud Computer

The Single-chip Cloud Computer (SCC) is a prototype chip multiprocessor with 48 Intel® IA-32 architecture cores supported on a 6 × 4 2-D mesh interconnect [127]. A 2-core cluster is connected to each of the mesh routers in the interconnect fabric and a total of four memory controllers, two on each side are attached to two opposite sides of the mesh. The SCC architecture supports a messaging-based communication protocol rather than hardware cache-coherent memory for intercore communication. The SCC supports 8 voltage and 28 frequency domains or islands. Voltage isolation and translation circuitry help to provide a clean interface between different voltage domains. The 2-D mesh itself is part of a single voltage and frequency domain. However, clock domain crossing FIFOs are used for clock synchronization at the mesh interface with core clusters that may be running at a different frequency. We discuss the SCC router and interconnect briefly below.

As shown in Figure 8.4, the SCC router represents a significant improvement in implementation of a 2-D on-chip interconnect over Teraflops processor chip. Contrasted with the 2-D mesh implemented for the Teraflops processor, this implementation is more tuned for a wider datapath required in a multiprocessor interconnect and is more latency, area and power optimized for such a width. It also targets a lower 2 GHz frequency of operation compared to the 5 GHz of its predecessor the Teraflops processor, yet with a higher-performance interconnect architecture. Some of the other features of this implementation are as follows:

- 8 virtual channels (VCs), 2 reserved VCs for request/response message classes (MCs), 6 performance VCs
- Flit buffer depth: 24 flits with a dedicated storage of 3 flits per VC
- Virtual cut-through switching
- Message-level arbitration done by a wrapped wave-front arbiter
- 3-cycle router pipeline plus a combined link traversal and buffer write stage (total 4 cycle zero-load latency per hop)
- Deterministic XY routing, implements route precomputation
- 2 request FIFOs, one per MC (round robin between classes)
- Prequalification by checking VC availability

In summary, on-chip interconnects for many-core processors span a continuum of design points as illustrated by the examples in this section. For processors with small enough core count, wide distinct rings offer the best design trade-off and are already in production. For many-core processors with 10s to 100s of cores, more scalable 2-D mesh topologies show promising characteristics and have been implemented in research prototypes. As the evolution of 2-D mesh implementation from Teraflops processor to Single-chip Cloud Computer demonstrated that significant opportunity for latency, bandwidth, area, and power optimizations exist and the designs of these interconnects will evolve to further tune these trade-offs. Moreover, additional features and capabilities can be incorporated to address the needs of emerging architectures and workloads and ways of addressing challenges introduced by process scaling (some of them addressed in previous chapters). In the next section, we will discuss the design of a 2-D on-chip interconnect for a processor with 100s of cores that addresses some of these requirements.

8.3 A Low-latency Flexible 2-D On-chip Interconnect

Tera-scale many-core processors present opportunities that enable high levels of parallelism to address the demands of existing and emerging workloads. These workloads may create hot-spots, jitters, and congestion during communication among various blocks such as cores, on-die storage elements, memory controllers, and I/O hubs. In this section, we first discuss a few example usage scenarios and their desired attributes, and then show how these would favor the use of a general-purpose interconnect. We then present a specific design of an on-chip interconnect, focusing on the optimizations made for latency, energy efficiency, and support for flexibility and adaptivity. Details of an FPGA prototype of this interconnect and illustrative performance results are then presented that indicate the benefits of various features.

8.3.1 Usage Scenarios

The on-chip interconnect architecture discussed here is targeted towards a general-purpose design that can meet the needs of different subsystems in a highly integrated and highly parallel architecture. The challenge of the underlying architecture is to be competitive with application-specific designs. A few example scenarios that are potential targets for tera-scale architecture are presented below.

Cloud Computing or a Virtualized Data Center

The aggregate compute capacity of a tera-scale processor can be partitioned and virtualized to provide cloud computing services to multiple applications sharing its resources. An environment to support this should allow dynamic allocation and management of compute, memory, and I/O resources with as much isolation between different partitions as possible. A large set of allocation and de-allocation of resources can create fragmentation that may not provide a clean and regular boundary between resources allocated for different purposes. The interconnection network bridging these resources should be flexible enough to allow such partitioning with high quality of service (QoS) guarantees and without causing undue interference between different partitions.

Scientific Computing

Scientific computing applications span the range from molecular dynamics to cosmic simulations, fluid dynamics, signal and image processing, data visualization, and other data and compute-intensive applications. These demanding applications are typically executed over large clusters of systems. The compute density per chip and energy per operation provided by tera-scale architecture can greatly enhance the capability of systems targeting this application domain. A typical design for this domain would be dominated by a large number of processing elements on-chip with either hardware or software-managed memory structures and provision for very high off-chip memory and system bandwidth. An interconnect designed for such applications should provide high bandwidth across all traffic patterns, even patterns that may be adversarial to a given topology.

Visual Computing

Visual computing applications are similar to scientific computing applications in some aspects. However, these applications may additionally have real-time requirements including bounded-time and/or bandwidth guarantees.

Also, different portions of an application such as artificial intelligence (AI), physical simulation and rendering have distinct requirements. This heterogeneity, reflected architecturally, would create topological irregularities and traffic hot-spots that must be handled to provide predictable performance for visual workloads.

Irregular Configurations

Cost and yield constraints for products with large numbers of cores may create a requirement for masking manufacturing defects or in-field failures of on-die components that in turn may result in configurations that deviate from the ideal topology of the on-chip interconnect. Another usage scenario that can create configurations that are less than ideal is an aggressive power-management strategy where certain segments of a chip are powered-down during periods of low utilization. Such scenarios can be enabled only when the interconnect is capable of handling irregular topologies with graceful performance degradation.

8.3.2 Desired Attributes

The on-chip interconnect for tera-scale architecture has to be designed keeping in mind the usage scenarios outlined above. Apart from the typical low-latency and high-bandwidth design goals, a flexible topology and predictable performance under a variety of loads are also essential. In this chapter we describe the mechanisms to provide these features.

Some of the main drivers for tera-scale architecture are the chip-level power and thermal constraints that have necessitated the shift from optimizing for scalar performance to optimizing for parallel performance. This shift has resulted in the integration of a higher number of relatively simpler cores on a chip instead of driving towards higher frequency and higher microarchitectural complexity. This trend also has implications for the on-chip interconnect. Due to the weak correlation of frequency with process technology generation, logic delay is becoming a diminishing component of the critical path, and wire delay is becoming the dominant component in the interconnect design. This implies that topologies that optimize for the reduction of wire delay will become preferred topologies for tera-scale architectures. Given the two-dimensional orientation of a chip, two-dimensional network topologies are obvious choices.

Another desirable attribute of an on-chip interconnect is the ability to scale-up or scale-down with the addition or reduction of processor cores and other blocks in a cost-performance optimal and simple manner, enabling the design to span several product segments. Based on the usage scenarios described at the beginning of this chapter, a latency optimized interconnect is critical for minimizing memory latency and ensuring good performance in a cache-coherent chip multiprocessor (CMP). In addition, support is required for partitioning and isolation, as well as fault-tolerance, based on the envisaged usage scenarios. Due to these considerations, two-dimensional mesh and torus, and their variant topologies are good contenders for tera-scale designs (see Chapter 4). The design presented in this chapter assumes the 2-D mesh and torus as primary topologies of interest and with support for some variations of the primary topologies to enable implementation flexibility.

Although on-chip wiring channels are abundant, shrinking core and memory structure dimensions and increasing numbers of agents on a die will put pressure on global wiring channels on the chip. This implies that wiring efficiency, i.e., the ability to effectively utilize a given number of global wires, will be an important characteristic of on-chip interconnect.

In order to support good overall throughput and latency characteristics with a modest buffering cost, our design assumes a buffered interconnect, based on wormhole switching [67] and virtual channel flow control [63]. It supports multiple virtual channels to allow different types of traffic to share the physical wires. The details of this design are discussed in the following sections.

8.3.3 Topological Aspects

Figure 8.5 depicts four on-chip interconnect topology options optimized for a CMP with the tiled modular design paradigm. All options are variants of 2-dimensional (2-D) mesh or torus topologies. In option (a), each processor tile is connected to a 5-port router that is connected to the neighboring routers in both X and Y dimensions. I/O agents and memory interfaces can be connected to the local port of the router or directly to an unused router port of a neighboring tile on the periphery of the chip. Option (b) depicts a 2-D folded torus with links connecting routers in alternate rows and/or columns to balance wire delays. Compared to option (a), average hop count per message is reduced at the expense of longer wires and the number of wires in the wiring channels is doubled for the same link width. In this particular example, more number of routers and links are needed to connect the peripheral devices. Option (c) shows a hybrid 2-D mesh-torus topology with wraparound links in X dimension only by exploiting the fact that all peripherals are now located along the Y dimension. Compared to option (a), the same number of routers are used but the number of wires is doubled in the X dimension. Option (d) shows yet another variation of hybrid mesh-torus with 2 cores sharing one router, resulting in a concentrated topology that requires fewer routers for interconnecting the same number of cores. To enable this topology, either an extra port is required in every router to accommodate the additional core in a tile, or the two cores in a tile share a single local port through a multiplexer and demultiplexer at the network interface.

The network topologies shown in Figure 8.5 and many others can utilize the same basic router design. The trade-off between different network topologies depends on the specific set of design goals and constraints of a given product. In other words, substantial topology flexibility can be achieved through minor design modifications. In the next subsection, we discuss the microarchitecture and pipeline of one such router in detail.

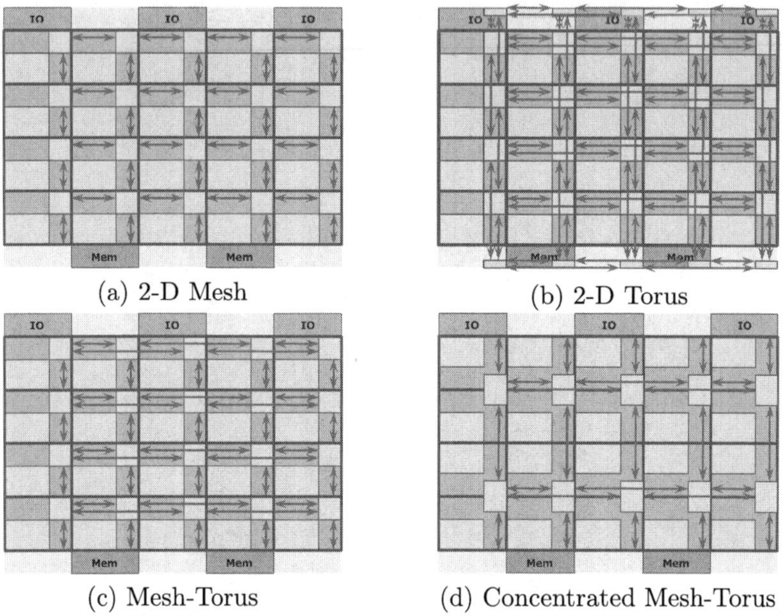

(a) 2-D Mesh (b) 2-D Torus

(c) Mesh-Torus (d) Concentrated Mesh-Torus

FIGURE 8.5: Examples of 2-D mesh and torus topology variants supported.

8.3.4 Router Microarchitecture and Pipeline

The principal component of the 2-D interconnect fabric is a pipelined, low-latency router with programmable routing algorithms that include support for performance isolation, fault-tolerance, and dynamic adaptive routing based on network conditions.

Our router uses a nonblocking 5×5 full crossbar using wormhole switching and flit-level credit-based flow control. The functionality in our adaptive router includes most of the standard functionality that one can expect to see in a wormhole switched router [69]. In the rest of this subsection, we will focus on several novel features that make our design for a many-core on-chip interconnect unique.

The router has a three-stage pipeline with local arbitration (LA), global arbitration (GA), and switch traversal (ST) stages. The LA stage at each input port fairly selects one candidate message request and presents it to the global arbitration (also known as switch arbitration) that occurs in the following clock. For each available output port, the GA stage grants that output port to one of potentially several input port requests selected in the previous cycle at the LA stage of input ports. Following the arbitration, the GA stage determines the crossbar configuration for switch traversal occurring in the following clock. The ST stage orchestrates the movement of the flits from the granted input ports to the corresponding output ports. In addition to the three pipe stages in the router, the link traversal (LT) stage drives a flit at the

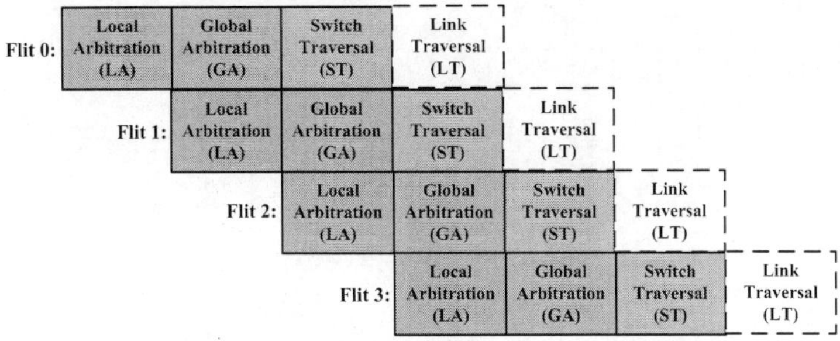

(a) Router pipeline under heavy traffic load

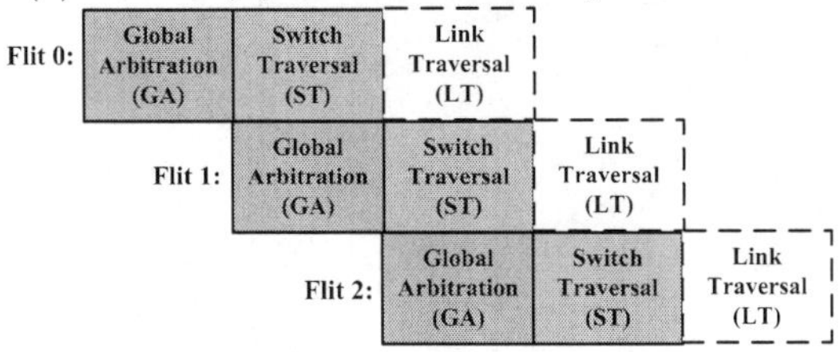

(b) Router pipeline under light traffic load

FIGURE 8.6: Overview of the router pipeline.

output port of an upstream router to the input port of a downstream router. It should be noted that the LT stage is the boundary between a router and a link and is not considered as a stage in the router pipeline per se. The bulk of the LA stage operations (except some bookkeeping) is skipped entirely when a flit arrives at a previously idle input port with no queued-up flits, thereby reducing the router latency under a light traffic load to just two cycles. In such a case, the flit (or message) proceeds directly to the GA stage as it is the only candidate from that input port. The pipeline is reconfigured automatically according to the traffic conditions. Figure 8.6 captures the router pipeline stages under heavy and light traffic conditions.

Figure 8.7 shows the key functions performed in each stage. The LA stage performs four major functions: input message processing, shared buffer management for flit write / insertion in to the flit buffer, virtual channel (VC) request prequalification, and input-port arbitration. The GA stage offers five major functions: switch arbitration, flit source selection (LA bypass or lack of it) and shared buffer management for flit-buffer read and removal from flit buffer, message route precomputation, VC allocation, and downstream router credit accounting. The ST stage performs two major functions: crossbar traversal and credit management for upstream router. And finally the LT stage moves flits from an upstream router to a downstream router. The detailed description of some of these functions follows.

Virtual Channels and Buffer Management

Our router architecture relies on virtual channel flow control [63] both to

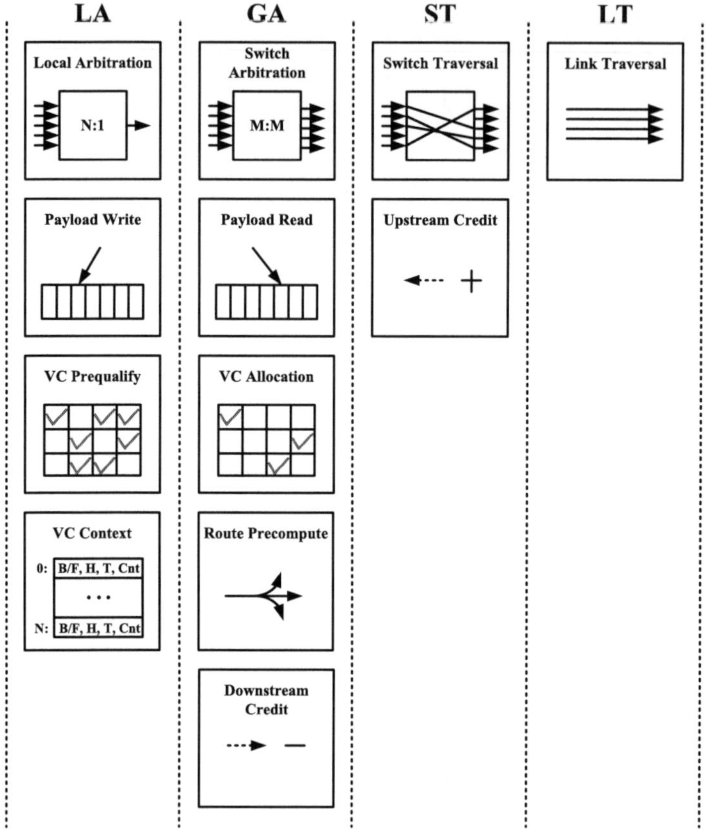

FIGURE 8.7: Key functional blocks in the router pipeline.

improve performance and to enable support for deadlock-free routing with various flavors of deterministic, fault-tolerant, and adaptive routing. The set of virtual channels (VCs) is flexibly partitioned into two logical sets: routing VCs and performance VCs. VCs are also logically grouped into virtual networks (VNs). VCs belonging to the same VN are used for message-class (MC) separation required by protocol-level MCs, but they use the same routing discipline. Routing VCs are also used for satisfying the deadlock-freedom requirements of particular routing algorithms employed. Each routing VC is associated with one and only one MC with at least one reserved credit. Performance VCs belong to a common shared pool of VCs. They can be used by any MC at a given time both for adaptive or deterministic routing schemes. A VC is used by only one message at any given time to manage design complexity and to ensure deadlock freedom for support of fully adaptive routing based on Duato's theory [84].

The number of VCs supported in a design is a function of design-time

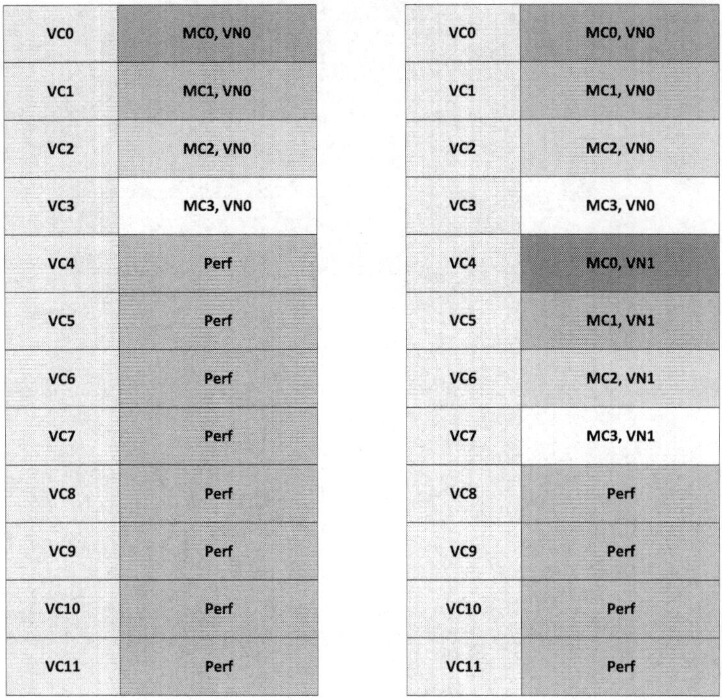

(a) 2-D mesh (b) 2-D torus

FIGURE 8.8: Example of mapping of virtual networks to virtual channels for 2-D mesh and 2-D torus topologies.

goals, area, and power constraints. Figure 8.8 depicts examples of mapping of the supported VCs into routing VCs belonging to specific message-classes and VNs required for supporting minimal deadlock-free XY routing algorithms for 2-D mesh and torus topologies, as well as the pool of performance VCs. The example configurations assume a total of 12 VCs and 4 MCs; the mesh requires a single VN (VN0) for deadlock freedom, whereas the torus requires 2VNs (VN0, VN1).

A single shared buffer at each input port [312] is used to support flexibility and optimal usage of message buffering resources with respect to performance, power, and area. The buffer is shared by all VCs, either routing VCs or performance VCs, at a port. Buffer slots are dynamically assigned to active VCs and linked lists are used to track flits belonging to a given message associated with a VC. A free buffer list tracks buffer slots available for allocation to incoming flits. The operation of the flit-buffer is illustrated in Figure 8.9.

Depending on the size of the buffer, a banked implementation is used where each bank can be individually put into a low-leakage state to save power during

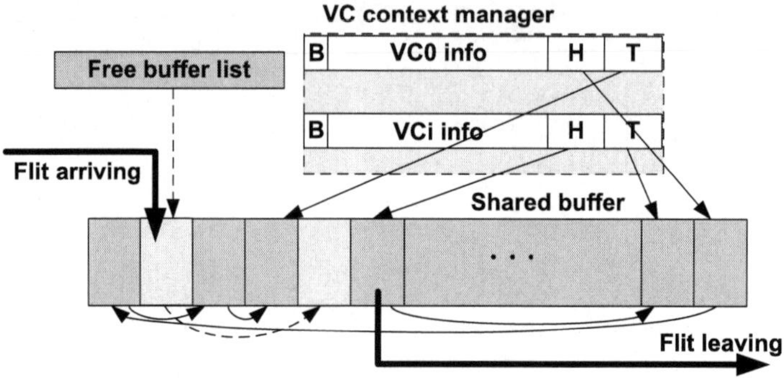

FIGURE 8.9: Shared flit buffer management across a set of virtual channels.

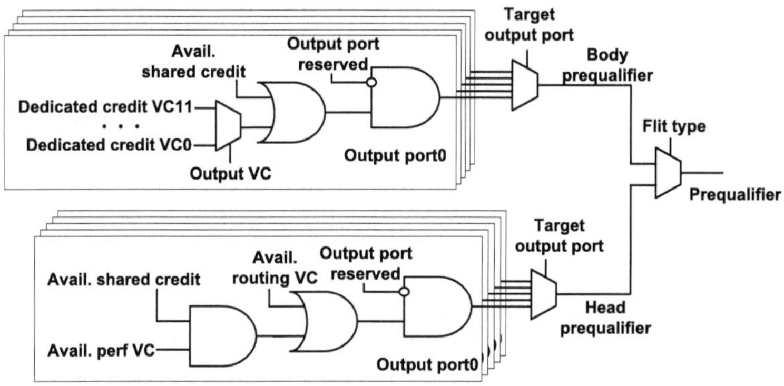

FIGURE 8.10: VC request prequalification.

low levels of buffer utilization. Active power-management strategies to trade off performance and power can also be implemented.

VC Request Prequalification

A request from each VC that has waiting flits is filtered using 3 conditions: target physical port availability, VC availability, and credit availability. We call this filtering approach, request prequalification and it is used to improve the success rate of switch arbitration in the GA stage, thereby yielding a high throughput design.

The prequalification conditions change slightly depending on whether a flit is a header or body flit. The details are shown in Figure 8.10. These conditions are updated in the GA stage at every clock. However, as the LA stage precedes the GA stage, it uses a one-cycle stale state for prequalification.

Flow Control

The router uses credit-based flow control to manage the downstream buffering resources optimally. It tracks available input buffer space in the downstream router at the output side of the crossbar, through handshaking between two parts: upstream credit management and downstream credit management. The flow control protocol is a typical scheme, except for the fact that it needs to handle the shared flit buffer resources and ensure reservation of at least one buffer slot resource for all routing VCs and for each active performance VC. Specifically, a routing VC always has one reserved credit, regardless of whether it is in use or not. A performance VC in use has one reserved credit. A performance VC not in use has no reserved credit. These reservations are required to guarantee resource deadlock freedom with wormhole switching. We adopt the following usage policies between reserved credits and shared credits: (a) reserved credit is used ahead of shared credits; (b) a returned credit is reserved for a particular VC if that active VC has no reserved credit; otherwise, the returned credit is put into the pool of shared credits.

Route Computation

Routing determines which path a message takes to reach its destination. We use a distributed routing scheme to select an output port at a given router that a message must take to move towards its destination. For minimal adaptive routing, up to two distinct directions may be permitted based on the region a destination node falls into. The routing decision (i.e., output ports and VN choices permitted) at each router is based on the current input port and VN a message belongs to, as well as on the destination address of the message. To support the flexible algorithms, two different options are supported in our design: a compressed table-based distributed routing (TBDR) [327] [93] and logic-based distributed routing (LBDR) [94] (see Chapter 5). Compared to the LDBR scheme, the TBDR scheme uses a 9-entry table per router providing more routing flexibility with higher storage overhead. More details can be found in [16].

The LBDR scheme uses a connection bit per output port and 2 turn-restriction bits. Different routing algorithms can be supported by setting appropriate turn restrictions. Our router architecture uses route precomputation [69] [97] for the route decision of neighboring routers, thereby removing route computation from the critical path of the router pipeline. As shown in Figure 8.11, it can be divided into 2 steps: (1) compute route tags based on message destination, identifying the target quadrant and (2) determine the output port based on the selected routing algorithm. In an adaptive routing scheme, this can imply that the message may have a choice of more than one output port towards its destination. We support up to two output port choices.

Arbitration Mechanisms

In our basic design, we use a two-phase round-robin flit-level arbitration scheme: input-port arbitration in the LA stage and switch arbitration in the GA stage. In the LA stage, at every clock, each message that passes prequalification arbitrates for the input port using a round-robin priority. The priority is adjusted only when a grant is accepted. To avoid any undesired bubbles or fragmentations in a message, the highest priority (HP) is set to the winner's

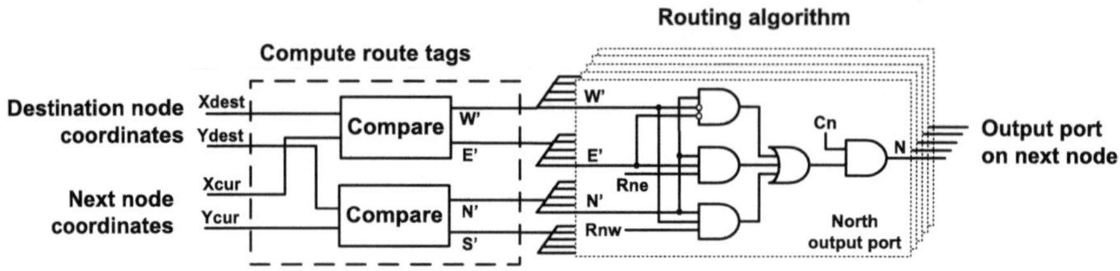

FIGURE 8.11: Route precompute (Cn is north port connection bit; Rne and Rnw are turn-restriction bits for northeast and northwest turn, respectively).

position if the grant is given to a nontail flit. Otherwise, the HP is set to the position immediately following the winner. Figure 8.12 shows an example of how priority rotation is done with 12 VCs. VCs 0, 1, 2, 4, 8, and 10 have no qualified requests; VCs 3, 5, 6, 7, 9, and 11 have qualified requests; and VC1 has the highest priority. VC3 will be the winner of this arbitration. If the grant to VC3 is accepted, the HP is set to VC3 provided the grant is to a nontail flit. Otherwise, the HP is set to VC4.

In the GA stage, switch arbitration is done by an array of arbiters using the round-robin priority, with one arbiter per output port. In every clock cycle, each output-port arbiter selects a single winning input port among one or more candidates from all input ports at every clock. A multiflit message may make a *multiflit tagged* request, and the arbiter grants the output port to such an input port for an additional cycle. Each arbiter speculates on the availability of a VC and a buffer credit on its associated output port to reduce timing critical paths. Misspeculation probability is low because most requests under high traffic conditions go through the LA stage prequalification having been queued at the input port. Under a low load, LA stage bypassing messages may not be prequalified but chances of finding available VC and buffer slots are high. VC allocation and credit checking occur in parallel with switch arbitration. A grant is accepted only if the speculation is confirmed correct. If the speculation is wrong, that output port will be idle for one clock and the failed request will trigger retry at its own input-port LA stage.

The priority rotation is done only if a grant is accepted. Similar to the priority rotation in the LA stage, each output-port arbiter adjusts its priority among the input ports depending on whether the grant is to a tail flit or not. An example is shown in Figure 8.13. Figure 8.13.(a) shows at the beginning of a clock, requests from different input ports arbitrate for desired output ports with current round-robin priority associated with each output port. Figure 8.13.(b) shows at the end of the same clock, which input requests have been granted and how the round-robin priority on each output port is adjusted for the next clock.

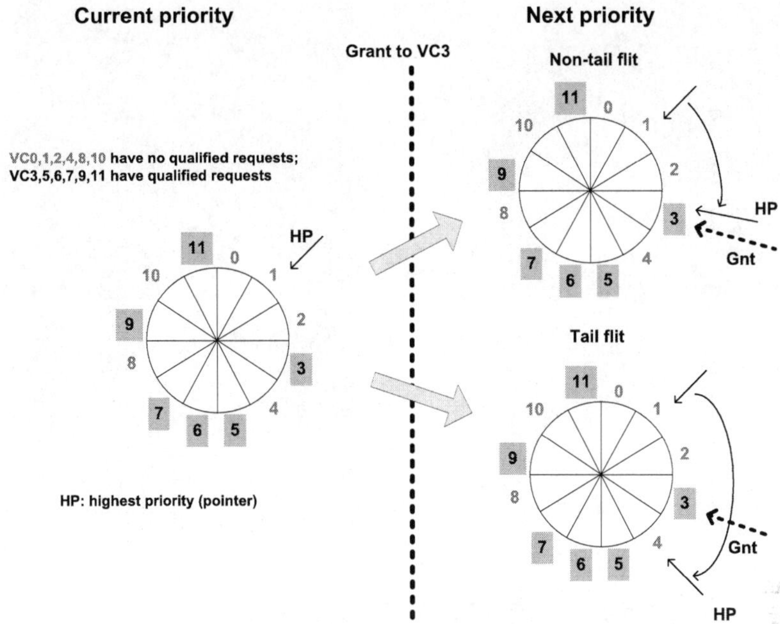

FIGURE 8.12: Arbitration priority rotation at LA stage.

8.3.5 Routing Algorithm Considerations

In this section we describe the support for various routing algorithms to enable a flexible, configurable, and adaptive interconnect, and we discuss the design implications.

Support for Multiple Routing Algorithms

The router architecture supports distributed routing wherein the subsets of the routing decisions are made at each router along the path taken by a given message.

In two-dimensional networks like mesh and torus, given a source node, the set of shortest paths to the destination fall into one of four quadrants. With the LBDR framework [94], any turn model-based routing algorithm [105] such as XY, west-first, odd-even [52], etc., can be implemented by setting the turn-restriction bits appropriately. For minimal adaptive routing, up to two distinct directions may be permitted based on the quadrant a destination node falls into. The routing decision (i.e., output ports and VN choices permitted) at each router is based on the current input port and VN a message belongs to, as well as on the desired destination. For each VN, we support the flexible algorithms with very economical storage of only a few bits per port or with an alternative small 9-entry table [16].

Minimal path deterministic routing in mesh and torus and partially and fully adaptive minimal path routing algorithms, such as those based on the

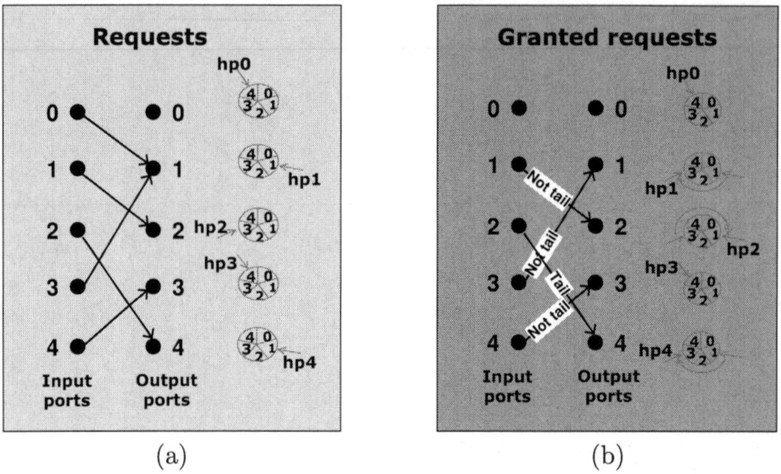

FIGURE 8.13: Arbitration priority rotation at GA stage.

	Message Classes	
Topology	Turn Model	Duato's theory
2-D Mesh	8	5
2-D Torus	12	9

TABLE 8.1: Minimum virtual channels required for fully-adaptive routing (Turn model v/s Duato's theory).

turn model [105], are supported. Our adaptive router architecture uses Duato's theory [84] to reduce the VC resource requirements while providing full adaptivity. Table 8.1 shows a comparison of the minimum number of VCs required to implement deadlock-free fully adaptive routing using the turn model versus one based on Duato's theory.

TBDR routing support also enables a deterministic fault-tolerant routing algorithm based on fault-region marking and fault-avoidance, such as in [36], as well as adaptive fault-tolerant routing algorithms [83]. Incomplete or irregular topologies caused by partial shutdown of the interconnect because of power-performance trade-offs can be treated in a manner similar to a network with faults for routing reconfiguration purposes.

A novel two-phase routing algorithm, which can use both minimal and nonminimal routing approaches, is also supported. It can be used to implement load-balancing and fault-tolerant routing in the interconnect.

In addition, the design also supports performance isolation among partitions on a mesh that may or may not have rectangular geometries. This is implemented through a hierarchical routing approach that helps isolate communication of each partition. This is purely a routing-based approach to provide isolation, and no additional resource management mechanisms, such as bandwidth reservation, are required with this approach.

Design Implications

Our flexible design had to have a minimal cost overhead in the definition phase. We outline next the various impacts of our design choices.

We maintain a high-performance router design by using a shared pool of VCs, as well as a shared buffer pool. The shared buffer pool reduces the overall buffer size and power requirements. Support for up to two output port choices with adaptive routing has little impact on the router performance. With the decoupled local and switch arbitration stages in the router pipeline, each message arbitrates for a single output port candidate after applying the prequalification filter and path selection. The criteria used could be based on several congestion prediction heuristics that use locally available information, such as resource availability and/or resource usage history (see [69] for examples of such criteria). Path selection can be implemented without a significant impact on the arbitration stages.

Support for the adaptivity and flexible routing configuration requires the use of configurable route tables. The LBDR framework provides us the desired flexibility with very small area and power overhead. These table storage optimizations, that are independent of the network size, considerably reduce the area and power requirements for supporting flexibility features in the on-chip interconnect.

8.3.6 FPGA Emulation Prototyping

We have developed a full featured register transfer language (RTL) implementation of the router using Verilog for the purpose of robust validation of the microarchitecture and design as well as to conduct a detailed performance characterization of the interconnect [61]. The larger goal of the interconnect prototyping effort is to have a robust interconnect that can then be interfaced to several production grade processor cores and their cache coherence protocol engines.

The two-cycle wormhole router microarchitecture with multiple MCs, virtual channels, and shared buffers has been implemented. Various microarchitectural parameters are configurable, including the number of MCs, performance and routing VCs, and buffer sizes. Along with each router in the prototype, a network node also implements a network interface (NI) block for message ingress and egress functionality as well as a synthetic traffic generator. Uniform random, transpose, bit-complement, and hot-spot traffic patterns are currently supported by the traffic generator along with several additional configurable parameters for controlling injection rates, MCs, and sizes. Various routing algorithms using programmable routing tables have been implemented including basic XY routing, turn model-based routing, load balanced and adaptive routing, fault-tolerant routing, and support for isolation of multiple partitions as well as support for mesh and torus topologies. Our prototype en-

Pipeline	2-cycle router pipeline with bypass support
Topology	2-D Mesh / Torus
Routing algorithm	Both hardwired routing and table-based routing Algorithms: East-Last, Odd-Even, Up-Down, South-Last, West-First, SR-Horizontal, SR-Vertical and XY. Fault-tolerant and hierarchical routing (for performance isolation) also supported.
Traffic Patterns	4 traffic patterns supported: uniform random, transpose, bit-complement, hotspot
# nodes per board	2×3 mesh per board (2 nodes per FPGA and 3 FPGAs are used for router fabric) 6 boards used to form a 6×6 mesh
Other	Configurable network size, buffer sizes, virtual networks, number of performance VCs

TABLE 8.2: Key features of the emulation framework.

ables both early design development and configuration exploration. The RTL has 120+ Verilog modules and approximately 32K lines of code. The RTL is synthesized to FPGA bit-streams using Synopsys SynplifyPro 9 and Xilinx ISE 10 tools.

Table 8.2 summarizes the key features implemented in the RTL. Figure 8.14 shows initial breakdown of look-up table (LUT) usage for each major router module after synthesis.

The FPGA prototype has been implemented on Intel's Many Core EMUlator (MCEMU) platform [203]. As shown in Figure 8.15, each MCEMU board has 5 Xilinx FPGA chips (Virtex 4 and Virtex 5 class). Our current interconnect emulation configuration supports six 5-port routers on each board in a 2×3 mesh configuration. Three FPGAs (numbered FPGA2, 3, 4 in Figure 8.15) are used for the network logic (2 nodes are implemented per FPGA, each node with a router, network interface, and traffic generator). FPGA1 contains Xilinx Rocket-I/O interface so 10 external links from each 2×3 mesh block on the FPGA board connect to up to 4 other neighboring MCEMU boards each with similar blocks of a 2×3 mesh. FPGA 5 supports the system interface to the host including control and the debug plane for each card.

A larger 2-D mesh or 2-D torus topology is built by suitable interconnections across multiple MCEMU boards. For example, we used 6 boards to build a 6×6 mesh/torus network, as shown in Figure 8.16.

Emulator control and visualization software system enables one to initialize a multiboard configuration with appropriate bit-streams and then configure and run multiple experiments with various microarchitectural parameters, routing algorithm tables, and traffic patterns. The software environment enables one to run each experiment for any given number of cycles after which a large array of performance-counter values that can be recorded. Registered

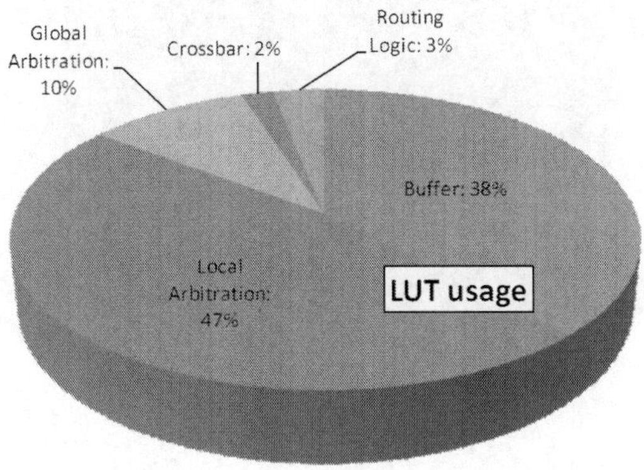

FIGURE 8.14: LUT usage of router modules (16 VCs and 24 flit buffers per input port for 144-bit link widths per direction per router port).

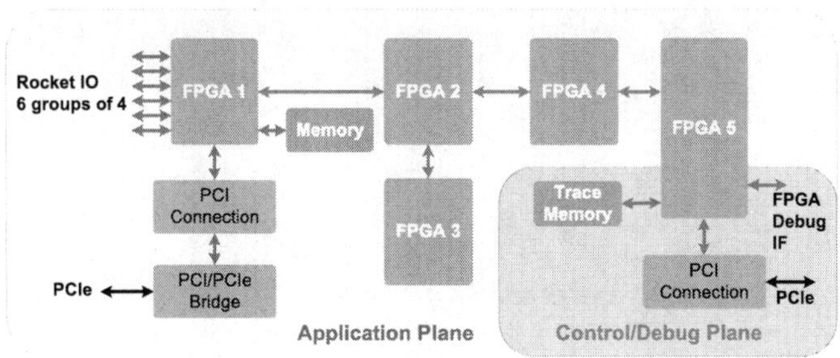

FIGURE 8.15: MCEMU FPGA board with key components and connectivity (from [203]).

values include number of injected and ejected messages, message latencies split by MCs, buffer utilizations, bypass and arbitration success / failure rates per port, etc. A custom graphical user interface (GUI) for control and performance visualization is used to run experiments interactively and graphically render performance data in real time for each experiment.

Extensive validation of our 2-D mesh RTL through this infrastructure over a wide range of parameters has led to the next phase of our prototyping effort that integrates RTL of multiple production processor cores and a cache coherence protocol with the 2-D mesh interconnect infrastructure to reliably execute a standard operating system and applications for several tens of billions of processor cycles.

(a) Physical configuration

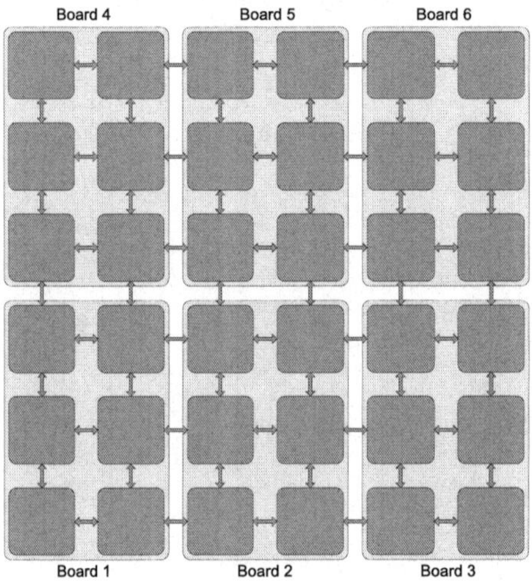

(b) Logical configuration (without showing wraparound links)

FIGURE 8.16: Example of an emulation system configuration (6×6 mesh/torus mapping onto 6 boards).

8.3.7 Performance Analysis

In this section, we present some representative performance results for different configurations and traffic patterns to demonstrate the effectiveness of the adaptive routing scheme described earlier in this section. However, before presenting the results on adaptive routing, we present the results to establish the optimal design parameters for typical operation.

Figure 8.17 shows the effect of buffer size and number of VNs on each input port on the network capacity for a uniform random traffic pattern on a 6×6, two-dimensional mesh network. This traffic pattern uses a mix of single flit and five-flit messages divided between two MCs, based on an expected distribution

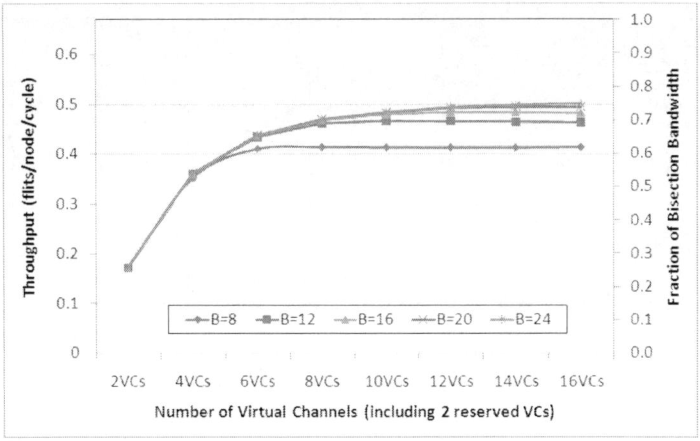

FIGURE 8.17: Effect of number of buffers and virtual channels on delivered throughput for a 6×6 mesh.

for cache coherency traffic. The target of each message is randomly selected with equal probability, and the injection rate is increased until the network is close to saturation. For different buffer depths and VC settings, the network saturates at different points. Each plot in Figure 8.17 represents different buffer depths and shows the delivered throughput in terms of flits accepted per cycle per node for increasing numbers of VCs. The plots indicate that a larger number of buffers and VCs result in increased throughput; however, the improvement tapers off beyond sixteen buffers and ten virtual channels. These results were obtained by using a deterministic dimension-order routing scheme as adaptive schemes are not beneficial for uniform random traffic patterns. We use 16 buffers and 12 VCs as the baseline to evaluate the effectiveness of an adaptive routing scheme against that of a deterministic scheme.

We use a traffic pattern that is adversarial to a two-dimensional mesh topology in order to illustrate the effectiveness of an adaptive routing scheme. Adversarial traffic for a given topology and routing scheme illustrates a worst-case scenario, which shows the extent of degradation in network performance for some traffic patterns. A transpose traffic pattern is a good example for a mesh with deterministic XY routing, since it results in an uneven traffic load across the links along the diagonal.

Figure 8.18 illustrates the source-destination relationship for a transpose operation. For this traffic pattern, a node labeled (i,j) communicates with node (j,i) and vice versa. This results in nodes in the upper triangle communicating with nodes in the lower triangle. The nodes on the diagonal do not generate any traffic. Figure 8.18 also illustrates the route taken when a deterministic XY routing scheme and some possible alternatives allowed by an adaptive routing scheme are used. Paths highlighted with thick lines do not follow

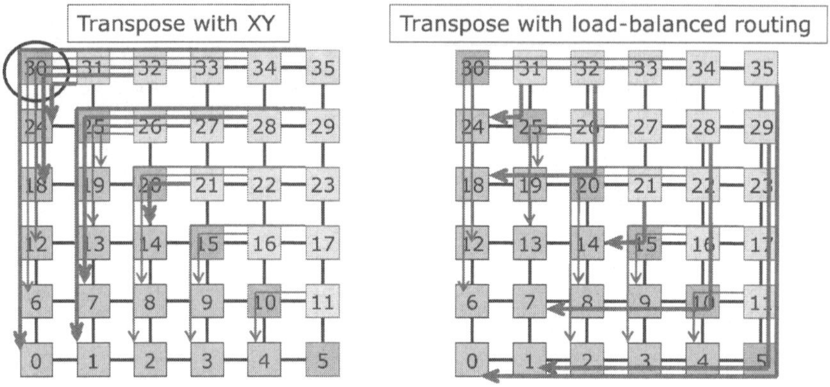

FIGURE 8.18: Traffic pattern for a transpose operation with deterministic and adaptive routing.

the XY routing scheme. There are many other paths possible between these source and destination nodes that are not shown in the figure. A deterministic routing scheme tends to concentrate the load on a few links in the network for this traffic pattern, and not utilize other alternative paths between source-destination pairs. An adaptive routing scheme allows more flexibility in path selection among multiple alternatives thereby avoiding congested routes and improving the overall capacity of the network.

The effect of path diversity in increasing the overall network capacity is illustrated through the load-throughput plot in Figure 8.19 for transpose traffic by using different routing schemes. "XY" represents deterministic XY routing, whereas "Adaptive" is an implementation of a minimal fully-adaptive routing scheme using Duato's theory. As shown in the plot, the network capacity is severely restricted for this pattern when a deterministic routing scheme is used. The adaptive routing scheme delivers much higher throughput for this traffic pattern.

When all the agents in the on-chip network are not of the same type, non-homogeneous traffic patterns can be created with the potential for transient hot-spots. One example of this is the traffic going to and generated from memory controllers, I/O interfaces, and system interconnect. Depending on the phases of execution or type of workload, some of these agents may be more heavily used than others and may become bottlenecks in the interconnect, thereby affecting other traffic that shares the path with congested traffic. Such scenarios can be approximated with a mix of traffic patterns between different sets of agents. To illustrate this scenario, we set up an experiment with a 6×6 mesh network where six agents at the periphery (three in the top row and three in the bottom row) are considered hot-spots with 30 percent of the traffic targeting these nodes with equal probability and the rest of the traffic targeting the remaining nodes with equal probability. Throughput delivered

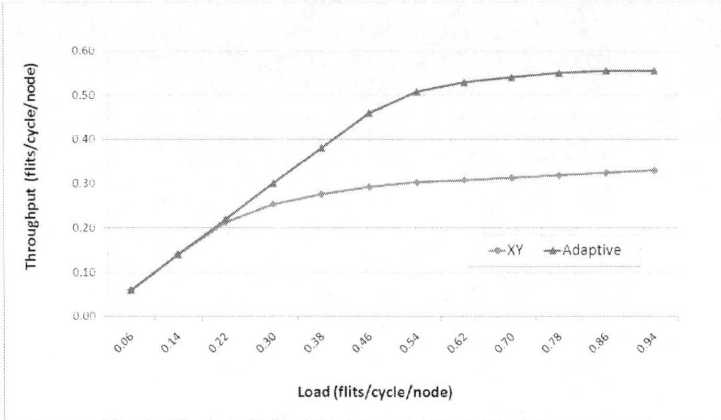

FIGURE 8.19: Network throughput with a deterministic and adaptive routing scheme for transpose traffic on a 6 × 6 mesh.

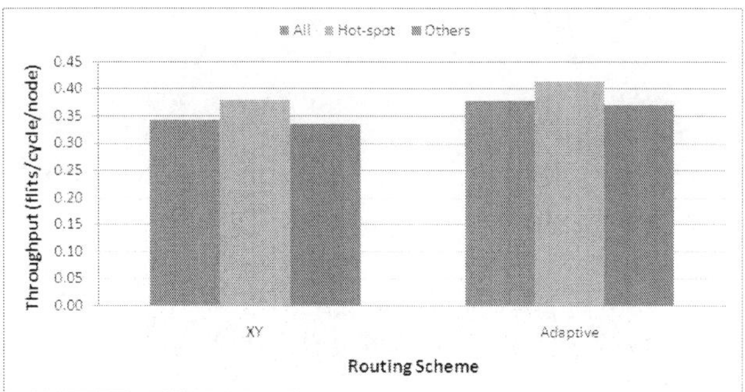

FIGURE 8.20: Network capacity with and without adaptive routing in a heterogeneous environment.

to each node in terms of flits per cycle was measured for all nodes combined, for hot-spot nodes, and for nodes excluding hot-spot nodes. The benefit of adaptive routing for this scenario is illustrated in Figure 8.20, which shows improved overall throughput as well as the throughput delivered to hot-spot nodes and other nodes excluding the hot-spot nodes.

The interconnection network can be reconfigured in the presence of faults in interconnection network links or switches by using fault-tolerant routing algorithms. Without support for fault-tolerant routing, the network may be rendered virtually unusable. (It might be able to function at a significantly diminished capacity by avoiding the use of a whole row and/or column of nodes containing a faulty node. However, we do not consider such highly degraded configurations valuable.)

Figures 8.21 and 8.22 illustrate network behavior in the presence of faults.

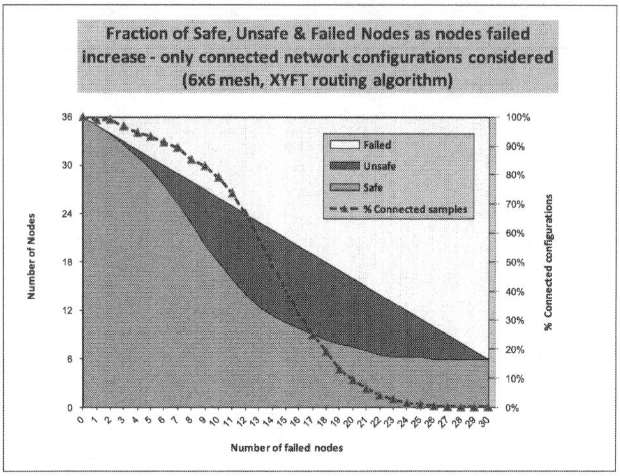

FIGURE 8.21: Profile showing an average number of safe (useable), unsafe, and failed node fractions as number of faults is increased for a 6×6 mesh with fault-tolerant XY routing. The averages are for randomly placed faults over a large number of samples and only configurations where all working nodes form a single connected cluster are considered. The fraction of such connected configurations over the total number of configurations for a given number of failed nodes is plotted alongside for reference.

These results are for a 6×6 2-D mesh network averaged over a large number of configurations, each with a given number of randomly placed faults. We have used the fault-tolerant XY routing algorithm described in [36] (XYFT). A node-marking algorithm first identifies faulty nodes and can turn off additional healthy nodes neighboring faulty ones that are unsafe for the XYFT algorithm. This creates "convex" or rectangular fault regions (with faulty and unsafe nodes) around which the XYFT algorithm can safely route in a deadlock-free manner. If all the remaining unmarked nodes (safe nodes) are all connected, the network has a working or "routable" configuration; else, a disconnected is considered unusable and discarded. Only connected network configurations are considered in the results shown.

Figure 8.21 shows the split-up of an average number of safe, unsafe, and faulty nodes in the network as the number of faulty nodes increases. When the number of faulty nodes (placed at random) is small, very few or no additional nodes need to be turned off by the node-marking algorithm. As the number of faults increases, additional nodes needs to be turned off, seen as the increase in the number of unsafe nodes. While the figure shows these trends for the connected configurations even for a very large number of faults, in reality, the region of interest is primarily for a small number of failures. The change in the percentage of connected configurations as a fraction of all possible configu-

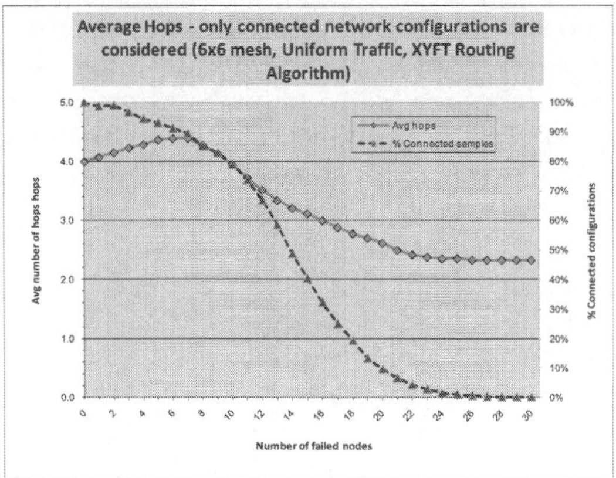

FIGURE 8.22: Average hops profile with increasing number of faults. The average includes only those configurations in which all safe nodes form a single connected cluster.

rations with an increasing number of faults is also shown in the same figure for reference. It can be noted that a very high fraction of configurations remain connected and routable with XYFT routing when the number of failed nodes is small. The percentage of connected and routable configuration deteriorates as the number of failed nodes increases.

Figure 8.22 shows that for the uniform traffic pattern, the increase in the average number of hops in the network is quite modest for a small number of faults. As the number of faults increases, that number plus the corresponding number of unsafe nodes together reduces the overall working cluster size compared to the size of the full network. This drop in working cluster size causes the average number of hops to decrease for larger fault counts.

The above results show that for a small number of failures—our primary region of interest—an overwhelming fraction of such configurations are routable, with little or no additional nodes required to be turned off. In this region of interest, there is only a modest increase in the average number of hops compared to the no-fault case leading to graceful performance degradation. The results demonstrate the value of fault-tolerant routing support in our network.

8.4 Summary and Conclusions

The number of cores in a processor is expected to scale from tens to hundreds, providing aggregated performance of over one trillion operations per second per chip. Such a tera-scale many-core processor will present opportunities that enable high levels of parallelism to address the demands of existing and emerging workloads at very affordable price points. Demanding workloads in a many-core processor will create transitory hot-spots, jitters, and congestion during communication among various blocks. In addition, effective approaches to address manufacturing defects, on-chip variation, and aggressive power management will be required. An adaptive and flexible on-chip interconnect capable of dynamically and gracefully responding to these requirements and conditions is a critical component of a tera-scale processor.

In this chapter, we first briefly reviewed the evolution of on-chip interconnects used in recent Intel products and research prototypes and then discussed new usage models and desired attributes for many-core processors. Following the lessons learned and new insights, we presented the details of our on-chip fabric and an adaptive router that supports various 2-D interconnect topologies. 2-D mesh and torus topologies provide good latency and bandwidth scaling for tens to more than a hundred cores. The rich connectivity in 2-D topologies enables multiple paths between source-destination pairs that can be exploited by adaptive routing algorithms.

The architecture presented has an aggressive low-latency router pipeline. We are also able to provide high throughput in the presence of adversarial traffic patterns and hot-spots, through the use of adaptive routing. Adaptive routing is supported without adversely impacting the router pipeline and it is supported through very economical and configurable routing table storage requirements. Our router also supports very efficient use of resources by making use of shared (performance) VC and buffer pools and suppresses undesirable message fragmentation.

Our fabric architecture and routing algorithms also support the ability to provide partitioning with performance isolation and the ability to tolerate several faults or irregularities in the topology, such as those caused by partial shutdown of processing cores and other components for power management. All the novel features and performance results presented in this chapter have also been implemented and validated using FPGA prototyping.

Acknowledgment

The work presented here has benefited from the contributions and insights of our colleagues Ching-Tsun Chou, Seungjoon Park, Roy Saharoy, Hariharan Thantry, Dongkook Park, Andres Mejía, and Gaspar Mora Porta as well as our former colleagues Jay Jayasimha, Partha Kundu, and the late David James.

We are also immensely thankful to Intel Labs leaders Joe Schutz, Justin Rattner, Jim Held, and Andrew Chien for their continued support and encouragement for our work on scalable on-chip interconnect fabrics through the tera-scale initiative.

Chapter 9

The TRIPS OPN: A Processor Integrated NoC for Operand Bypass

Paul V. Gratz

Texas A&M University, USA

Stephen W. Keckler

The University of Texas at Austin, USA

9.1 Introduction

With continued process technology scaling, wire delay and design complexity has imposed significant constraints on traditional, superscalar processor design [6]. Achieving high performance in superscalar processors requires complex, dynamic operand bypass networks to route the output of producer instructions to the input of consumer instructions in a timely fashion. Traditionally, operand bypass networks in superscalar processors were implemented with ad-hoc buses and point-to-point networks; however, as smaller process technologies led to greater numbers of instructions in flight and increased wire delays, the unfeasibility of building a traditional, ad-hoc point-to-point operand network has limited the scalability of superscalar processors. This chapter describes the design and implementation of a scalable, Network-on-Chip (NoC) based, operand bypass network as an alternative to traditional operand bypass networks.

9.1.1 Interconnect in Superscalar Processors

Instruction-level parallelism (ILP) characterizes the number of instructions within a program that may execute simultaneously given the dependence relationships between those instructions. Superscalar, out-of-order processors, such as the Intel i7 [43] and IBM Power7 [140], leverage ILP to achieve performance. In these processor architectures, a program's sequential instruction stream is searched to find a set of independent instructions that are ready to execute at a given time. Independent instructions issue to the Arithmetic-Logic Units (ALUs) dynamically, depending on the availability of operands for those instructions. As instructions complete, they produce operands that will be consumed by the next set of independent instructions.

The number of available ALUs in a superscalar processor defines the maximum number of instructions it may execute at a time, providing an upper bound on the ILP that a processor may exploit. Assuming a given program has available ILP, one way to increase performance is to increase the number of ALUs in the processor. Increasing the number of ALUs in a traditional superscalar processor, however, implies scaling the operand bypass network used to route operands between inputs and outputs of the ALUs, register file, and load/store units.

An operand bypass network shuttles instruction operands between the register files, load/store units, and the ALUs. Each of these units has a unique function. The register files provide an architecturally visible, local storage for operands and are the initial source for most ALU operations. The register files also maintain the architected state of committed instructions in the face of exceptions. The load/store units provide an interface to the memory system to fill and spill the register file according to the program's needs. The ALUs perform the actual work of the processor, computing on operands provided by the register file, or fed back from the output of the ALUs themselves. Because the function of the units are different, their communication needs are different.

Low latency in the operand bypass network, particularly between the outputs of the ALUs back to their inputs is critical to achieve performance in superscalar processors [128]. Delay in routing an operand from the execution of a producer instruction to its consumption by a consumer instruction reduces the effective exploitable ILP. Operand bypass networks must also provide enough bandwidth to convey all the operands produced and consumed each cycle, a factor of the total number of ALUs and available program ILP. Finally the operand bypass network must scale to match the number of ALUs in the system. Performance gain through provisioning of additional ALUs is only feasible if the bypass network can be scaled to match.

9.1.2 Scalable Operand Bypass Networks

In current superscalar processor microarchitectures, the bypass network is composed of specialized buses, crossbars, and other ad-hoc interconnects. Some processor designs must incorporate extra pipeline stages to accommodate the wire delay global bypass wires require [302], detracting from their potential performance. The complexity of the wire routing and electrical design of these specialized interconnects increases quadratically with the number of ALUs that smaller process technologies allow.

NoCs enjoy a scaling advantage relative to buses and fully connected crossbars since network wire lengths between adjacent switches can be kept short and unidirectional. NoCs also enable better pipelining of data between nodes and greater aggregate bandwidth than buses. Finally, design complexity is bounded since a switch is designed once and replicated for use wherever needed. This chapter presents an NoC-based operand bypass network as a scalable alternative to traditional operand bypass networks.

9.1.3 Operand Bypass NoC Characteristics

In contrast to the network traffic on chip-multiprocessor (CMP) and system-on-chip (SoC) interconnect, operand bypass traffic has several unique characteristics. NoCs in CMPs and many SoCs primarily carry memory system traffic such as cache line fills and spills and cache coherency traffic. Memory system packets range in size from single flit fill requests and coherence traffic to multiflit, cache-line-sized fill reply and spill request packets. Operand bypass networks convey operands that are sized to match the register width of the processor, often 32 or 64 bits. Thus, packet sizes in an operand bypass NoC can contain only a single flit. Despite the smaller packet size, bandwidth scales directly with the program's exploited ILP and therefore may be higher than memory system bandwidth. Because operand bypass networks provide forwarding between producing and consuming instructions, latency is critical to processor pipeline performance. CMP and SoC traffic latency may be less critical to system performance because of greater tolerance of longer memory or cross-chip interconnection delays. Unlike the typically homogeneous CMP architectures, operand bypass networks may also interconnect a set of heterogeneous units, such as register files, primary cache banks and ALUs, each unit having different network traffic characteristics.

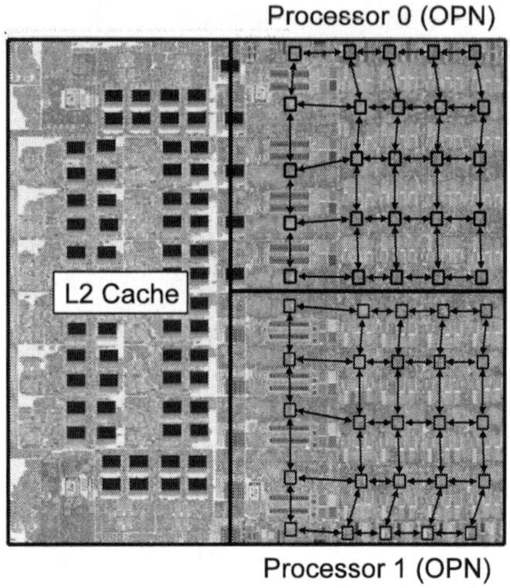

FIGURE 9.1: TRIPS chip plot with operand networks highlighted (from [111] ©2007 IEEE).

These characteristics imply a different network design than a CMP or SoC NoC. Operand bypass network switches must be very lightweight and fast to reduce latency, leveraging the single flit packet nature of the traffic to remove the need for virtual channels and multiflit packet accounting. Furthermore, an operand network may be tightly integrated with a processor's pipeline to reduce packet generation latency, by taking advantage of information available before the full data payload has been computed.

9.1.4 The TRIPS Processor

This chapter describes and evaluates the design and implementation of the TRIPS prototype processor's operand network (OPN). Figure 9.1 shows a plot of the TRIPS prototype processor chip, which was fabricated in a 130 nm application-specific integrated circuit (ASIC) process. On the right side of the figure are the two processors, each with its own separate operand network, as indicated by the superimposed meshes. Each processor's OPN is a 5×5 dynamically routed 2-D mesh network with 140-bit links. The OPN connects a total of 25 distributed execution, register file, and data cache tiles. Each tile is replicated and interacts only with neighboring tiles via the OPN and other control networks. The OPN subsumes the role of several traditional microprocessor interconnect buses, including the operand bypass network, register file read and write interface, and the L1 memory system bus.

The remainder of this chapter is organized as follows. Section 9.2 describes related work in operand bypass networks. Section 9.3 introduces the TRIPS processor microarchitecture. Section 9.4 describes the design and implementation of the TRIPS OPN, highlighting where the OPN differs from typical NoC. Section 9.5 presents an evaluation of the network's performance under different loads and explores the sensitivity of processor performance to OPN latency and bandwidth. Section 9.6 summarizes the material discussed in this chapter.

9.2 Related Work

Operand bypass networks were introduced more than forty years ago and have been prevalent in every processor since. As increased integration and clock rates have exposed the wire-delay bottleneck, processors with more and distributed ALUs have been implemented. This section employs a taxonomy of operand bypass networks to categorize various operand bypass network designs from industry and academia.[1]

9.2.1 Operand Bypass Network Taxonomies

Taylor et al. [315], Pinkston and Shin [253] and Sankaralingam et al. [279] all present useful taxonomies of operand bypass networks. Taylor categorizes operand networks based on whether the assignment, transport, and ordering are each either static or dynamic. Similarly, Pinkston and Shin [253] demonstrate how trends in semiconductor technology are leading to partitioned microsystem architectures and provide a taxonomy that categorizes microsystem architectures based on how they are partitioned. Their insight being that the trend toward partitioned architectures is driving the adoption of NoCs.

Sankaralingam et al. develop the Routed Inter-ALU Network (RIAN) taxonomy to categorize operand networks based on the following characteristics: network organization–point-to-point (P) vs. broadcast (B); network architecture–single hop (S) vs. multihop (M); and switch control–static (S) vs. dynamic (D). These characteristics together form a three-letter acronym which represents the network's configuration: <{P,B}, {M,S}, {D,S}> [279]. We will use the RIAN taxonomy to categorize the related work in operand bypass networks. Table 9.1, shows a summary of the networks discussed in Section 9.2.2.

---

[1]Portions of this and the following sections reprinted, with permission, from "Implementation and Evaluation of a Dynamically Routed Processor Operand Network," P. Gratz, K. Sankaralingam, H. Hanson, P. Shivakumar, R. McDonald, S.W. Keckler, and D. Burger, 1st International Symposium on Networks-on-Chips (NOC), pp. 7–17. Copyright 2007 IEEE.

Processor	Organization	Architecture	Switch Control
	Point-to-point(P) vs. Broadcast(B)	Single-hop(S) vs. Multihop(M)	Static(S) vs. Dynamic(D)
IBM System 360/91	B	S	D
Intel i7	B	S	D
IBM Power7	B	S	D
Alpha 21264	B	M	D
M-Machine	P	S	D
MIT Raw	P	M	S
Tilera Tile64	P	M	S
TRIPS OPN	P	M	D
Wavescalar	P	M	D

TABLE 9.1: Routed inter-ALU network taxonomy (from [279] ©2003 IEEE).

9.2.2 Operand Bypass Network Implementations

The first operand bypass network was introduced with the IBM System 360/model 91 to avoid delaying the sequential execution of dependent instructions [321]. This bypass network employed a simple broadcast bus (common data bus) that linked each ALU output to each ALU input. Thus, an instruction could receive an operand directly from a preceding instruction's output without the delay of passing it through the register file. Using the RIAN taxonomy of Table 9.1, this network's dynamic, single-hop broadcast would be represented as <B,S,D>. Most modern superscalar processors, including the Intel i7 [43] and the IBM Power7 [140] would similarly be represented by <B,S,D>.

The development of deeply pipelined superscalar architectures drove increases in the complexity and latency of bypass networks. Deeper pipelines increased the number of stages in which an instruction could produce a result or consume an operand. Wider pipelines increase the bypass bus network complexity quadratically with the number of ALUs because of the full connectivity between ALU outputs and ALU inputs. This N^2 complexity scaling is not viable beyond a small number of ALUs. The Alpha 21264 architecture divided its four integer ALUs into two clusters to reduce the complexity of its bypass network [151]. Operands produced within one cluster are available for use in the same cluster in the next cycle, but they must pay a single cycle penalty to be used in the other cluster. The RIAN taxonomy characterizes this form of partitioned architecture as <B,M,D> in Table 9.1 because the bypass network is no longer single hop to all ALUs.

Other machines have sought to reduce communication latency between processors through cross-processor register-to-register communication. The M-Machine is represented by a <P,S,D> in Table 9.1 because it employed an on-chip cluster crossbar switch to connect the register bypass networks for three processors; an instruction writing to a remote register injects its result into the switch, which delivers the data to a waiting instruction on a remote processor [148].

The MIT RAW processor took this strategy further, using a 4×4 mesh network to interconnect its processor tiles between execution units [333]. The integration of the RAW network into the local bypass network of each execution unit reduced the latency of operand passing between units to three cycles. One interesting feature of RAW is that network routing arbitration and ordering are statically determined, making a <P,M,S> network in the RIAN taxonomy. While this strategy simplifies the switches, a compiler or programmer must generate a routing program that executes concurrently with the application program. In addition to the statically routed network, RAW also implemented a network for load/store traffic which is dynamically routed due to the difficulty of compiler-based memory disambiguation.

The TRIPS OPN described in this chapter is also integrated directly with the execution unit. However, to allow for out-of-order instruction execution and uncertain memory delays, the OPN switches are dynamic, resulting in a RIAN taxonomy of <P,M,D>. We also employed additional routing optimizations to reduce the per-hop latency to one cycle.

The WaveScalar processor has a similar philosophy and execution model as TRIPS, but uses a hierarchy of interconnection networks to pass operands between processing elements [310]. Operands are broadcast within the eight processing elements making up one domain. Although operands pass through a crossbar switch to travel between the four domains that make up a cluster, operands traveling to another cluster traverse a 2-D mesh network similar to the TRIPS OPN.

9.3 TRIPS Processor Overview

TRIPS is a distributed processor consisting of multiple tiles connected via multiple NoCs. Figure 9.2 shows a tile-level diagram of the processor with its OPN links. The processor contains five types of tiles: each execution tile (ET) contains ALUs and reservation stations, each register tile (RT) contains a portion of the processor's register file, each data tile (DT) contains a portion of the level-1 data cache, each instruction tiles (IT) contains a portion of the level-1 instruction cache, and the global control tile (GT) orchestrates instruction fetch, execution, and commit. In addition, the processor contains several control networks for implementing protocols such as instruction fetch, completion, and commit in a distributed fashion. Tiles communicate directly only with their nearest neighbors to keep wires short and mitigate the effect of wire delay.

ISA and execution model: TRIPS is an explicit datagraph execution (EDGE) architecture, an instruction set architecture with two key features:

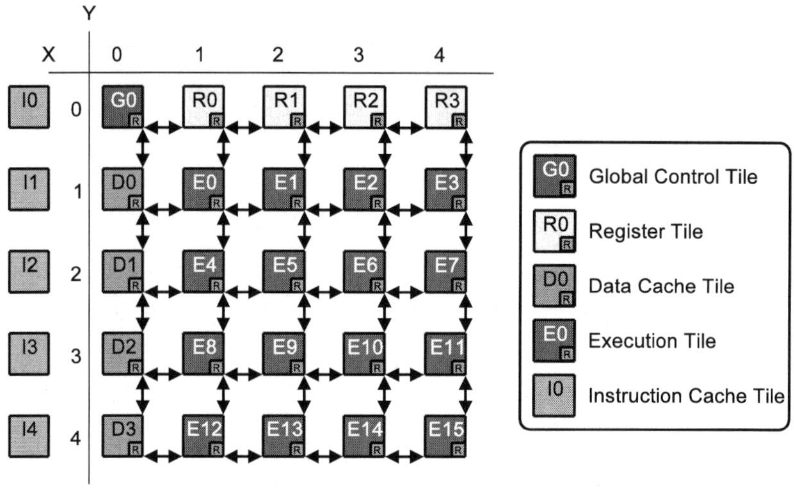

FIGURE 9.2: Block diagram of the TRIPS processor core with tiles and OPN network connections (from [111] ©2007 IEEE).

(1) the hardware fetches, executes, and commits blocks of instructions, rather than individual instructions, in an atomic fashion; and (2) within a block, instructions send their results directly to other instructions waiting to execute, rather than communicating through a common register file [44]. The compiler is responsible for constructing blocks, which can contain up to 128 instructions. Since basic blocks typically contain only a handful of instructions, the TRIPS compiler uses techniques such as predication, loop unrolling, and function inlining to create large hyperblocks. After hyperblock formation, a scheduler maps the block onto the fixed array of 16 execution units, with up to 8 instructions per ET. The scheduler is aware of the topology of the ETs and attempts to minimize the distance between dependent instructions along the program's critical path. The scheduler determines where an instruction will execute and encodes this in the program binary, but the hardware executes instructions in dataflow order based on when an individual instruction's operands arrive.

Block execution: Processing a TRIPS block requires four phases: fetch, execute, complete, and commit. To fetch a block, the GT transmits a fetch request to each of the ITs using the TRIPS global dispatch network (GDN). Each IT then retrieves a portion of the block (32 instructions) from its instruction cache bank and delivers them to preallocated reservation stations in the ETs and RTs. An instruction waits in its reservation station until all of its operands have arrived before it can execute. Block execution is instigated by special register read instructions that fetch block inputs from the RTs and deliver them to waiting instructions via the OPN. Instructions within the block then execute in dataflow order. Load and store instructions compute their addresses in the ETs, which are then transmitted to one of the DTs to access the data cache. Addresses are interleaved across the DTs on cache-line boundaries (64 bytes). Register outputs are transmitted back to the RTs where they wait in write queues before updating the architecturally persistent register file banks.

When all of the RTs and DTs have received all of the register writes and stores for the block, they signal the GT via a separate protocol network called the GCN (global control network). When the GT receives completion notifications from all DTs and RTs, the block is complete. If the block has not caused any exceptions, the GT signals to the DTs and RTs that the block can commit. The DTs then update the cache with the store values from the store buffers and the RTs update the register file banks with the contents of the write queues. When all of the state of the block has committed, a new block may be mapped into its place for execution. The TRIPS processor allows up to 8 blocks in-flight and executing simultaneously: one nonspeculative block and up to seven speculative blocks. Complete details of the TRIPS microarchitecture can be found in [278].

During block execution, the TRIPS operand network (OPN) delivers operands among the tiles. The TRIPS instruction formats contain target fields indicating to which consumer instructions a producer sends its values. At runtime, the hardware resolves those targets into coordinates to be used for network routing. An operand passed from producer to consumer on the same ET can be bypassed directly without delay, but operands passed between instructions on different tiles must traverse a portion of the OPN. The TRIPS execution model is inherently dynamic and data driven, meaning that operand arrival drives instruction execution, even if operands are delayed by unexpected or unknown memory latencies. Because of the data-driven nature of execution and because multiple blocks execute simultaneously, the OPN must dynamically route the operand across the processor.

9.4 OPN Design and Implementation

The operand network (OPN) is designed to deliver operands among the TRIPS processor tiles with minimum latency. While tight integration of the network into the processor core reduces the network interface latency, two primary aspects of the TRIPS processor architecture simplify the switch design and reduce routing latency. First, because of the block execution model, reservation stations for all operand network packets are preallocated, guaranteeing that all OPN messages can be consumed at the targets. Second, all OPN messages are of fixed length, one flit broken into header and payload phits.

Control phit		Data phit	
Field	bits	Field	bits
Valid	1	Valid	1
Type (LD/ST/etc.)	4	Type (normal/null/exception)	2
Block ID	3	Data operation (access width)	3
Destination node	6	Data payload	64
Destination instruction	5	LD/ST Address	40
Source node	6		
Source instruction	5		

TABLE 9.2: Breakdown of bits for OPN control and data phits (from [111] ©2007 IEEE).

9.4.1 OPN Design Details

The OPN is a 5×5 2-D routed mesh network as shown in Figure 9.2. Buffer management is performed via stop&go signaling, meaning that the receiver informs the transmitter when the available buffer space falls below a threshold set to guarantee in-flight flits will not be dropped despite round-trip control signaling latency. Packets are routed through the network in Y-X dimension-order with one cycle taken per hop. A packet arriving at a switch is buffered in an input First-In First-Out (FIFO) queue prior to being launched onward towards its destination. Due to dimension-order routing and the guarantee of the consumption of messages, the OPN is deadlock-free without requiring virtual channels. The absence of virtual channels reduces arbitration delay and speeds routing.

Each operand network message consists of a control phit and a data phit. The control phit is 30 bits and encodes OPN source and destination node coordinates, along with identifiers to indicate which instruction to select and wakeup in the target ET. The data phit is 110 bits, with room for a 64-bit data operand, a 40-bit address for store operations, and 6 bits for status flags. Some of the fields, such as the source identifiers, are not strictly necessary for operand network functionality, but we included them in the prototype for debugging and monitoring. Table 9.2 shows a breakdown of all of the bits in the data and control phits.

The data phit always trails the control phit by one cycle in the network. The OPN supports different physical wires for the control and data phit so one can think of each OPN message consisting of one flit split into a 30-bit control phit and a 110-bit data phit. Because of the distinct control and data wires, two OPN messages with the same source and destination can proceed through the network separated by a single cycle. The data phit of the first message and the control phit of the second are on the wires between the same two switches at the same time. Upon arrival at the destination tile, the data phit may bypass the input FIFO and be used directly, depending on operation readiness. This arrangement is similar to flit-reservation flow control, although

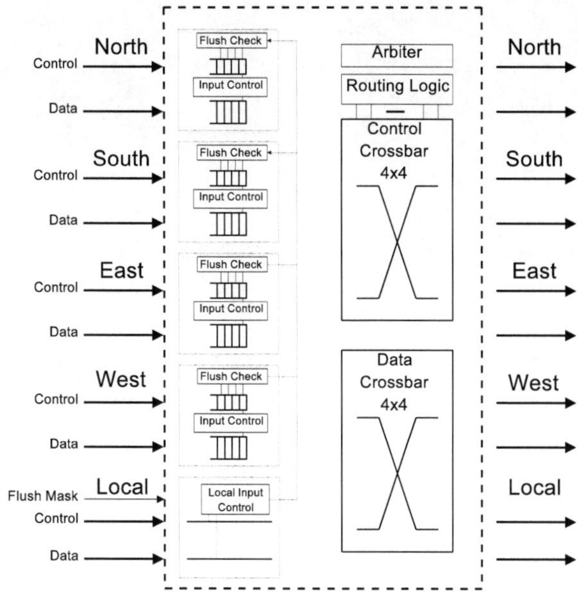

FIGURE 9.3: OPN switch microarchitecture (from [111] ©2007 IEEE).

here the control phit contains some payload information and always moves exactly one hop ahead of the data phit [250]. In all, the OPN has a peak injection bandwidth of 175 GB/sec when all nodes are injecting packets every cycle at its designed frequency of 400 MHz. The network's bisection bandwidth is 70 GB/sec measured horizontally or vertically across the middle of the OPN.

Figure 9.3 shows a high-level block diagram of the OPN switch, with five inputs and five outputs, one for each ordinal direction (N, S, E, and W) and one for the local tile's input and output. The ordinal directions inputs each have two four-entry deep FIFOs, one 30 bits wide for control phits and one 110 bits wide for data phits. The local input has no FIFO buffer. The control and data phits of the OPN packet have separate 4×4 crossbars. All arbitration and routing is done on the control phit, in round-robin fashion among all incoming directions. The data phit follows one cycle behind the control phit in lock step, using the arbitration decision from its control phit.

9.4.2 OPN/Processor Integration

ET/OPN datapath: Figure 9.4 shows the operand network datapath between the ALUs in two adjacent ETs. The instruction selection logic and the output latch of the ALU are both connected directly to the OPN's local input port, while the instruction wakeup logic and bypass network are both

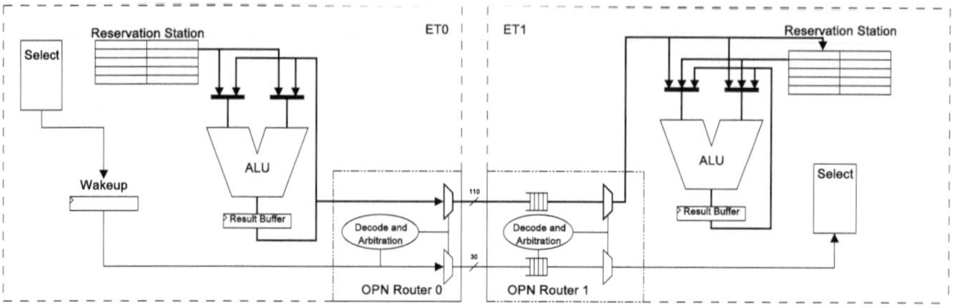

FIGURE 9.4: Operand datapath between two neighboring ETs (from [111] ©2007 IEEE).

connected to the OPN's local output. The steps below describe the use of the OPN to bypass data between the ALUs.

- Cycle 0: Instruction wakeup/select on ET 0
  - ET0 selects a ready instruction and sends it to the ALU.
  - ET0 recognizes that the instruction target is on ET1 and creates the control phit.
- Cycle 1: Instruction execution on ET0
  - ET0 executes the instruction on the ALU.
  - ET0 delivers the control phit to switch FIFO of ET1.
- Cycle 2: Instruction wakeup/select on ET1
  - ET0 delivers the data phit to ET1, bypassing the FIFO, and depositing the data in a pipeline latch.
  - ET1 wakes up and selects the instruction depending on the data from ET0.
- Cycle 3: Instruction execution ET1
  - ET1 selects the data bypassed from the network and executes the instruction.

The early wakeup, implemented by delivering the control phit in advance of the data phit, overlaps instruction pipeline control with operand data delivery. This optimization reduces the remote bypass time by a cycle (to one cycle) and improves performance by approximately 11% relative to a design where the wakeup occurs when the data arrives. In addition, the separation of the control and data phits onto separate networks with shared arbitration and routing eliminates arbitration for the data phit and reduces network contention relative to a network that sends the header and payload on the same wires in successive cycles. This optimization is inexpensive in an NoC due to the high wire density.

The OPN employs round-robin arbitration among all of the inputs, including the local input. If the network is under load and chooses not to accept the control phit, the launching node captures the control phit and later the data phit in a local output buffer. The ET will stall if the instruction selected for execution needs the OPN and the ET output buffer is already full. However, an instruction that needs only to deliver its result to another instruction on the same ET does not stall due to OPN input contention. While OPN contention can delay instruction execution on the critical path of program execution, the scheduler is effective at placing instructions to mitigate the distance that operands must travel and the contention they encounter.

Selective OPN message invalidation: Because the TRIPS execution model uses both instruction predication and branch prediction, some of the operand messages are actually speculative. On a branch misprediction or a block commit, the processor must flush all in-flight state for the block, including state in any of the OPN's switches. The protocol must selectively flush only those messages in the switches that belong to the flushed block. The GT starts the flush process by multicasting a flush message to all of the processor tiles using the global control network (GCN). This message starts at the GT and propagates across the GCN within 10 cycles. The GCN message contains a block mask indicating which blocks are to be flushed. Tiles that receive the GCN flush packet instruct their switches to invalidate from their FIFOs any OPN messages with block-identifiers matching the flushed block mask. As the invalidated packets reach the head of the associated FIFOs they are removed. While we chose to implement the FIFOs using shift registers to simplify invalidation, the protocol could also be implemented for circular buffer FIFOs. A collapsing FIFO that immediately eliminates flushed messages could further improve network utilization, but we found that the performance improvement did not outweigh the increased design complexity. In practice, very few messages are actually flushed.

9.4.3 Area and Timing

The TRIPS processor is manufactured using a 130 nm IBM ASIC technology and returned from the foundry in September 2006. Each OPN switch occupies approximately 0.25 mm^2, which is similar in size to a 64-bit integer multiplier. Table 9.3 shows a breakdown of the area consumed by the components of an OPN switch. The switch FIFOs dominate the area in part because of the width and depth of the FIFOs. Each switch includes a total of 2.2 kilobits of storage, implemented using standard cell flip-flops rather than generated memory or register arrays. Utilizing shift FIFOs added some area overhead due to extra multiplexors. We considered using the library-generated,

Component	% Switch Area	% E-Tile Area
Switch input FIFOs	74.6%	7.9%
Switch crossbar	20.3%	2.1%
Switch arbiter logic	5.1%	0.5%
Total for single switch	–	10.6%

TABLE 9.3: Area occupied by the components of an OPN switch (from [111] ©2007 IEEE).

Component	Latency	% Path
Control Phit Path		
Read from instruction buffer	290ps	13%
Control phit generation	620ps	27%
ET0 switch arbitration	420ps	19%
ET0 OPN output mux	90ps	4%
ET1 OPN FIFO muxing and setup time	710ps	31%
Latch setup + clock skew	200ps	9%
Total	2.26ns	–
Data Phit Path		
Read from output latch	110ps	7%
Data phit generation	520ps	32%
ET0 OPN output mux	130ps	8%
ET1 switch muxing/bypass	300ps	19%
ET1 operand buffer muxing/setup	360ps	22%
Latch setup + clock skew	200ps	12%
Total	1.62ns	–

TABLE 9.4: Critical path timing for OPN control and data phit (from [111] ©2007 IEEE).

hard-macro Static Random Access Memory (SRAM) arrays instead of flip-flops, but the area overhead turned out to be greater given the small size of each FIFO.

A single OPN switch takes up approximately 10% of the ET's area and all the switches together form 14% of a processor core. While this area is significant, the alternative of a broadcast bypass network across all 25 tiles would consume considerable area and is not feasible. We could have reduced the switch area by approximately $\frac{1}{3}$ by sharing the FIFO entries and wires for the control and data phits. However, the improved OPN bandwidth and overall processor performance justifies the additional area.

We performed static timing analysis on the TRIPS design using Synopsys Primetime to identify and evaluate critical paths. Table 9.4 shows the delay for the different elements of the OPN control and data critical paths, matching the datapath of Figure 9.4. We report delays using a nominal process corner, which we obtained by scaling our worst-case process corner delays by a factor of $\frac{2}{3}$. A significant fraction of the clock cycle time is devoted to overheads such as flip-flop read and setup times as well as clock uncertainty (skew and jitter). A custom design would likely be able to drive these overheads down. On the logic path, the control phit is much more constrained than the data phit due to switch arbitration delay. We were a little surprised by the delay associated with creating the control phit, which involves decoding and encoding. This path could be improved by performing the decoding and encoding in a previous cycle and storing the control phit with the instruction before execution. We found that wire delay was small in our 130 nm process given the relatively short transmission distances. Balancing switch delay and wire delay may be more challenging in future process technologies.

9.4.4 Design Optimizations

We considered a number of OPN enhancements but chose not to implement them in the prototype to simplify the design. One instance where performance can be improved is when an instruction must deliver its result to multiple consumers. The TRIPS ISA allows an instruction to specify up to 4 consumers, and in the current implementation, the same value is injected in the network once for each consumer. Multicast in the network would automatically replicate a single message in the switches at optimal bifurcation points. This capability would reduce overall network contention and latency while increasing ET execution bandwidth, as ETs would spend less time blocking for message injection. Another optimization would give network priority to those OPN messages identified to be on the program's critical path. We have also considered improving network bandwidth by replicating the operand network by replicating the switches and wires. We examine this optimization further in Section 9.5.5. Finally, the area and delay of our design was affected by the characteristics of the underlying ASIC library. While the trade-offs may be somewhat different with a full-custom design, our results are relevant because not all NoC connected systems will be implemented using full-custom silicon. Our results indicate that such ASIC designs would benefit from new ASIC cells, such as small but dense memory arrays and FIFOs.

9.5 OPN Evaluation

In this section, we evaluate the behavior of the OPN on statistical and realistic network workloads, using our operand network simulator to model the OPN hardware. We characterize the operand network message workload and show that injection is not distributed evenly across the nodes, due to the TRIPS execution model and scheduler optimizations. Finally, we examine the sensitivity of program performance and operand network latency to OPN bandwidth and latency parameters.

9.5.1 Methodology

The OPN simulator is a custom network simulator configured with the operand network design parameters. It can inject messages using different traffic patterns, including random and bit-reversal, with variable injection rates. It can also accept a network trace file that specifies source nodes, destination nodes, and injection timestamps. We obtained realistic workload traces from an abstract TRIPS processor performance estimator (*tsim-cyc*), which runs compiled TRIPS programs. This simulator models TRIPS block execution at a high level, but employs a simple analytical performance model without accurate OPN contention estimation. Nonetheless, this simulator estimates the performance of the TRIPS hardware to within 25%. The high simulation speed of tsim-cyc allows us to obtain traces for long running programs. However, the message injection times only approximate those that will be seen in hardware. For a more detailed analysis, we also used our low-level simulator (*tsim-proc*) which accurately models all aspects of a TRIPS processor core, including network contention. This simulator has been validated for accuracy against the TRIPS RTL and hardware. Unfortunately, the speed of this simulator prevents analysis of large programs.

Our realistic workloads include programs from the Embedded Microprocessor Benchmark Consortium (EEMBC) [338] and SPEC2000 [118] benchmark suites. The 30 EEMBC benchmarks are small enough to run to completion on both tsim-cyc and tsim-proc. The 19 SPEC CPU2000 benchmarks were run with the Minne-SPEC [163] reduced input set, but were still too long-running for tsim-proc. The SPEC benchmarks were run to completion (50 million cycles for the shortest benchmark), or for 300 million cycles after program warmup. The traces include 2–70 million operand messages, depending on the benchmark.

9.5.2 Synthetic Statistical Loads

Interconnection networks are typically evaluated by examining their performance on stochastically generated workloads. Two common workloads are bit-reversal and uniform random traffic. In bit-reversal, each node exchanges packets with a node on the opposite side of the network. The random traffic model randomly chooses source and destination pairs from among all the TRIPS core tiles. Both traffic models inject packets at a uniform random distributed rate. Figure 9.5(a) shows the offered vs. accepted rate for both of these types of traffic. The offered rate is the rate at which packets are generated, while the accepted rate is the throughput of the network. In these diagrams, the offered and accepted rates are shown as a percentage of the peak injection bandwidth. For bit-reversal traffic, the accepted rate tracks the

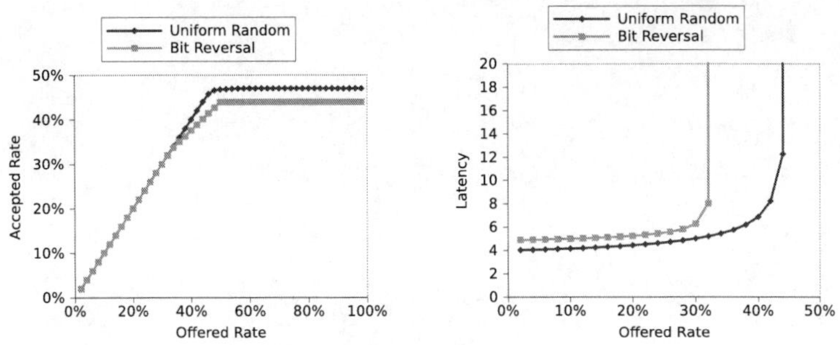

(a) Offered vs. accepted rate for random and bit-reversal traffic.

(b) Offered rate vs. average latency for random and bit-reversal traffic.

FIGURE 9.5: Synthetic statistical loads on the TRIPS OPN (from [111] ©2007 IEEE).

offered rate until the 33% mark, after which the accepted rate increases at a slower rate before finally leveling off at 44%. For random traffic the accepted bandwidth tracks the offered bandwidth up to approximately 46% before leveling off to a maximum of 47%. These are typical curves for this type of 2-D mesh network.

Figure 9.5(b) shows the average measured packet latency in cycles for increasing offered rate. The average latency for bit-reversal traffic gradually increases from around 5 to 8 cycles for offered rates of 1% to 32%. The latency then increases exponentially as the network becomes saturated. Similarly, for random traffic the latency increases from about 4 to 7 cycles for offered rates from 1% to 40% before increasing dramatically. This diagram shows that 32% and 40% are the saturation offered rates for bit-reversal and random traffic, respectively.

9.5.3 OPN Traffic Trace Analysis

Previous work examining the TRIPS OCN [110] showed that real benchmark generated traffic in NoCs would not be modeled well by traditional synthetic loads. We perform a similar analysis for the OPN using network traces generated from tsim-cyc and characterize the network workload.

Variation in application offered rate: Figure 9.6 shows the average offered rate for various SPEC CPU2000 benchmark traces generated from tsim-cyc. For each application's trace, we derived the offered rate by dividing the number of total messages by the product of the cycle count and the number of injecting nodes (25 for the 5 × 5 network). While the offered rates vary widely, from under 1% for `twolf` to almost 14% for `mcf`, the average offered rate is well below the saturation threshold range of 30–40% for bit-reversal and uniform random. The magnitude of the offered rates correlate to the degree of ILP that the TRIPS compiler has exposed to the processor. A benchmark

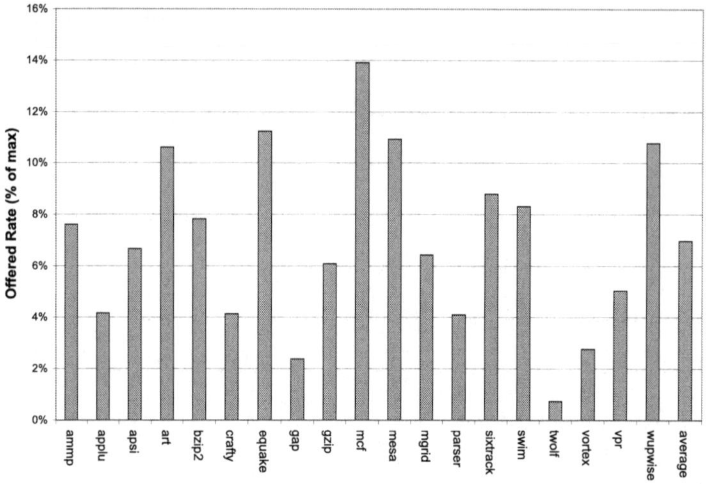

FIGURE 9.6: Average offered rates in SPEC CPU2000 benchmark traces (from [111] ©2007 IEEE).

with more exposed ILP will have more operations occurring simultaneously and will therefore generate more operands each cycle than a benchmark that has long dependency chains and lower ILP.

Average packet hop distance: Figure 9.7 shows the average number of hops, or switch traversals, from source to destination for OPN packets from various SPEC CPU2000 benchmark traces. These values are generated by averaging the Manhattan distance from source to destination for each packet in the trace of each benchmark. Because the OPN is a single cycle per hop network, this distance also represents a best case routing delay for each packet in the absence of any network contention.

The figure shows that average message distance across all benchmarks is about 2.1 hops, with little variance among the benchmarks. The hop counts are low by the design of the TRIPS system. The TRIPS compiler statically maps instructions to particular nodes, purposely placing each instruction near where its operands are produced. Because the compiler is also balancing locality with parallelism and load balance of instructions across the ALUs, not all operand communications can be between adjacent ETs.

Variation in offered rate by source: While the overall average offered rate of the OPN is low, that metric does not accurately capture hot spots in the network. Figure 9.8 shows the average offered rate for each individual OPN node as a percentage of the peak offered rate of one message per cycle, averaged across all SPEC CPU2000 benchmarks. The X-Y plane of the graph matches the layout of the 5 × 5 operand network and the different shades highlight the different tile types. The per-tile offered rates vary widely, from

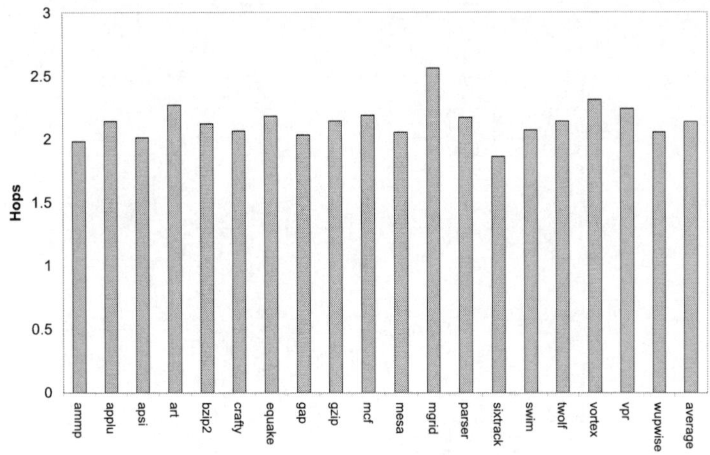

FIGURE 9.7: Average number of hops from source to destination for various SPEC CPU2000 benchmarks (from [111] ©2007 IEEE).

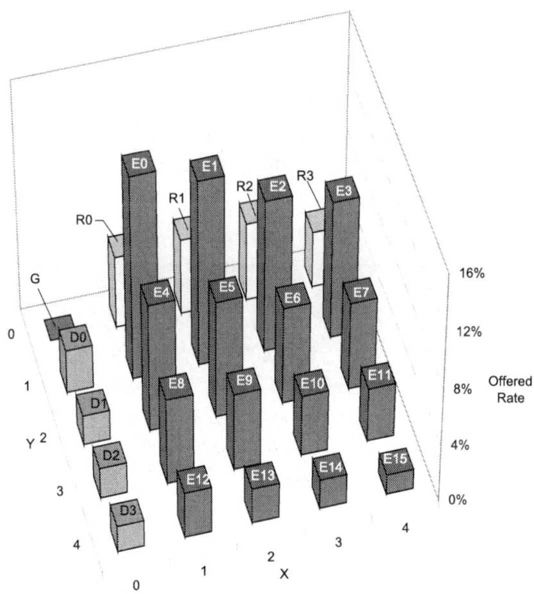

FIGURE 9.8: Offered rates for SPEC CPU2000 benchmarks broken down by sources (from [111] ©2007 IEEE).

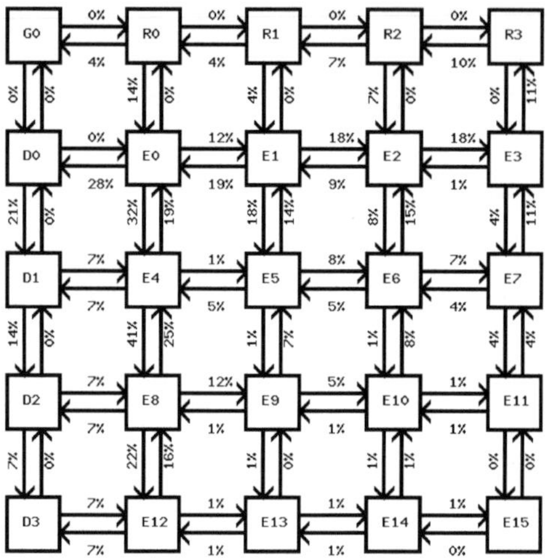

FIGURE 9.9: Link utilization for `mesa` SPEC CPU2000 benchmark (from [111] ©2007 IEEE).

a low of 0% for GT and 2.6% for ET15 to a high of 16.7% for ET0 at the upper left. The disposition of offered rates reflects the heterogeneity of processor tiles, the GT injects no traffic and is only an end node for traffic. The RTs and DTs each show similar offered rates to tiles of the same type, differing from GT and ETs, respectively. The disposition of offered rates among ETs reflects the TRIPS compiler's instruction placement optimizations that attempt to minimize operand routing distance. The scheduler preferentially places instructions near the register file and data cache tiles to reduce block input and output latency.

Our analysis shows that even though average offered rate is low, applications can easily create network injection hot spots that may approach local saturation, producing higher than expected transmission latencies. The compiler schedules instructions to more evenly distribute the network traffic; however, such optimizations must be balanced against the effect of increasing the average source to destination hop count.

Variation in link utilization: Hot spots also form when many messages must pass through the same link. Figure 9.9 labels each OPN link with the link occupancy percentage for the `mesa` SPEC CPU2000 benchmark. We choose to show the data for one benchmark instead of averaging across all of them because of the variance across the benchmarks. The southbound link between E4 and E8 has a high utilization of 41% and many other links are in the 15–20% range. High link utilization has a disproportionately large effect on latency due to congestion and limits performance for this benchmark. Our

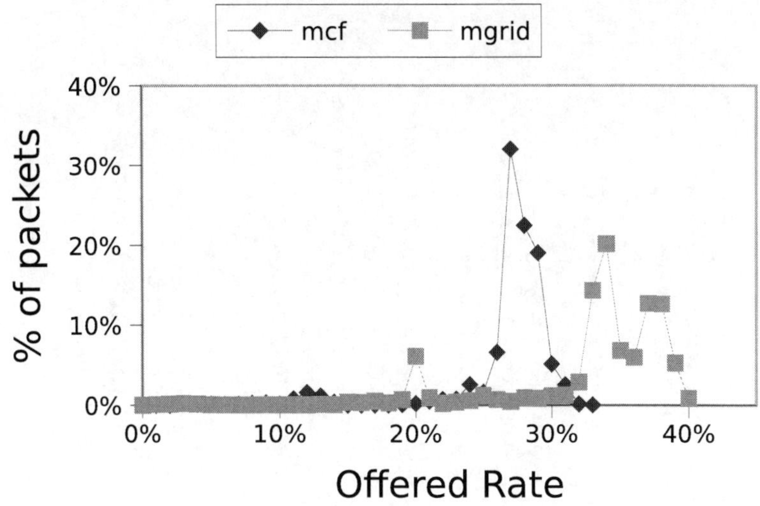

FIGURE 9.10: Distribution of offered rates measured as a percentage of packets injected at a given offered rate (from [111] ©2007 IEEE).

experience shows that other benchmarks place a maximum load of only 5–10% on any link.

Traffic burstiness: In addition to load variability across applications and network nodes, offered traffic can vary over time. TRIPS naturally has traffic bursts because a block begins execution through the injection of many register values from the top of the network. To measure burstiness, we examined the trace at 1000 cycle intervals, counted the number of messages in each interval, and computed the offered rate for the interval. Figure 9.10 shows a histogram of the offered rates for two SPEC CPU2000 benchmarks, `mcf` and `mgrid`. The X-axis shows the histogram buckets at 1% intervals, while the Y-axis shows the fraction of all messages that fall into each bucket. The figure shows that `mcf` has a relatively stable offered rate centering around 29% for most packets, meaning that the network is evenly loaded over time. Conversely, `mgrid` shows more diversity in its offered rates with significant numbers of packets clustered around 20%, 34%, and 38%. Based on these results, we conclude that the traffic of `mgrid` has more bursts than that of `mcf`, and likely has spikes in latency for critical operands traversing the network. Such bursts may motivate lightweight network designs that tolerate and spread traffic in response to varying loads.

9.5.4 Network Simulator-based Analysis

We applied the traces to the OPN trace-driven simulator to examine how the network performs under load. The inherent weakness of trace-driven network simulation is the lack of a feedback loop between the network simulation

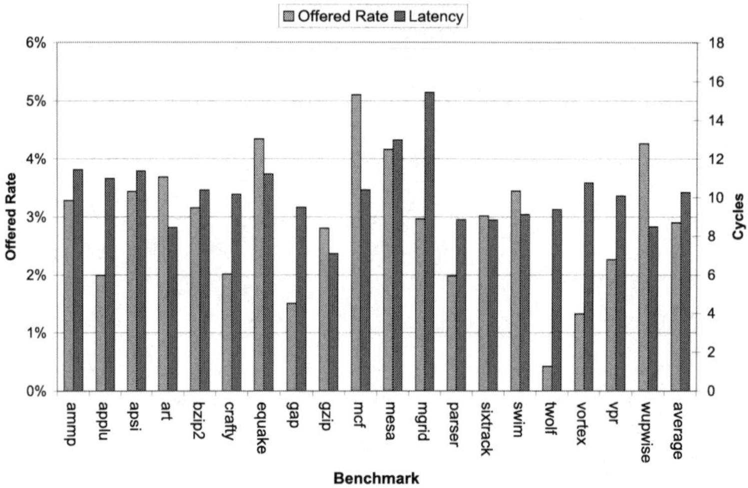

FIGURE 9.11: Offered rates and latencies for SPEC CPU2000 benchmarks from the OPN network simulator (from [111] ©2007 IEEE).

and trace generation. In the real TRIPS processor, network congestion will throttle instruction execution, in turn throttling the offered rate. To bound this difference, for each message we tracked the instruction block to which it belongs and ensures that messages from only eight consecutive blocks are considered for injection at any one time. These eight correspond to the one nonspeculative and seven speculative blocks that can execute simultaneously. This approach represents a reasonable compromise that keeps the processor and network simulators separate. The lack of intrablock throttling places some excess stress on the network, giving additional insight on the load if the network were ideal and noncontented.

Offered rate and latency: Figure 9.11 shows the average offered rate and latency for the SPEC CPU2000 on the OPN simulator. Compared to the results in Figure 9.6, the offered rates are significantly lower because of the block-level throttling. The benchmarks that had the highest offered rates show offered rates that are reduced by as much as two thirds.

The right bar for each benchmark shows the average message latency for each benchmark. In general, benchmarks with higher offered rates show higher average latencies, but certain benchmarks show the reverse. For example, `mcf` has the highest offered rate at 5.1% while it has a fairly average latency of about 10 cycles. Conversely `mgrid` has a fairly average offered rate of around 3% but the highest average latency at 15.5 cycles. This dichotomy can be attributed to the burstiness in the traffic and high utilization of particularly hot links. The OPN simulator shows that the average latencies are high, ranging from 6 to 15 cycles. Although throttling will prevent actual OPN latencies

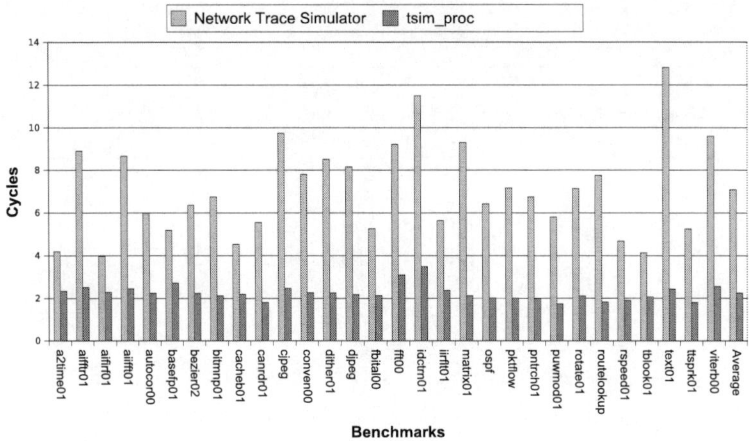

FIGURE 9.12: EEMBC benchmark latency from the high-level tsim-cyc traces versus the detailed processor model tsim-proc (from [111] ©2007 IEEE).

from reaching these levels, the measured latencies highlight where OPN network performance improvement has a direct affect on processor performance.

Network throttling: To examine the impact of throttling on latency, we used the cycle-level simulator tsim-proc. Because tsim-proc is approximately 300 times slower than the tsim-cyc simulator we used to generate traces, we chose the shorter EEMBC 2.0 suite of embedded system benchmarks. Figure 9.12 shows the average latencies of OPN packets as measured in both tsim-proc and the OPN network trace driven simulator. While the network simulator shows an average latency of 7 cycles, tsim-proc shows only 2.25 cycle, again due to throttling from instruction dependences in the program. This can be a little deceiving because throttling manifests as stalls in the execution tiles (ETs) rather than in the network. Thus, for network research, trace-based simulation still provides good insight into network behavior, but one must take care when analyzing system performance based on network performance.

9.5.5 Operand Network Sensitivity Studies

Packet End-To-End Latency: The TRIPS prototype is designed to support one-cycle communication latency between adjacent ETs. Speculative injection of the operand message header, early wakeup of the consumer, and bypassing directly from the network input limit the latency of operand network transmission. Each additional hop in the network costs only one cycle. To examine the sensitivity of performance to latency, we simulated two alternate designs. The first emulates an architecture that does not have early wakeup

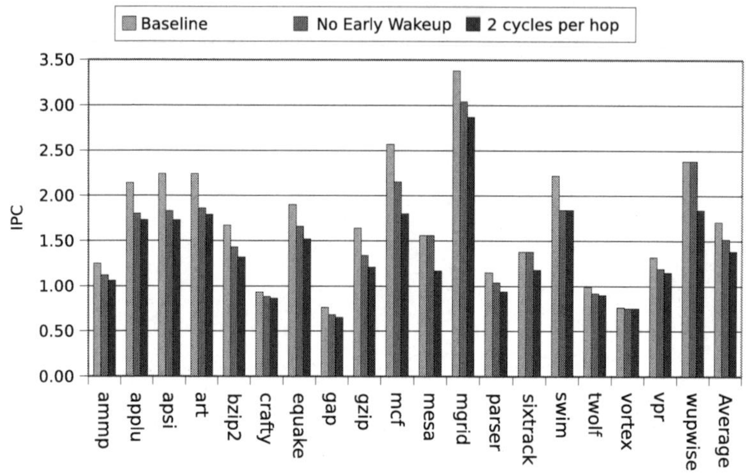

FIGURE 9.13: Comparison of baseline OPN versus an OPN without the early wakeup and an OPN that consumes two cycles per hop (from [111] ©2007 IEEE).

and thus requires one additional cycle for every operand transmission. The second emulates a two-cycle-per-hop network to model slower switches and wires. Figure 9.13 compares the IPC (instructions-per-clock) of the TRIPS processor core for the different design points. Without early wakeup, processor performance drops by about 11%; a two-cycle per hop network decreases IPC by 20%. Thus, performance of TRIPS is quite sensitive to OPN latency.

Bandwidth: A simple way to improve the performance of a network is to increase its bandwidth. Typically one would increase the bit-width of the network's interfaces to decrease the number of flits per message, network occupancy, and message injection and extraction latency. Because the OPN already has single-flit packets, increasing the link-width will not affect network occupancy or processor performance. Another way to improve the performance of a network is to decrease the network diameter by using higher-radix switches and a more highly interconnected topology. This approach is not a good fit for the OPN for two reasons. First, increasing the radix of the switches increases the logical complexity of the switches, possibly to the point of becoming the TRIPS core's critical timing path. Second, as shown in Figure 9.7 the average hop distance for packets on the OPN is just over 2, so increasing the network topology dimensions would not decrease the end-to-end latency of a large fraction of the injected messages. Furthermore, the higher logical complexity of high radix switches would require either more pipeline stages in the switch or a slower system clock to meet timing. The extra latency these changes imply would mitigate the benefits of a smaller diameter network.

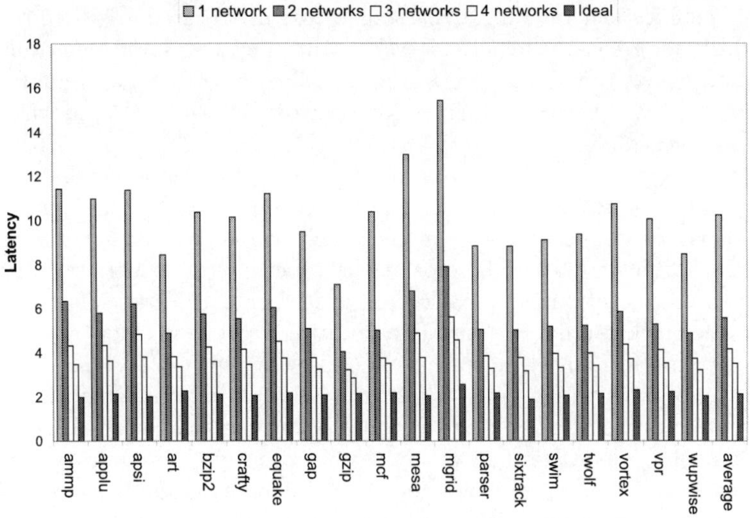

FIGURE 9.14: Average packet latency for SPEC CPU2000 benchmarks with 1, 2, 3, and 4 OPN networks (from [111] ©2007 IEEE).

As an alternative, we investigated replicating the network links and switches as a means to increase the effective bandwidth of the network and reduce contention. We simulate a simple scheme in which nodes inject packets into each network in a round-robin fashion. If a network is blocked due to congestion, the injecting node skips it until the congestion is alleviated. Figure 9.14 shows the average packet latency from the OPN trace simulator for the SPEC CPU2000 benchmarks with 1 (the current OPN configuration), 2, 3, and 4 networks interconnecting the nodes of the OPN. The expected latency without contention is shown as "Ideal." The biggest improvement in latency occurs between 1 and 2 networks, almost halving the average latency. However, replication comes at a cost of doubling the area consumed by the network.

9.6 Summary

In this chapter, we presented the design, implementation, and evaluation of the TRIPS OPN. The TRIPS OPN is an NoC that interconnects the functional units within the TRIPS processor core. The OPN replaces an operand bypass bus and primary memory system interconnect in a technology scalable manner. The tight integration between the OPN and the processor core elements enables fast operand bypassing across distributed ALUs, providing opportunity for greater instruction-level concurrency. Our implementation and fabrication shows that such a network is feasible in terms of area and delay, and that the network design provides good performance for the traffic provided by real applications.

We used synthetic benchmarks along with static traces generated from SPEC CPU2000 traffic to evaluate the performance of the OPN under different loads. We found that the offered traffic varied widely across multiple applications and across the heterogeneous processor tiles. Our experiments confirm the expectation that distributed processor performance is quite sensitive to network latency, as just one additional cycle per hop results in a 20% drop in performance. Increasing the link width in bits does not help this network since the messages already consist of only one flit. Replicating the network to improve bandwidth and reduce latency is promising as increasing the wire count in NoCs is not prohibitively expensive. Network switch area (particularly switch buffers), however, is not insignificant and these costs must be balanced with network performance benefits.

We expect that fine-grained networks will increase in importance, initially as memory oriented networks for CMPs and SoCs, but ultimately in support of finer grained communication and synchronization. Further research is needed to re-examine standard multichip interconnection network architectures with respect to the constraints and opportunities of NoCs. In addition to network latency and area, we expect network power, efficiency, and quality of service to be critical. We also expect NoCs to provide new opportunities in other aspects of distributed system and processor design.

Acknowledgment

We thank our design partners at IBM Microelectronics, and Synopsys for their generous university program. This research was supported financially by the Defense Advanced Research Projects Agency under contracts F33615-01-C-1892 and NBCH30390004, NSF instrumentation grant EIA-9985991, NSF CAREER grants CCR-9985109 and CCR-9984336, IBM University Partnership awards, grants from the Alfred P. Sloan Foundation and the Intel Research Council.

Part III

Upcoming Trends

Chapter 10

Network Interface Design for Explicit Communication in Chip Multiprocessors

Stamatis Kavadias

Foundation for Research and Technology–Hellas (ICS-FORTH)

Manolis Katevenis

Foundation for Research and Technology–Hellas (ICS-FORTH)

Dionisios Pnevmatikatos

Foundation for Research and Technology–Hellas (ICS-FORTH)

10.1 Network Interface Evolution and Outlook

Scalable on-chip networks define a new environment for network interfaces (NI) and communication of the interconnected devices. On-chip communicating nodes are usually involved in a computation framework, whose efficiency critically depends on the organization of resources as well as on the performance of the communication architecture. The principle of device networking was inspired by off-chip network architectures, such as those in the Local Area Network (LAN)/ Wide Area Network (WAN) or personal computer (PC) cluster domains, which however evolved in a different direction in light of their distinctive features, with complex protocols, robustness in the face of varying traffic patterns and latency not being a primary concern.

From the perspective of the network interface, supercomputer and multiprocessor systems targeting parallel computations have a lot more in common with an on-chip setting. For instance, NI-CPU proximity enables communication and synchronization optimizations that can potentially improve computation efficiency and are suitable for integration in a chip multiprocessor (CMP). Furthermore, as the number of cores per chip will increase beyond a few tens, the need for a scalable on-chip interconnection network and bandwidth efficient mechanisms will become more pronounced. Communication mechanisms employed in supercomputers can then be matched to the needs of chip multiprocessor (CMP) systems and adopted appropriately.

This chapter focuses on network-on-chip (NoC) interfaces for CMPs, and sheds light on architecture-level design techniques that can be applied to largely integrated CMP systems, that utilize a scalable on-chip interconnect. Systems-on-chip are a more resource-constrained application domain, and as such design techniques for their network interfaces may be radically different from those illustrated in this chapter. The interested reader is referred to [27] for further details on network interfaces for multiprocessor Systems-on-Chip.

It should be observed that designing a suitable interface for the scalable on-chip interconnection network in a CMP is tightly related with the programming model. Therefore, this chapter also addresses some programming model related issues and relevant synchronization support. Interprocessor communication (IPC) can be *implicit*, when the "address" supplied by the program does not identify the physical location where the communicated data should be sent or placed, or it can be *explicit* when it does. Implicit communication is supported by cache coherence, and is easier for the programmer, but it may not allow for performance optimizations in those cases where the programmer, compiler, or runtime system, can manage locality better than coherence hardware. It is resonable to expect that both types of IPC will be useful for different functions and parts of a computation, as future CMP systems will become increasingly distributed.

The place where explicit communication mechanisms are implemented is traditionally called the "network interface." Even though coherent caches in the past distributed shared memory systems and provided an interface to the interconnection network, that was not called a network interface partly because it was transparent to software and was considered part of the cache. In addition, these systems usually provided a bus that connected the processor's cache, the directory with main memory, and a network interface to the remote access network, because of node design and integration issues. Note that network interface-based explicit communication was not regarded as competitive to cache-based communication, partly because the performance of scalable networks lagged behind intranode buses.

This chapter follows this traditional identification of network interface mechanisms with explicit communication mechanisms. Although the interface of *any* device to a network is literally and actually a "network interface," since this chapter refers to the design of network interfaces for explicit communication in CMPs, a more contextualized terminology is adopted. The term "network interface" is used only to denote the support for explicit communication, and the more general term "NoC interface" is employed to address any kind of interface to the NoC, regardless of its support for explicit or implicit transfers. What is meant with the selection of these terms is that a network interface (that is one that supports explicit communication mechanisms) includes a NoC interface, just as any device connected to the NoC, but may also provide more advanced functions than simple packetization and unpacketization of data.

Caches provide a NoC interface, and have been optimized for an on-chip environment long before the multicore challenge became the focus of processor chip architecture. Although the hardware cost of caches is relatively small and the processor's view of a consistent memory system based on caches has been studied extensively, the implicit communication style supported by cache coherence may not scale well in terms of latency and energy efficiency, and the hardware cost of coherence directories is not negligible. On-chip network interfaces provide a competitive choice, supporting explicit communication.

Two main alternatives have been considered in the literature as well as implemented in contemporary CMPs, for the network interface, depending on the way communication targets are identified. The first is to directly name the destination *node* of communication, possibly also identifying one of a few queues or registers of a specific processor, but *not* any memory address. This results in a network interface tightly coupled to the processor, and may provide advantageous end-to-end communication latency, but usually necessitates receiver software processing for every message arrival.

The second approach is to place NI communication memory inside the processor's normal address space. In this case, communication destinations are identified via memory addresses, which include node identifier (ID) information, and the routing mechanism may be augmented or combined with address translation. The resulting NI has the flexibility of larger send and receive on-chip space, that is directly accessible from software and allows the use of more advanced communication mechanisms. Challenges raised in the design of such a network interface include NI control register virtualization, the possible interactions with the address translation mechanism, and potentially managing the costs of sharing with an application the memory of the NI, which is a critical resource in an on-chip setting.

Both types of NoC interfaces will be described, with an emphasis on the second type, which can support additional communication and synchronization mechanisms than simple messaging, and presents a more challenging design target. Because interprocessor synchronization is very important for the efficiency of parallel processing of small amounts of work (*fine-grain* parallel computation), CMP network interfaces may provide special support for it. To this end, a case study is presented concerning the support for task syn-

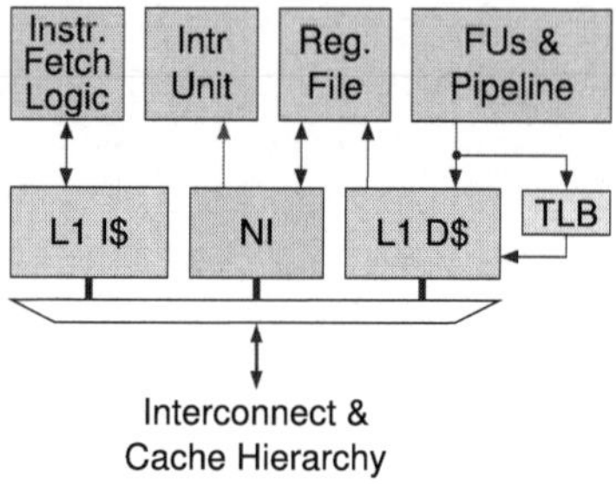

FIGURE 10.1: NI directly interfaced to processor registers.

chronization and scheduling in the Scalable computer ARChitecture (SARC) research prototype [141].

10.2 Processor-integrated Network Interfaces

The on-chip network can be directly interfaced to the processor pipeline by mapping it to processor registers (see Figure 10.1). With this type of network interface, destinations can only be determined by their node ID, and no naming or addressing mechanism is provided to identify memory of a remote processor or network interface. Any NI buffers are *private* to the sending and receiving nodes, and are managed by hardware. Restricted local access to such buffers may be provided, through processor privileged operations, to support context switches.

Only a send-receive style of communication is supported with this type of network interface. Processors send and receive messages directly from a subset of their registers, which essentially act as NI registers. Hence, communication operands appear directly in these registers, while the compiler cannot use them for general purposes. The processor is occupied during message send proportionally to the transfer size, and may *block* because of NoC or receiver occupancy. Messages are usually of a small, fixed maximum size.

Receive-side processing is normally required for every communication event (i.e., message). The processor can block on reading from an *empty* register, or, user-level reception interrupts may be used. Alternatively, an asynchronous privileged interrupt can be used to remove incoming data from the NI, and allow subsequent processing by software. This mechanism has the overhead of copying incoming traffic, and is typically only used for overflow handling. Furthermore, privileged reception interrupt costs in out-of-order processors can be very high because of outstanding memory operations, unless a special mechanism that squashes noncommitted instructions is provided for asynchronous interrupts.[1]

Multiword messages are usually atomic in the sense that individual words of the message are presented to the receiver as a unified entity, and thus message transfer appears as a single transaction. For the initiation, an explicit message send instruction is usually provided, that posts the whole message towards its destination once it is completed. This type of interface minimizes transfer latency, because communication is directly initiated from and arrives to *private* processor registers, avoiding processor interaction with any intermediate device. Bandwidth is limited to at most one processor word per cycle and potentially reduced by transfer initiation overheads.

In past multichip multiprocessor systems, iWarp [39] interfaced the network directly to processor registers in support of *systolic communication* and in the MIT J-Machine [65] similar mechanisms were used for fast remote handler dispatch. More recently, this approach has been used in CMPs, in the MIT Multi-ALU Processor (MAP) [148] to exploit fine-grain parallelism and provide concurrent event handling via multithreading, and in Tilera's TILE64 chip (see Chapter 7 and [339]) in support of operand networks. In the Imagine stream processor [142] the network is mapped in the stream register file for intercluster communication. Processor-interfaced *asynchronous direct messaging (ADM)* has also been proposed, for fast task synchronization support in CMPs [274].

In order to provide decoupling of the sending and receiving processors with this type of NI, some buffer space is usually provided at receivers. To keep the buffering requirements low and avoid dropping messages, the NI must provide a mechanism for *end-to-end flow control*, that will prevent or retry message transmission when no space is available at the destination. In case protected processor context switches must be supported, the newly scheduled thread should not have access to the buffer contents for the previously scheduled thread. To provide a clean receive buffer on context switches, either supervisor software must have access to the buffer contents, or a hardware mechanism should support copying of received messages to off-chip memory.

In addition, the NI must guarantee the delivery to software of the packets stored in its receive buffers, so that transactions complete and NI buffers are freed. When this is not guaranteed, a form of protocol deadlock specific to message passing systems can take place. If two hardware threads continuously send messages to each other, without removing arriving messages from the network interface, receive buffers will eventually become full. Network buffers will then fill up because of backpressure, and the threads will, eventually, deadlock when no more messages can be injected into the network. To prevent erroneous or malicious software from this kind of deadlock, when the NI receive buffer is almost full, some mechanism must guarantee its contents are copied

[1]Modern processors provide mechanisms to enter the kernel that do not change the context and have a much smaller overhead. These mechanisms, though, are processor synchronous.

to off-chip storage. This can be done, again, either by an automatic hardware facility, or by a high priority system thread that is scheduled-in via a receive buffer overflow interrupt [193]. It is also possible to allow user-level interrupt handlers to process incoming messages, but such handlers must be restricted from actions that may cause them to block, e.g., by acquiring locks or sending messages [332].

Regardless of the mechanism that guarantees software delivery of messages, buggy or malicious sender software could be filling receive buffers faster than the receiver could do any useful processing of incoming traffic. Once receive buffers become full, backpressure will incrementally clog the network with packets, interfering with the performance of other threads. There are two possible approaches to this problem: automatic off-chip buffering and end-to-end flow control. Providing an automated facility to copy messages off-chip may require costly rate-matching on-chip buffers to inhibit affecting other threads, and requires an independent or high-priority channel for the transfer to off-chip storage.

Alternatively, the NI may provide some form of end-to-end flow control that can keep the buffering requirements low. Providing some guaranteed receive side buffering per sender and acknowledged transfers for end-to-end flow control would be viable only in small systems. One possible solution is provided by limited buffering at the receiver for all senders, and sender buffering for retransmission on a negative acknowledgment (NACK). Such sender-side buffers should also be flushed on a context switch, if message ordering is required. Hardware acknowledgments (ACKs and NACKs) for end-to-end flow control require a network channel independent to the one used for messages, to avoid deadlock.

Because software usually requires point-to-point ordering of messages, if the NoC does not provide ordered delivery of packets (e.g., because of using an adaptive or other multipath routing scheme), the NI must provide reordering at receivers. Keeping the cost of reordering hardware low in this case, probably requires that the amount of in-flight messages per sender is kept very low. Even if the network supports in-order packet delivery, the NI may still need to provide some support for point-to-point ordering when the end-to-end flow control mechanism involves NACKs and retransmissions [274].

10.3 NI Integration at Top Memory Hierarchy Levels

Explicit interprocessor communication can also be supported with a network interface integrated at one of the higher levels in the memory hierarchy, close to the processor. This is usually some kind of scratchpad memory—also called a *local store* in the literature. Parts (a) and (b) of Figure 10.2 show two different placements for memory hierarchy integration of the NI. In part (c) of

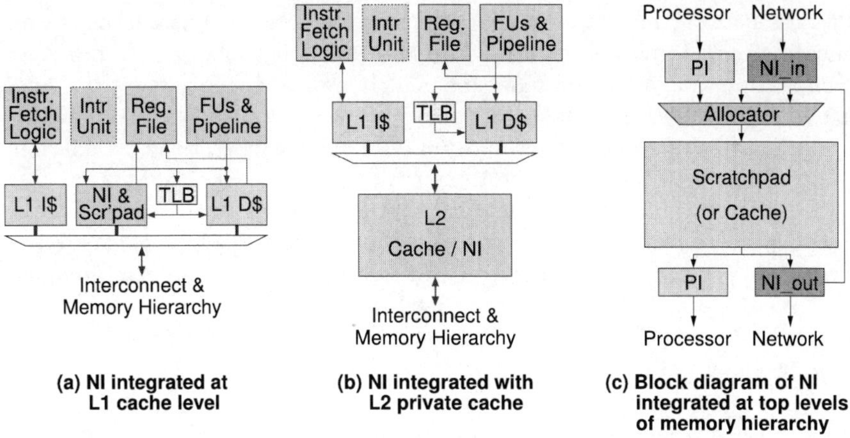

FIGURE 10.2: Microarchitecture of NI integrated at top memory hierarchy levels.

the same figure, a block diagram of the internal arrangement of NI resources is provided.

In Figure 10.2.(a), an NI supporting scratchpad memory at the level of the L1-cache is depicted. Figure 10.2.(b) shows an NI integrated with a private L2 cache, using portions of cache memory for scratchpad and other functions (see the next section) and exploiting common functionality. In both of these parts of the figure a detailed processor interface is shown. The TLB distinguishes cacheable and scratchpad accesses and can prevent the L1 data cache from responding to scratchpad acceses. In the case of L2-cache level integration, local scratchpad may or may not be cacheable in the L1. If L1 caching is allowed for local scratchpad memory, the L1 may need to be invalidated on remote scratchpad accesses.

Figure 10.2.(c) abstracts away the processor interface and shows the internal organization of the NI. An allocator arbitrates scratchpad accesses initiated from the processor interface (PI), from incoming network traffic (NI_in), or for outgoing communication (NI_out). Incoming network traffic is buffered independently, per network channel in NI_in. The NI_out schedules transfers to the different subnetworks as required, accessing scratchpad for local source regions (e.g., for remote direct memory access (RDMA)). Transfer requests from the processor may not require access of local scratchpad (e.g., remote scratchpad loads and stores), but go through the allocator to access NI_out.

A less aggressive placement of the network interface, further from the processor, at some level of the memory hierarchy shared by a group of cores, is also perceivable. In this case, processor access of the NI can be uncached (as in most network interfaces for off-chip communication), or may exploit coherence mechanisms [222]. In this case, the NI would be utilized for explicit transfers among parts of the hierarchy belonging to different processor groups. This organization, though, is not studied here.

With a network interface integrated at top memory hierarchy levels, communication operations always have their source, destination, or both in memory. Scratchpad memories provide sufficient buffering for bulk transfers, and thus the NI may support RDMA (get and put) or copy operations. For RDMA one *local address* operand is necessarily used for the source or the destination of the transfer, whereas copy operations require *global addresses* for both the source and the destination. It is natural to provide access to the local scratchpad with processor loads and stores. When the processor supports global addresses (i.e., processor addresses are wide enough), or if remote scratchpad memory can be mapped in its address space, direct load/store access to remote scratchpads can also be provided. In any case, atomic multiword messages can be supported as well.

At least one of the source and destination operands of communication operations always resides in memory shared by processors other than the one initiating a transfer. For this reason, synchronization is required, both before the transfer and for its completion, with the processor(s) that may be accessing the shared operands concurrently. Conversely, because application data can be placed in NI memory, communication can occur asynchronously with computation (i.e., without occupying the processor), both for transmission and for reception. The aforementioned communication mechanisms, with the exception of scratchpad loads and stores, can provide such decoupling of NI communication operations and processor computation, effectively allowing software to overlap multiple transfers with each other and with computation.

The latency of short transfers is generally increased compared to processor-integrated NIs, because transmitted operands may need to be written to the local scratchpad before departure, and received operands must be accessed from the scratchpad after synchronization, as opposed to sending and receiving from processor registers directly. On the contrary, as long as scratchpad memory and network resources are not busy, NI bulk transfers can exploit the full NoC bandwidth, which may oftentimes be two or four processor words per cycle.[2]

Moreover, it is possible for this type of NI to also support blocking scratchpad loads to minimize reception overhead and reception interrupt mechanisms to allow arbitrary processing to occur asynchronously (not shown in Figure 10.2), as provided with synergistic processing elements (SPE) channels in the CellBE [138]. Normally, though, polling must be used to detect operand reception for communication operations other than scratchpad loads.

In multichip multiprocessors this type of NI was exploited for the sender-side of the AP1000 [288] to transfer cache lines, and there are a few recent CMP designs that also take this approach. These include the CellBE [138], from IBM, Sony, and Toshiba, implementing eight SPEs with private scratchpad memories inside a global address space accessible via coherent RDMA; Intel's

[2]This results in reduced latency for bulk transfers that exceed some minimum size, provided that scratchpad memory bandwidth is equal or exceeds that of the NoC.

Single-chip Cloud Computing (SCC) experimental 48-core chip [127], which provides 8 KB/core on-chip message passing buffers (MPB), accessible by all cores via loads and stores, as a conceptual shared buffer inside the system address space; and the cache-integrated NI for the SARC European project [255] implemented in a multicore FPGA-based prototype [141, 147], which supports all the communication mechanisms referenced above as well as a set of synchronization primitives.

NIs integrated at top memory hierarchy levels provide explicit communication and synchronization, but are related to caches in that NI communication memory is also the application memory. With this kind of NI, instead of separate processor and NI memory resources, on-chip memory is shared and better utilized, obviating the redundant copy of communication operands that was common in traditional off-chip network interfaces with dedicated memory. With memory hierarchy integration there are no dedicated buffering resources managed exclusively in hardware. As a result, producer-consumer decoupling and flow control, usually supported by the NI, need to be arranged under software control.

Load and store accesses to local or remote scratchpad memory and to NI control registers must utilize the processor translation lookaside buffer (TLB) for protection and to identify their actual target. The destination of these operations can be determined from the physical address or by enhancing the TLB with explicit locality information, i.e., with extra bits identifying local and remote scratchpad regions. In the case of a scratchpad positioned at the L1-cache level, as the one illustrated in Figure 10.2.(a), TLB access may necessitate deferring scratchpad stores locally, as is usual in L1 caches for tag matching.

Although scratchpad memory can be virtualized in the same way as any other memory region, the system may need to migrate its contents more often to better utilize on-chip resources. For example, scheduling a new context on a processor may need to move application data from scratchpad to off-chip storage, in order to free on-chip space for the newly scheduled thread. Migration of application scratchpad data is a complicated and potentially slow process, because any ongoing transfers destined to this application memory need to be handled somehow, for the migration process to proceed. Furthermore, mappings of this memory in TLBs throughout the system need to be invalidated. To facilitate TLB invalidation, the NI should provide a mechanism that allows the completion of in-progress transfers without initiating new ones.

In a parallel computation, consumers need to know the availability of input data in order to initiate processing that uses them. Reversely, producers may need to know when data are consumed in order to replenish them, essentially managing flow control in software. For these purposes the NI should provide a mechanism for producers and consumers to determine transfer completion. Such a mechanism is also required—and becomes more complicated—if the on-chip network does not support point-to-point ordering. Write transfers that require only a single NoC packet can be handled with simple acknowledg-

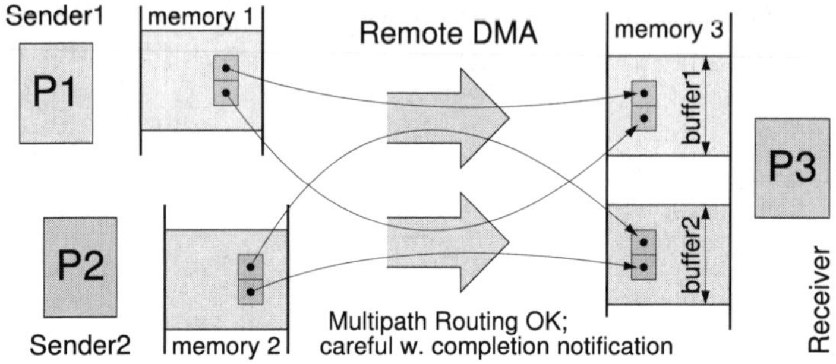

FIGURE 10.3: One-to-One Communication using RDMA: single writer per receive buffer.

ments, but multipacket RDMA is more challenging. Furthermore, generalized copy operations, if supported, may be initiated by one NI and performed by another, e.g., node A may initiate a copy from the scratchpad of node B to the scratchpad of node C.

To illustrate some of these points consider Figure 10.3, which illustrates a typical use of write RDMA with multiple parallel transfers in progress. For multiple senders (P1 and P2), to be sending to a same receiver (P3), the receiver must have set up separate memory areas where each transfer is taking place. Figure 10.3 also illustrates that RDMA works well even if the network uses adaptive or multipath routing, which can drastically improve network performance. Under such circumstances, packets may arrive out-of-order; however, each packet carries its own destination address, so their data can all be written into the proper places, regardless of arrival order. Completion detection can no longer be made based on when the "last" word has been written to the highest address; instead, the number of arriving bytes must be counted (assuming no duplicates ever arrive), and compared to the number of expected bytes. An elegant mechanism that provides transfer completion notification in such an environment was proposed for the SARC NI and will be presented later in this chapter.

Let us go back to the properties of a network interface integrated at the top memory hierarchy level. Since NI communication memory coincides with application memory, software delivery of write transfer operands is implied, as long as NoC packets reach their destinations. Nevertheless, the NI needs to guarantee that read requests can always be delivered to the node that will source the response data without risking network deadlock. Networks guarantee deadlock-free operation as long as destinations sink arriving packets. The end nodes should be able to eventually remove packets from the network, regardless if backpressure prevents injection of their own packets. Because nodes sinking read requests need to send one or more write packets in reply, reads "tie" together the incoming and the outgoing network, effectively not sinking the read, which may lead to what is called *protocol deadlock*.

To remedy this situation the NI may use for responses (writes in this case) a network channel (virtual or physical) whose progress is *independent* to that of the channel for requests (reads). As long as responses are always sunk by NIs, the request channel will eventually make progress without deadlock. Alternatively, reads may use the same network channel with writes, but they need to be buffered at the node that will send the data. The reply of the read will then be posted by that node, similar to a locally initiated write transfer, when the network channel is available. Finally, for transfer completion detection, the NI may need to generate acknowledgments for all write packets. These acknowledgments must also use an independent network channel and NIs must always be able to sink such packets.

Since bulk transfers are offloaded to the NI while the processor can continue computation and memory access, a weak memory consistency model is implied. When posting a bulk transfer to the NI, subsequent memory accesses by the processor must be considered as concurrent to NI operation (i.e., no ordering can be assumed for NI and processor operations) until completion information for the bulk transfer is conveyed to the processor. To provide such synchronization, the NI must support *fences*[3] or other mechanisms, to inform software of individual transfer completion or the completion of transfer groups.

When remote scratchpad loads and stores are supported, these accesses may need to comply with a memory consistency model [221]. For example, for *sequential consistency* [175] it may be necessary to only issue a load or store (remote or local) after all previous processor accesses have completed, sacrificing performance. With a consistency model like *weak ordering of events*, fences are required together with special synchronizing accesses [86]. In addition, the NI may need to implicitly support ordering for load and store accesses to the same address, so that the processor expected order of operations is preserved and read-after-write, write-after-read, or write-after-write hazards are avoided.

Additional complexities arise because the network interface resides outside of the processor environment, in the memory system. The network interface may need to tolerate potential reordering of load and store operations to NI control registers by the compiler or an out-of-order processor. For example, the NI should be able to handle a situation where the explicit initiation of a message send operation and the operands of the message arrive in an unexpected order. Finally, out-of-order processors may issue multiple remote scratchpad loads, for which the NI needs to keep track of the destination processor register. This, in turn, requires buffering of the outstanding operations, in a structure similar to cache miss status holding registers (MSHR).

A final thing to note is that in the presence of cacheable memory regions,

[3]In the context of the processor, fences or "memory barriers" are instructions that postpone the initiation of subsequent (in program order) memory operations, until previous ones have been acknowledged by the memory system and thus are completed. Similarly, the network interface can provide operations that expose to software the completion of previously issued transfers.

all NI communication mechanisms may need to interact with coherence directories, which, in turn, will need to support a number of protocol extensions. To avoid complicating directory processing, it may be preferable that explicit transfers to coherent memory regions are segmented at destination offsets that are multiples of cache-line size and are aligned according to their destination address (i.e., use destination alignment explained later in this chapter).

10.4 NI Control Registers and Virtualization

Virtualization of the network interface allows a *hypervisior* (e.g., the OS) to hide the physical device from a *guest* (e.g., a thread), while providing controlled access to a virtual one. The virtual device provides direct access to the physical one, without requiring hypervisor intervention in the usual case. The virtual device must be accessible in a protected manner to prevent guests that belong to different protection domains from interfering with each other. Protection is managed by the hypervisor, and allows control on how sharing of the physical device is enforced. Virtualization can allow parallel access to the physical device by multiple guests, without the need for synchronization.

The mechanisms that provide synchronization-free sharing of the physical device and delegate control of the sharers to the operating system, are implemented in hardware. For example, a mechanism commonly used for virtualization is memory mapping of device resources. Access control provided by address translation and protection hardware, allows the required OS control over the memory mapped resources. In addition, virtualization usually requires special hardware support, to allow OS handling of context swaps. The OS must preserve the state of the switched out thread, including any communication state, so that the thread can later resume. It must also guarantee that, after the context switch is completed, the physical device allows unobstructed use by other processes or threads.

Traditionally, NI control registers are used to provide multiword or block transfer descriptions to the NI, which cannot be programmed with a single instruction. Such descriptions may refer to messages and RDMA or copy operations. In addition, special NI control registers can support data and thread synchronization, providing atomic operations, blocking processor access, receive-side queuing, and configuration of communication event interrupts.

The control registers of a processor-integrated NI are a subset of the processor registers. Because send operations are processor synchronous, only a single set of such registers is adequate for the supported transfers of short messages and scalar operands. Virtualization of the NI requires that any send and receive buffers associated with these control registers are brought to a consistent state and freed when another thread is scheduled on the processor, and can later be restored to that state when the initial thread is rescheduled.

In addition, communication destinations can be virtualized by identifying the *destination thread* instead of the destination processor. This can be supported by a translation mechanism similar to address translation, i.e., a hardware *thread translation table* in the NI that maps threads to processors and is filled by the OS on misses. The NI must be aware of the thread currently scheduled on the processor, for which it can accept messages, and a mechanism is required to handle messages destined to threads other than the scheduled one. For example, Sanchez et al. [274] have proposed the use of NACKs and *lazy invalidation* of the corresponding entry in the thread translation table of the source processor.

A network interface integrated at the top memory hierarchy levels makes more extensive and demanding use of communication control and status registers, especially when asynchronous bulk transfers are supported. It is advantageous to bring NI control registers close to the processor, preferably in a processor-private part of the hierarchy, to reduce initiation latency for transfers and to allow uncontended local access and communication progress monitoring. In addition, NI registers need to be virtualized to provide low-overhead protected access, avoiding interaction with the operating system. Finally, communication overlap benefits from multiple sets of control registers, when multiple simultaneous outstanding transfers are supported.

Programmable accelerators usually relax virtualization requirements, advocating a computation model in which tasks are processed to completion. As a result, they may provide limited support for protected local access to their NI resources, or limited support for migrating partial results from scratchpad memory and communication state of ongoing computation for a context switch. For example, even though the CellBE provides support for SPE context migration, the SPEs cannot perform a full context switch on themselves.

To provide NI virtualization and transfer overlap capability, a number of copies of the control and status registers is needed. Memory mapping of a NI register set in the address space of a process, allows low overhead, user-level NI access. The number of register sets defines the number of processes that can access the NI concurrently. In case more processes need simultaneous access to the NI, the OS will need to resort to frequent context switches to timeshare the use of the register sets. In addition, the number of control registers per set limits the number of concurrent transfers available to each process. The cost of virtualization is proportional to the total number of control register copies.

An alternative is to provide configurability of scratchpad memory, so as to allow programmable control "registers" inside its address space. This is feasible by providing a few tag bits per *scratchpad line* (block). The tag bits are used to designate the varying memory access semantics required for the different *line types* (i.e., control "registers" and normal scratchpad memory). The resulting memory organization is similar to that of a cache and can sup-

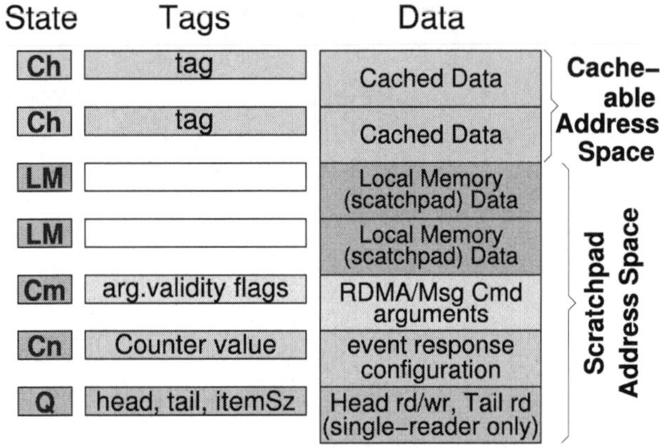

FIGURE 10.4: State bits mark lines with varying access semantics in the SARC cache-integrated network interface (from [147] ©2010 ACM, used with permission). Top to bottom: (Ch) Normal cached data lines; (LM) Normal scratchpad memory lines; (Cm) Command buffers (communication control/status "register"); (Cn) Counter synchronization primitives (support atomic add-on-store operations); (Q) Single and Multiple Reader Queue descriptor synchronization primitives. Cacheable and scratchpad address spaces are identified from the physical address provided via normal address translation.

port virtualization via address translation and protection for accesses to the scratchpad memory range.

This design allows a large number of communication "registers" to be allocated inside the (virtual) address space of a process. The NI can keep track of outstanding operations by means of a linked list of these communication control "registers" formed inside scratchpad memory. Alternatively, the total number of scratchpad lines configured as communication "registers" may be restricted to the number of outstanding jobs that can be handled by a fixed storage NI *job list*. Such a job list processes in first-in first-out (FIFO) order transfer descriptions associated with NI control "registers"—potentially allowing recycling of descriptions of transfers in progress. Scaling the size of such a job list structure, and thus the number of supported outstanding transfers, results in low hardware complexity increase, in contrast to outstanding transfers for a cache that require a transaction buffer or miss status holding register (MSHR) fully associative structure.

To support migration of scratchpad regions that may include lines marked as NI "registers," the operating system may read and record tag bits of the region at migration time. Alternatively, lines marked as NI "registers" may be recorded by the OS at the time of their allocation. To optimize the migration process the OS may restrict NI "register" allocation inside special scratchpad pages.

The SARC cache-integrated NI exploits this approach to provide configurable use of cache lines as cache, scratchpad, NI control registers, or synchronization primitives (see Figure 10.4). Because scratchpad lines are located in memory space, processors can only access them going through address translation and protection. Thus, each process can freely access, in user-mode, its own special "registers," independent of and asynchronously to other processes.

Virtualization of cache-integrated or scratchpad-attached NIs supporting RDMA or copy operations, also requires that address arguments passed to control registers are given in virtual—rather than physical—space, and are protection-checked. Although a few solutions have been proposed to the NI address translation problem [116, 50, 282], processor proximity of a CMP NoC interface may simplify the situation. A TLB (or memory management unit (MMU)) structure can be implemented in the NI to support the required functions. This TLB would be placed in NI_out of Figure 10.2. Updates of address mappings may use memory mapped operations by a potentially remote processor handling NI translation misses, as is the case of the Cell BE where SPE translation misses are handled by the power processing element (PPE). Alternatively, access to a second port of the local processor's TLB may be provided.

10.5 Communication and Data Synchronization Mechanisms

Explicit communication mechanisms supported by CMP network interfaces can provide direct point-to-point transfers and thus offer efficient communication and minimize energy per transfer. Bulk explicit transfers can be used for macroscopic software prefetching and to overlap multiple transfers and computation, while messages and direct loads and stores to scratchpad can provide low latency signaling and common data access. It is thus essential for the functions of a CMP network interface to be accessible at the application level, without OS intervention, to exploit the latency and bandwidth advantages of on-chip transfers.

Because software needs to actively manage the ordering of operations, explicit communication can become crabbed in the case of processor-asynchronous handling of bulk transfers by the NI, or when the on-chip network does not support ordering. To simplify the handling of operation ordering the NI should support both efficient and straightforward transfer completion detection. In addition, in the context of producer-consumer communication, detection of data arrival by the consumer should be both flexible and fast to allow efficient fine-grain interactions. We call this special case of transfer completion detection by the consumer *data synchronization.*

10.5.1 Communication Mechanisms

Messaging-like mechanisms can be used for scalar operand exchange, atomic multiword control information transfers, or, combined with a user-level interrupt mechanism at the receiver, for remote handler invocation. In processor-integrated NIs, message transmission directly uses register values. An explicit *send* instruction or an instruction that identifies the setting of the final register operand is used to initiate the transfer. At the receiver the message is delivered directly to processor registers.

In the case of a network interface integrated at the top memory hierarchy levels, the message must be posted to NI registers and the transfer can be initiated either explicitly via an additional control register access or implicitly by NI monitoring of the transfer size and the number of posted operands. At the receiver the message can be delivered in scratchpad memory or in NI control registers used for synchronization purposes.

Loads and stores to local or remote scratchpad regions can be started after TLB access, depending on the consistency model. Remote stores can use write-combining to economize on NoC bandwidth and energy. In this case, the processor interface depicted in parts (a) and (b) of Figure 10.2 would include an additional path from the combining buffer to the NI. Although it is possible to acknowledge stores to scratchpad to the processor immediately after TLB access, scratchpad accesses must adhere to the program dependences as discussed in the previous section. Remote loads can be treated as read direct memory acces (DMA) requests, or exploit an independent network channel for their single-packet reply, avoiding the possibility of protocol-level deadlock. In order to support vector accelerators, the NI may also provide multiword scratchpad accesses.

When the NI resides below the L1 cache level (see Figure 10.2.(b)), scratchpad memory may be L1 cacheable. Coherence of scratchpad and cached copies may be kept by hardware or software. For example the SARC cache-integrated network interface resides in a private L2 cache, as in the case of Figure 10.2.(b). Scratchpad memory "locked" in the L2 can be cached only in the local L1, which is write through and is invalidated appropriately on remote writes to scratchpad. In the case of the Single-chip Cloud Computing (SCC) Intel experimental chip, per core message passing buffers (scratchpad memories) can be cached without coherence in L1s throughout the chip. The L1s support a single cycle operation that allows software to rapidly invalidate all scratchpad copies locally cached.

The remote direct memory access (RDMA) mechanism implements *get* and *put* bulk transfers to and from the local scratchpad. The direction of the transfer (read or write) is explicit in RDMA, which requires that software is aware of both local and global addresses. Specifying an RDMA transfer initiated by a remote NI is not fully supported in the usual case. For example, the SPE DMA engine of the Cell processor utilizes *local store* and *effective* addresses and corresponding commands, while for transfers initiated by the

Power (or peripheral) processor element (PPE) each SPE has a separate DMA queue where get and put operations can be placed remotely.

Write DMA commands that include the source and destination addresses of the transfer, as well as its size and opcode, need to be buffered in appropriate NI control registers to keep track of operation progress. A large write DMA transfer should be segmented and processed in multiple iterations. This will avoid blocking other processor traffic (messages or remote stores) while the DMA is in progress, and will allow interleaving segments of multiple outstanding DMAs for parallel progress of transfers toward potentially different destinations.

For read DMA the receiving NI must be able to remove the incoming request from the NoC and generate the appropriate reply. In this case, not blocking the input channel while sending a multipacket response can be accomplished by breaking the read request in multiple read packets so that each requires a "short" reply. Alternatively, the NI can buffer read DMA requests until they are fully processed, which obviates the need for separate read and write network channels, but limits the number of concurrent reads the NI can support depending on the amount of buffering provided.

With this second approach, if the available buffer space is exceeded, the NI needs to *drop* superfluous requests, record the event and notify the nearby processor of the error condition (e.g., with an interrupt). The SARC network interface employs a software-provided buffer in scratchpad memory for read requests, delegating to the software the responsibility to provide a buffer large enough for its needs. An alterative approach for a read request exceeding the available buffer space at the NI hosting the read's source region, would be to employ some kind of negative acknowledgment that notifies the error condition to the initiating node.

When scratchpads can only be accessed via a global address space, it is more natural not to limit the sources and destinations of communication mechanisms to the local scratchpad. This precludes get and put operations (RDMA), and the congruent mechanism provided by the NI supports a *copy* operation. Copy operations are more general than RDMA because they allow the specification of transfers that have both remote source and destination addresses. Notification of the initiating node for transfer completion in hardware is more complex in this case.

RDMA and copy operations that allow arbitrary changes of data alignment may be dealt with in three different ways, as illustrated in Figure 10.5. First, it is possible to send write data in network packets keeping their *source alignment*, as shown in Figure 10.5.(a). This implies that packets may have "padding" both at the beginning (before useful data) and at the end. In addition, a barrel shifter is required at the receiver to provide the operation requested alignment. Second, if we choose to send packets with their requested *destination alignment*, as shown in Figure 10.5.(b), then "padding" may also be required both at the beginning and at the end, and a barrel shifter must be placed at the source node.

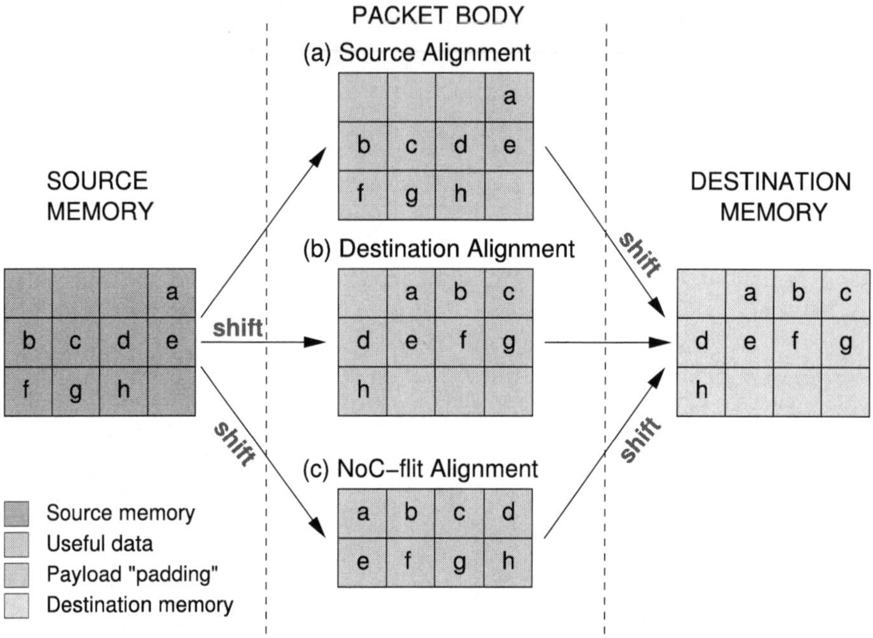

FIGURE 10.5: Three alternatives of handling transfers with arbitrary alignment change.

Third, we may want to minimize the number of NoC flits transferred, as shown in Figure 10.5.(c), in which case we need a barrel shifter at the source node to align the transmitted data to the NoC flit size boundary and another barrel shifter at the destination node to fix the requested destination alignment. In this case, packets can only have "padding" after useful data. This third approach is more expensive, requiring two barrel shifters per node, and cannot reduce the amount of transferred data by more than a single NoC flit. Restricting the supported alignment granularity reduces the cost of the barrel shifter circuit, but complicates software use of RDMA or copy operations.

Transfer pipelining is important for both latency and bandwidth efficiency of transfers. For instance, NoC request scheduling can overlap with NoC injection of a previous packet. To keep the latency of communication mechanisms to a minimum, it is important that the network interface implements cut-through for both outgoing and incoming packets. When the maximum packet size does not exceed the width of the receiver's memory, cut-through implementation for the incoming path can speculatively advance the tail pointer of the NoC interfacing FIFO for packet reception, until the correct cyclic redundancy check (CRC) is computed at the end. Alternatively, for NIs integrated at the top of the memory hierarchy, providing separate CRC for the packet's header and body allows writing incoming packets if the destination address is correct before checking if the data CRC is correct.

10.5.2 Data Synchronization

Detection of data reception by a consumer (i.e., data synchronization) is necessary before a computation on communicated data can start. For this purpose NI designers usually optimize data synchronization combining it with data reception. This is true for send-receive style communication and consumer-initiated transfers. Reversely, in the case of producer-initiated transfers, individual transfer completion information enables the initiating node to synchronize with the consumer as required, and enforce point-to-point ordering when it is not supported by the NoC or the NIs. The common element of these mechanisms, that enables producer-consumer interactions and allows exploitation of NI-initiated transfers, is transfer completion detection. Providing this type of information, without compromising NI scalability and communication performance, may be a challenging goal for network interface design.

There are three basic mechanisms for application software to detect message reception:

1. *Blocking receive operations.* This first mechanism is based on NI support for blocking read access to NI registers, until a message arrives. A blocking "receive" operations can provide the lowest reception latency when the operation is issued before message arrival. Software must express correctly an order of issuing send and receive operations that does not cause deadlock (e.g., two nodes exchanging values should not both do a blocking "receive" before sending across their values). Blocking operations can be combined with processor transition to a low power state for energy efficiency.

2. *Polling on NI registers or scratchpad memory.* The second mechanism avoids blocking by means of "peek" operations on NI status registers or scratchpad memory, where a message or an acknowledgment is expected. This provides nonblocking reception handlers and allows computation to proceed while waiting for a message. Polling should be local to reduce reception overhead and to avoid congesting the network. The potential downside of this mechanism is that for purposes other than event processing, the appropriate frequency of polling is very difficult to assess effectively and may introduce unnecessary software overhead.

3. *Interrupting the destination processor.* The third mechanism invokes a user-level interrupt that forces the execution of a message reception handler. Dispatch of the appropriate handler can be automated with hardwired information in special message types [65]. User-level interrupts have lower overhead than privileged ones, but event handling with interrupts is not straightforward. The user needs to disable interrupts when synchronous processing atomicity must be provided [42]. In addition, the user-level interrupt mechanism may require that the handler removes the received message from NI dedicated storage without blocking (e.g., by sending messages or acquiring locks) [332]. Nevertheless, this

last restriction can be relaxed by providing a higher priority privileged interrupt mechanism that handles NI resource overflow and underflow transparently.

Processor-integrated NIs usually support point-to-point ordering transparently, in order to preserve program semantics of operand passing and be programmed more naturally, while data synchronization is provided by the message reception mechanisms discussed above. NIs integrated at top memory hierarchy levels can provide the same data synchronization mechanisms for messaging by exploiting load and store accesses to scratchpad or to NI "registers."

Considering read-write instead of send-receive communication style, similar mechanisms can be provided for RDMA operations through NI control registers, both on the producer and the consumer side. These mechanisms though, require some additional support to detect transfer completion, whose complexity can vary depending on a number of factors: (i) the number of outstanding accesses supported, (ii) whether copy semantics are supported for RDMA, and (iii) whether the NoC supports point-to-point ordering. A large number of outstanding accesses requires a lot of state (and a lengthy transfer ID) to track their completion. Copy semantics require that more than two nodes may be involved in the transfer and its completion. Lastly, with an unordered network point-to-point completion signals cannot be grouped per involved node and need to be counted appropriately.

In addition, with the read-write communication style (i.e., for load and store as well as for RDMA-copy operations), the memory consistency model interferes with synchronization. A weak consistency model must be assumed since bulk multipacket transfers would be difficult or inefficient to handle otherwise, especially with an arbitrary size. In this case, the NI may provide fence operations that guarantee the completion of all previously initiated transfers. Separate fence operations can be supported for each mechanism to provide more flexibility. For example, a simple fence mechanism for scratchpad stores only requires a counter at the initiating NI that keeps track of the arithmetic sum of departures minus acknowledgments.

Individual operation completion can also be exposed to software, providing *explicit transfer acknowledgments*. This can be supported by allowing software to specify an acknowledgment address for each communication operation, at which an acknowledgment value will be deposited upon transfer completion. This mechanism is sufficient for messaging over an unordered network, but multipacket RDMA or copy operations require additional support to confirm that all associated packets have been delivered to their destination.

The SARC cache-integrated NI provides an elegant mechanism for individual copy operation *completion notification*, based on explicit acknowledgments and synchronization *counter* primitives, which we briefly describe. Counters are NI control "registers" allocated in tagged scratchpad blocks (see Figure 10.4). They support an atomic *add-on-store* function that adds to the counter contents values stored at offset zero of the associated scratchpad line. They

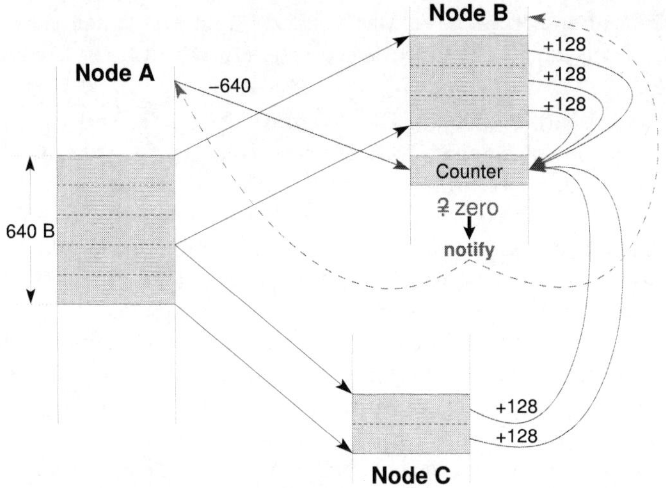

FIGURE 10.6: Copy operation completion notification in the SARC research prototype.

also provide configurable *event response* functionality for which arguments are set in the rest of the line. The event response mechanism *triggers* when the counter becomes zero and sends up to four notifications (stores) towards addresses specified by software in the counter line. A single word as data for the notifications as well as a reset value for the counter are also supplied by software within this line.

The SARC network interface requires that for every write packet received, an acknowledgment packet is sent to an address provided in the write, carrying as data the size of the write packet. At their destination, acknowledgments are treated as writes, but do not generate further ACKs. For completion notification, a copy operation needs to specify an acknowledgment address that is the address of a counter, and will be carried in all packets generated for the transfer. The counter must be configured in advance to send completion notifications to the intended nodes (addresses).

Figure 10.6 shows an example of the mechanism's operation. Node A initiates a copy operation for 640 bytes residing locally to be transfered to a destination scratchpad region that is scattered to nodes B and C (the destination region crosses a page boundary and the two pages reside in different nodes). The acknowledgment address of the copy operation points to a preconfigured counter at node B, initialized to zero. The transfer is broken in five 128-byte NoC packets, three sent to node B and two to node C. Node A also writes to the counter the opposite of the total transfer size (-640). For each packet of the copy operation that arrives at its destination, an acknowledgment is generated, writing the value 128 to the counter that is actually incremented by the same amount. The counter accumulates the values of all the acknowledgments and the opposite of the total transfer size sent by node

A, and will become zero once all of them have arrived. When that happens, the counter sends notifications to the preconfigured addresses, which in the example reside at nodes A and B.

Several points are evident in this example. First, observe that this completion notification mechanism does not require network ordering—the order the values are accumulated in the counter is not important. Second, the copy operation of the example could be initiated equivalently by node B, or C, or by a fourth node instead of node A. The only difference would be the additional read request packet sent from the initiating node to node A before the illustrated communication is triggered.[4] This means that the completion notification mechanism is suitable for general copy operations and is not restricted to simpler point-to-point get and put operations. Finally, observe that completion notification for an arbitrary number of user-selected copy operations can use a single counter, provided that the opposite of the aggregate size of all transfers is written to the counter with a single operation to prevent notification triggering for subsets of the transfered data.

The network interface can support synchronization mechanisms that *multiplex* data from multiple senders (producers) in a NI queue at the receiver (consumer), when ordering among producers is not important. This allows the consumer to avoid searching for arriving data. In addition, multiple such queues can be provided, to allow a receiving node to *demultiplex* messages, according to some categorization. For example, messages from the same sender, or messages carrying a stream of data, or messages of some specific type, can be automatically categorized by hardware in order to be processed in a uniform or orchestrated manner. The first property of such synchronization support (multiplexing of senders) is natural in processor-integrated NIs, which queue messages from multiple senders. Its second property (message demultiplexing in categories) is natural in memory hierarchy integrated NIs, which demultiplex arriving packets to different addresses.

Tilera's TILE64 multicore chip [339] (Chapter 7) supports a small number of queues in a processor integrated NI. Demultiplexing is enabled via hardware supported message tags specified by the sending node. Queues at the receiver allow software settable tags for tag matching in hardware, and an additional queue is provided for unmatched messages. The SARC prototype provides similar functionality allowing an arbitrary number of *single-reader queues* to be allocated in scratchpad memory, and supporting multiple queue item granularities. Each queue operates as a single point of reception, in support of many-to-one communication that does not require independent per sender buffering. While SARC network interface requires software handling of flow control and the receiver reads data from scratchpad, Tilera's processor-integrated NI provides flow control in hardware and data are found in registers.

A SARC NI single-reader queue is illustrated in Figure 10.7. A queue

[4] The write of the opposite of the total transfer size (-640) can actually be done by any node.

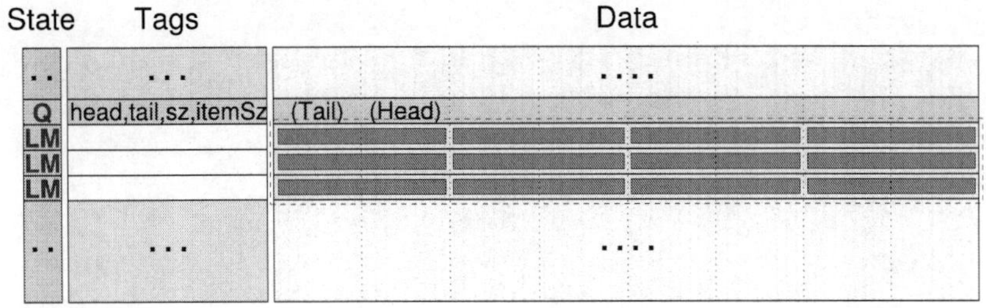

FIGURE 10.7: Illustration of a single-reader queue in scratchpad memory of the SARC cache-integrated NI. Double-word queue item granularity is depicted.

descriptor special scratchpad line is allocated adjacent to and preceding the scratchpad range that forms the queue body. The queue descriptor is denoted with (Q) in its state bits, whereas normal scratchpad lines are marked (LM) and a dashed line frame is drawn around the queue body.

Multiple queue item granularities are supported and all queue metadata are kept in the queue descriptor tag, including queue and item size configuration as well as queue head and tail pointers. Read access to the tail pointer and read/write access to the head pointer is provided for software via the first and second word of the descriptor block, intended for polling and dequeue from the local processor. Local or remote writes at block offset zero of the queue descriptor atomically enqueue an item in the queue, and multiword enqueues must use messages. Queue descriptor and body alignment as well as item size restrictions are enforced to optimize the implementation.

10.6 Case Study: Task Synchronization and Scheduling Support in the SARC Network Interface

A task (or job) is a part of a larger computation, that is intended to run in parallel with other tasks in a multiprocessor system. A task can be an instance of a function, a loop iteration (or a number of iterations), or simply a block of code. Tasks should be selected so that there are other independent tasks that can be executed in parallel.[5] Usually, a job has a set of (input) dependences from other jobs and a set of jobs that depend on its completion. We will refer to conveying the resolution of these dependences among tasks

[5]Alternatively, tasks can be selected so that they only have a few dependences with each other and can provide sufficient execution overlap. Here we do not consider such tasks to simplify the discussion of task synchronization support.

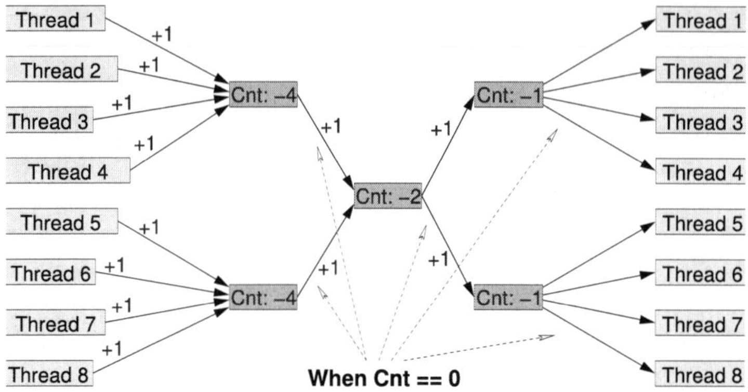

FIGURE 10.8: A barrier based on SARC network interface counters.

as task synchronization. In an analogy between serial and parallel execution, one can say that true data dependences in serial code correspond to data synchronization in parallel code and control dependences correspond to task synchronization. *Task scheduling* is the selection of a processor (or hardware thread) to run a task once its input dependences have been resolved.

The synchronization counters of the SARC network interface can provide efficient task synchronization support. For example, a counter can be allocated for a task (T) waiting input dependences and initialized to the opposite of the number of these dependences. As tasks on which T depends complete, they signal to T the resolution of input dependences. When all dependences are resolved for T, its counter will become zero and automatically send a notification to a scheduling task (in scratchpad memory), or a network interface queue, so that T can be scheduled.

Counters can also be combined to construct scalable and efficient *barriers* for global task synchronization. Figure 10.8 shows how counters are combined to form two trees with a single counter at their common root, in a barrier for eight threads. The tree on the left accumulates arrivals and the tree on the right broadcasts the barrier completion signal. In the figure, threads are shown to enter the barrier writing the value one to counters at the leafs of the arrival tree. The counters of both trees are initialized to the opposite of the number of expected inputs.

For the arrival tree, when counters become zero they send a single notification with the value one, that is propagated similarly towards the root of the tree. When the root counter triggers, the barrier has been reached. Counters in the broadcast tree, other than the root one, expect only a single input and generate multiple notifications, propagating the barrier completion event to the next tree level, until the final notifications are delivered to all the threads. Triggering counters are automatically re-initialized and ready for the next barrier episode.

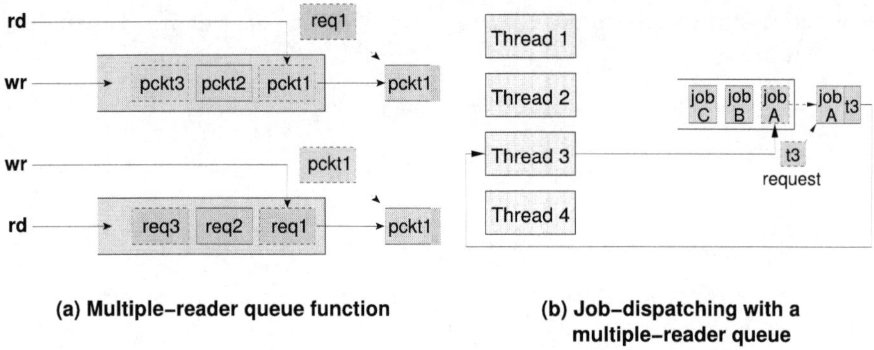

FIGURE 10.9: (a) Read and write access semantics to a multiple-reader queue (from [147] ©2010 ACM, used with permission) and (b) automatic job-dispatching exploiting the functionality of a multiple-reader queue.

The SARC network interface also provides *multiple-reader queues* in support of task scheduling. Multiple-reader queues are constructed in scratchpad memory similarly to single-reader queues of Figure 10.7, but do not provide access to the head and tail pointers of the queue. Instead, write and read type messages are used for enqueue and dequeue operations respectively, and one maximum size message item is supported per scratchpad line in the queue body.

Read and write access to multiple-reader queues is *symmetric*, as illustrated in Figure 10.9(a). Both read messages (requests) and write messages (data) are buffered when arriving at a queue that is empty or if messages of the same type of the arriving one are pending. Conversely, both reads and writes result in a dequeue when messages of the opposite type are found in the queue. In essence, multiple-reader queues *match* reads and writes in time, regardless of the order of their arrival. The data of the write are then sent automatically to the response address provided in the read.

Figure 10.9(b) shows an example of how multiple-reader queues can be used for task scheduling. Job IDs are initially enqueued to the multiple-reader queue. When one of the threads 1 through 4 of the figure becomes available, it dequeues a task ID from the multiple-reader queue. Although job-dispatching illustrated in this example is one variant of task scheduling, there are others that are not supported as efficiently by multiple-reader queues.

Multiple-reader queues can also be used to implement efficient locks, accessed directly with messages, as a general mechanism for task-synchronization. This can be done by initially writing a single lock-token in the queue. Threads may contend to dequeue the lock-token by reading from the queue. The first one to arrive at the queue will get the token directly, while other reads will be buffered. The thread that *acquires* the lock-token enters its critical section and on exit writes (*releases*) the token back to the queue so that it can be matched and forwarded to the next requestor pending in the queue.

10.7 Conclusions and Future Perspectives

In order for multicore architectures to provide further performance scaling for applications, efficient ways to exploit the ever-increasing chip resources must be discovered. To improve the performance of a single application using multiple cores, architects need to optimize parallel processing and the communication architecture. However, a good matching between the system architecture and the programming model is key to the success of chip multiprocessor platforms.

Explicit communication supported by network interfaces allows better understanding of hardware functions, and thus could ease reasoning about communication performance and optimization of application behavior. On the contrary, coherence mechanisms are transparent to the programmer, which makes interpreting performance difficult and platform dependent. In addition, only a limited number of optimizations can be implemented in hardware, and even if hardware support for customized protocols is provided, customizing implicit communication mechanisms to an application should be at least as difficult as programming with explicit communication mechanisms.

Caches natively support mechanisms for locality management that simplify the task of programming. Nevertheless, recent research indicates that sharing a cache may have limited effectiveness for more than a few cores and data movement may be inefficient in that context [131, 114]. Cache coherence addresses data replication and migration in a unified framework, but fixes the locality management strategy in the protocol. In addition, coherent caching over a multistage NoC requires a considerable amount of on-chip state for the directories, and introduces communication overheads [45] because of the round-trip nature of caching and the directory indirection required.

On-chip explicit transfers over a scalable NoC may have better latency and energy properties, by avoiding directory accesses [144]. In addition, latency and bandwidth characteristics of explicit communication mechanisms and their implementation can scale better than those pertaining to cache-coherent transfers. These advantages might very soon make the difference since they become more pronounced for a larger number of cores.

The current state-of-the-art for CMP network interfaces provides promising support for distributed computations [339, 127], efficient synchronization [147], and probably automated parallelism exploitation. It also offers the most efficient approach for handling latency critical tasks [274] and may be able to provide performance guarantees. More research is required though, in order to efficiently manage the limited on chip resources, and to better support data migration in general purpose systems.

Although implicit communication supported by coherent caches provides an easier target for programmers, explicit communication should allow performance advantages in many cases. While several traffic patterns for fine granularity communication are supported efficiently by caches through numerous coherence protocol optimizations, it is reasonable to expect that many new mechanisms can be developed for explicit communication and synchronization in the years to come, thus making them an appealing choice for an increasing number of platforms and applications.

Chapter 11

Three-dimensional On-chip Interconnect Architectures[1]

Soumya Eachempati

Pennsylvania State University, USA

Dongkook Park

Intel Labs, Intel Corporation, USA

Reetuparna Das

Intel Labs, Intel Corporation, USA

Asit K. Mishra

Pennsylvania State University, USA

Vijaykrishnan Narayanan

Pennsylvania State University, USA

Yuan Xie

Pennsylvania State University, USA

Chita Das

Pennsylvania State University, USA

11.1 Introduction

Design of Chip Multiprocessors (CMP) / multicores and System-on-Chip (SoC) architectures by exploiting the increasing device density in a single chip is quite complex mainly because it needs a multiparameter (performance, power, temperature, and reliability) design space exploration. The core of this design lies in providing a scalable on-chip communication mechanism that can facilitate the multiobjective design space trade-offs. The Network-on-Chip (NoC) architecture paradigm, based on a modular packet-switched mechanism, can address many of the on-chip communication design issues, and thus, has been a major research thrust spanning across several design coordinates. These include high performance [158, 159, 225, 251], energy-efficient [334, 298, 178], fault-tolerant [87, 82, 243], and area-efficient designs [178, 194, 247]. While all these studies, except a few [159, 178, 194, 247], are

[1]This chapter is an extended version of "A Novel Dimensionally-Decomposed Router for On-chip Communication in 3-D Architectures," by J. Kim, C. Nicopoulos, D. Park, R. Das, Y. Xie, V. Narayanan, M. Yousif, C. R. Das, published in the proceedings of the 34th annual International Symposium on Computer Architecture [157] (in Sections 11.2.1 – 11.2.4), and "MIRA: Multilayered On-chip Interconnect Router Architecture" by D. Park, S. Eachempati, R. Das, A. K. Mishra, Y. Xie, N. Vijaykrishnan, and C. Das, published in the International Symposium of Computer Architecture, 2008, [242] (in Sections 11.2.5 – 11.6).

targeted for 2-D architectures, we believe that the emerging 3-D technology provides ample opportunities to examine the NoC design space.

3-D integration has emerged to mitigate the interconnect wire delay problem by stacking active silicon layers [33, 152]. 3-D integrated chips offer a number of advantages over the traditional 2-D design [323, 260, 2]: (1) shorter global interconnects; (2) higher performance; (3) lower interconnect power consumption due to wire-length reduction; (4) higher packing density and smaller footprint; and (5) support for the implementation of mixed-technology chips. In this context, several 3-D designs, from distributing different logical units among different layers to splitting a unit (such as a processor) into multiple layers, have appeared recently [260]. In order to build 3-D CMPs, where the logical units are split among multiple layers, the NoC that forms the communication medium also needs to be designed to be split among multiple layers. In this chapter, we will first discuss the various 3-D architectural design options for NoCs and also illustrate a novel design that moves from the traditional design options to build a 3-D stacked NoC router,[2] called Multilayered On-chip Interconnect Router Architecture (MIRA). The following sections describe four types of routers extended to the third dimension.

11.2 3-D Router Design

The on-chip network design is severely power- and area-constrained, while at the same time is required to provide low latency and high bandwidth. Consequently, these trade-offs need to be evaluated for each new technology. Among all the router components, the crossbar design is the most important for extending to three dimensions. Extension to three dimensions results in higher number of ports and results in larger crossbars. The crossbar area, power, and performance scale poorly with higher number of ports. As a result, we need to pay close attention to the crossbar design. In the following sections, we will see that the crossbar design undergoes progressive improvement for building a 3-D NoC.

11.2.1 3-D Baseline (3DB)

The first router design is the straightforward extension of the 2-D router into three dimensions (see Figure 11.1). Such a 3-D router requires a larger crossbar (7×7) where the extra ports are for the up-down direction, along with the associated buffers and arbiters.

In such a router design, vertical movement from a layer to the next layer is

[2]In this chapter, we use the term "router" instead of switch. Switch and crossbar switch are used interchangeably.

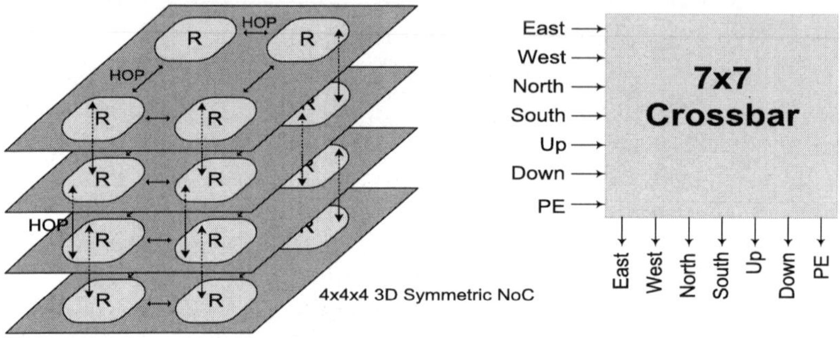

FIGURE 11.1: 3-D Baseline (3DB) NoC (from [157] ©2007 ACM).

a hop. We call this router the 3-D baseline (3DB) or symmetric NoC router. The 3DB router does not distinguish between the vertical short links (which are of the order of a few microns) and the longer horizontal links (of the order of a few mm). Using 3DB routers, a traditional 2-D NoC can be extended to three dimensions in a symmetric fashion where interlayer and intralayer communication is identical, i.e., hop-by-hop traversal. In this design, the main advantage of using 3-D is the decreased hop count due to smaller network size in each layer. Yet, the latency benefits of using 3-D integration are smaller due to the multihop interlayer traversal. The area and power of the crossbar scales inefficiently with higher number of ports. Thus, the increased size of the crossbar switches and increased number of components eats into the power savings obtained by decreased hop count. In addition, the size of buffer space and the complexity of routing logic and arbitration logics (in both virtual channel allocation (VA) and switch allocation (SA)) also increase due to these additional ports.

11.2.2 3-D NoC-bus Hybrid Router Design

In this design (see Figure 11.2), the inherent asymmetry in delays in the 3-D architecture between the fast (short) vertical interconnects and the (longer) horizontal interconnects is exploited. Given the very small interlayer distance, single-hop communication across all layers is in fact feasible. Consequently, a shared link can provide single hop traversal between two (non-adjacent as well) layers and thus, this design uses a shared bus along the vertical dimension. The incorporation of bus results in an additional port to the generic 2-D 5×5 crossbar, instead of the two ports in the 3DB case. The bus link has its own dedicated queue, which is controlled by a central arbiter. Flits from different layers need to arbitrate to gain access to the shared bus. Figure 11.2 illustrates the vertical via structure. It depicts the usefulness of large via pads between layers to cope for the misalignment issues during fabrication process. As a result, these larger via pads ultimately limit the ver-

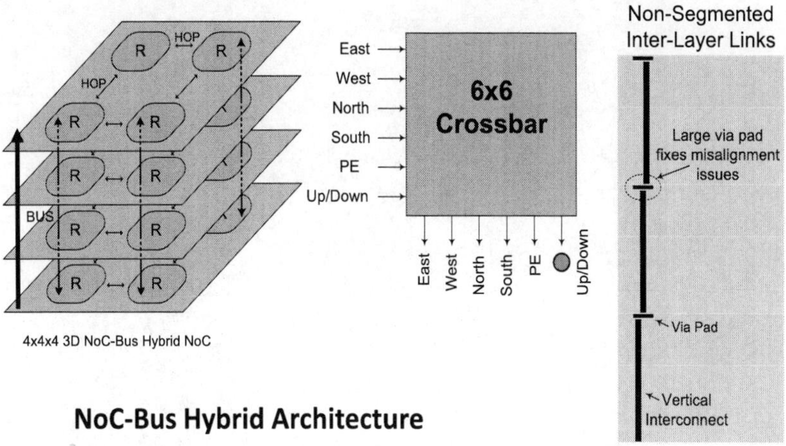

FIGURE 11.2: Left: 3-D NoC-bus hybrid design; Right: Interlayer via structure (from [157] ©2007 ACM).

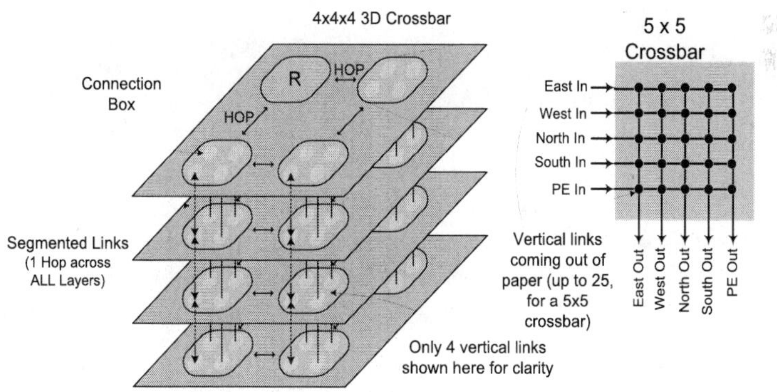

FIGURE 11.3: True 3-D architecture (from [157] ©2007 ACM).

tical via density, i.e., the bus width in the vertical direction. Even though, this architecture improves upon the 3DB design, it is plagued by the inherent serialization of the bus limiting the concurrency. Under high network load, the shared bus medium can lead to high contention. The single hop vertical communication improves the performance in terms of overall latency but the interlayer bandwidth takes a hit.

11.2.3 True 3-D Router Design

Moving beyond the previous options, we can envision a true 3-D crossbar implementation, which enables seamless integration of the vertical links in the overall router operation. Figure 11.3 illustrates such a 3-D crossbar layout. It

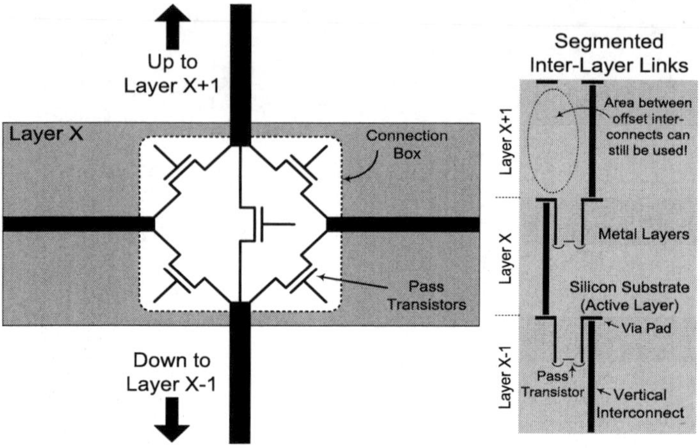

FIGURE 11.4: Connection box and interlayer via layout (vertical link segmentation) (from [157] ©2007 ACM).

should be noted that the traditional definition of a crossbar in the context of a 2-D physical layout is a switch in which each input is connected to each output through a single connection point. Instead of building a complex true 3-D crossbar, a connection box that can accommodate the interconnection of an input port to an output port through more than one connection points is chosen. The vertical links are now embedded in this new crossbar design and extend to all layers. As a result, no additional physical channels need to be dedicated to the interlayer communication and the crossbar size is retained at 5×5.

Interconnection between the various links in a 3-D crossbar would have to be provided by dedicated connection boxes at each layer. The internal configuration of such a connection box (CB) is shown in Figure 11.4. The vertical link segmentation also affects the via layout as illustrated in Figure 11.4. While this layout is more complex than that shown in Figure 11.2, the area between the offset vertical vias can still be utilized by other circuitry, as shown by the dotted ellipse in Figure 11.4. Hence, the 2-D crossbars of all layers are physically fused into one single three-dimensional crossbar. Multiple internal paths are present, and a traveling flit goes through a number of switching points and links between the input and output ports. Moreover, flits reentering another layer do not go through an intermediate buffer; instead, they directly connect to the output port of the destination layer. For example, a flit can move from the eastern input port of layer 1 to the southern output port of layer 3 in a single hop.

Yet just like a coin has two sides, there is a downside to this approach. Adding a large number of vertical links in a 3-D crossbar to increase NoC connectivity results in increased path diversity. While this increased diversity

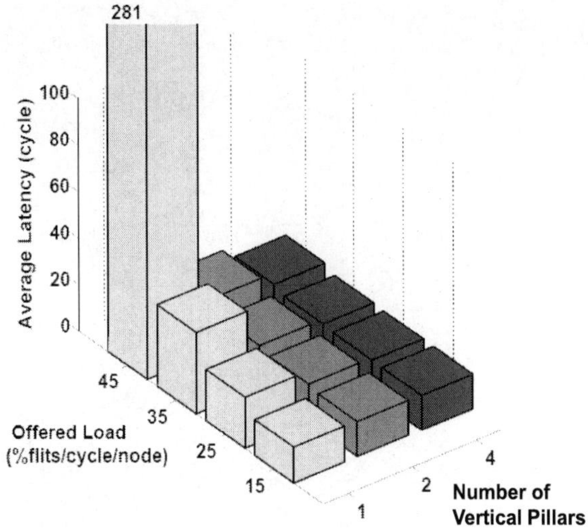

FIGURE 11.5: Latency sensitivity to number of pillars for true 3-D router design (from [157] ©2007 ACM).

may initially look advantageous, it actually leads to a dramatic increase in the complexity of the central arbiter, which coordinates interlayer communication in the 3-D crossbar. The arbiter would now require excessive number of control signals to enable so many interconnections. Alternatively, if static connections are assigned, the exploration space is still daunting in deciding how to efficiently assign those paths to each source-destination pair. In addition, such a 3-D crossbar requires connection boxes at each crosspoint per layer and the number is proportional to P^2xL where P is the radix of the crossbar/number of physical ports and L is the number of layers. For example, for a four-layer mesh router, the number of connection boxes would be 100. Such a large number of connection box poses high-control complexity. Thus, a small number of vertical links is only supported. Sensitivity analysis (shown in Figure 11.5) of the number of pillars with varying load was conducted. It was found that there is a small latency increase when you use 2 pillars as opposed to the 4 pillars case. Thus, a small number of vertical links are sufficient to achieve good latency characteristics. Moreover, the redundancy offered by the full connectivity is rarely utilized by real-world workloads, and is, in fact, design overkill [159].

11.2.4 3-D Dimensionally-decomposed Router Design

To understand this design, we first explain it in the context of the 2-D router and then extend it to the 3-D design. A traditional 2-D router has a 5×5 crossbar that connects the four cardinal directions and the connection

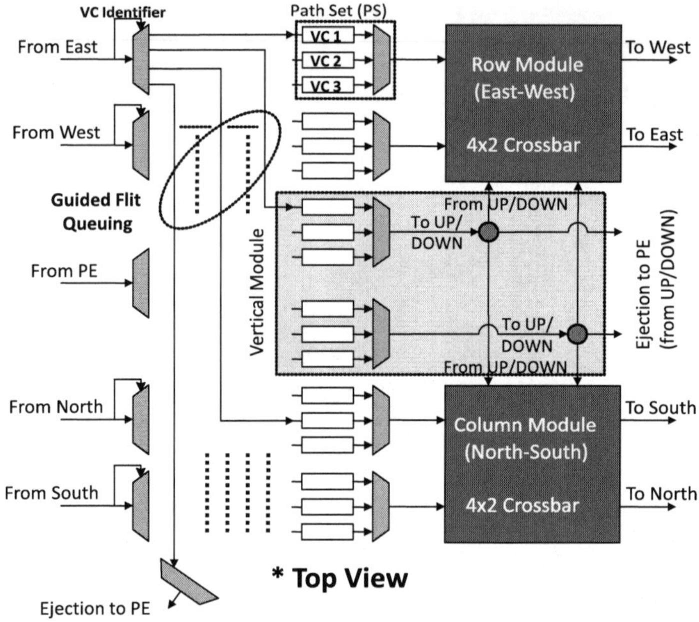

FIGURE 11.6: 3-D dimensionally decomposed router architecture (from [157] ©2007 ACM).

to the local processing element (PE). It has been shown in [158] that by using some simple switching process called *guided flit queuing*, the incoming traffic can be decomposed into East-West traffic (moving along X direction) and North-South traffic (moving along the Y direction). Segregating the traffic directionally allows breaking down the crossbar into smaller crossbars and the isolation of the traffic in two submodules called *Row module* and *Column module.*

Along the same lines, the traffic in 3-D NoC can be split into three classes, X direction, Y direction, and Z direction (interlayer communication). An additional module is required for traffic in the third dimension called the *vertical module.* Further, connecting links between vertical and row/column modules are required. Consequently, such a router requires smaller 4×2 crossbar design resulting in a power efficient design. Figure 11.6 shows the 3-D dimensionally decomposed NoC router design. More extensive reading about this design can be found in [159].

Having introduced the various design paradigms, let us now compare them in terms of different metrics. Figure 11.7 compares the average packet latency and network power consumption across all of these four router designs for a representative set of benchmarks chosen from scientific and commercial workloads. As seen from the figure, the dimensionally decomposed router design

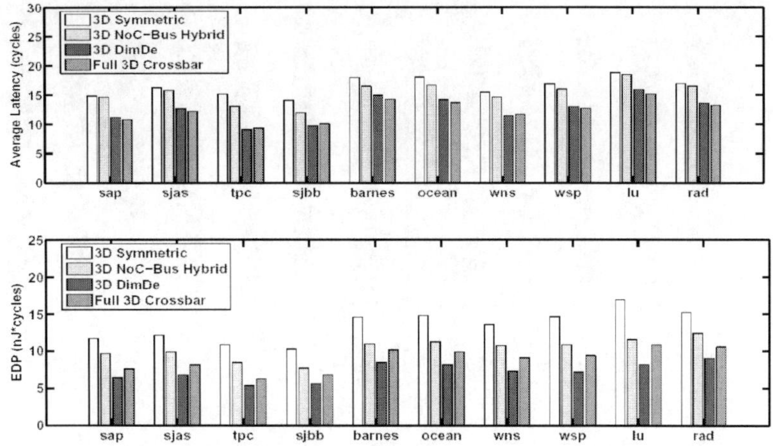

FIGURE 11.7: Average packet latency and power consumption (from [157] ©2007 ACM).

is the most energy efficient and equivalent in packet latency to the true 3-D crossbar.

11.2.5 Multilayered Router Design

The design of a multilayered router is based on the concept of dividing a traditional 2-D NoC router along with the rest of the on-chip communication fabric into multiple layers, with the objective of exploiting the benefits of the 3-D technology in enhancing the design of the router micro-architecture for better performance and power conservation. This multilayer NoC design is primarily motivated by the observed communication patterns in a Non-Uniform Cache Architecture (NUCA)-style CMP [23, 153]. Such a NoC adopts a different design philosophy compared to all the design options discussed so far.

The NoC in a NUCA architecture supports communication between the processing cores and the second-level cache banks. The NUCA traffic consists of two kinds of packets, data and control. The data packets contain certain types of frequent patterns like all 0's and all 1's as shown in Figure 11.8 [8]. A significant part of the network traffic also consists of short address / coherence-control packets as shown in Figure 11.9. Thus, it is possible to selectively power down the bottom layers of a multilayer NoC that have redundant or no data (all 0 word or all 1 word or short address flits), helping in energy conservation, and subsequently mitigating the thermal challenges in 3-D designs. Furthermore, such a multilayered NoC design complements a recent study, where a processor core is partitioned into multiple (four) layers [260]. It was shown that by switching off the bottom three layers based on

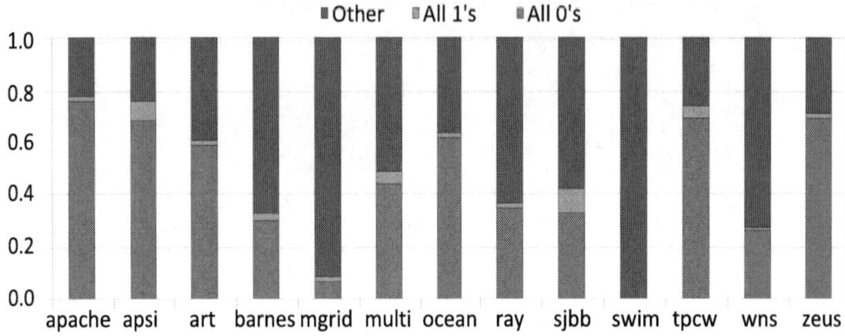

FIGURE 11.8: Data pattern breakdown (from [242] ©2008 IEEE).

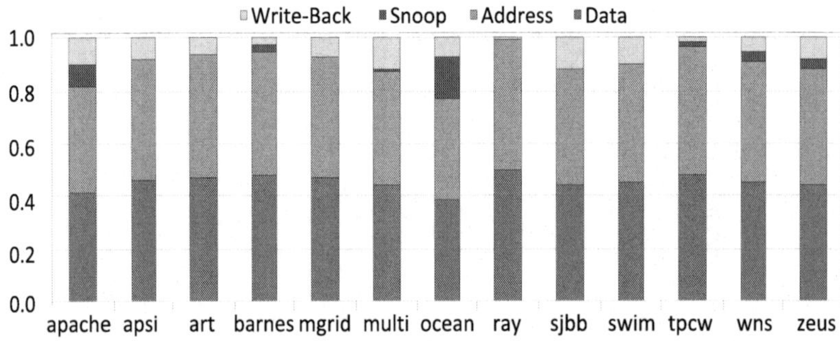

FIGURE 11.9: Packet type distribution (from [242] ©2008 IEEE).

the operand manipulation characteristics, it is possible to achieve significant power savings. These savings in power result in minimizing the thermal impacts compared to a standard 3-D stacking. If future CMP cores are designed along this line, the underlying interconnect should also exploit the concept of a stacked multilayered NoC architecture, as explained here.

The rest of the chapter describes two 3-D multilayered designs for a mesh interconnect:

- 3-D multilayered router (3DM): In this approach, the design implications of splitting the router components such as the crossbar, virtual channel allocator (VA) and buffer in the third dimension, and the consequent vertical interconnect (via) design overheads are analyzed. The multilayered design renders several advantages in terms of reduced crossbar size and wire length with respect to the 3DB design.

- 3-D multilayered router with express paths (3DM-E): The saving in chip area in 3DM approach can be used for enhancing the router capability, and it is the motivation for this design, which is called a 3-D router with express paths (3DM-E). These express paths between nonadjacent nodes reduce the average hop count, and help in boosting the performance and power behavior.

Due to the structural benefits obtained by utilizing the third dimension, the link traversal stage can be squeezed into the crossbar stage without increasing the cycle time, leading to reduced latency. Thus, the 3DM-E shows the best performance with up to 51% latency reduction and up to 42% power savings over the 2-D baseline router. This design is discussed extensively from Section 11.4 onwards.

11.3 Related Work

The prior work can be categorized into two subsections: NoC router architectures and 3-D integration techniques.

11.3.1 NoC Router Architectures

Due to the resource-constrained nature of an NoC router, many researchers have focused on two major themes; improving the performance and reducing power consumption. Performance can be improved by smart pipeline designs with the help of advanced techniques such as look-ahead routing [96] and path speculation [251, 225]. Also, dynamic traffic distribution techniques [158, 160] can be used to reduce the contention during the switch arbitration, reducing overall latency. These architectures can also save the crossbar power consumption by decomposing a monolithic crossbar into smaller subcrossbars. Recently, [230] proposed a shared buffer design that achieves improved performance by dynamically varying the number of virtual channels. This can improve buffer utilization of a router and consequently maintain similar performance even with half the buffer size, significantly reducing area and power consumption.

3-D on-chip interconnection architectures have been investigated [159, 178, 306]. An NoC-bus hybrid structure for 3-D interconnects was proposed in [178], and [159] proposed a dimensionally decomposed crossbar design for 3-D NoCs. A layer multiplexing and demultiplexing scheme that replaces one-hop per layer was proposed and significant power and performance benefits were shown. This scheme was implemented on a 3-D baseline and a randomized partially minimal oblivious routing algorithm was used [309]. Radix of the router is important for the power efficiency and diameter of the network determines the worst-case zero load delay. Higher radix leads to small diameter and high power. Xu et al., proposed a low radix, low diameter 3-D architecture by using long links within a layer and by leveraging 3-D technology [345].

Understanding and modeling through-silicon-vias (TSVs) is essential for designing 3-D integrated circuits. A synthesis approach for 3-D application specific power and performance efficient NoC was presented in [226]. TSVs are very important for supporting vertical links in 3-D NoC. The low yield of TSVs can have a significant impact on the design of 3-D NoCs. Fault tolerant design using TSVs for 3-D NoC links was studied in [185]. The design implications of different switch radices using TSVs in 3-D was studied in [182]. Mesosynchronous communication for 3-D NoC was proposed in [183].

The design of on-chip caches can take advantage of the characteristics of 3-D interconnection fabrics, such as the shorter vertical connections [350] and increased bandwidth between layers. A CMP design with stacked memory layers was proposed in [152], where the authors claim to remove the L2 cache and connect the CPU core layer directly to the dynamic random access memory (DRAM) layer via wide, low-latency interlayer buses. In addition, [259] has proposed multiple bank uniform on-chip cache structure using 3-D integration and [181] performed an exploration of the 3-D design space, where each logical block can span more than one silicon layer. However, all these approaches assume that planar 2-D cores are simply distributed on multiple layers and then a 3D-aware interconnection fabric connects them. None of them have considered the actual 3-D design of the interconnect spanning across multiple layers. On the other hand, as proposed in [260], processor cores can benefit from three-dimensional multilayered design, and the 3DM routers as explained in this chapter can directly work with such true 3-D cores.

11.3.2 3-D Integration Techniques

In a three-dimensional chip design, multiple device layers are stacked on top of each other and connected via vertical interconnects tunneling through them [71, 220]. Several vertical interconnect technologies have been explored, including wire bonding, microbump, contactless (capacitive or inductive), and TSV vertical interconnect [72]. Among them, the TSV approach offers the best vertical interconnection density, and thereby, has gained popularity. Wafers can be stacked either Face-to-Face (F2F) or Face-to-Back (F2B) and both have pros and cons. The F2B approach is more scalable when more than two active layers are used. In this chapter, the F2B approach with TSV interconnects is assumed.

The move to a 3-D design offers increased bandwidth [180] and reduced average interconnection wire length [136], which leads to a saving in overall power consumption. Also, it has been demonstrated that a 3-D design can be utilized to improve reliability [194]. However, the adoption of a 3-D integration technology faces the challenges of increasing chip temperature due to increasing power density compared to a planar 2-D design. The increased temperature in 3-D chips has negative impacts on performance, leakage power, reliability, and the cooling cost. Therefore, the layout of 3-D chips should be carefully designed to minimize such hotspots. To mitigate the thermal challenges in 3-D designs, several techniques, such as design optimization through intelligent placement [108], insertion of thermal vias [55], and use of novel

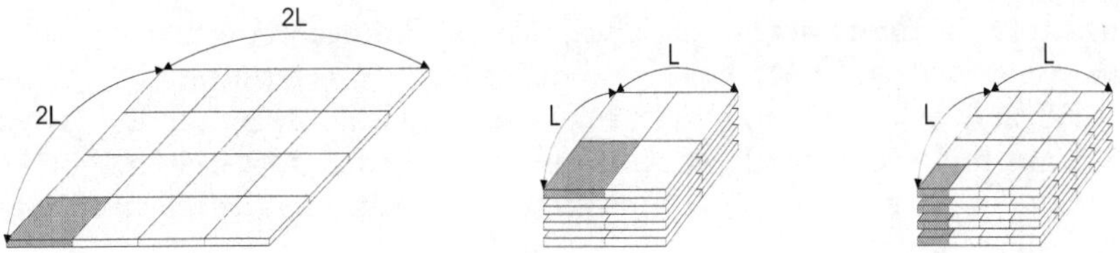

(a) Baseline 2-D (2DB) (b) Baseline 3-D (3DB naive approach) (c) Multi-layered 3-D (3DM)

FIGURE 11.10: Area comparison of 2-D & 3-D architectures with 16 nodes; 4×4 in 2DB/3DM and $2 \times 2 \times 4$ in 3DB (from [242] ©2008 IEEE).

cooling structures [32] have been proposed. The 3DM / 3DM-E architectures described in this chapter can minimize the formation of hotspots by placing highly active modules closer to the heat sink to avoid severe thermal problems.

11.4 3-D NoC Router Architectures

On-chip routers are major components of NoC architectures and their modular design makes them suitable for 3-D architectures as well. Such a 3-D design of an NoC router can help reduce the chip footprint and power consumption, leading to an optimized architecture. In this section, we investigate the two 3-D design options for NoC routers that are using mesh interconnects: (1) a 3-D multilayered router (3DM); and (2) a 3-D multilayered router with express paths (3DM-E). Microarchitectural details of each 3-D router design approach are described in the following subsections.

11.4.1 A Baseline 3-D Router (3DB)

Towards integrating many nodes into a 3-D chip, a naive approach is to group the nodes into multiple layers and simply stack them on top of each other as shown in Figure 11.10 (b), which shows 4 layers stacked together, each with 4 nodes, totaling 16 nodes. This is the same baseline 3-D (3DB) architecture as described in Section 11.2.1.

The nodes considered here can be any type of intellectual property (IP) blocks (e.g., CPUs or cache banks). However, if a node consumes significant power by itself, for instance a CPU node, it is not desirable to stack them on top of each other since such a design would significantly increase the on-chip temperature due to higher power density. Therefore, a good design approach would be to put all the power-hungry nodes in the top layer (which is closer to the heat sink), while accommodating the other relatively low-power consuming nodes in the lower layers. Thus, if all nodes are power-hungry, then the naive approach may not be a good design choice. For this study, we use both CPU and cache nodes and place the CPU nodes only on the top layers to avoid thermal problems. In comparison to the earlier router architectures [158, 159], the distributed router architecture described in the following section is a different design style.

11.4.2 A Multilayered 3-D Router Architecture (3DM)

Puttaswamy and Loh [260] proposed a 3-D processor design, where the design of individual functional modules spans across multiple layers. Although such a multilayer stacking of a processor is considered aggressive in current technology, we believe that such stacking will be feasible as the 3-D technology matures. Such a multilayered processor would benefit from a similar on-chip network that is based on a 3D multilayered router (3DM) designed to span across the multiple layers of the 3-D chip. Logically, the 3DM architecture is identical to the 2DB case (see Figure 11.10(a)) with the same number of nodes albeit the smaller area of each node and the shorter distance between routers (see Figure 11.10(c)). Consequently, the design of a 3DM router does not need additional functionality as compared to a 2-D router and only requires distribution of the functionality across multiple layers.

We classify the router modules into two categories, separable and nonseparable, based on the ability to systematically split a module into smaller sub modules across different layers with the interlayer wiring constraints, and the need to balance the area across layers. The input buffers, crossbar, and interrouter links are classified as separable modules, while arbitration logic and routing logic are nonseparable modules since they cannot be systematically broken into submodules. The following subsections describe the detailed design of each component.

11.4.2.1 Input Buffer

Assuming that the flit width is W bits, the buffers can be placed onto L layers, with $\frac{W}{L}$ bits per layer. For example, if $W = 128$ bits and $L = 4$ layers, then each layer has 32 bits starting with the least significant bits (LSB) on the top layer and most significant bits (MSB) at the bottom layer.

Typically, an on-chip router buffer uses a register-file type architecture and it is easily separable on a per-bit basis. In this approach as shown in Figure 11.11(b) and Figure 11.11(c), the word-lines of the buffer span across L layers, while the bit-lines remain within a layer. Consequently, the number of interlayer vias required for this partitioning is equal to the number of word-lines. Since this number is small in the case of on-chip buffers (e.g., 8 lines for 8 buffers), it makes it a viable partitioning strategy. The partitioning also results in reduced capacitive load on the partitioned word-lines in each layer

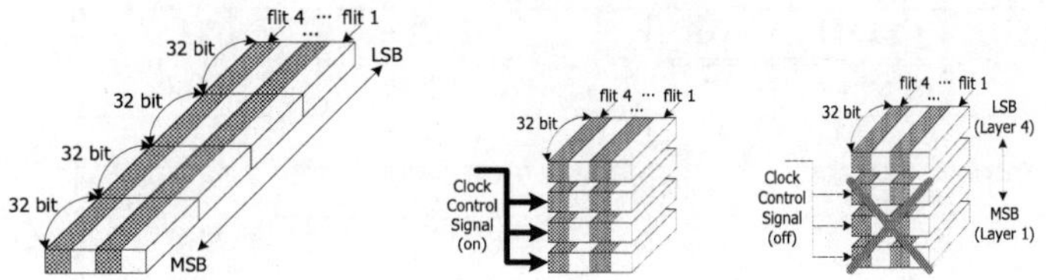

(a) 2DB, 3DB (b) 3DM (full) (c) 3DM (partial)

FIGURE 11.11: Input buffer distribution (from [242] ©2008 IEEE).

as the number of pass transistors connected to it decreases. In turn, it reduces the driver sizing requirements for the partitioned word-lines [323].

Since buffers contribute about 31% of the router dynamic power [334], exploiting the data pattern in a flit and utilizing power saving techniques (similar to the scheme proposed in [260]) can yield significant power savings.

The lower layers of the router buffer can be dynamically shutdown to reduce power consumption when only the LSB portion has valid data [342]. We define a short-flit as a flit that has redundant data in all the other layers except the top layer of the router data-path. For example, if a flit consists of 4 words and all the three lower words are zeros, such a flit is a short flit. The clock-gating is based on a short-flit detection (zero-detector) circuit, one for each layer. The overhead of utilizing this technique in terms of power and area is negligible compared to the number of bit-line switching that can be avoided. Figure 11.11(b) shows the case, where all the four layers are active and Figure 11.11(c) depicts the case, where the bottom three layers are switched off.

11.4.2.2 Crossbar

In the 3DM design, a larger crossbar is decomposed into a number of smaller multibit crossbars positioned in different layers. As shown in Figure 11.12, the crossbar size and power are determined by the number of input/output ports (P) and flit bandwidth (W) and therefore, such a decomposition is beneficial. In the 2DB case, P=5 and the total size is 5W $\times$ 5W, whereas in the 3DM case with 4 layers, the size of the crossbars for each layer is (5W/4) $\times$ (5W/4). If we add up the total area for the 3DM crossbar, it is still four times smaller than the 2DB design.

We use the matrix crossbar for illustrating the ideas in this chapter. However, such 3-D splitting method is generic and is not limited to this structure. In this design, each line has a flit-wide bus, with tristate buffers at the cross points for enabling the connections from the input to the output. In the 3-D designs, the interlayer via area is primarily influenced by the vertical vias required for the enable control signals for the tristate buffers (numbering P $\times$ P) that are generated in the topmost layer and propagated to the bottom layers.

Area(μm^2)	2DB	3DB	3DM*	3DM-E*
RC	1,717	2,404	1,717	3,092
SA1	1,008	1,411	1,008	1,814
SA2	6,201	11,306	6,201	25,024
VA1	2,016	2,822	2,016	3,629
VA2	29,312	62,725	9,770	41,842
Crossbar	230,400	451,684	14,400	46,656
Buffer	162,973	228,162	40,743	73,338
Total area	433,628	760,416	260,829	639,063
Total vias	0	W=128	2P+PV+Vk	2P+PV+Vk
Via overhead per layer	0%	0.4%	1.6%	0.6%

TABLE 11.1: Router component area. *Maximum area in a single layer. k: buffer depth in flits per VC (from [242] ©2008 IEEE).

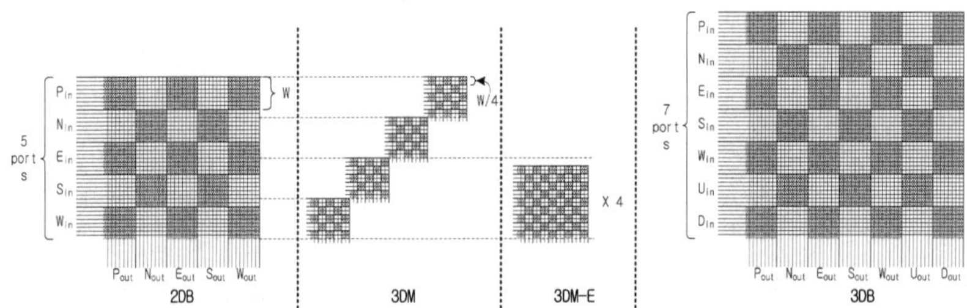

FIGURE 11.12: Crossbar distribution/relative area comparison (from [242] ©2008 IEEE).

The area occupied by these vias is quite small (see Table 11.1) making such a granularity of structure splitting viable.

11.4.2.3 Inter Router Link

Inter router links are a set of wires connecting two adjacent routers, and therefore, they can also be distributed as in the crossbar case above. Assuming that the link bandwidth is W and the number of layers is L, the cross-section bandwidth across L layers is W × L in the 3DB case as shown in Figure 11.13(a). To maintain the same cross-section bandwidth in the 3DM case for fair comparison, this total bandwidth should be distributed to multiple layers and multiple nodes. For example, if we assume 4 layers (L=4), the 3DB architecture has 4 separate nodes (A, B, C, and D), with one node per layer, as shown in Figure 11.13(b), whereas in the 3DM case, we have only 2 separate nodes (A and B in Figure 11.13(c)), since now the floor-plan is only half of the size of a 3DB node. Consequently, in the 3DB design, 4 nodes

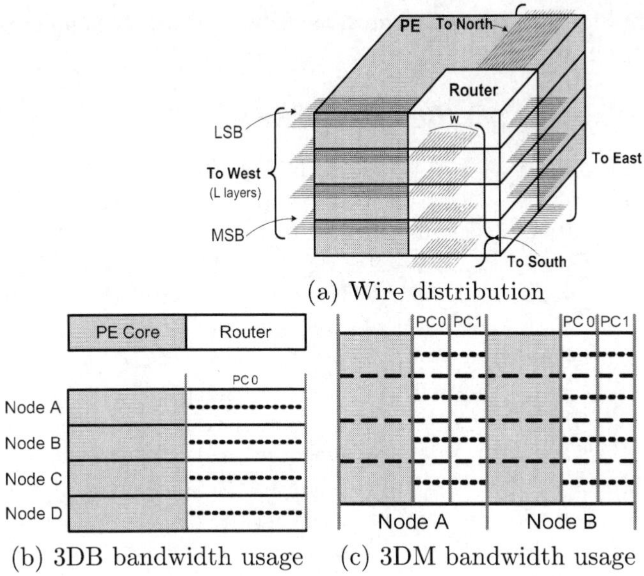

FIGURE 11.13: Inter router link distribution (from [242] ©2008 IEEE).

share $4 \times$ W wires; while in the 3DM design, 2 nodes share $4 \times$ W wires. This indicates that the available bandwidth is doubled from the perspective of a 3DM node. For example, node A in Figure 11.13(c) has $4 \times (W/2) = 2 \times W$ wires available, which doubles the bandwidth over the 3DB case. This extra bandwidth can be used to support topologies with higher connectivity, such as adding one more physical channel (PC) , as will be explained later in this chapter.

11.4.2.4 Routing Computation (RC) Logic

A physical channel (PC) in a router has a set of virtual channels (VCs) and each VC is associated with a routing computation (RC) logic, which determines the output port for a message (packet); however, a RC logic can be shared among VCs in the same PC, since each PC typically takes at most one flit per cycle. Hence, the number of RC logic blocks is dependent on the number of VCs (or PCs if shared) per router. Since the RC logic checks the message header and is typically very small compared to other logics such as arbiters, it is best to put them in the same layer where the header information resides; this avoids feeding the header information across layers, thereby eliminating the area overheads from interwafer vias. In this chapter, we fix the number of VCs per PC to be 2 and evaluate the increase in RC logic area and power. The choice of 2 VCs is based on the following design decisions: (i) low injection rate of NUCA traffic (ii) to assign one VC per control and data traffic, respectively (iii) to increase router frequency to match CPU frequency, and (iv) to minimize power consumption. However, this technique is not limited to this configuration.

11.4.2.5 Virtual-channel Allocation (VA) Logic

The virtual channel allocation (VA) logic typically performs a two-step operation [63]. The first step (VA1) is a local procedure, where a head flit in a VC is assigned an output VC. If the RC logic determines the output VC in addition to the output PC, this step can be skipped. In this case, we assume that the RC logic assigns only the output PC and thus requires P $\times$ V V:1 arbiters, where P is the number of physical channels and V is the number of virtual channels. The second step (VA2) arbitrates among the requests for the same output VC since multiple flits can contend for the same output VC. This step requires P $\times$ V PV:1 arbiters. As the size of VA2 is relatively large compared to VA1, the VA1 stage arbiters is placed entirely in one layer and distribute the PV arbiters of the VA2 stage equally among different layers. Consequently, PV interlayer vias are required to distribute the inputs to the PV:1 arbiters on different layers. It should also be observed that the VA complexity for 3DM is lower as compared to 3DB since P is smaller. Hence, 3DM requires smaller number of arbiters and the size of the arbiters is also small (14:1 vs. 10:1).

11.4.2.6 Switch Allocation (SA) Logic

In this design, since the SA logic occupies a relatively small area, it is kept completely in one layer to help balance the router area in each layer. Further, the SA logic has a high-switching activity due to its per-flit operation in contrast to the VA and RC logics that operate per-packet. Hence, the SA logic is placed in the layer closest to the heat sink.

11.4.2.7 3DM Router Design Summary

In summary, the 3DM router design has the RC logic, the SA logic and the VA stage1 logic in the layer closest to the heat sink and the VA stage2 logic is distributed evenly among the bottom three layers. The crossbar and buffer are divided equally among all the layers. The pitch size for the through-silicon via (TSV) is assumed to be 5 $\times$ 5 μm^2 in dimension (based on technology parameters from [134]). The resulting area of each of the modules is shown in Table 11.1. Note that in the 3DM design via overhead is less than 2%.

11.4.3 An Enhanced Multilayered 3-D Architecture (3DM-E)

The extra wire bandwidth available in the 3DM architecture (refer to Section 11.4.2.3 and Figure 11.13(c)) can be used to support an additional physical port per direction. This extra physical port can be used for purposes such

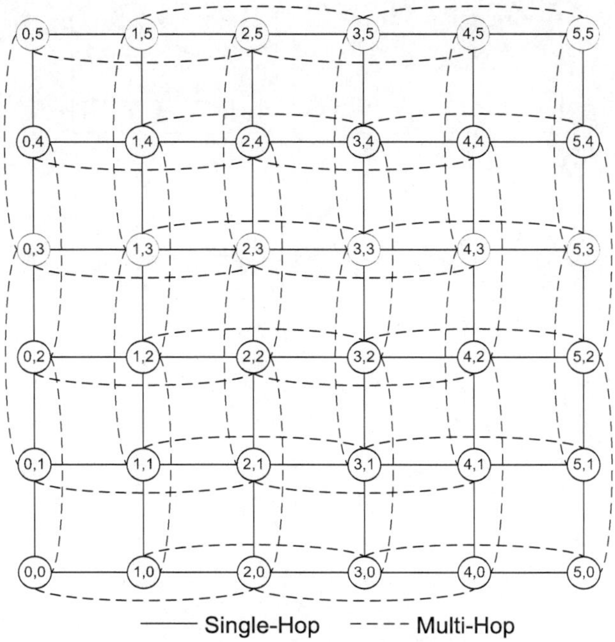

FIGURE 11.14: A 6 × 6 MESH topology with multihop links (from [242] ©2008 IEEE).

as quality of service (QoS) provisioning, for fault-tolerance, or for express channels. In this chapter, the extra bandwidth is used to support multihop express channels as shown in Figure 11.14 to expedite the flit transfer [62]. The express topology (3DM-E) requires each router to support 9 physical ports (4×2 ports on cardinal directions and one port to the local node). Consequently, it requires additional buffers and a larger crossbar in comparison to the 3DM design. Since 3DM-E components are distributed across multiple different layers, the area in a single layer is still much smaller than that of the 2DB and 3DB designs. For example, the crossbar size in a single layer is $7W \times 7W$ for 3DB and $(9W/4) \times (9W/4)$ for 3DM-E. Overall, the router area for the 3DM-E case is 2.4 times that of the 3DM case and 0.7 times that of the 2DB case.

11.4.4 Design Metrics

As discussed earlier, the 3DM architecture has inherent structural benefits over the traditional 2DB and the baseline 3-D cases. These benefits can be leveraged for better performance and reduced power consumption. In this section, we will understand these benefits and study the architectural enhancements that are born to exploit the structural benefits.

Wire delay per mm	Inverter delay	Inter router link length (mm)	
		2DB	3DM
254 ps	9.81 ps	3.1	1.58

TABLE 11.2: Link properties (from [242] ©2008 IEEE).

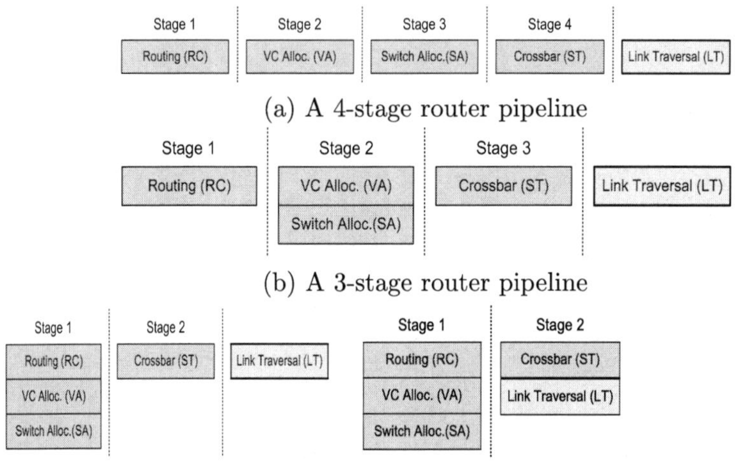

(a) A 4-stage router pipeline

(b) A 3-stage router pipeline

(c) A 2-stage router pipeline (d) 3DM router pipeline

FIGURE 11.15: Router pipeline (from [242] ©2008 IEEE).

11.4.4.1 Performance Optimization

The performance of a router depends on many factors such as traffic patterns, router pipeline design, and network topology. Among these, we have less control over traffic patterns compared to router pipeline design and network topology. Therefore, optimizing the router in terms of these two aspects will lead to improved performance.

A typical on-chip router pipeline consists of four stages, RC, VA, SA, switch traversal (ST), and one inter router link-traversal stage (LT) as shown in Figure 11.15(a). Many researchers have proposed techniques to reduce the router pipeline using techniques such as speculative SA (Figure 11.15(b)), look-ahead routing (Figure 11.15(c)).

In the 3DM router using 4 layers, the distance between two adjacent nodes is halved as compared to the 2DB or 3DB cases, leading to reduced inter router link delay. Also, the crossbar length is shortened by 1/4 as described in the previous section. This reduces the crossbar wire delay, which is a significant portion of the crossbar delay, and enables the combination of ST and LT stage together (see Figure 11.15(d)). Therefore, basically, each hop of the transfer will take one less cycle than the comparable designs using 2DB/3DB. The viability of this design is demonstrated for a 90 nm router based on the switch

	XBAR (ps)	Link (ps)	Combined Delay (ps)	ST and LT combined
2DB	378.57	309.48	688.05	No
3DM	142.86	154.74	297.60	Yes
3DM-E	182.85	309.48	492.33	Yes

TABLE 11.3: Delay validation for pipeline combination (from [242] ©2008 IEEE).

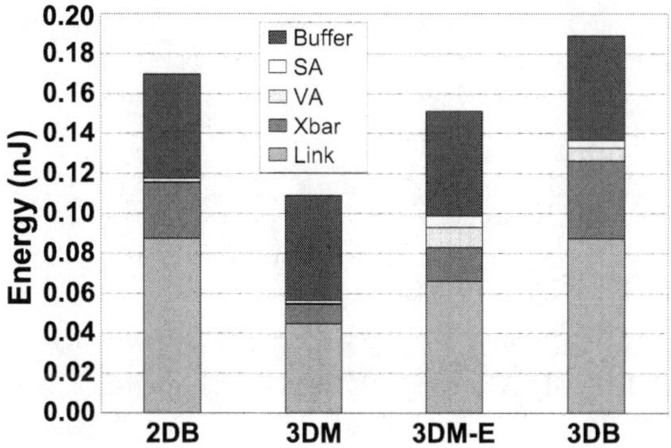

FIGURE 11.16: Flit energy breakdown (from [242] ©2008 IEEE).

design from [172] scaled to 90 nm and a link with optimal buffer insertion using parameters from [95]. The design parameters are shown in Table 11.2, and the resulting delays are shown in Table 11.3 for a $2GHz$ router that has a maximum per stage delay of $500ps$.

11.4.4.2 Energy Behavior

The dynamic energy breakdowns of the different router designs were evaluated using the Orion power model [335], and are reported in Figure 11.16. The 3DM design exhibits the lowest energy consumption due to reduced dimensions (and associated capacitance) of its structures. The biggest savings for 3DM comes from the link energy due the length reduction as described earlier. Further, decomposing the crossbars to smaller parts provides significant energy savings. A 35% reduction in energy is observed for the 3DM case over 2DB. In contrast to 3DM, the energy consumed by 3DB is higher primarily due to the increased number of ports to support communication in the vertical dimension. In the 3DM-E design, the extra ports to support the express links incur energy overheads compared to the 3DM design.

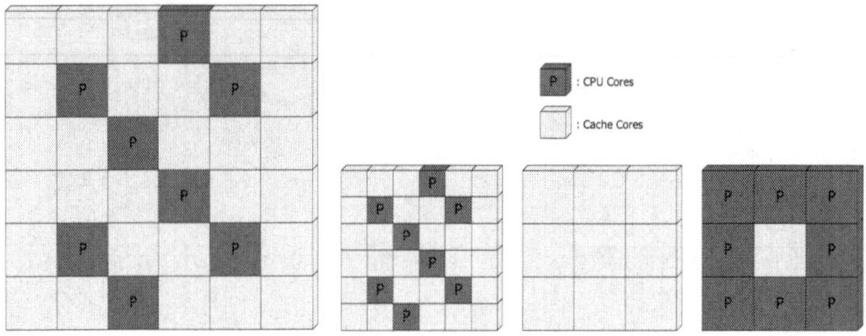

FIGURE 11.17: Node layouts for 36 cores. Left: 2DB (single layer), Middle left: 3DM(-E) layers 1–4, Middle right: 3DB layer 4, Right: 3DB layers 1–3 (from [242] ©2008 IEEE).

11.5 Performance Evaluation

In this section, an in-depth evaluation of the six architectures 2D-Base (2DB), 3D-Base (3DB), 3D-Multilayer (3DM) with switch traversal and link traversal combined into one stage, 3DM without switch traversal and link traversal combining ((3DM(NC)), 3DM with Express paths (3DM-E), and (3DM-E (NC)) with respect to average latency, average power consumption, and thermal behavior is presented. An in-house cycle-accurate NoC simulator (NoX) is used for the performance analysis. The simulator simulates the router pipeline and adopts wormhole switching. X-Y deterministic routing algorithm is used for all experiments, and analyze the metrics with both synthetic workloads (uniform random injection rate and random spatial distribution of source and destination nodes) and application traces. The energy consumption of the router modules are obtained from Orion [335] and fed into the cycle-accurate NoC simulator to estimate overall power consumption. For the thermal analysis, HotSpot 4.0 [294] is used. All the cores and routers are assumed to operate at 2 GHz. For fair comparison, the bisection bandwidth is kept constant in all configurations.

11.5.1 Experimental Setup

11.5.1.1 Interconnection Network Configuration

For these experiments, a 36-node network configuration is used. Four layers in all 3-D cases were used. Out of the 36 cores, 8 cores are assumed to be processors and other 28 cores are L2 caches. For the CPU, a core similar to Sun Niagara [169] is assumed and use SPARC ISA in the Simics simulation [196]. Each L2 cache core is 512 KB, and thus, the total shared L2 cache is 14 MB.

For the 2DB, 3DM, and 3DM-E cases, a 6 × 6 2-D MESH topology is configured, where the processor cores are spread in the middle of the network as shown in Figure 11.17 (left) and Figure 11.17 (middle left). Also, in 3DM and 3DM-E, we assume that the processor cores can be implemented in a multilayered (four layers in this study) fashion as proposed in [260]. In these experiments, we assume that all four layers in each processor and cache core statically consume the same amount of power. In the 3DB case, a 3 × 3 × 4 topology is formed and place most of the cache cores in the bottom three layers, while all the processor cores and one cache core are placed in the top (4th) layer so that more active cores stay closer to a heat sink, as shown in Figure 11.17 (middle right and right).

11.5.1.2 Cache Configuration

The memory hierarchy for these experiments consists of a two-level directory cache coherence protocol. While each core has a private write-back first-level (L1) cache, the second-level (L2) cache is shared among all cores and split into banks. These banks are interconnected via the NoC routers. The cache coherence model includes a detailed timing model of the MESI protocol with distributed directories, where each bank maintains its own local directory and the L2 caches maintain inclusion of L1 caches. The memory model is implemented as an event-driven simulator to speed up the simulation and have tractable simulation time. The simulated memory hierarchy mimics static non-uniform cache access (SNUCA) [23, 153] and the sets are statically placed in the banks depending on the low-order bits of the address tags. The network timing model simulates all kinds of messages such as invalidates, requests, responses, write-backs, and acknowledgments. The memory traces were generated by executing the applications on the Simics full system simulator [196]. The memory configurations for these experiments are summarized in Table 11.4. The workloads used are:

- TPC-W: We use an implementation of the TPC-W benchmark [58] from New York University. It consists of two tiers—a JBoss tier that implements the application logic and interacts with the clients and a MYSQL-based database tier that stores information about items for sale and client information—under conditions of high consolidation. It models an online book store with 129,600 transactions and 14,400 customers (tpcw).
- Java Server Workload: SPECjbb. SPECjbb2000 [300] is a Java-based benchmark that models a 3-tier system. We use eight warehouses for eight processors.
- Static Web Serving: Apache. We use Apache 2.0.43 for SPARC/Solaris 10 with the default configuration. We use SURGE [19] to generate web requests. We use a repository of 20,000 files (totaling 500 MB) and simulate 400 clients, each with 25 ms think time between requests (apache).

Private L1 Cache: Split I and D cache, each cache is 32 KB and 4-way set associative and has 64 bit-lines and 3-cycle access time
Shared L2 Cache: Unified 14 MB with 28 512 KB banks and each bank has 4 cycle access time (assuming 2 GHz clock)
Memory: 4 GB DRAM, 400 cycle access time. Each processor can have up to 16 outstanding memory requests.

TABLE 11.4: Memory configuration (from [242] ©2008 IEEE).

- Static Web Serving: Zeus. Zeus [351] is the second web server workload we used. It has an event-driven server model. We use a repository of 20,000 files of 500 MB total size, and simulate with an average 400 clients and 25 ms think time (zeus).
- SPEComp: We used SPEomp2001 [14] as another representative workload (apsi, art, swim mgid).
- SPLASH 2: SPLASH [343][51] is a suite of parallel scientific workloads (barnes, ocean).
- Multimedia: Mediabench. We use eight benchmarks (cjpeg, djpeg, jpeg2000enc, jpeg2000dec, h263enc, h263dec, mpeg2enc, and mpeg2dec) from the Mediabench II [326] suite to cover a wide range of multimedia workloads. We used the sample image and video files that came with the benchmark.

11.5.2 Simulation Results

11.5.2.1 Latency Analysis

We start with the performance analysis by measuring the average latency for the six configurations with the synthetic Uniform Random traffic (UR), simulated Uniform Random traffic with the NUCA layout constraints (NUCA-UR-Traffic), and six Multiprocessor Traces (MP-Traces). The results are shown in Figure 11.18 (a) through Figure 11.18 (c). The UR traffic represents the most generic case, where any node can make requests to any other nodes with uniform probability. This does not capture any specific layout of the cores and is the baseline configuration for all of our comparisons. From the performance perspective for UR traffic, 3DM-E is the best architecture since it has the least average hop counts as shown in Figure 11.18 (d). The simulation results corroborate this; 3DM-E has about 26% saving on an average over 3DB, 51% over 2DB, and 49% over 3DM at 30% injection rate. Also, the pipeline combination (merging ST and LT stages to a single stage) in 3DM and 3DM-E reduces the latency significantly before the network saturates. Thus, the pipeline combination makes 3DM/3DM-E architectures attractive at low

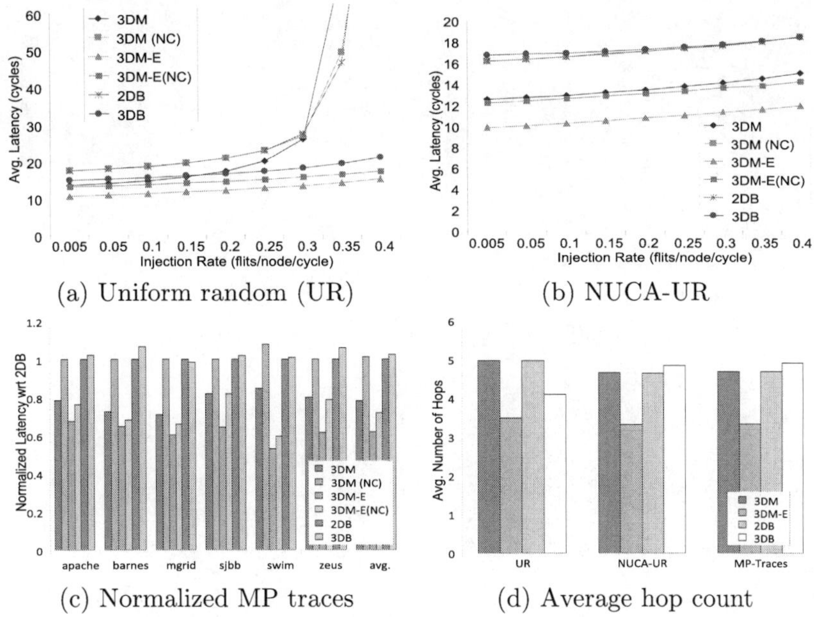

(a) Uniform random (UR) (b) NUCA-UR

(c) Normalized MP traces (d) Average hop count

FIGURE 11.18: Average latency results (from [242] ©2008 IEEE).

injection rates, which is typical of NUCA networks. Also, for 3DM-E, since the average latency is lower as pointed out before, the network saturates at higher injection rates compared to other architectures, making it more robust even in the saturation region.

NUCA traffic is typically different from uniform random traffic in a sense that the source and destination sets are constrained. A CPU needs to communicate only with cache nodes and a cache node only needs to communicate with CPU nodes. Thus, the average hop counts will be layout dependent. Hence, we should have different results with MP traces. To capture this layout specific traffic pattern, we also run experiments that model the request-response type bi-modal traffic, where the eight CPU nodes generate requests to the 28 cache nodes with uniform random distribution. Every request is matched with a response to a CPU. The results are shown in Figure 11.18 (b).

Interestingly, 3DB performance takes a hit for layout dependent traffic simulations. This can easily be explained by analyzing the traffic patterns between different layers. Because of thermal constraints, all the CPU nodes need to be placed in the top layer and most of the cache nodes in the lower layers (refer to Figure 11.17 (c)). Thus, the average hop count increases as most requests go from the top layer to the bottom layers and similarly, from the bottom layers to the top layer for responses. The trend can be seen in Figure 11.18 (d), which shows the average hop count for the three kinds of simulations. 3DM-E has the minimal hop count, while both 2D and 3DM

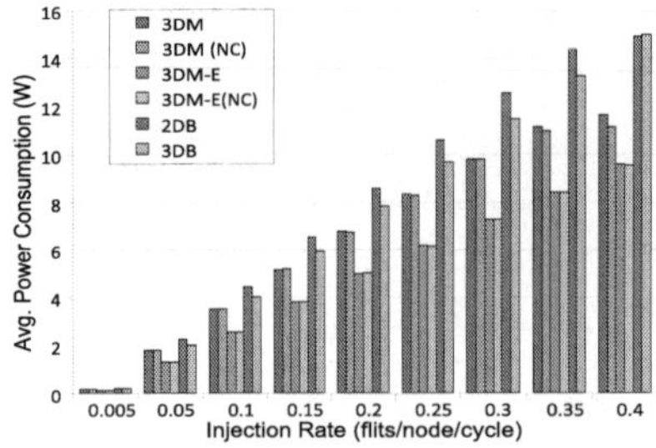

(a) Uniform random (UR)

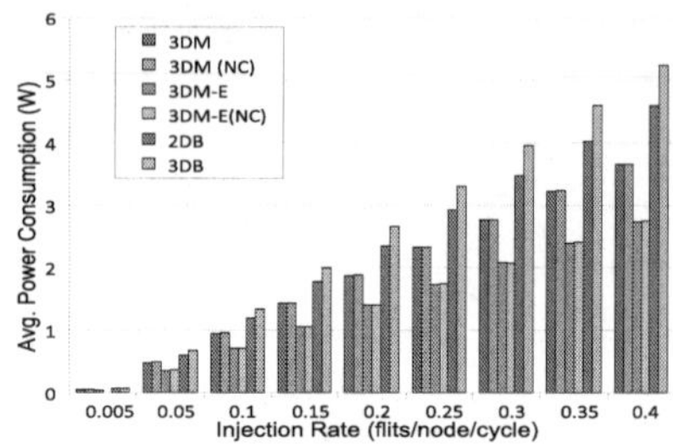

(b) NUCA-UR

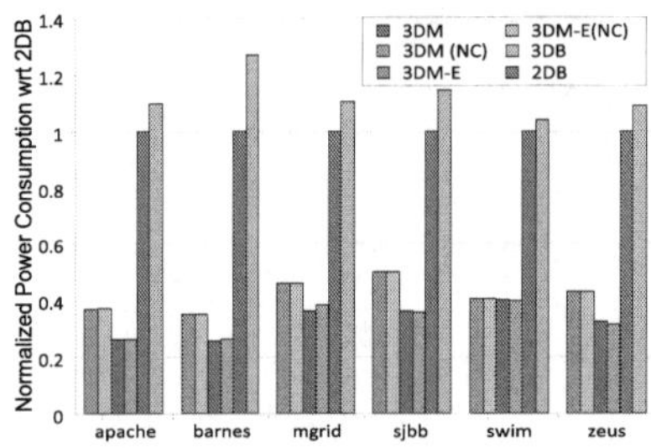

(c) Normalized MP traces

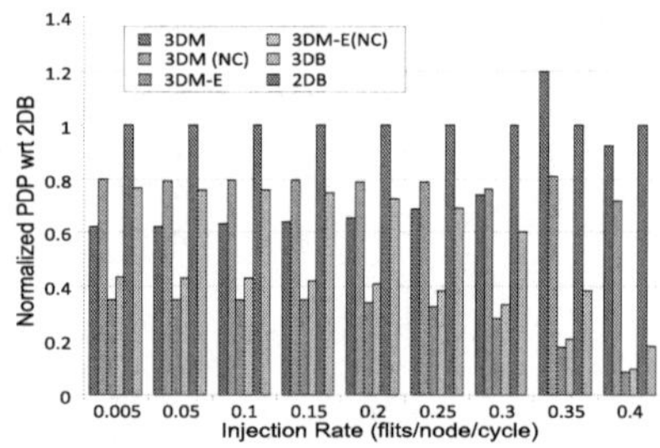

(d) Normalized PDP UR

FIGURE 11.19: Average power consumption results (from [242] ©2008 IEEE).

have the same hop counts as expected. All these three architectures are almost agnostics to the traffic patterns, while the 3DB configuration suffers with NUCA-UR and multiprocessor (MP) traces. MP trace simulation results present similar trend in latency behavior. 3DM-E performs around 38% better than 3DB and 2DB on an average. 3DM exhibits on an average 21% and 23% lower latency than 3DB and 2DB architectures, respectively. As expected, the 2DB and 3DM (NC) configurations have similar performance since they have the same logical network layout. Pipeline combination gives 3DM latency reduction up to 14% over 3DM (NC), and 3DM-E results around 23% latency reduction compared to 3DM-E (NC). Thus, the pipeline combination, possibly due to smaller crossbars and shorter wires in these 3-D architectures, helps in performance improvement.

11.5.2.2 Power Analysis

Figure 11.19(a) shows the average power consumption of the six architectures with 0% short flits. This result shows purely the power improvements of designing the self-stacked multi-layered router without any impact of layer shutdown techniques. The 3DM, 3DM-E designs show lower power consumption than 2DB and 3DB. This can be explained by looking at the per flit energy in the case of 3DM, 3DB, and 2DB as shown in Figure 11.16. Due to reduced router footprints, crossbar size, and link length reduction, the 3DM

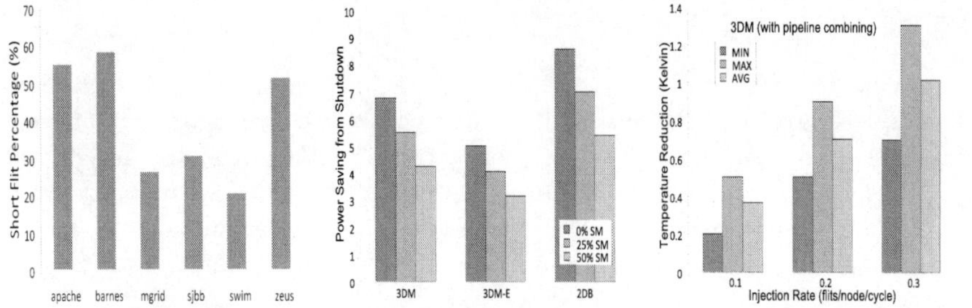

(a) Short flit percentage (b) Power saving techniques (c) Temperature reduction

FIGURE 11.20: Power and temperature simulation results (from [242] ©2008 IEEE).

router has lower power consumption than the 2DB and 3DB architectures. In the 3DM-E case, although the individual router power is higher due to increased radix, because of the reduced number of hops via express channels, the overall power consumption has decreased, on an average, by 37% and 42% over the 3DB and 2DB, respectively. 3DM design offers around 22% and 15% power savings over 2DB and 3DB, respectively. Unlike the latency case above, in general, the pipeline combining does not have significant impact on power consumption.

The power saving due to the layer shutdown technique is also evaluated. Figure 11.20 (b) shows the power savings with 25% and 50% layer shutdown. We save up to 36% power when 50% of the flits are short (32 bits and thus, use only one layer) on an average for the 3DM/3DM-E/2DB configurations. This demonstrates the potential of the shutdown technique. Comparing 3DM-E with 50% short flits with 2DB (0% short flits), we observe a power reduction of 63%. Similarly comparing 3DM with 50% short flits with 2DB (0% short flits), we see a power reduction of 50%. The MP traces (Figure 11.20 (a)) show up to 58% short flits and on an average 40% of the flits are short. This results in significant power savings (Figure 11.19 (c)); 3DM-E and 3DM both reduce power consumption around 67% and 70% with respect to 2DB and 3DB, respectively, with no layer shut down in the base cases. The power savings obtained with traces are due to the structural benefits of the self-stacked routers and due to the layer shutdown techniques. 3DB exhibits the worst power behavior because of increased hop count / latency per flit.

As a combined metric of both performance and power consumption, we measured the power delay product (PDP), which is the product of delay and power consumption. The normalized PDP result with respect to 2DB is shown in Figure 11.19 (d), and it shows that 3DM-E and 2DB are the best and worst choices, respectively. At lower injection rates, all 3DM-based techniques performed better than the 2DB and 3DB architectures.

11.5.2.3 Thermal Analysis

For the thermal analysis, we use the Hotspot simulator [294] that uses the complete layout of the entire chip along with the power numbers for each component. We use the power numbers from the SUN Niagara design (90 nm) [169] for the processors and power numbers for the cache memory from CACTI [314]. In these experiments, a processor core consumes 8 W and a 512 KB cache bank consumes 0.1 W. The network power numbers for the six configurations are obtained from Orion and then, fed into our NoC simulator. The NoC simulator generates power traces for Hotspot. For the multilayered configurations, the processor and memory powers are divided equally among the four layers.

The shutdown technique can be applied to all four architectures, but the 3DM and 3DM-E architecture would benefit more from reduced overall temperature since lower layers will not be active for short messages, thereby reducing the overall power density. Figure 11.20 (c) shows the temperature difference between 50% short message and 0% (none) short message cases in the 3DM (with pipeline combining) at three different injection rates. The temperature drops up to 1.3K and on an average, we get up to 1K temperature reduction. Also, as the injection rate increases, we tend to get more temperature reduction. We speculate that this is due to increased number of flit activities in the router, which triggers more activities in separable modules (buffer, crossbar, inter router links), where we can benefit from short flits. Although the temperature saving is small, it is reasonable since the average power each router consumes is relatively small. Overall chip temperature will depend heavily on CPU/cache core power consumption rather than router power.

11.6 Conclusions

As the 3-D technology is envisioned to play a significant role in designing future multicore / SoC architectures, it is imperative to investigate the design space of one of the critical components of multicore systems, the on-chip interconnects, in the 3-D setting. In this chapter, we explain the various 3-D router designs and focus on the design of a multilayered 3-D NoC router architecture, called MIRA, for enhancing the performance, energy efficiency, and thermal behavior of on-chip interconnects. The design of the MIRA router is based on the observation that since a large portion of the NUCA communication traffic consists of short flits and frequent patterns, it is possible to dynamically shut down the bottom layers of the multilayer router to optimize the power consumption while providing better performance due to smaller footprints in the 3-D design. Two 3-D design alternatives, called 3-D multilayer (3DM), and

3DM with Express channels (3DM-E) are discussed along with their area and power analysis. The MIRA routers were compared to the baseline 3-D router. Comparison study of the dimensional decomposed 3-D router and MIRA architectures can also provide further insights. We find that the 3-D multilayer approach is quite promising and this is confirmed by the experimental results. The experimental results show that the 3DM and 3DM-E designs can outperform the 3-D base case and the 2-D architectures in terms of performance and power (up to 51% reduction in latency and 42% power savings with synthetic workloads, and up to 38% reduction in latency and 67% power savings with traces). These benefits are quite significant.

The next step in this course of study is the combination of the multilayered 3-D processor architecture as proposed in [260] with the 3-D router architecture presented in this chapter. This would give a complete picture of multilayered 3-D CMPs.

Acknowledgment

The authors would like to thank Chrysostomos Nicopoulos and Jongman Kim for their help and support. This work is supported in part by NSF grant 0702617, EIA 0202007, CCF-0429631, CNS-0509251.

Chapter 12

Congestion Management and Process Variation

José Flich

Technical University of Valencia, Spain

Federico Silla

Technical University of Valencia, Spain

Carles Hernández

Technical University of Valencia, Spain

Mario Lodde

Technical University of Valencia, Spain

The Network-on-Chip (NoC) concept has been profoundly researched in the last decade, since the concept arose. Many research groups and individuals have focused on NoCs, from many different points of views and providing many useful approaches. Given the wide spectrum of domains that NoCs can be applied to, and the wide number of architecture and design techniques that can be considered, two prominent publications appeared trying to provide steering directions to the community. The first one [238] evolved from a workshop organized in 2006, promoted by the National Science Foundation, where the new challenges arising were identified. The second publication [199] emerged in 2009. In both cases new problems are identified and potential solutions and research directions are provided.

In [238], the main research challenges pointed out by the attendees to the workshop were:

- **NoC technology and circuits**. How will current and future manufacturing technologies influence the design of NoCs? In this regard, power consumption and required area are important concerns, as current designs require a large fraction of the chip power and area budget.

- **NoC microarchitecture**. What should the microarchitecture of NoCs look like in order to meet the upper-level requirements, mainly message latency and NoC area and power? Other important research challenges are programmability, scaling NoCs to new technologies, and managing reliability and process variability.

- **NoC architecture**. What should be the combination of network topology, routing algorithm, flow control mechanism, etc, that delivers the best performance in this scenario?

- **Computer-aided design tools for NoCs**. How should the design tools be in order to easily design this kind of networks?

- **Evaluation and driving applications**. What should be the benchmarks that NoC designers use and what will be the applications running on top of these networks?

Regarding [199], the main research problems that NoC designers should address in the near- and mid-term were:

- **Application modeling and optimization**. In this regard, NoC designers should consider modeling traffic in the network in order to understand what are the problems in a NoC in order to carry out realistic performance analysis. Since using real-life applications is costly in terms of time, a set of realistic benchmarks should be devised. On the other hand, application mapping and scheduling should also be considered, as different mapping and scheduling mechanisms deliver different performance and, therefore, the analysis carried out in the designed network could be biased.

- **NoC communication architecture and optimization**. Many NoC characteristics must be considered in order to design a network that best fits the requirements of applications. For example, the routing algorithm and the switching technique should be seriously taken into account. Also, quality-of-service and congestion control are related with application performance as may greatly influence it. Other technological concerns like power and thermal management, or reliability and fault tolerance should be considered. In this direction, process variation is an important source of uncertainty that sometimes requires to be managed as if a fault had occurred. It may also require additional architectural and/or technological approaches.

- **NoC design**. Technological issues also influence NoC performance. For example, the design of the network topology or the switch to be deployed are primary concerns. It is also important the design of the links as well as the overall floor-planning and NoC layout, as these decisions also will later influence the performance of the applications using the NoC.

- **NoC evaluation and validation**. Once the network has completely been designed, it must be tested and evaluated.

One clear issue that both studies address for the community is the need to know the behavior of the application working on top of the analyzed system. Traffic characteristics must be exploited in order to produce the best NoC. In this sense, in chip-multiprocessor (CMP) systems the cache coherence protocol is the entity that creates the traffic, thus influences the NoC behavior. In this chapter we provide traffic analysis for different cache coherence protocols, thus providing some light to the real traffic characteristics.

Another important issue is congestion management. In [199] congestion is linked to Quality of Service (QoS) support. Differentiating flows with different quality-of-service metrics becomes a need in future systems. Indeed, latency will be highly impacted by the lack of QoS support. Linked with this, congestion control mechanisms, mostly at the network interface, have been identified as a different way to guarantee QoS, by avoiding end nodes to oversubscribe resources.

As a key challenge identified in [238] and [199], process variation is related with the tremendous integration scales used nowadays, that use such a small feature size that some degree of unpredictability in manufactured devices arises. Process variability is caused because current manufacturing processes are no longer able to perfectly translate designs into real devices. There are several reasons for this, like small deviations introduced in the photolithographic process, or random variations in the number of atoms of the dopants used. Additionally, capacitance and resistance variations introduced as a consequence of varying wire dimensions because of defects in the chemical metal planarization process, are another source of uncertainty in current and future chips. The main consequence of process variation is that manufactured devices present delay characteristics that do not exactly match the parameters established at the design phase. For this reason, process variation arises as one of the most important challenges to be tackled in new on-chip system architectures starting from 65 nm manufacturing technologies down to 16 nm ones [238].

The goal of this chapter is to shed some light on the pros and cons of some of the research challenges pointed out in the two publications mentioned above that have not been explicitly addressed in other chapters of the book. More specifically, this chapter addresses congestion management techniques for NoCs and the impact of process variation in technology on the performance of applications. To do this, we will first provide analysis of real traffic obtained by executing real applications in different-modeled CMP systems. The goal is to formally identify the kind of traffic we can expect, and to see how performance may change if the entire system is not taken into account in the simulated system.

12.1 Traffic Analysis for CMP Systems

In this section we provide an analysis on the traffic requirements for NoCs in the domain of CMP systems. CMPs typically implement a cache coherence protocol on top of the network (shared-memory CMP system). The protocol guarantees coherency and consistency when sharing memory blocks between multiple cores (processors). Indeed, the protocol keeps the coherency among different copies of the same memory block replicated through different L1 and/or L2 cache memories.

In the last years, research on on-chip interconnects have relied on synthetic benchmarks where assumptions on the running applications, the operating system, and the cache coherence protocols have been obviated or simplified. The performance of the network, however, largely depends on other important design choices. Indeed, the cache coherence protocol (in a shared-memory CMP system) is the main driver for the network. There is a need to jointly evaluate both environments in a unified simulation platform. In the other direction, research work in cache coherence protocols have usually assumed simplified network models thus potentially leading to conclusions that obviate the network complexity. In addition, by the joint co-design of both the cache coherence protocol and the on-chip interconnect better solutions can be achieved, each component being fine-tuned for the other.

The goal of the section is to provide an initial evaluation on the effects of the cache coherence protocols on on-chip network performance. The ultimate goal is to see how the network affects the final performance of the entire CMP system. Indeed, there are many cache coherence protocols and implementations that may end up in different performance levels. To do this, we have performed a thorough evaluation of NoC behavior with different applications

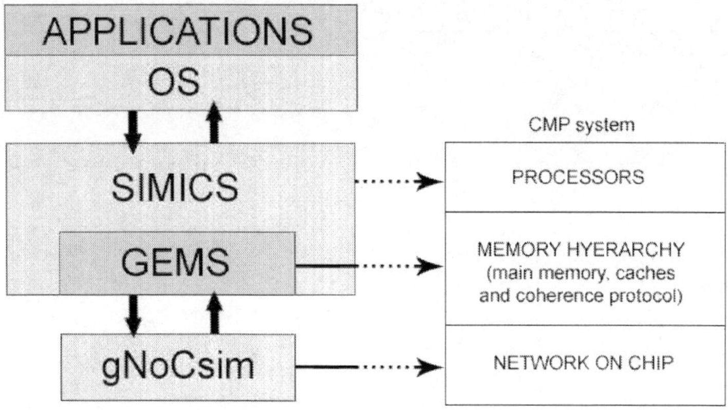

FIGURE 12.1: Simulation platform for on-chip network simulation in shared-memory CMP systems.

running on a tiled CMP system with several cache coherence protocols and chip organizations.

12.1.1 Simulation Platform

The complete simulation environment has the structure shown in Figure 12.1. The performance of a CMP system (and its on-chip network) has been modeled by using the full system simulator Virtutech Simics 2.2.18 [196]. Simics is extended with the Multifacet General Execution-driven Multiprocessor Simulator (GEMS) 2.1 [201] to model memory timing. GEMS is composed of a set of modules that allow detailed simulation of multicore systems; the module we used, Ruby, simulates in detail the whole memory hierarchy (caches, main memory, and memory controllers) of a wide range of multicore systems; it also provides a specification language that describes the cache coherence protocol used to keep coherence in the cache memory hierarchy; furthermore, it models the interconnection network, providing a set of standard interconnects and describes a custom network, although we modeled the NoC with gNoCsim, a more detailed network simulator.

In our simulations Simics models 32 UltraSPARC-III processors running SUN Solaris 8 operating system; Ruby is used to accurately calculate the delays introduced by memory hierarchy. When a Ruby traffic source, as a cache memory or a memory controller, generates a message, GEMS passes it to gNoCsim. gNoCsim then performs a cycle-by-cycle simulation of the flits advancing through the network and wakes up the Ruby destination node as the message is delivered.

GEMS has been extended with the gNoCsim network simulator to better model the NoC. Indeed, gNoCsim models an on-chip network using the

GEMS parameters	
Cache block size	64 bytes
L1 cache	64K instructions + 64K data
	4 ways, 2 hit cycles
L2 cache	512K, 16 ways, 4 hit cycles
gNoCsim parameters	
Topology	4x4 2-D mesh
Switch	4-stages switch (gNoC switch)
Switch latency	4 cycles
Switching technique	virtual cut-through
Routing algorithm and implementation	XY and LBDR
Flit size	2 bytes
Packet size	5 flits
Virtual networks	4
Virtual channels	1

TABLE 12.1: GEMS and gNoCsim parameters.

gNoC switch design described in the Appendix. gNoCsim is a flit-level cycle-accurate network simulator that can run in stand-alone mode or act as a slave driven by external traffic sources (in this case the GEMS module). gNoCsim models wormhole (WH) and Virtual Cut-Through (VCT) switching techniques, provides tree-based broadcast support and implements different routing algorithms and implementations, like the Logic-Based Distributed Routing (LBDR) (see Chapter 5). The main parameters used for GEMS and gNoCsim configuration are shown in Table 12.1.

Notice that virtual cut-through switching has been used. Indeed, this is needed to guarantee deadlock-free routing when tree-based broadcast communication is permitted within the network (see Chapter 5 for a description of the problem). This leads to the need of defining the packet size and possibly to packetize large messages generated by the cache coherence protocols. The small flit size has been selected in order to achieve an efficient and compact virtual cut-through switching mechanism [268].

The system we considered has 16 tiles, organized as shown in Figure 12.2, with 4 cores on each tile; each core has its private L1 cache and an L2 cache bank that can be private or shared depending on the coherence protocol. We assumed to allocate 32 cores to the running application, and evaluated the performance with two different core allocation policies: a uniform allocation policy, on which the application runs in two cores on each tile, and a nonuniform allocation policy, on which the number of cores allocated to the application on each tile varies from zero to four. Figure 12.3 shows the two policies where applications run on white cores. The goal of such policies is to analyze the impact the core distribution of the application has in terms of traffic dis-

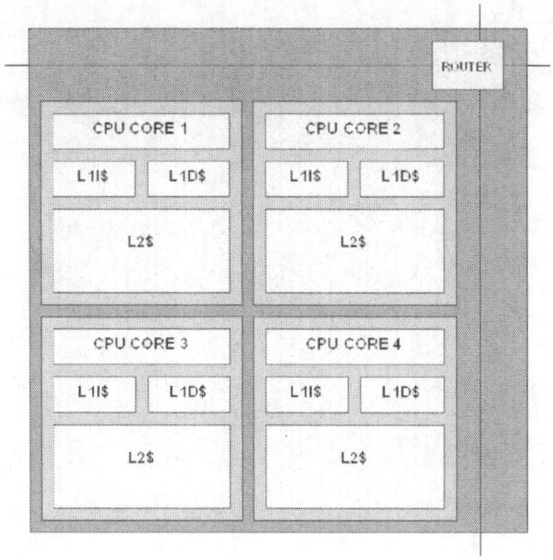

FIGURE 12.2: Tile structure.

tribution and execution time of the application. The memory controllers are located in the tiles marked with MC.

12.1.2 Cache Coherence Protocols

Prior to describing the different cache coherence protocols evaluated, we need to differentiate between systems with shared cache structures and systems with private cache structures. Let's consider a CMP system composed of four tiles, each one with a processor, and a two-level cache memory. While the L1 cache is usually private for each processor, the L2 can be private or shared. Figure 12.4 shows a system with private L2 caches: each processor has its own L2 cache block and in case of a L2 miss, a request is sent to the memory controller. Later, the request will be forwarded to another L2 cache or to memory. In Figure 12.5 a system is shown with shared L2 caches: the four L2 blocks are shared by all processors, and each cache block is mapped on a particular L2 bank. In case of a L1 miss, a request is sent to that bank, that can be located in the same tile or in one of the other tiles. This way the four banks of the L2 cache are part of a single L2 cache physically distributed among the tiles.

Three different cache coherence protocols have been considered: Directory [20], Token [200], and Hammer [239], each one generating a different kind of network traffic. All protocols use MOESI states in private caches; a cache block in states M (modified, dirty) or E (exclusive, clean) is held only by one core, which can read or write on it; if a subsequent read request is issued by another core, the block is provided by the core that holds the data: in

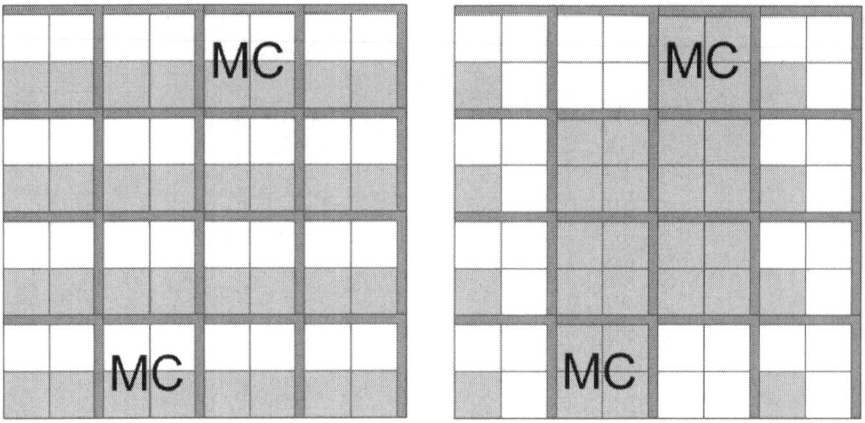

FIGURE 12.3: Uniform and nonuniform allocation.

this core the block's state becomes O (owned) while in the receiving core the block's state will be S (shared). The owner core provides the block to further requestors. Blocks in I (invalid) state are stale or not present: a read or write request for these blocks ends up in a cache miss.

In the Directory protocol directory information is associated with each cache block, including information about which core, if any, has a valid copy of the block, what's the current state of the block, and whether or not the block is dirty. In case of a miss in the private cache a request is sent to the home node, which forwards it to the owner core or sends the block to the requestor depending on the block's state. If L2 caches are private, the home node is located in the memory controller; if L2 caches are shared, each L2 bank is the home node to a given subset of cache blocks. Since the directory has to be stored in the home node, space overhead is the main drawback of this protocol. Cache misses generate low network traffic: in case of a read miss, two or three messages are sent: the request message to the home node, eventually a forwarded request to the owner core, and finally a message with the block; in case of a write miss, more messages may be sent since the home node has to invalidate the cores that share the block, if any.

The Token protocols avoid using space to store the directory information by associating with each cache block a given number T of tokens, usually equal to the number of cores in the system. One of the T tokens, called the Owner token, marks the owner core. A core can read a block if it has a valid copy in his private cache and at least one token; the core is allowed to write on the block only if it has all the T tokens assigned to that block. This way a core cannot write on a block while another one is reading it. In case of a private cache miss, a request is broadcasted to all other cores. If it is a write request, all the cores that have at least one token answer by sending all the tokens they have; to lower network traffic, only the core who has the Owner token sends

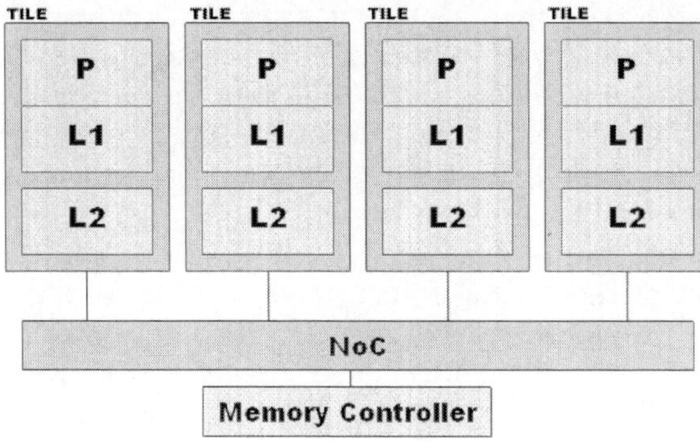

FIGURE 12.4: CMP system with private L2 caches.

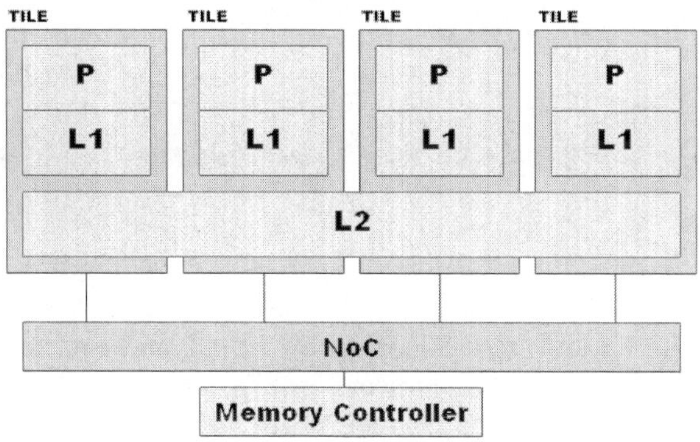

FIGURE 12.5: CMP system with shared L2 caches.

the block to the requestor; in case of a read request, the cores who have more than one token (usually only one core, the one who holds the Owner token) answer by sending both a token and the block. Space overhead is much lower in token protocols than in directory protocols but, since a broadcast request is performed each time a private cache miss occurs, a large amount of network traffic is generated.

In the Hammer protocol no additional information associated with the cache block is needed. When a cache miss occurs, a request is sent to the home node (which is located in the memory controller if L2 are private or in an L2 bank if L2 are shared, as in the Directory protocol). The home node broadcasts the request to all other cores and sends an additional offchip request to the main memory. The cores answer by sending the block or, in case they do not have a valid copy, an acknowledgment message; so, if the system has N cores, the requesting core has to wait for N messages: N-1 data or acknowledgment messages from the other cores and one data message from the memory. If at least one core sends a data message, the memory's data is considered stale and ignored. The Hammer protocol generates even more traffic than the Token protocol, since broadcast operations are performed at every cache miss and all cores have to answer.

The applications we used in our simulations are part of the SPLASH-2 suite [343], a set of eight parallel applications written to allow evaluations of shared memory systems. We show average results for all the applications and more detailed ones for all of them, specially for : FFT, LU, Raytrace and FMM. Notice that other applications, like PARSEC [31] could be suitable for this analysis.

12.1.3 Performance Results

We have simulated three different network scenarios for each application. First of all, applications have been tested on a CMP system with an on-chip network with a fixed one cycle delay and infinite bandwidth. This scenario is referred to as an *ideal network case*. Then, we evaluate the applications on a CMP system with the network with parameters shown in Table 12.1 and with no broadcast support. Indeed, a unicast-based broadcast approach has been used where if a message has D destination nodes, D unicast messages are sent sequentially. This method generates a large amount of traffic when broadcast-oriented protocols are used, such as the Token or Hammer, however the complexity of the network gets simplified. Finally, a scenario where a tree-based broadcast support in the network exists has been evaluated. In that case, the Signal Bit-Based Multicast (SBBM) method described in Chapter 5 is used. In that scenario, if a message has D destination nodes and D is greater than a threshold T, a broadcast message is sent to all nodes. If not, D unicast messages are sent. We have assumed T=1, thus generating a broadcast message for each message with more than one destination node. Notice that multicast support is not considered and the entire chip is flooded instead. Each protocol uses four virtual networks to avoid protocol-level deadlocks and one virtual channel per virtual network.

Figure 12.6 shows the normalized (to the worst case) average execution time of eight applications (FFT, LU, Raytrace, OCEAN, VOLREND, RADIOSITY, WATERSP, and FMM) executed in the CMP system varying coherence protocols, L2 sharing degrees (private L2 caches or shared L2 caches), core allocation (homogeneous or heterogeneous), and network implementation (ideal network, no broadcast support, or tree-based broadcast support). With an ideal network, the average execution time of applications do not depend on core allocation policy, since any message can reach any node with one cycle delay. The small differences in execution time are due to coherence protocols (since some protocols generate more messaging than others). The Directory and Hammer with private L2 caches experience higher execution times since

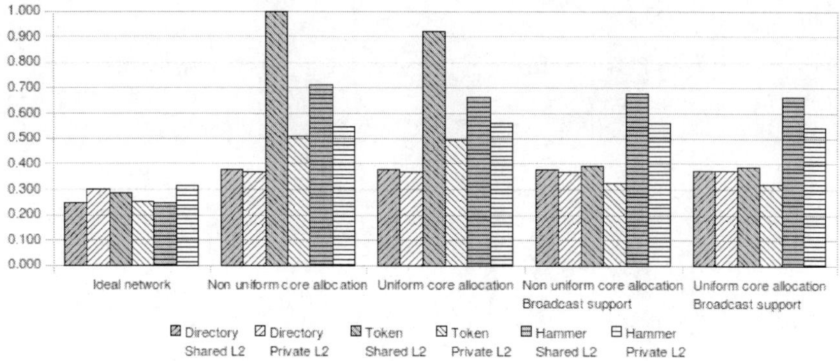

FIGURE 12.6: Average execution time under different CMP systems and networks.

when a miss in L2 cache occurs a request is sent to the home node and then the home node forwards the request to the Owner (Directory) or broadcasts the request to all nodes (Hammer). In the Token protocol with private L2 caches instead, in case of a L2 cache miss, a broadcast is performed directly by the cache, without passing through the home node. However, notice that differences are quite slim. Figure 12.7 shows the results obtained for every application. Although results differ (slightly) from application to application, in all the cases the cache coherence protocols have similar behavior and show similar trends.

Now let's focus on a CMP system when the network is accurately modeled. Execution time grows when the system is simulated with a real network. Let us consider first the case of a network without broadcast support (sets two and three of results in Figure 12.6). Taking as a reference the ideal network case, the average execution time of applications increases from 60% (Directory with private L2) to more than 370% (Token with shared L2). Execution time growth is bigger when L2 caches are shared. Indeed, using a shared configuration, when an L1 miss occurs a request is sent to the L2 bank where the requested block is mapped, which can be located in any other tile; then a response is sent back to the L1 cache (in case of L2 hit) or a request is sent to the memory controller; unless the home L2 is located in the same tile, in the best case (home L2 located in a close tile) a request message has to cross at least two switches and three links and a response has to cross back to the same path. In the worst case (home L2 located in the furthest tile in the system) the messages have to cross 7 switches and 8 links in both directions. If L2 caches are private, however, an L1 cache miss is resolved without sending any message through the network (as in case of an L2 cache hit), so execution times are not affected so much by network latency. Notice that this is the usual trend

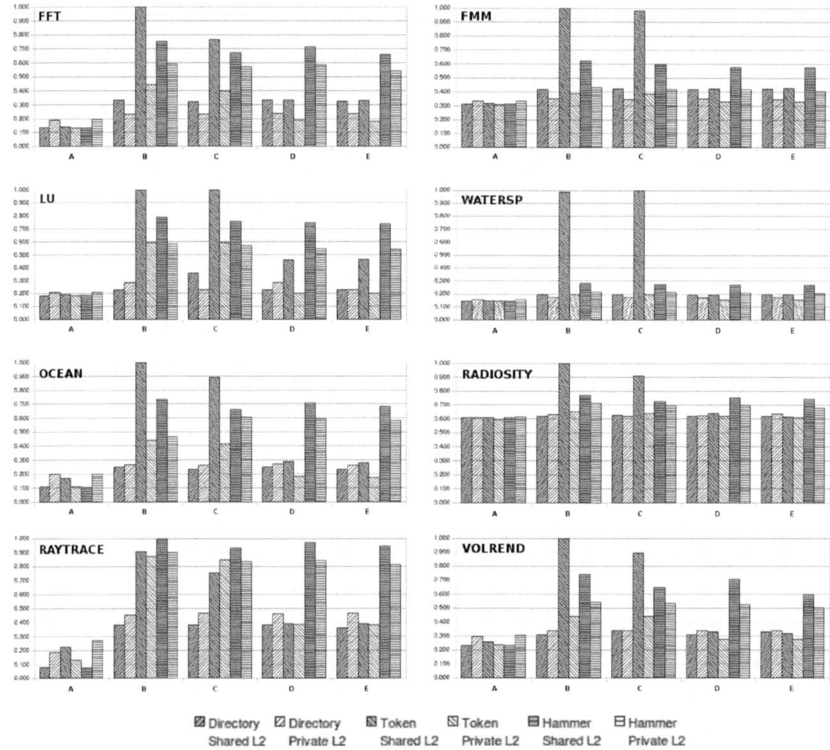

FIGURE 12.7: Execution times with: (a) Ideal network, (b) Nonuniform core allocation, (c) Uniform core allocation, (d) Nonuniform core allocation with broadcast support, (e) Uniform core allocation with broadcast support.

(configurations with shared L2 banks perform worse than configurations with private L2 memories). Indeed, configurations with private L2 caches benefit for the small working sets of applications that usually fit caches, thus processors having all its data locally within the tile and not going outside through the network. Figure 12.7 shows the particular numbers for every application.

Let us now focus on the core allocation policy (still with no support for tree-based broadcast communication). As we can deduce from the figures (second set of results vs. third set of results), uniform core allocation policy tends to lower execution time of applications. Reductions are however, nonuniform. A uniform core allocation policy tends to lower average network latencies thus leading to faster communication exchanges in the coherence protocol. This improvement is higher when L2 caches are shared for the reasons discussed above but can be considerable with private L2 caches too. Figure 12.8 shows how the average execution time improves when passing from a nonuniform allocation

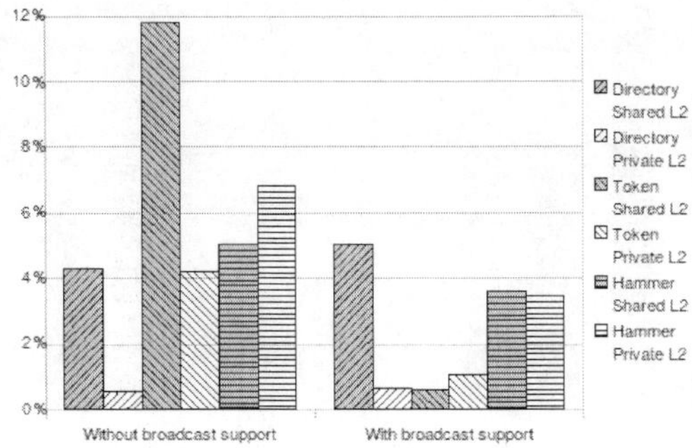

FIGURE 12.8: Average improvement in execution times due to uniform allocation of cores.

policy to a uniform one. In the absence of broadcast support, the Token and Hammer protocols perform better with a uniform core allocation, while the Directory protocol does not improve its performance; with broadcast support only the Hammer protocol has noticeable improvements. In all the evaluated cases, execution time savings can be as large as 12% (for Token shared L2) but on average savings are 3% only. This might indicate also that the network is not highly stressed (we will see this later).

Previous commented results relied on a unicast-based broadcast communication support. Now, we focus our attention on when the tree-based broadcast is supported. As we might expect, when adding broadcast support at the network level performance of applications are further improved for protocols that rely on many broadcast messages. Figure 12.9 shows the percentages of multicast/broadcast and unicast messages generated and received for each coherence protocol. Directory-based protocols use almost only unicast messages, especially if L2 caches are shared, so the improvements when network broadcast support is available are quite low. Indeed, if we compare execution time in Figures 12.6 and 12.7 we see marginal differences.

On the other hand, the Token-based protocols improve significantly their performance by the use of a network-level broadcast facility. Every L1 cache miss generates a request that is broadcasted to all other nodes: more than 95% of total received messages are multicast/broadcast messages (20% and 50% of the injected messages), and the applications' execution time in some cases is cut by 50% (as an example the FFT for the Token protocol with private L2 caches). The Hammer protocol also receives a large set of broadcast traffic at destinations, between 61% and 66% of the total, however in general network broadcast support does not seem to improve the performance as much as we would expect given the amount of broadcast traffic received. Indeed, the percentage of injected broadcast messages are much lower (less than 5%).

Tuning the broadcast threshold for each particular coherence protocol can lead to a performance improvement. Figure 12.10 shows the normalized ex-

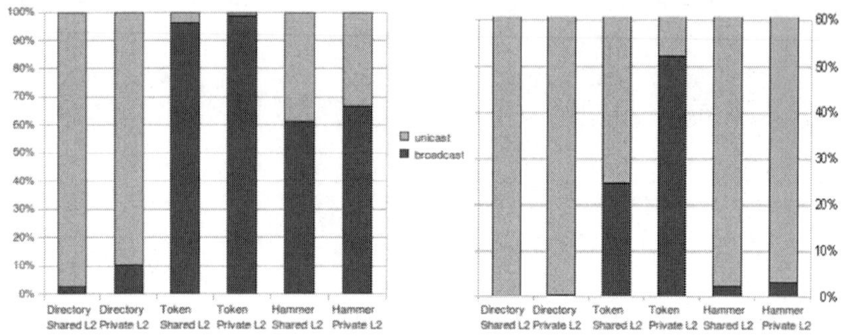

FIGURE 12.9: Broadcast vs. unicast message ratio. Computed at destination (left plots) and at source (right plots) end nodes.

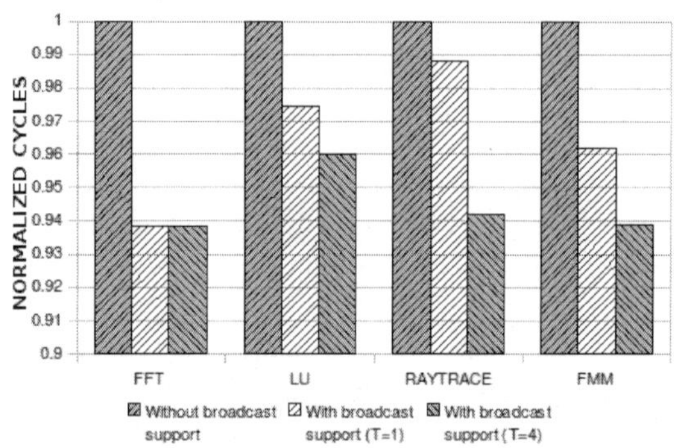

FIGURE 12.10: Execution time with hammer protocol using different broadcast thresholds (bars are normalized to the worst case of each protocol).

ecution time when the Hammer protocol is used. For each application, the first bar represents the execution time without broadcast support, while the second and third bars represent the execution time with broadcast support for two different thresholds to trigger the broadcast mechanism (T set to 1 and 4). As shown, using a higher threshold slightly improves the Hammer protocol performance since it avoids flooding the network with broadcast operations when an small set of destinations needs to be reached (messages with two or three destination nodes are not broadcasted to all nodes). Execution time reduction ranges from no reduction at all (FFT application) up to 5% reduction for RAYTRACE.

Figure 12.11 shows the number of allocated flits in the on-chip network by every simulated case (excluding the ideal network case and assuming the T parameter set to 1). We can observe the large reductions in traffic when moving from a NoC system with no broadcast support to a NoC with built-

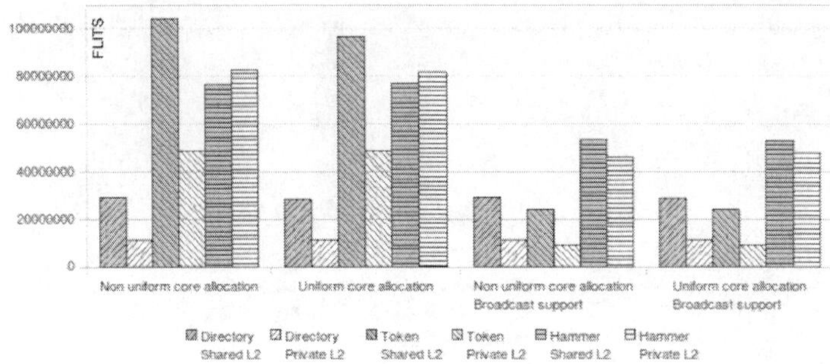

FIGURE 12.11: Allocated flits for the FFT application.

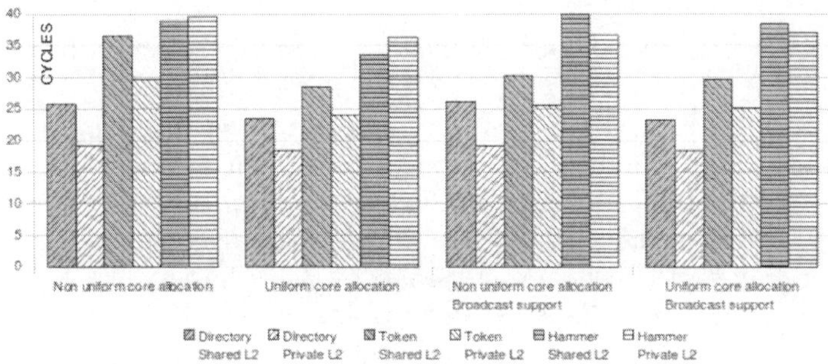

FIGURE 12.12: Flit latency for the FFT application.

in broadcast support. Both Token and Hammer largely benefit and reduce the network traffic by average factors of 5 and 1.6, respectively. Also, notice how directory-based protocols do not benefit. Additionally, the use of private L2 configurations also tends to reduce network traffic, since L2 hit accesses are always satisfied within the tile. In both, Directory-based and Token-based traffic is reduced to half when using private L2 configurations.

Figure 12.12 shows the average flit latency for the different cases. Depending on the case the flit latency can be as much as doubled (comparing Directory with private L2 caches vs. Hammer with private L2 caches).

Finally, Figure 12.13 shows the average network throughput achieved by each scenario. Protocols with high demand on collective communication (Token and Hammer) exhibit a very large network throughput. However, this effect is due to the internal flit replication the network is performing, thus an injected flit reverts into 32 flits received (one per destination), thus having a much larger throughput. Anyway, the tree-based broadcast mechanism increases network throughput. Besides this effect, we can observe that the network is not highly loaded (on average for the entire run of the applications). Indeed, on average 3 flits per cycle are obtained, which is very low when com-

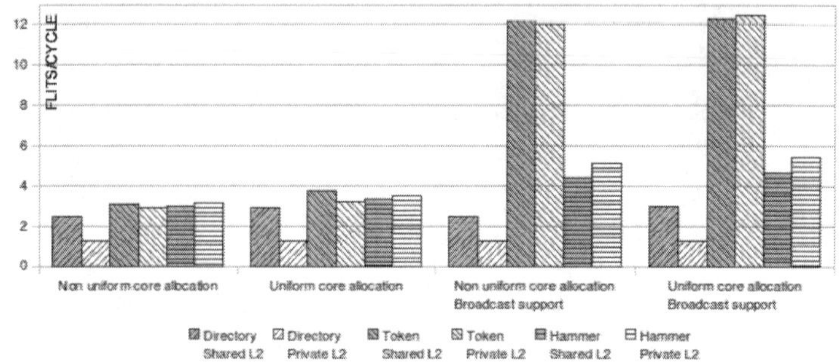

FIGURE 12.13: Flit throughput for the FFT application.

pared with the theoretical network bandwidth at the bisection of the 4×4 mesh.

As an overall conclusion in this section we can state that the cache coherence protocol exhibits a large variation in performance when the network is accurately modeled and different built-in mechanisms are used. The evaluation tries to highlight the necessity to evaluate all the system parameters, thus not relying only on the network evaluation (without taking into account the cache coherence protocol, the operating system, and the application). Indeed, as we have observed, the network (in the evaluated scenarios) does not congest and usually is very lightly loaded. These results should be carefully put in the context of the evaluated scenarios and the applications we have chosen.

12.2 Congestion Management

Congestion Management (CM) has been researched for decades in high-performance (off-chip) interconnection networks. Congestion occurs when messages contend persistently for the same set of resources and there is an excess of demand for the offered resources in the network. The most straightforward example comes when two messages reaching the same switch through different input ports contend for the same output port at the same time. Since not enough bandwidth exists for both messages one of the messages wins the arbitration and is forwarded through the output port, while the other message waits. This is mostly known as *contention* where the blocked message suffers an additional latency due to the other message. However, if instead of two single messages we have a flow of messages also contending for the same output port, then contention reverts to *congestion* since both flows will be impacted by each other. Indeed, congestion can be seen as a contention situation that lasts for a large amount of time.

Typically, congestion has been seen as a problem to avoid. Indeed, there are different solutions that either focus their attention on preventing the appearance of congestion, or try to detect and eliminate the congestion. The first series of techniques can be referred to as *proactive techniques* and are typically found in networks where messages are injected only if there is a 100% guarantee that no congestion spot will be created (avoidance-based techniques) [347, 30], or in networks where paths taken by messages are restricted (prevention-based techniques) [125, 284].

The second set of solutions can be referred to as *reactive techniques* and there are many proposed solutions. In these kinds of solutions mechanisms are proposed to detect the appearance of congestion within the network (typically buffer utilization and/or link utilization). To complement such detection mechanism, techniques to eliminate the congestion spots are also used. The most widely used mechanisms are injection limitation techniques [319, 161, 64, 21] where end nodes contributing (or supposedly contributing to the congestion) limit their injection rate. These techniques have, however, as their main limitation scalability, since the reaction time for a congestion situation is directly proportional to the distance from the congestion point to the end nodes.

The problem with the congestion effect can become quite serious from the performance point of view of the network. The reason is the spreading effect of the congestion spot. Indeed, congestion may spread quickly over part of, or even the entire network, thus collapsing the entire system. Indeed, in Figure 12.14 we can observe the typical plots for a network where accepted traffic and average message latencies are shown against the injected traffic rate. As traffic injection increases, messages start to experience contention levels thus increasing the average message latency. However, this is not a major issue until the network enters the maximum throughput (defined as the maximum injection rate where traffic does not saturate). Near that point we can observe an exponential increase of the message latency. Indeed, messages are enqueued at the end nodes (possibly at the network interfaces). Moreover, as we can see, the accepted network traffic may also lessen. In that situation the network is totally collapsed.

The effect of congestion can be better viewed in Figure 12.15(a). In that figure the generated traffic rate is shown together with the accepted traffic rate. The modeled network is a bidirectional multistage network (BMIN) with 64 end nodes connected through 3 switch stages, each switch with eight bidirectional ports. Uniform traffic is injected during the entire simulation process (6 ms). Suddenly, hot spot traffic is added during 300 μs where eight end nodes send traffic to the same destination. Thus, a congestion situation is created within the network. As can be viewed, throughput severely degrades due to the hot spot traffic by 70%. However, in addition, the network throughput stays low for quite a long period, during that time congestion is still present within the network. Only when congestion subsides the network recovers its normal throughput. During congestion situation messages have a very large latency, as plotted in Figure 12.15(b). Three orders of magnitude increase are

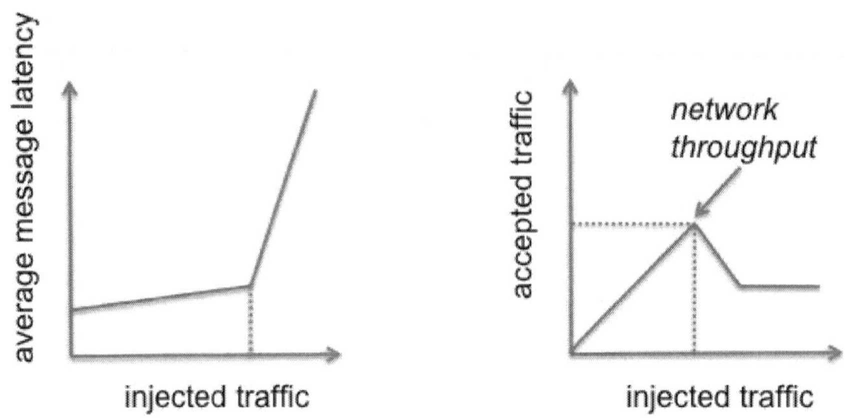

FIGURE 12.14: Typical performance graphs for interconnection networks.

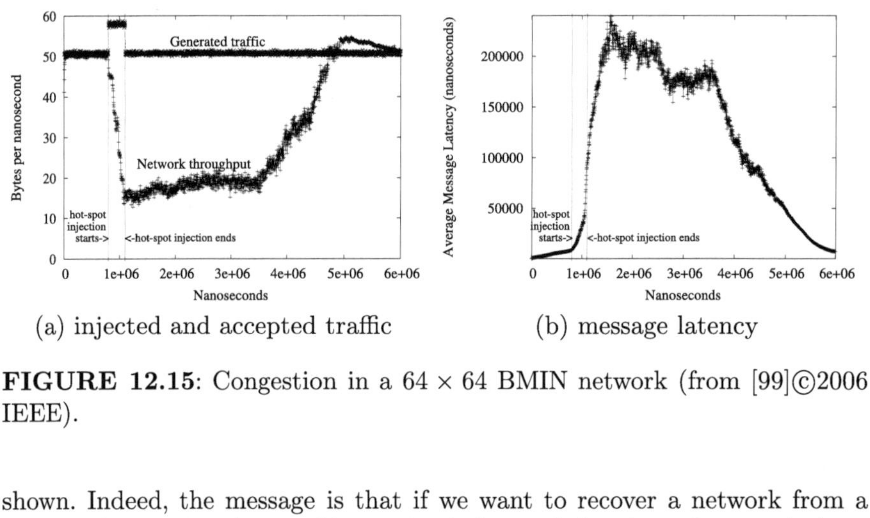

(a) injected and accepted traffic (b) message latency

FIGURE 12.15: Congestion in a 64 × 64 BMIN network (from [99]©2006 IEEE).

shown. Indeed, the message is that if we want to recover a network from a congested situation (its working point is beyond the throughput point) we need to reduce injection into the network to a far lower injection rate, as we need the network to slowly drain congested messages.

Although proactive and reactive mechanisms have been thoroughly researched in the literature for decades, recently a different view of the congestion problem was addressed [99]. Indeed, in this new view congestion is not seen as a problem by itself. If many sources request the same set of resources the flow control mechanism will work and will regulate the flows. Indeed, flows contributing to the congestion will advance at their maximum allowable rate thanks to the flow control mechanism. Congestion occurs because no more bandwidth is available. Thus, the real and single solution to the problem would be to add more resources to the system so to avoid the lack of bandwidth (this is related to network overdimensioning). The real problem with

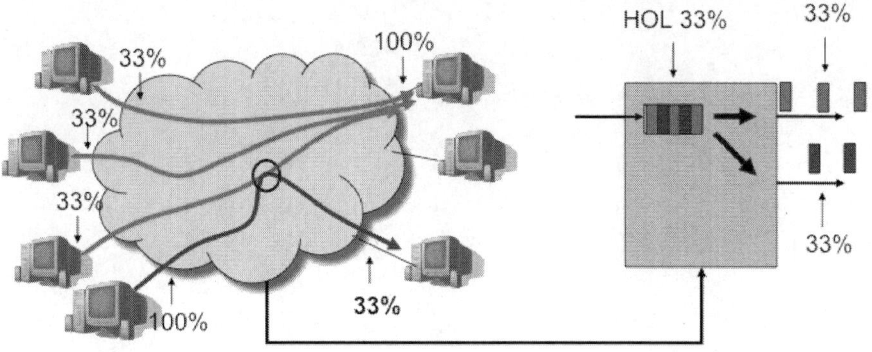

FIGURE 12.16: Example where HoL blocking spreads the congestion.

congestion is the effect it causes to other flows that are not contributing to the congestion. Figure 12.16 shows an example where three flows are building a congestion tree. Since all of them request the same output port, a congestion tree is formed (because of the flow control mechanism) and they adapt their injection rates naturally to 33%. However, notice that there is a flow, not contributing to the congestion but requesting the same resource as one of the congested flows. Because the congested flow is moving at a slower pace the noncongested flow will be impacted. The reason is the use of the same resource (indeed they are using the same buffer) and the *head-of-line* blocking effect between both flows. As the figure depicts, messages of both flows are mixed in the same queue. Congested messages will stay blocking the queue during 66% of the time (due to the congestion) and 33% of that time the message will be blocking the advance of the noncongested flow. From that moment on, the noncongested flow converts into a congested flow, thus spreading the congestion.

If a mechanisms solves (eliminates) all of the Head-of-Line blocking (HoL) effect caused by congestion trees, the congestion becomes harmless within the network and there is no need for eliminating the congestion. Indeed, the regional explicit congestion notification (RECN) solution [99] advocates for such a solution. RECN dynamically detects and allocates queues when detecting congestion. Congested flows are allocated in separate queues, eliminating the HoL blocking effect they might cause to noncongested flows. With less than eight queues per input port, RECN is able to totally eliminate the HoL blocking of congestion, thus getting rid of its effects. Figure 12.17 shows the ideal and expected performance numbers of a network without the HoL blocking effect of congestion. As can be seen, the network can enter the congestion zone (beyond network throughput point), however, accepted traffic keeps high and maximum. Thus, there is no need to drain the network to leave from a congestion situation. Figure 12.18 shows the performance results when RECN is applied to a multistage network in the presence of a congestion situation.

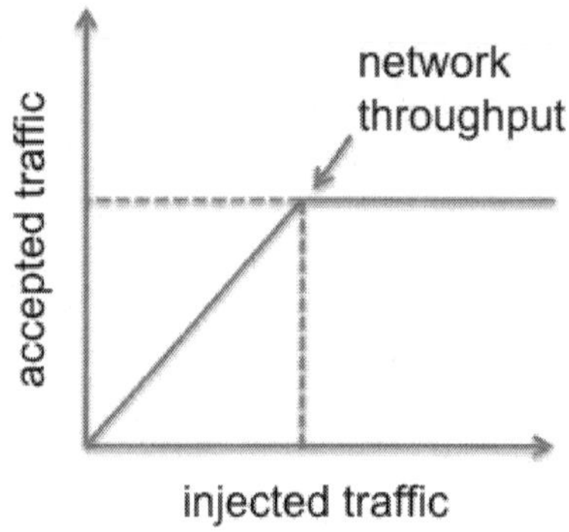

FIGURE 12.17: Expected performance when HoL blocking effect is avoided.

VOQ_{net} (virtual output queueing at network level) devotes as many queues per input port as destinations in the entire system (does not scale, but avoids completely HoL blocking) and VOQ_{sw} (virtual output queueing at switch level) devotes as many queues per input port as output ports has the switch (it scales in resources but does not avoid HoL blocking completely).

Obviously, on-chip networks is a different domain where dynamic techniques like RECN may be difficult to implement. Moreover, RECN relies on virtual cut-through switching (where differentiating congestion from contention is much easier than in wormhole switching). However, there are approaches that may help in the on-chip domain where simpler and deterministic mechanisms are used (to reduce or minimize the HoL blocking effect). One such example is Destination-Based Buffer Management (DBBM) [229] where queues are statically assigned to different disjoint sets of destinations, thus flows pertaining to destinations in different sets will never experience HoL blocking between them (congestion is localized into a subset of destinations only).

There is not much prominent research in congestion management for on-chip networks. Indeed, most of the existing work relies on previous well-established techniques from off-chip networks (mostly high-performance interconnects). However, in the last decade several works have been published. In [232] a buffer-less switch is presented where deflective routing is used when needed. To increase the maximum tolerable load of the network, the Proximity Congestion Awareness (PCA) technique is added, where switches use load information from neighboring switches, called *stress values*, for their own switching decisions, thus avoiding congested areas. Similarly, in [109] Regional Congestion Awareness (RCA) has been proposed to improve load balance when using adaptive routing. RCA relies on regional information in order to take better routing paths so as to avoid congestion spots in the neighborhood of the switch. Up to 71% of latency reductions are reported with no impact on switch delay and negligible impact on area.

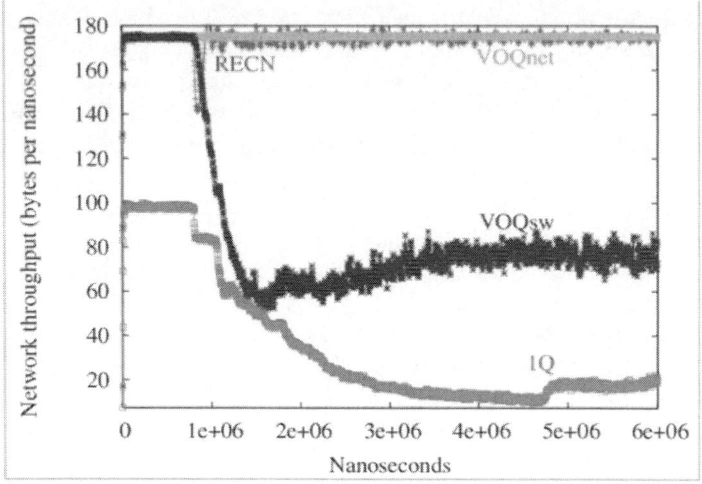

FIGURE 12.18: Network performance when RECN is used.

Dyad [129] indirectly tackles congestion by switching from a deterministic routing algorithm to an adaptive one. The method, however, does not resolve congestion (i.e., the alternative paths might also be congested) as it only tries to avoid it. In [15] both data plane and control plane are separated so as to avoid congested data from affecting control data. Network interface statistics are used as a congestion measure. In [234], a prediction-based flow-control strategy for NoCs is proposed where each switch predicts future buffer fillings to detect future congestion problems. The buffer filling predictions are based on a switch model. The switch buffer filling information is used for toggling the sources.

Congestion-Controlled Best-Effort (CCBE) is proposed in [328], where congestion levels in the NoC are measured (link utilization) and informed to a Model Predictive Controller (MPC). The goal is to still keep the guarantees in bandwidth and latency for the established connections while the network is congested. The proposal is made for embedded systems. Finally, in [235] a predictive closed-loop flow control mechanism is proposed where the packet injection rate is limited at traffic sources in order to regulate the total number of packets in the network.

Another issue with on-chip networks and the congestion/HoL blocking effect is the use of overdimensioning. It is well known that there is bandwidth in excess within a chip, so one could think congestion is a minor issue (if any). However, as power consumption is becoming the driving factor when designing a chip, and because power reduction comes mainly by switching off unused components, congestion may become a hot topic in the future. One clear example is a chip where, due to low traffic utilization, most of the network components are switched off, or their bandwidth offerings are reduced (for instance, a link is downgraded from 1 GBps to 128 KBps). In that scenario, a sudden increase on network demand would be seen as a congestion spot around the slow link. Indeed, an efficient power management technique should come with an appropriate method to deal with congestion within the network.

As a conclusion for congestion management in NoCs, we can expect the coming years will have an increase in the importance of researching and developing mechanisms and techniques that deal with on-chip congestion. However, currently it seems the available bandwidth within the NoC makes the problem still an academic one. Future challenges, like different network domains (DVFS, virtualization) and the need of aggressive power-saving techniques, may require an effective congestion management technique to keep maximum network performance and to avoid the significant system degradation a congested situation could cause. Notice, however, that entire full system modeling is a requirement to better assess the degree of congestion the system under test will have.

12.3 Variability in NoCs

The sources of variability in NoCs have been partially analyzed in many recent works. For example, the main sources of process variation in NoC links are identified in [217]. In summary, variability sources can be divided into front-end and back-end ones. The front-end phase of the integrated circuit (IC) fabrication process is related to the steps involved in the creation of devices whereas the back-end stage comprises steps involved in the wiring definition. Front-end process variation can be further decomposed into systematic and random components. Systematic variation is caused by deviations in the photolithographic process like small imperfections in the lenses used. Random variation is caused, for example, by dopant fluctuations. Systematic variation is characterized by a spatial correlation, meaning that differences in neighboring areas are expected to be low. On the contrary, random variation may cause different operation characteristics in adjacent areas. It is commonly accepted that random and systematic variations are uncorrelated [280]. Additionally, spatial correlation depends on some ρ parameter dependent on the exact manufacturing process. The ρ parameter represents the fraction of the chip that is correlated. Typical values for the ρ parameter range from 0.5 (half of the chip size) to 1 (the entire chip) according to [280].

Regarding back-end variation sources, they are mainly capacitance and resistance variations due to defects introduced by the chemical metal planarization process in metalizations, for example. The way back-end variation affects a NoC depends on the exact dimensions of the metalizations considered. For example, variations introduced by the chemical metal planarization process in logic areas, like a switch, are negligible because wires connecting logic gates are quite short. On the other hand, in the case of links, where metalizations are much longer, the influence of back-end variability may be more noticeable. Nevertheless, as shown in [121], when links are designed for minimizing power consumption, delay variation due to back-end variation is less than 0.1% because in that scenario link delay is dominated by repeater delay and, therefore, variations in wire resistance have a negligible contribution to the resulting link delay.

The consequences of variability in NoCs have been analyzed in several studies. For example, an analytical expression of timing variability from the variation of the parameters involved in the interconnect delay is provided in [217]. However, this study does not consider the spatial features of variability, thus ignoring that link delay variation depends on the location of the link in the die. Other studies on variability focus on the impact of within-die variation in cores, not considering variations in links [280][304], although links in the NoC are also affected by variability as shown in [121] and [217].

The impact of process variation in NoCs is also analyzed in [231]. However, this study does not consider delay variations between switches as a consequence of their exact location in the die. Additionally, this study simplistically models random variability as a percentage of the nominal delay, thus not considering other studies that show that random variations strongly depend on the critical path depth [132] and the size of devices [121]. Another drawback of [231] is that it states that variability does not considerably affect NoC links because it is possible to take advantage of the large slack present in links in order to compensate link delay variations. However, as shown in [121], this is only true when links are designed to present minimal delay at the expense of a larger power consumption [49].

12.3.1 A Variability Model for NoCs

A new model for variability in NoCs has been recently presented [121][120]. This model provides accurate data on the effects of process variation both in links and switches. Additionally, it considers the spatial features of variability and precisely models random variation, thus being noticeably more accurate than previous works [217][280][304][231]. Nevertheless, the new model does not consider the influence of back-end variability because it is assumed that NoC links are designed for minimizing power consumption and, therefore, delay variation due to back-end variation is less than 0.1%, as mentioned above. In the following sections, the model is presented. A detailed description of the model can be found in [120].

12.3.1.1 Modeling Front-end Systematic Variation

The systematic component of front-end variation is strongly related to the photolithographic process. Lens aberrations may lead to an important systematic spatial nonuniformity of L_{eff} over the reticle field [237], that can be as high as 12% for 45 nm processes, according to [95]. However, it is not enough knowing the maximum percentage of variation in L_{eff}. It is required to know how variations in L_{eff} are spatially distributed in the exposure field as well as how variations in L_{eff} influence variations in link and switch delay.

In order to model the spatial nonuniformity of L_{eff} as well as its correlation, Gaussian Random Fields [115] along with the spherical model proposed in [280] can be used. This model depends on the ρ parameter mentioned above. The R tool [1] can be used to implement the Gaussian Random Fields with the spherical model. To do so, the chip surface has to be discretized using, for example, a 1000×1000 square matrix. This will allow computing L_{eff} maps for the chip.

Once the L_{eff} values are computed for the entire chip surface, we can proceed to compute how variations in L_{eff} influence variations in link and switch delay, that is, we can compute the delay for every gate in the switches, or repeaters in the links, depending on their exact location in the die. As this computation also depends on random variation, we postpone how to perform it until Section 12.3.1.3, thus revisiting how to model random variations first.

12.3.1.2 Modeling Front-end Random Variation

The main source of random variation in NoCs is threshold voltage variation due to Gaussian Random Dopant Fluctuations (RDF) caused because the number of dopant atoms that fit into the transistor channel area gets smaller as technology scales down. This number is in the range of a few tens atoms for 45 nm down to around 10 atoms for 16 nm, on average. Therefore, a few atoms more or less considerably matters. Thus, Random Dopant Fluctuations refer to the variations in the number of dopant atoms in the transistor channel. According to [149], RDF will increasingly affect deep submicron technologies by causing variations in the threshold voltage. For example, for a 45 nm technology, this variation in threshold voltage, referred to as $3\sigma_{V_{th}}$, is 40% according to [95]. However, it is not enough to know the value for $3\sigma_{V_{th}}$. It is required to connect that value to the size of the particular transistor being analyzed. Thus, to compute $\sigma_{V_{th}}$ for a given device, the formulas provided in [13] can be used.

12.3.1.3 Introducing Process Variation into Switches and Links

In order to introduce process variations in switches (see [120] for a deeper description), the L_{eff} values previously computed with the R tool in Section 12.3.1.1 will be mapped to individual logic gates at each of the switches in the network after postlayout synthesis. Then, the delay of each switch cell is computed according to the value of the L_{eff} assigned to it and, simultaneously, the effect of random variation on that cell. Then, a static timing analyzer like PrimeTime could be used in order to compute the delay of the switch. The output of this process will be the delay data for each of the switches in the network, individually computed taking into account their position in the die.

Parameter	Values
Core	1 GHz, in-order, 1 thread
L1 inst cache	32 KB, 4-way
L1 data cache	32 KB, 4-way
L2 cache	128 KB, 8-way

TABLE 12.2: Tile configuration.

In order to introduce process variations in links, the methodology is quite different. The reader could refer to [121] for a complete description. In this case, links are simulated with SPICE using the Predictive Technology Model (PTM) for the technology used during the synthesis of the chip. Again, the effect of variability on links is computed taking into account their location in the die. With the location of each link and the number of repeaters in it, it is possible to know the exact L_{eff} value for each repeater of each link in the network thanks to the L_{eff} maps computed in Section 12.3.1.1. On the other hand, random variations, as in the case of the switches, are introduced varying the threshold voltage taking into account the size of repeaters. As can be seen, every wire of every link in the network is individually analyzed in order to assess the delay characteristics for each link in the NoC.

12.3.2 Analyzing Variability in NoCs

This section presents how process variation affects delay in a typical NoC. For doing so, a test bench CMP NoC is first introduced. Then, the model presented in the previous section is applied to that NoC.

12.3.2.1 On the Design of a NoC-based CMP

In the following, a common case NoC-based CMP is presented to be used as a case study for applying the previous variability model presented. This CMP, aligned with current proposals from industry [126][266][320] and academia [70][353], includes 64 cores interconnected by an 8×8 bidimensional mesh, and uses a tiled approach [352], option widely accepted for designing multicore processors [126][353][266][320].

Each tile includes a general purpose CPU with its associated private L1 instruction and data cache banks, and a fragment of a distributed L2 shared cache. The tile also includes part of the coherency protocol directory, which is distributed among them. Finally, in order to provide connectivity among the different tiles, they also include a switch connecting them to the NoC. Table 12.2 summarizes the characteristics of the tile used in this study.

In order to interconnect the 64 tiles in the CMP, an 8×8 2-D mesh NoC was designed and synthesized in a 45 nm technology. This NoC will be used as a test bench for the variability model. The switch (gNoC switch, see the Appendix) designed for this NoC implements wormhole switching with five

module	area (mm^2)	critical path (ns)	gates	critical path depth
IB	$3.08 * 10^{-3}$	0.58	1205	8
RT	$8.91 * 10^{-5}$	0.30	41	4
VA-SA	$1.09 * 10^{-3}$	0.74	576	12
XB	$4.47 * 10^{-3}$	0.43	2103	7

TABLE 12.3: Area, delay, and number of gates for the switch modules.

virtual channels. That number of virtual channels was selected so that support for the underlying coherency protocol is provided. The switch is a pipelined input buffered wormhole switch with five stages: input controller (IB), routing (RT), virtual channel allocator and switch allocator (VA-SA), crossbar (XB), and link traversal (LT). Link width is set to 8 bytes. Flit size is also set to 8 bytes. Input buffers can store four flits. A Stop&Go flow control protocol has been deployed. Additionally, the routing stage has been implemented to support the XY routing algorithm. Table 12.3 summarizes the delay, area, and number of gates for each of the modules of the switch. After postlayout, the switch clock cycle is 0.99 ns.

Regarding links, as they are usually long interconnects, they will present a considerable capacitance and resistance. To efficiently deal with them while not increasing power consumption, repeaters are used. Actually, repeater insertion is an efficient method to reduce interconnect delay and signal transition times. In our case, provided that link length is 2.4 mm and switch delay is equal to 0.99 ns, links have been designed with a supply voltage equal to 1.1 V and consisting of 5 repeaters of size 4. Additionally, links are placed in metalization layers M4 and M5. Using this link configuration, a nominal link delay of 0.79 ns has been obtained by using SPICE and the PTM model for 45 nm [325].

12.3.2.2 Variability in the Switch

In order to analyze the effects of variability in the gNoC switch, we applied the variability model to 100 instances of the 8×8 mesh NoC and studied how variability modifies the operating frequency of each of the switches of the network and also each of their modules. From the 100 NoC instances analyzed, 50 of them were produced using a value equal to 1 for the ρ correlation parameter while the other 50 were produced with ρ set to 0.5. Figure 12.19 shows the probability distribution function (*pdf*) of the relative operating frequency with respect to their own nominal frequency (Table 12.3) of each stage of the switch in two scenarios: when only systematic variation with correlation 1 is considered (Figure 12.19(a)) and when only random variation is taken into account (Figure 12.19(b)).

Figures 12.19(a) and 12.19(b) show that systematic variation has a larger influence in the operating frequency than random variability. This is due to the fact that random variability differently affects two adjacent components.

Thus, random variability, as will be shown later, may even be reduced or canceled as the number of gates in a chain of logic increases [121], as it can be seen for the IB, and the XB modules. Notice that despite the differences in the *pdf* of the different stages, all the *pdf* have the same nature. They can be seen as an addition of different peaks that are spread out due to variability. This is more noticeable for random variability. Each peak in the *pdf* represents different paths in a stage. These paths become alternatively the critical path in their respective stage. The fact that the critical path changes due to variability will increase the difficulty to estimate the influence of the variability in a presynthesis design, because not only the nominal critical path has to be taken into account.

Furthermore, Figures 12.19(a) and 12.19(b) show that variability (both systematic and random) influence different stages in different ways. However, the different kinds of variability affect the same stage in the same manner by generating the same peaks. Then, the properties of the implementation of each stage are critical in order to understand how variability affects it. For a small and fast module, like the RT module, with a small number of gates in its different paths, variability affects considerably. Small changes due to variability have great impact. On the opposite end, a slow module with a large number of gates in its paths (VA-SA module) is slightly affected. Notice that, as the number of paths is small in the VA-SA module, it only has one main peak. In between these two ends, we can find the IB and the XB modules. Their main characteristic is that as their number of paths is bigger and these paths are similar in length (they are regular modules), the number of paths that may become the critical path increases. This is the reason for the two peaks in the graphs for these modules.

Figures 12.20(a) and 12.20(b) show the probability distribution function of the relative operating frequency of the switch with respect to its own nominal frequency when only systematic variation with correlation 1 is considered (Figure 12.20(a)) and when only random variation is taken into account (Figure 12.20(b)). When only random variability is considered, the influence of this variability is smaller than when only systematic variability is considered. Then, when only random variability is considered, it is the VA-SA stage that determines the operating frequency in all cases since this stage is slower than the others. On the other hand, when only systematic variability is considered, the operating frequency for the switch could also be determined by others. This can be seen in Table 12.4. This table shows the correlation between the operating frequency of a switch and the operating frequency of the stages of that switch. Note that the correlation of the VA-SA stage is the highest in all cases because this stage is the one that fixed the operating frequency of the switch in almost all the cases. Note, however, that the correlation when only systematic (sys) variability is considered is higher than when only random (rnd) variability is taken into account. When only systematic variability is considered, this correlation is high even for modules other than the VA-SA one. This is due to the fact that all the components in the switch are affected

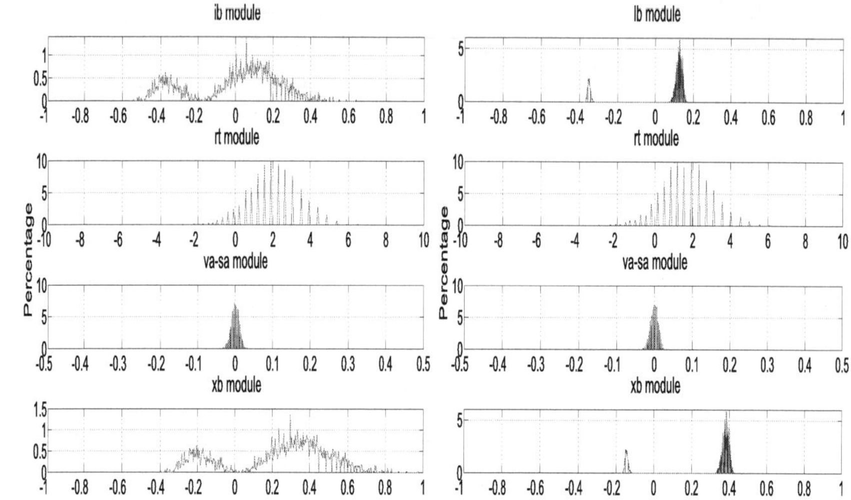

(a) Only systematic L_{eff} variations considered ($\rho = 1$).

(b) Only random V_{th} variations considered.

FIGURE 12.19: Frequency variation in the stages of the pipelined switch.

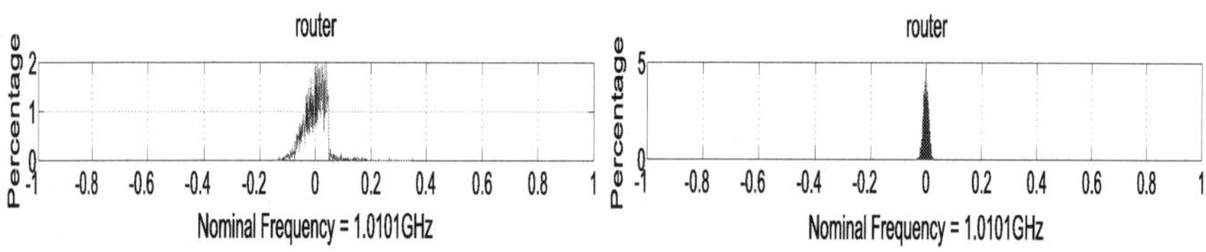

(a) Only systematic L_{eff} variations considered ($\rho = 1$).

(b) Only random V_{th} variations considered.

FIGURE 12.20: Frequency variation for the switch.

by variability in a similar way. This means that if the frequency of one stage is reduced because of systematic variability, then the frequency of other stages will probably be also reduced. Therefore, systematic variability can be seen as a biasing variability that causes that the critical path does not change among the switches so often. This effect is not present when only random variability is considered due to the nature of random variability, that affects adjacent components differently.

Figure 12.21 shows the probability distribution function of the switch operating frequency when systematic and random sources of variation are simultaneously considered. This figure, and Table 12.4, show that there exists small differences in frequency between high and low correlations.

	variability				
stage	($\rho = 1$)	($\rho = 0.5$)	rnd	($\rho = 1$)+rnd	($\rho = 0.5$)+rnd
IB	0.9955	0.8789	0.1178	0.9732	0.8745
RT	0.9922	0.7644	0.2354	0.9549	0.6879
VA-SA	1.0000	0.9998	1.0000	0.9788	0.9902
XB	0.9933	0.9765	0.1715	0.9644	0.9712

TABLE 12.4: Correlation between stage delay and switch delay.

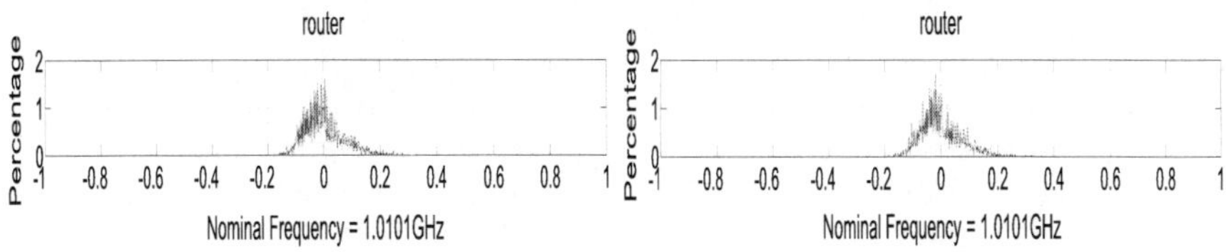

(a) Systematic correlation of variability set to 1. (b) Systematic correlation of variability set to 0.5.

FIGURE 12.21: Frequency variation in the switch as a consequence of both systematic and random variations.

12.3.2.3 Variability in Links

Figures 12.22 and 12.23 show link operating frequency variation as a consequence of systematic L_{eff} variation, random V_{th} variations, and both random and systematic variations, respectively. Figure 12.22(a) shows that when considering only systematic variations, link operating frequency varies between 1.15 GHz and 1.4 GHz, for both values of ρ, despite that the nominal frequency was 1.26 GHz. The exact value of correlation does not introduce significant differences, as shown by the σ parameter ($\sigma = 6.62$ and $\sigma = 6.58$ for $\rho = 1$ and $\rho = 0.5$, respectively). When only random variations are analyzed, variations in wires move practically in the same range of systematic variations. The upper plot of Figure 12.22(b) shows the maximum achievable operating frequency of all wires of each link in the network. However, as all wires of a given link have to work at the same frequency, the slowest wire will cause a slight operating frequency slowdown. This is shown in the bottom plot of Figure 12.22(b), where the mean frequency is reduced to 1.21 GHz and the frequency variation is reduced to 5%. This effect is similar to the behavior analyzed by Bowman et. al in [41]. In that work it is shown that when the number of critical paths increases, the mean delay increases and the standard deviation decreases. In the case of links, a higher number of wires per link will cause a higher frequency slowdown but also a reduction in the σ of the link operating frequency. Finally, Figure 12.23 shows that when random and systematic variations are simultaneously considered, the average link operating frequency is reduced as a consequence of random variations. The mean values of link operating frequency are 1.22 GHz and 1.21 GHz ($\rho = 1$ and

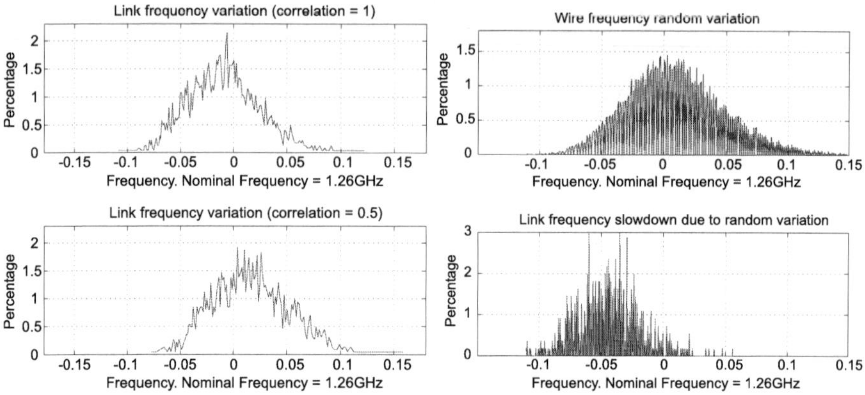

(a) Link frequency variation as a consequence of systematic L_{eff} variations

(b) Link frequency variation as a consequence of random V_{th} variations

FIGURE 12.22: Link frequency variation.

$\rho = 0.5$). However, frequency variation of links is almost the same than when considering only systematic variations, $\sigma = 5.5$ and $\sigma = 5.8$ for $\rho = 1$ and $\rho = 0.5$, respectively.

12.3.3 Consequences of Variability in GALS-based NoCs

This section analyzes the consequences of variability in GALS-based NoCs (Global Asynchronous Locally Synchronous). In these systems, each of the components of the network can be clocked at the maximum frequency it allows. For this analysis, the 8×8 CMP test bench network has been simulated and performance metrics, like link utilization, message latency, and network throughput have been collected. Figure 12.24 shows two examples of the 8×8 CMP NoC when affected by variability. The left side of Figure 12.24 shows the maximum frequency achievable for each switch and link when the ρ parameter is set to 0.5. The right side shows such frequencies when ρ is set to 1.

The traffic pattern used is intended to synthetically emulate the coherent traffic present in a CMP chip, and is the composition of two different types of traffic. The first one, which accounts for 60% of the overall network traffic, follows a uniform destination distribution and tries to emulate cache-coherent traffic between cores. The second traffic is intended to emulate memory accesses targeted to on-chip memory controllers. We have simulated a CMP chip having four memory controllers located at both sides of the chip: two of them on the right side and the other two on the left side. Additionally, the two memory controllers on the right are directly connected to the 4 upper or 4 lower switches in the right edge of the network, respectively. The memory controllers on the left side follow the same interconnection pattern. In this configuration, every core sends 30% of the messages it generates to the mem-

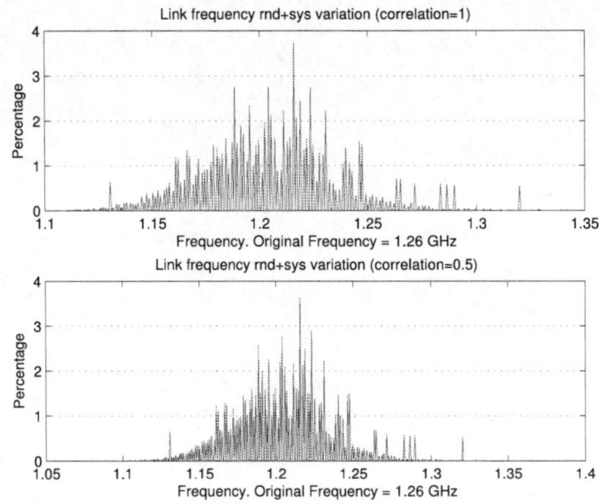

FIGURE 12.23: Link frequency variation as a consequence of both systematic and random variations.

ory controller closer to it and the remaining 10% to other memory controllers, with a uniform distribution.

Figure 12.25 shows the average message latency versus received traffic for several of the chips analyzed. Low and high correlation values are considered ($\rho = 0.5$ and $\rho = 1.0$). As can be seen, in the presence of variability (curves labeled "Chip #n") the network is able to manage almost 20% less traffic than in the absence of process variation (curve labeled "Nominal"). Moreover, average message latency is increased by 23% even for low traffic loads. On the other hand, it is interesting to notice that the overall performance of the network in the presence of variability is almost independent of the exact characteristics of that variability. This is shown by all the "Chip #n" curves being almost overlapped. Nevertheless, when the correlation of the manufacturing process is lower, we can see more differences in network performance.

One of the reasons for the differences in performance shown in Figure 12.25 is the lower average network bandwidth caused by random variation in links (note that systematic variation does not cause a reduction in the aggregated network bandwidth). However, this reduction in network bandwidth does not completely explain the plots in Figure 12.25. An important contribution to that performance reduction is shown in Figure 12.26, that displays link utilization for all the links in the network when it is close to saturation both with and without process variation. As can be seen, process variation not only causes lower link utilization but also larger differences in link utilization, as shown in Figure 12.26(b), where a few links present a much larger utilization than others (on the right end) or a much lower utilization (left end). This uneven distribution of link utilization leads to the performance loss in Figure 12.25.

Figure 12.27 shows a similar study but instead of focusing on link uti-

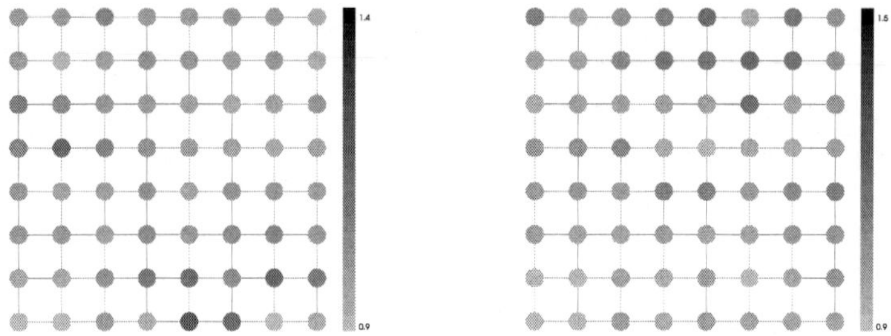

(a) $\rho = 0.5$ (b) $\rho = 1$

FIGURE 12.24: Operating frequency distribution in a NoC in the presence of process variation (random and systematic).

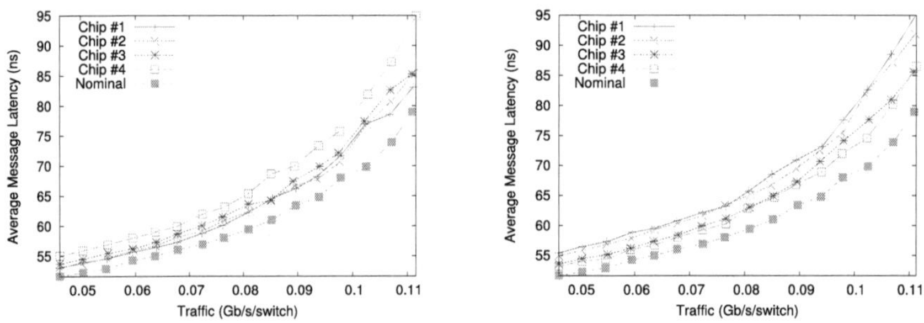

(a) $\rho = 0.5$ (b) $\rho = 1$

FIGURE 12.25: Network performance in the presence of process variation.

lization, it displays the variation of buffer utilization with the flit injection rate. Plots labeled as "nominal" represent the network working at the nominal frequency whereas plots labeled as "corr=1" represent the average results of all chips with high correlation. Results for low correlation ($\rho = 0.5$) are not shown as they are very similar to high correlation results. This figure shows how average buffer utilization decreases in a NoC when switches and links work at different frequencies. More concretely, in the ideal network, buffer utilization is around 50% while in a network with process variations the average buffer utilization is around 40%. Additionally, it is possible to see how in a GALS-based NoC in the presence of process variations the standard deviation of buffer utilization is much greater than in a fully synchronous network where buffer utilization is more uniformly distributed. This behavior is better shown in Figure 12.28 where the spatial distribution of buffer utilization is represented for an injection rate of 0.11 flits/cycle. In this figure the plot of the left represents the nominal network and the right plot is a chip presenting variation of low correlation ($\rho = 0.5$). Note that the frequency distribution of this chip was shown in Figure 12.24(a). As shown in Figure 12.28, in a

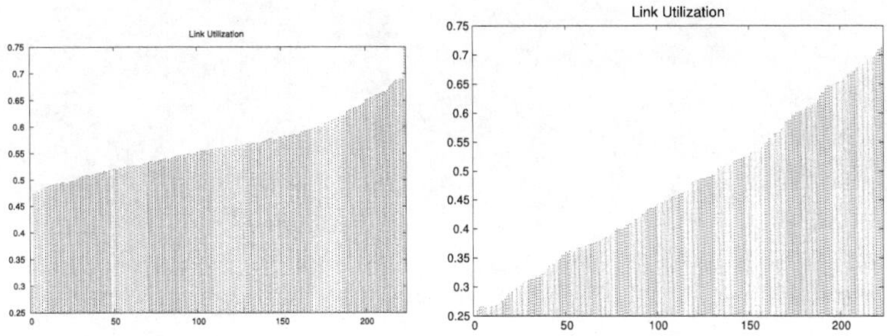

(a) Nominal network (b) Network with variability

FIGURE 12.26: Link utilization in the presence of process variation.

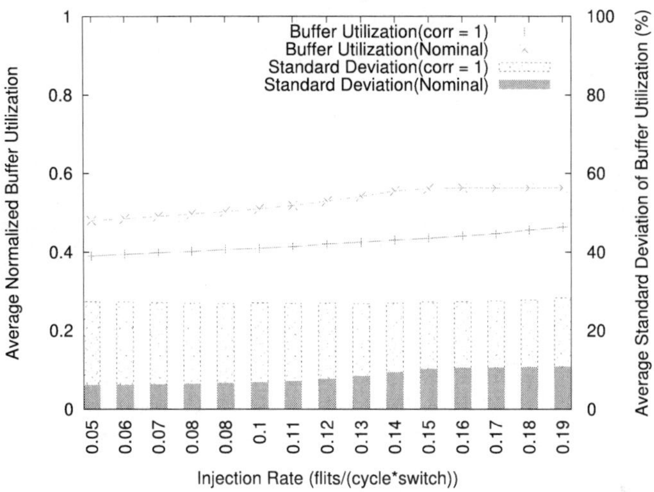

FIGURE 12.27: Buffer utilization in a 64-clock domain CMP.

GALS-based NoC in the presence of variations, there will exist hot spots as a consequence of the nonuniformity of the operating frequency distribution. This behavior causes in the first instance a decrease in the average buffer utilization, as previously shown, and in the last instance a reduction in network performance as shown in Figure 12.25.

12.3.4 Consequences of Variability in NoC-based CMPs

According to the results in previous sections, the traditional synchronous design technique is not feasible anymore because NoC clock frequency should be lowered to match the frequency of the slowest component in the network, noticeably reducing network performance. This fact is widely known by NoC architects. As a solution, GALS systems have been proposed [236]. However,

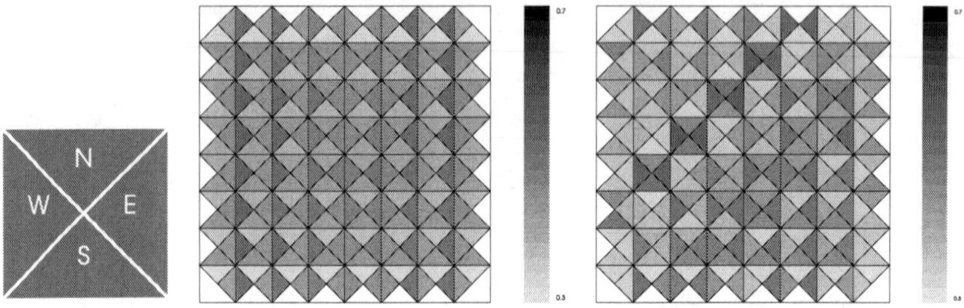

(a) A switch (b) Nominal network (c) Network with PV ($\rho = 1$)

FIGURE 12.28: Buffer utilization in the presence of process variations.

speed-up (%)	4 cores		8 cores		16 cores	
	Mean	Max	Mean	Max	Mean	Max
freqmine	8.1	20	3.2	20.2	-	-
swaptions	8.7	17.1	6.4	18.7	3.9	13.1
blackholes	9	16.8	5.8	17.9	2.8	15.9
canneal	7.8	16.2	5.2	16.2	3.6	14.7

TABLE 12.5: Speed-up achieved using a variability-aware mapping.

note that this is a technological patch that has several architectural consequences. The immediate one is that slower components quickly become bottlenecks, as shown in the previous section. Therefore, in order to avoid these bottlenecks to noticeably impact application performance, the mapping algorithm that selects the cores to be used by an application should be improved by considering variability data. This improvement may be based on two different observations. The first one is that front-end systematic variations cause different delays for different components in the network, but those differences are quite small for adjacent switches and links, due to correlation. Figure 12.24 showed the resulting operating frequency for switches and links after being affected by variability. As could be seen in that figure, neighboring switches and/or links present similar operating frequencies.

The second observation is that many-core chips are usually devoted to several applications simultaneously, by means of virtualization mechanisms. This virtualization layer assigns cores to applications on demand and usually isolates traffic among the applications so that traffic from one application does not traverse a NoC region assigned to another application. When assigning cores to applications, if no variability data is taken into account, we may assign to an application some cores belonging to a fast area of the chip and some other cores belonging to a slow area. In this case, communication among cores using such an assignment would end up running at the frequency of the slowest switch or link because of the bottlenecks mentioned before, causing that the fastest resources are underutilized. On the opposite end, it would be worth

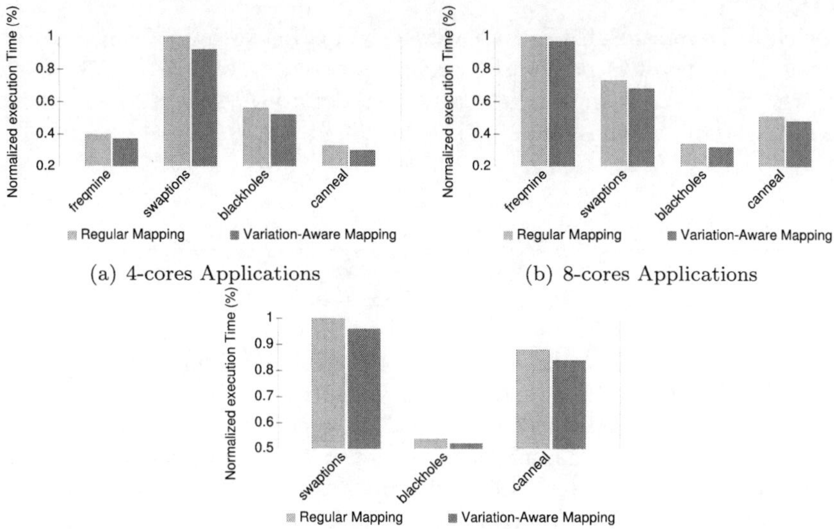

(a) 4-cores Applications

(b) 8-cores Applications

(c) 16-cores Applications

FIGURE 12.29: Average normalized execution time of applications mapped in isolation.

it to assign cores to applications taking into account variability information, thus assigning to a given application cores that are interconnected by links and switches working at similar frequencies. This assignment would not waste faster resources because those resources would not be mixed with slower ones.

In order to characterize the performance of such core assignments, the execution of 4 different applications of the PARSEC benchmark suite [31] on 50 instances of the 8×8 CMP test bench have been simulated. The benchmarks used are freqmine, swaptions, canneal, and blackholes running with 4, 8, and 16 parallel threads.

Figure 12.29 shows the average performance of different mapping policies. Light gray bars represent a regular mapping policy, that is, without taking into account variability data, whereas dark gray bars represent a variation-aware application mapping. The variation-aware mapping strategy selects the most uniform possible region containing the highest frequency core available, when the application requests the mapping. The most uniform region is the region where the frequency variation of switches and links forming it is the lowest.

Table 12.5 shows the speed-up achieved when using the variation-aware assignment policy. For each application, the mean and maximum speed-up values are provided. In the case for 16-core applications, the mapping possibilities are restricted due to the larger size of the regions. Additionally, frequency uniformity of the regions decrease because they span to a larger area of the chip. This translates into a lower speed-up, between 2.8% and 3.9% on average. However, it is possible to achieve a speed-up of 14.9% that occurs when the regular mapping assigns the application to the slower cores and the variability-aware mapping strategy is able to find a uniform fast region.

12.4 Conclusions

On-chip networks have clearly arose as the interconnection choice for multicore chips. Because of their many similarities with off-chip networks, which in some way could be seen as their large counterpart ancestors, on-chip networks have rapidly evolved inheriting many of the techniques and mechanisms previously proposed for the off-chip domain.

However, the different application domains (off-chip vs. on-chip) determine that most of the previously devised mechanisms must be reconsidered and/or evaluated. In this regard, prominent researchers from all over the world met in a workshop organized in 2006 and identified many new challenges for this new scenario. Additional challenges have been also considered by other researchers.

All the new challenges accompanying on-chip networks require a tremendous effort to be done by the research community, either in academia or in industry. Some of those challenges have been already addressed in previous chapters in this book. In this chapter we have selected three of the remaining challenges and have provided a brief insight to them. Namely, traffic characterization in CMPs, congestion management, and process variation.

The traffic characterization carried out in this chapter shows that the on-chip network in a CMP system may not be highly loaded. However, an accurate full-system model is of crucial necessity to really assess the final performance either of a particular on-chip network architecture or a cache coherence protocol. Depending on the network implemented, a particular cache coherence protocol can behave differently. With respect to congestion management in on-chip networks, not much work exists dealing with congested NoCs. However, congestion may arise as a need for new challenges, like DVFS domains and aggressive power-saving techniques.

Finally, the analysis presented in this chapter regarding process variation clearly shows variability is becoming an increasing concern for chip manufacturers. On one hand, if process variation is not faced, this phenomenon may cause an important decrease in yield, as many of the manufactured chips would not reach the required timing constraints. This would negatively affect the price of processor devices, as many of them should be discarded. Nevertheless, even if additional efforts are carried out in order to face process variation, thus not decreasing yield, the performance of manufactured chips will become less predictable, as shown in this chapter.

Acknowledgment

This work was supported by the Spanish MEC and MICINN, as well as European Commission FEDER funds, under Grants CSD2006-00046 and TIN2009-14475-C04. It was also partly supported by the project NaNoC (project label 248972), which is funded by the European Commission within the Research Programme FP7.

Appendix

Switch Models

Davide Bertozzi
University of Ferrara, Italy

Simone Medardoni
University of Ferrara, Italy

Antoni Roca
Technical University of Valencia, Spain

José Flich
Technical University of Valencia, Spain

Federico Silla
Technical University of Valencia, Spain

Francisco Gilabert
Technical University of Valencia, Spain

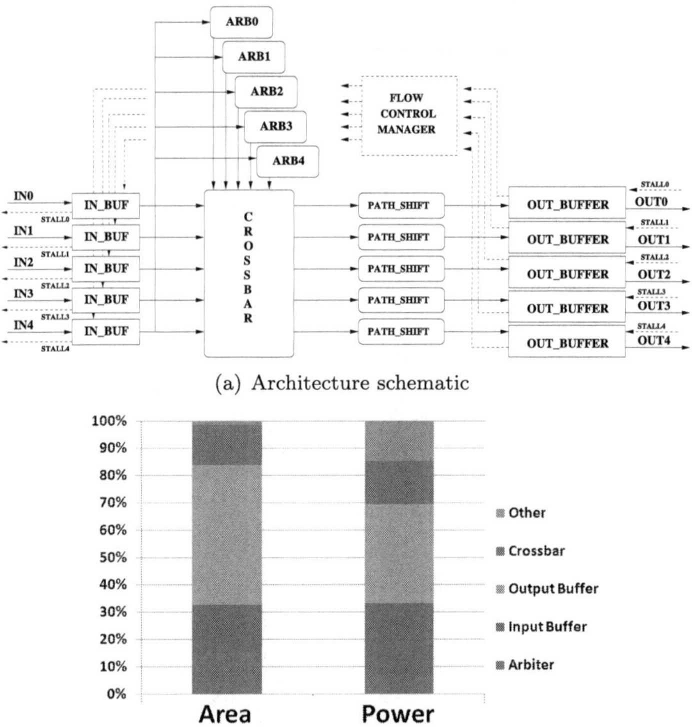

(a) Architecture schematic

(b) Area and power breakdown

FIGURE A.1: Baseline VC-less xpipesLite switch.

A.1 Case Study: The xpipesLite Switch Architecture

The xpipesLite network-on-chip architecture has been conceived for the resource-constrained Multiprocessor system-on-chip (MPSoC) domain, hence features a high degree of parametrization and compact implementation [305]. In contrast to typical switches targeting the Chip Multiprocessor (CMP) domain, the xpipesLite switch is fully synthesizable (no full custom design techniques), is unpipelined and achieves frequencies that peak at around 1.5 GHz for the fastest configurations. In the CMP domain, switches targeting the multi-GHz operating speed are not uncommon.

In this book, the xpipesLite switch has been used as an experimental NoC platform to validate concepts in Chapters 4, 6, and partly in Chapter 5 (for the MPSoC-related part). The switch is conceived as a soft macro from the ground up, and the possibility of design-time tuning of parameters such as flit width, number of I/O ports, buffer size, and flow control policy makes it suitable to instantiate application-specific Networks-on-Chip (NoC) targeting low-end MPSoCs. At the same time, suitability for the regular interconnect fabrics of high-end MPSoCs is preserved.

The baseline architecture of the single-cycle xpipesLite switch is illustrated in Figure A.1(a). Buffers are present both at the input and the output ports, although the switch behaves purely as an input-queued switch. The availability of buffer slots is signaled using a stall&go backpressure protocol [257]. It is a very simple realization of an on/off flow control protocol. It requires two control wires: one going forward and flagging data availability, and one going backward and signaling either a condition of buffer filled (stall) or of buffer free (go). Stall/go can be implemented with distributed buffering along the link; namely, every link pipeline stage can be designed as a two-stage First-In First-Out (FIFO). This is the reference link pipelining implementation considered in Chapter 4.

Buffers are needed both at inputs and outputs in order to break the timing path across switch-to-switch links. Implementation efforts of NoC topology layouts show that in 65 nm technology the delay of interswitch links, due to the increasing role of Resistor-Capacitor (RC) propagation delay [124], causes a significant performance drop for most regular NoC topologies depending on the intricacy of their connectivity pattern [102, 187].

Figure A.1(b) illustrates the area and power breakdown of a 5×5 xpipes-Lite switch synthesized for maximum performance on a 65 nm STMicroelectronics technology library. Due to the maximum speed required, combinational circuits are inferred into area-expensive gate level netlists. Overall, input and output buffers dominate both area and power breakdowns. The critical path of the switch starts from the finite state machine of the switch input buffer, goes through the arbiter, the crossbar selection signals, some header processing logic and finally includes any other clocking overhead (check Figure A.1(a)). Under tight delay constraints the crossbar takes about 15% of the total area, while input and output buffers altogether take about 68%.

Power was measured post-place&route with a 50% switching activity on all input ports. The contribution named "other" coincides almost entirely with the clock tree power. In relative terms, the arbiter's contribution to total power is lower than to total area, while clock tree plus input/output buffers consume approximately 70% of total power. This breakdown reflects typical conditions for wormhole switches in the MPSoC domain (e.g., [18]), while at network level the clock tree typically plays a greater role in determining total power [187].

A.1.1 Extension for Virtual Channel Support

The above baseline switch is augmented with virtual channels by following conventional design techniques. For the sake of simplicity, the focus of this section is restricted to *statically allocated Virtual Channels (VC)* and to *de-*

terministic routing algorithms, and a reference VC switch architecture deeply optimized for this case is derived [101].

The switch input port receives the virtual channel identifier (ID) together with the flit from the upstream switch. This ID is used to select the virtual channel where arriving flits must be stored. Also, a stall signal is generated by each virtual channel and propagated upstream to the attached output port to notify availability of buffer space on a per-VC basis. Each virtual channel implements its own buffering space and a very simple decoding logic that reads the target output port for each packet from the routing field in the header.

Switch allocation is performed immediately after the flit arrives, and the routing information is used to identify the intended switch output port. VCs are assigned nonspeculatively after switch allocation: the winning VC that is granted access to a given output port automatically reserves the VC with the same ID at that output port. This is because VCs are statically allocated. As will be clarified shortly hereafter, it can never occur that a VC is granted access to an output port and the intended VC at that port is occupied.

Switch allocation is performed with a separable input-first allocator. Since allocation requires 2 stages of arbitration we call it the *multistage architecture.* One rule that is enforced during switch allocation is that a flit, either head or body flit, can only win the arbitration in the first stage if it requests an output VC that has free buffer space and is not in use by another input VC. In practice, the first stage arbiter filters the requests for nonfree output virtual channels. This way, it is not possible to waste a cycle by selecting a winner in switch allocation that will find its target virtual channel reserved or with no space. To provide fairness among all the input virtual channels, if the winner of the VC arbiter does not win the port (second-stage) arbitration, it receives the highest priority in the virtual channel arbiter. This guarantees that the last winner will be proposed again as soon as possible.

This architecture limits the amount of virtual channels of each output port that can be allocated in a single cycle to 1. But since the outputs of the crossbar remain equal to the number of input ports even if several output virtual channels of the same output port were assigned in the same cycle, only a single flit from an input VC would be able to reach that output port at each cycle. So there is no performance penalty.

A.1.1.1 Virtual Channel Overhead

By inferring a 5×5 VC-less switch and a corresponding multistage switch architecture in the same 65 nm technology library with Synopsys Physical Compiler, the critical path delay and area results in Table A.1 were obtained. 2 VCs were considered for the multistage architecture.

The multistage realization of a VC switch incurs a delay overhead of 20%, associated with the more complex arbitration. For the sake of fair area comparison, the VC-less switch was resynthesized to match the same delay of the multistage one. This relaxation of delay constraints enabled the logic synthesis

Architecture	Critical Path (ns)	Area (normalized)
VC-less Switch (MAX Perf.)	0.98	0.52
M.Stage VC Switch (MAX Perf.)	1.2	1
VC-less Switch (Relaxed)	1.2	0.43

TABLE A.1: Overhead for VC support.

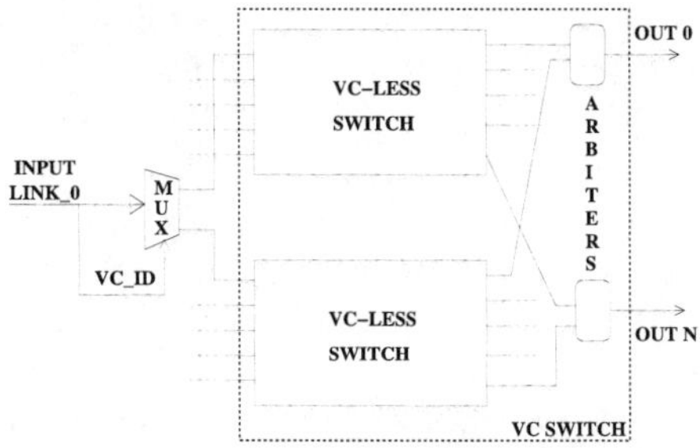

FIGURE A.2: Multiswitch implementation of a VC switch.

tool to infer the same logic functions with a more compact gate-level netlist, in practice moving the design point along the performance-area optimization curve. This is a well-known principle of logic synthesis [73] and therefore holds in general, while the amount of achieved area savings depends on the specific design, on the set of cells available in the technology library, and on the optimization techniques implemented by the synthesis tool at hand, resulting in different combinational logic implementations.

When focusing on the crossbar, the synthesis tool tends to implement it under loose delay constraints by means of a tree of smaller multiplexers that combine more than two inputs at each level of the tree. As the constraint is made tighter, two relevant optimizations are used. On the one hand, an AND-OR tree of gates is used, leveraging specific compound cells. On the other hand, the driving strength is exploited as much as possible to meet the predefined delay target. To the limit, the compound cells are broken down into individual logic cells, with selective optimization of the driving strength.

By looking at area numbers of the relaxed netlist, it can be observed that twice the area of a VC-less switch accounts for only 86% of the area of the multistage switch and that the relaxation resulted in almost 10% area savings with respect to the netlist synthesized for maximum performance. This result suggested the novel VC switch implementation technique hereafter illustrated, and which was categorized in Chapter 2.

A.1.2 VC Switch Implementation by Switch Replication

An alternative VC switch architecture consists of replicating not just buffers per channel, but rather the entire baseline VC-less switch as many times as the intended number of virtual channels. Replicated switches then share the same physical input and output links, similar to what conventional VCs do, but with the main difference that in the new implementation VCs have their own access to a replicated crossbar and the first stage of arbitration can be finally removed, as illustrated in Figure A.2.

This solution will be denoted as the *multiswitch VC implementation.* The underlying principle is simple: Instead of replicating buffering resources inside a switch, the idea is to replicate the baseline VC-less switch without impacting its internal critical path.

Similar to the multistage architecture, also this solution requires an additional stage of *link arbitration* in order to multiplex the outputs of the baseline VC-less switches into the same physical output links connecting to downstream switches. As Figure A.2 indicates, this stage is cascaded to the replicated VC-less switches it arbitrates on a flit-by-flit basis while the arbiters of the replicated switches keep arbitrating at the packet level.

Interestingly, delay of this arbitration stage does not add up to that of the VC-less switches to determine the critical path, since they are separated by a retiming stage (the switch output buffers). In practice, *the critical path of the multiswitch architecture is the same of a VC-less switch*, since it does not make use of a multistage arbiter. However, one might argue that this comes at the cost of replicating more physical resources (e.g., the crossbars).

At this point, a basic principle of logic synthesis comes into play and leads to opposite conclusions. When comparing the multistage with the multiswitch VC implementations, this latter has less functions on the critical path, hence potentially resulting in a more area/power-efficient gate-level netlist after logic synthesis. In fact, the multiswitch architecture certainly provides a higher maximum speed than the multistage one. However, if we require the two architectures to be aligned to the speed of the slowest one (the multistage), then combinational logic of the multiswitch design can be inferred with relaxed delay constraints and therefore thoroughly optimized for area and power. In practice, a different design point along the area-performance optimization curve is inferred. This fact stresses the importance of logic-based optimization of NoC switch components, as illustrated in Chapter 3.

Now, the issue is to determine whether the area savings achieved by logic synthesis are enough to compensate for the larger amount of hardware resources that are instantiated in the multiswitch architecture, especially the replicated crossbars. Please observe that the multistage and the multiswitch architectures can be designed to instantiate the same overall amount of buffering resources: N VC queues in the multistage switch are equivalent to a single queue in N replicated switches.

A.1.2.1 Reassessment of Virtual-channel Overhead

Two 5×5 switches were synthesized, placed, and routed with 2 and 4 virtual channels, while link width was varied from 32 to 64 bits. The designs were

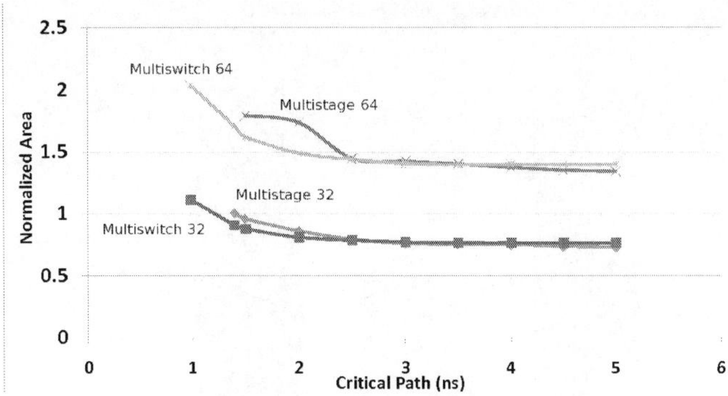

(a) Area comparison: 2 Virtual channels

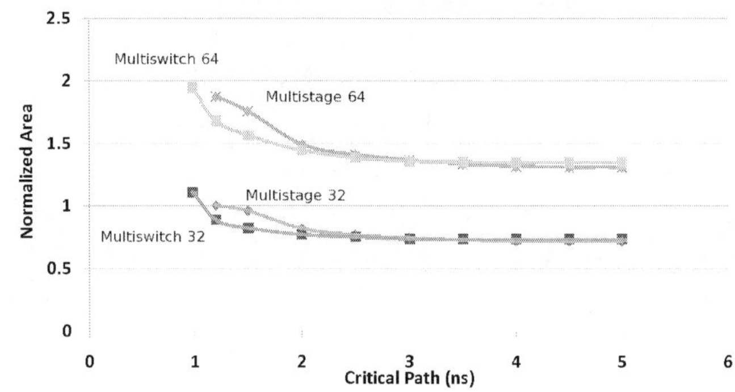

(b) Area comparison: 4 Virtual channels

FIGURE A.3: Area scalability as a function of the target delay constraint.

synthesized at their maximum performance first, then the delay constraint was gradually relaxed, thus getting area/critical path curves. For 2 VCs, such curves are illustrated in Figure A.3(a).

We observe the following:

1. The multiswitch architecture can achieve a higher speed than the multistage one since it implements less control functions on the critical path. Therefore, the physical synthesis tool can reduce the area of this design while relaxing its performance constraint. It is then possible to match the same maximum speed of the multistage architecture, while incurring a lower area, since the area scalability process for the internal combinational logic (e.g., the crossbar) is very effective. Area savings in almost all cases amount to 10%.

2. When operating at a lower speed, both gate-level netlists can be optimized for the relaxed timing constraint and therefore save area. This optimization process saturates around a cycle time of 3 ns for the mul-

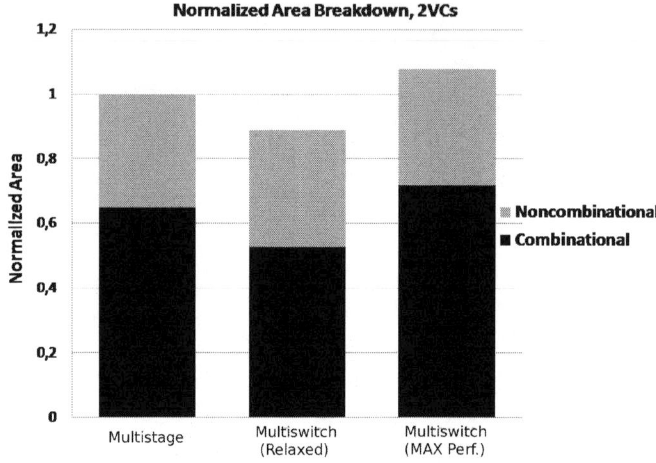

FIGURE A.4: Area breakdown of 32-bit multiswitch (max. and relaxed perf.) and multistage VC implementations.

tiswitch, while the multistage can still be optimized until 4.5 ns. As a consequence, there is a target cycle time (3 ns) beyond which the multistage architecture actually saves area. Apart from the unrealistically low operating speed at the break-even point, the area savings from there on are marginal (maximum of 5% at 200 MHz for 4 VCs).

3. While it is true that replicating a 64-bit crossbar is in principle more expensive, it has to be considered that the multistage architecture employs additional MUX-based logic: one to select one VC per input port and one de-MUX to send the crossbar output to one output VC. In practice, the multistage has a larger overall MUX-based logic. So, by increasing the width of the crossbar data path, the multistage architecture is more impacted. The overall balance is again in favor of the multiswitch solution (area-wise) even at 64 bits.

Figure A.3(b) reports the same area scalability curves for 4 VCs. The trend is very similar to the 2 VCs case.

Let us now better detail how the area optimization process operates in the multiswitch architecture (2nd and 3rd bars in Figure A.4). When the timing constraint is relaxed, area of the noncombinational circuits remains almost unchanged. In contrast, we observe that gate-level netlist transformations during logic synthesis enable a significant reduction of combinational logic area (crossbar, arbiters, multiplexers, buffer control logic). For the crossbar, what actually happens when a lower delay is required is that driving strength of gates is largely increased, and complex logic cells (like 4-input multiplexers) are decomposed into simpler and individually tunable logic gates. Understand-

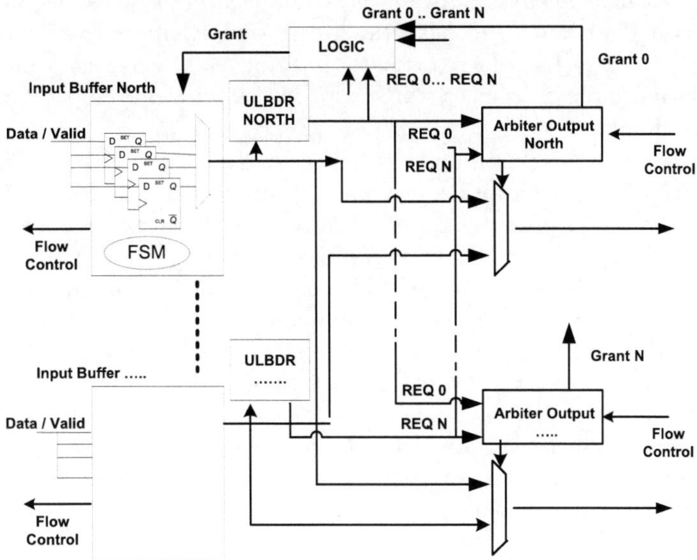

FIGURE A.5: Logic-based distributed routing in the xpipesLite switch.

ing of these concepts is straightforward after reading Chapter 3. When comparing 1st and 2nd bars in Figure A.4, it is clear that the capability to relax performance and optimize area of the multiswitch gate-level netlist enables a higher area efficiency.

By comparing Figure A.3(a) and Figure A.3(b), we also notice that for multiswitch implementations, the maximum performance is always the same regardless the number of virtual channels, while for the multistage case the addition of one virtual channel incurs an 8% degradation of the maximum speed (around 16% for 2VCs and so on). The multiswitch architecture thus avoids the well-known degradation of maximum speed with the number of VCs since it keeps adding resources in parallel without impacting the critical path. This holds until the critical path moves from the switch internal logic to the switch-to-switch link. At that point, multistage and multiswitch architectures would feature the same degradation since they have the same link architecture (including the same link arbitration logic).

A.1.3 Evolution to Distributed Routing

Natively, xpipesLite implements source-based routing: the routing information is embedded into the head flit of a packet by the network interface, which checks the transaction address against a Look-Up-Table (LUT). The length of the routing path field in the header depends on the maximum switch radix and hop count in the network instance at hand. Next, the switch evolution for the support of distributed routing is described, with reference to a logic-based distributed routing (LBDR) strategy as introduced in Chapter 5. To make the description more focused, a xpipesLite switch with only input buffers is hereafter assumed.

In the original switch, the allocator checks the target output port from the head flit of the packet and compares it with its own output port ID, thus generating a *match* signal in case of correspondence. In the modified switch variant, this task is offloaded from the allocator since it is now on the burden of LBDR logic. This time, the head flit contains destination switch coordinates and not routing bits. However, the packet length is not affected, since in any case the reference architecture places this kind of information in the head flit of the packet and payload is never allowed to mix with the header. LBDR logic is illustrated as a single box for each input port in Figure A.5. It is interesting to observe that LBDR keeps the modular design style of the switch architecture. The output signals from the LBDR modules represent exactly the *match* signals that the allocator used to compute by itself before and which indicate that the packet from a given input port requires a specific output port. The allocator now just has to discriminate between competing requests. LBDR logic therefore fits nicely into the xpipesLite switch architecture. Moreover, the used two-phase allocator requires input LBDR modules to activate only one routing option each, therefore for certain routing algorithms the filtering logic described in [94] would be needed. Please note that all LBDR variants compatible with wormhole switching and illustrated in Chapter 5 can be implemented with this architecture template.

A.1.4 Evolution to Virtual Cut-through Switching

This switch can then be evolved to virtual cut-through (VCT) switching to support LBDR with forks as well as multicast communication (Chapter 5). The signals used and the architecture schematic are the same as in Figure A.5, just the meaning of flow control signals and the arbiter behavior change. First of all, the basic stall/go flow control protocol has to be evolved to credit-based flow control. In fact, stall/go would have been acceptable only in case all packets were of the same length, which we find is not appropriate for the MPSoC domain. If packets exhibit variable length (e.g., reads vs. writes, variable number of write/read burst beats, etc.), then the router arbiter needs to know the number of available slots in the downstream buffer before granting a new packet head. Therefore, the $flowcontrol$ signals in Figure A.5 should now be used as $credits$. An input buffer asserts a credit high when it has a grant from the arbiter and it has valid flits to send.

The arbiter behavior has to be modified as well. A port arbiter (say for the N output port) performs round-robin arbitration among all inputs with $valid$ asserted and presenting a $headflit$. Say that input N is the winner. Then, the arbiter compares its counter value (denoting the number of free slots in the

Switch Variant	Area	Cycle Time
LBDR (wormhole)	1	1
LBDR (VCT)	1.12	1.21

TABLE A.2: VCT overhead (normalized to wormhole switching quality metrics).

downstream input buffer) with the packet length from the *Ninputport*. If it is larger, then *grant* is asserted, thus enabling switch traversal to all of the winning packet flits. If there is no space downstream for the entire packet, the grant is kept low.

In LBDR with forks and deroute bits, packets can be forked through two output ports. When this happens, the LBDR logic asserts two match signals heading to two different port arbiters. When *both* of them assert their *grant* signals, a unique *grant* is sent to the requesting input buffer, as illustrated in Figure A.5. One of the packets will reach the destination. The other one will reach a router where the LBDR logic will not provide a valid match signal. In that situation, the *grant* signal is set by default to asserted, thus the packet will be forwarded to the crossbar that is not configured for the input port, thus the packet will be filtered. The input buffer is not aware that no arbitration has been performed for the forked packet, and the *grant* signal is kept asserted. Therefore, it will also correctly generate a *credit* to the upstream router, since buffer slots are cleared. This is the way the misrouted forked packet is silently discarded.

Postsynthesis area and cycle time results for the LBDR switches with wormhole and VCT are showed in Table A.2. All switches implement the same amount of buffering (4 slots per input port) and are synthesized for maximum performance. Clearly, the evolution to VCT comes with a performance drop by 21%. The breakdown of the critical timing path in the VCT switch is also of interest. In both cases the routing function takes around 24% of the total critical path, while the arbitration, input buffer control logic, and crossbar take 35%, 22%, and 19%, respectively. By looking at the area results, the VCT switch variant is more than 10% larger than the wormhole switch, clearly due to the more complex port arbiters and to the need to support full credit-based flow control. To make this relatively more complex circuit faster, the synthesis tool tried to speed up the crossbar at the cost of further increased area.

A.2 gNoC: A Switch Design for CMP Systems

gNoC is a baseline switch design targeted for CMP systems, where high frequencies are usually required. However, it is not the intention of gNoC

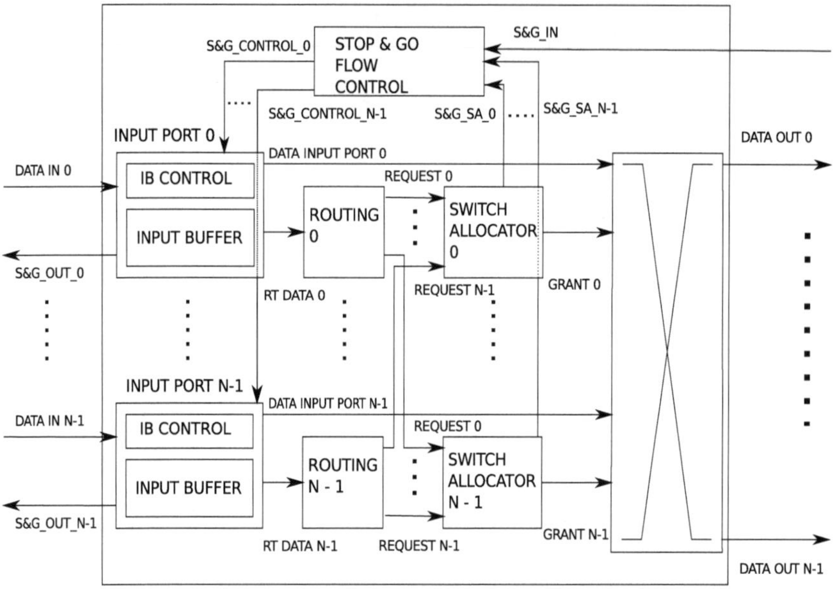

FIGURE A.6: gNoC switch schematic.

to reflect current state-of-the-art switches for CMPs. Instead, gNoC must be seen as a canonical academic implementation of a basic switch design. The real purpose of gNoC is to help in the evaluation of different techniques proposed and presented in the book, mainly routing designs in Chapter 5 and process variation analyses in Chapter 12.

The main features of the switch is its pipelined design and its support for virtual channels. Furthermore, gNoC has been built as a modular switch where the main parameters that define the switch can be tuned at design time. Parameters such as flit width, link width, number of input/output ports, and number of virtual channels, can be set at synthesis time. However, the gNoC switch instantiation used throughout this book is designed with five input and five output ports, so that four ports are intended to provide connectivity with the neighboring switches in a 2-D mesh and the fifth port connects to the local computing core.

The baseline design relies on wormhole switching but virtual cut-through switching has also been implemented to support broadcast (details of the changes required for the support of broadcast communication are described in Chapter 5). The Stop&Go flow control protocol has been deployed in order to control the advance of flits between adjacent switches.

In the following sections we describe the gNoC baseline switch design. Figure A.6 shows the main components of the baseline switch design. In this Figure, gNoC does not implement virtual channels (we later enhance the switch with support for VCs).

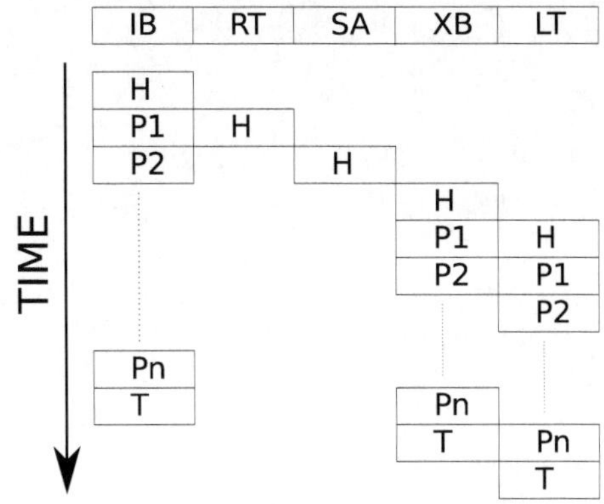

FIGURE A.7: Pipeline stages when forwarding a message through a gNoC switch.

A.2.1 Pipeline Organization

The gNoC switch follows the standard pipelined design with five stages: input buffer (IB), routing (RT), switch allocator (SA), crossbar (XB), and link traversal (LT). Note, however, that as the last stage does not belong to the switch, the gNoC design is actually a four-stage pipelined switch design.

A standard pipelined design as gNoC is a nonoptimized design from a point of view of latency because no stages are performed in parallel and therefore the latency for traversing the pipeline is five cycles. However, a pipelined switch design like gNoC allows each stage to be almost independent from the others, thus allowing for a simpler analysis of the impact each component has over the entire switch. This feature is very interesting from an academic point of view. Figure A.7 shows the pipeline when different flits of a single message are processed by the switch. Notice the first flit is a header flit followed by payload flits of the same message. The header flit crosses all the stages while the remaining flits cross only the IB, XB, and LT stages.

The gNoC implementation used throughout this book implements buffering of flits at the input ports only. In this implementation, the IB stage can store up to four flits at each input port. This number has been chosen in order to guarantee the round-trip time of the link and the delay in the flow control (Stop&Go) mechanism. The link delay is set to one cycle. Therefore, bubbles between flits are avoided and thus, maximum throughput per flow is achieved. Furthermore, notice the IB stage not only stores the incoming flits but also the flits that are in transit across the switch (e.g., flits crossing the switch), which are not removed from the input buffer until they completely leave the switch. Therefore, those flits that are at different stages of the switch such as the RT, SA, or XB stages are kept in the IB buffer, thus being unnecessary to replicate those flits across the switch. For this purpose, control signals distribute the basic control information required by each module. In this way, there is

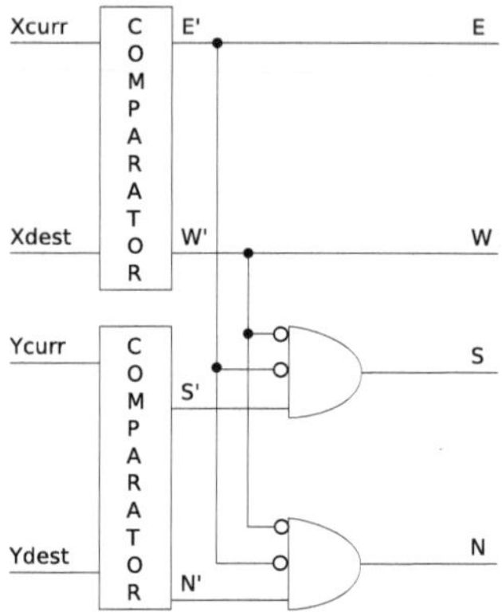

FIGURE A.8: Simple DOR implementation in gNoC switch design.

no need to add pipeline registers within the switch for storing the message flits, but only control pointers. More specifically, a set of control read/write pointers at each input buffer guarantees the proper operation of the switch.

Instead of using memory macros, buffers in the IB stage have been designed using FIFO registers. More specifically, they have been implemented using a Push-In-MUX-Out register [208] as its power consumption is lower than that of a conventional shift register. Implementing a Push-In-MUX-Out register forces the IB stage to implement a write pointer and several read pointers used to store the incoming flits and later retrieve the stored flits. The write and read pointers also make the implementation of the Stop&Go flow control easier, since those pointers reflect the input buffer occupancy, as it will be explained later.

The routing stage (RT) has been designed as a modular block, in order to support, with no major modifications, the several routing algorithms described in Chapter 5. Nevertheless, the Dimension-Order-Routing (DOR) algorithm is the one used in the baseline design. Additionally, distributed routing has been supported in gNoC, instead of using the source-based routing approach. Notice that the RT stage is replicated on every input port. The reason is that a global routing unit for the whole switch would increase the wiring complexity of the switch. Nevertheless, the DOR algorithm (and other options like LBDR) are small enough to allow replication through all the input ports, thus being a more cost-effective option than the increased wiring associated with an implementation relying on a global routing unit.

The RT stage is fed by the flit header of the message that includes its destination (in coordinates of the 2-D mesh). The switch has an internal register with its coordinates and a small logic is used to implement DOR routing. Figure A.8 shows the logic for the DOR routing implementation in the RT stage. The routing proposals from Chapter 5 also rely on current and destination switch coordinates.

The SA (switch allocator) stage is in charge of deciding the flits that get access (through the crossbar) to a given output port. A switch allocator module has been implemented for each output port. The allocators are independent from each other (arbitration decisions from one allocator do not influence another allocator's decision). The SA module receives from each RT module (one per input port) the requests for the output port. The SA module determines which input port will be connected to its own output port. As a result, the SA module generates control signals that program the crossbar and the FIFO of an IB stage. Each SA module has been designed using a round-robin arbiter according to [289]. The decision of designing a round-robin arbiter has been made due to its simplicity, that allows the SA stage to have a low delay [289]. Note that there exist more advanced arbiter designs (see Chapter 3). However in a switch with no virtual channels and with deterministic routing, the round-robin arbiter guarantees maximal matching. If an output port receives at least one request, that output port will grant one of them. Also, an input port will only raise one request.

In [69] there exists a discussion on the convenience of performing a flit-level arbiter or a message-level arbiter. In a flit-level arbiter each flit of a message has to compete for an output port. In contrast, in a message-level arbiter, when the header of a message obtains an output port, the connection remains set for the whole message. Hence, only the header of a message will compete for an output port. In the first case, a fair arbitration can be achieved [69], meaning that all messages of all input ports have the same chances to be selected at every arbitration cycle. As a disadvantage, flits from different messages will share the same buffer at the IB stage in the downstream switch, thus increasing the complexity of the control logic that determines the proper operation of the input buffer. In contrast, a message-level arbiter may lead to unfair situations [69], but allows a simpler arbiter design and an easier IB stage design as flits from different messages are not mixed in the same downstream buffer. In addition, in a message-level arbiter power consumption will be much lower since arbitration is performed message by message rather than flit by flit. gNoC implements a message-level arbiter.

A.2.2 Flow Control

The flow control mechanism is an important part of a switch. The gNoC switch implements a simple Stop&Go flow control. The operation of the Stop&Go algorithm is simple: when the input buffer of the IB stage is going to be full, a Stop signal is sent back to the previous switch, which is injecting flits to this IB. When the IB reaches a level where new flits can be stored, a Go signal is sent back. Notice that both signals can be seen as a physical signal where the value of this physical signal determines the Stop or Go status of an IB. Thus, the IB status—determined by the write/read pointers—is enough to generate the flow control information. However, the Go signal generation and the round-trip time of a link determine the minimum IB size needed in order to not introduce any bubble between data flits. In order to minimize this dependency between the flow control mechanism and the IB size, the gNoC switch generates the Stop&Go signal using the current IB and the SA status. In this way, once the SA determines that a flit is winning an output resource, as this flit will leave the switch in the next cycle, then we can anticipate the IB status, thus anticipating the generation of the Go signal by one cycle.

A.2.3 Virtual Channel Support

Some changes have been introduced to the baseline switch presented above in order to support virtual channels. At each input port the number of input buffers has been incremented, thus providing buffering support for V virtual channels. As all the virtual channels are devoted the same numbers of resources, a static partition of the overall buffer at the input port is feasible, thus simplifying the design. Note that a multiplexer has to be introduced at the input port to map an incoming flit into the proper buffer (associated to the virtual channel the flit is using). In order to reduce the complexity of the control logic associated to each input port, each virtual channel has its own RT module as can be seen in Figure A.9(a). Notice that an alternative could be to have only one RT module for each input port.

The basic functionality of the switch with virtual channel support determines that a message mapped to a virtual channel of an input port can access any virtual channel at any output port. That forces the resource allocator of the switch to be complex (see a complete discussion in Chapters 2 and 3). To deal with this issue, in the gNoC switch, the SA stage is extended to support the virtual channel allocation stage (VA-SA). Indeed, we may think of this stage performing both actions in parallel. The first action deals with assigning a free virtual channel to a header flit that requests a free virtual channel of an output port. The second action deals with each output port to arbitrate among all the virtual channels of all the input ports that request that output port. The way to implement the VA-SA stage has been widely studied (see Chapter 3). gNoC implements a VA-SA stage that is divided into two simple arbiters. First, there is an arbiter at each input port that arbitrates among all the virtual channels of that input port. That functionality has been implemented with a round-robin arbiter identical to the one implemented in the SA stage previously described. After performing this arbitration, only a single header flit from each input port is competing for an output port. The second task of the VA-SA implemented in the gNoC switch design is to arbitrate among all the input ports for an output port resource. Note that this second task is identical to the SA stage described above. As a side effect, by implementing the VA-SA stage in two parts reduces the matching capabilities of the VA-SA stage as only one header flit from an input port may compete for an output port, thus not achieving maximal matching. However, splitting the VA-SA stage into two simple arbitration processes reduces its latency with respect to more complex arbiter designs that perform maximal matching among all the

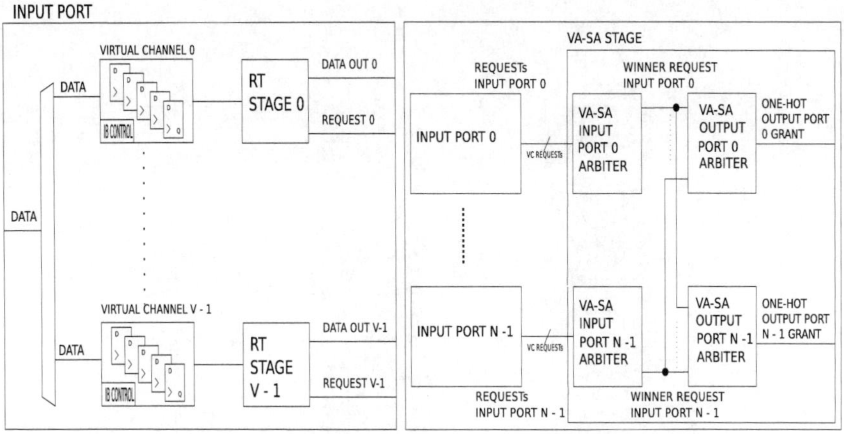

(a) Input port with virtual channels.

(b) VA-SA stage.

FIGURE A.9: gNoC switch support for virtual channels.

module	area (um^2)	critical path (ns)
IB	3113.45	0.55
RT	124.26	0.32
SA	337.88	0.75
XB	1975.6	0.52
gNoC	17651	0.75

TABLE A.3: Area and delay of the switch modules and the complete gNoC switch.

virtual channels of all input ports. An scheme of the VA-SA stage can be seen in Figure A.9(b).

A.2.4 Switch Implementation

Table A.3 shows the area and latency (critical path) of the gNoC switch design without virtual channels and five input and outputs ports. Flit size has been set to 4 bytes. Input buffer size is set to four flits. The switch has been implemented using the 45 nm technology open source Nangate [179] with Synopsys DC. The wire model used is 5K-hvratio-1-1. We have used M1-M4 metallization layers to perform the Place and Route with SoC Encounter.

Notice that the SA stage is the one with the highest latency, thus becoming the bottleneck of the switch. This large latency is due to the complexity of the arbiter (although being a round-robin arbiter) when compared to the other stages, and the management of flow control signaling in the same stage. For the area results note that the IB stage is the stage with major area requirements. The area computed for the IB and RT stages must be replicated for each

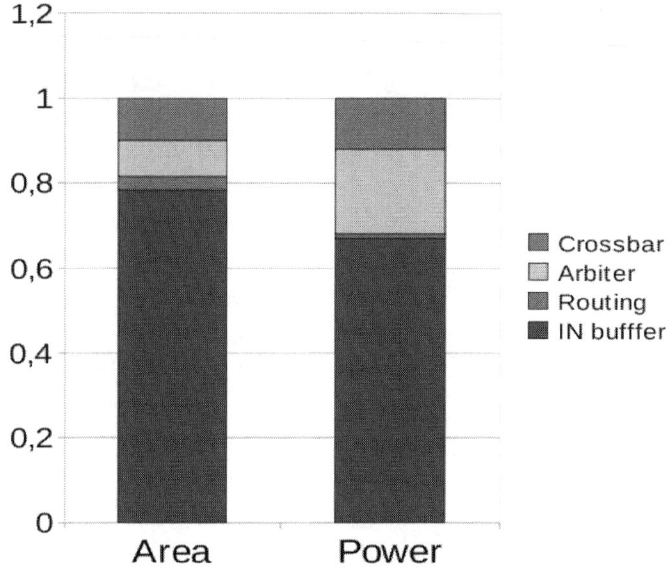

FIGURE A.10: Area and power of the gNoC switch.

input port, while the area computed for the SA stage must be replicated by the number of output ports. The final entry in the table reflects the total area and delay of a 5-port gNoC switch design. Figure A.10 summarizes the normalized area of each stage (computing all the modules) of a gNoC switch. The figure also shows the normalized power consumption of each stage. Note that the IB stage consumes almost 70% of the overall power budget of the switch.

When introducing virtual channels the area and latency of the switch are, obviously, incremented. Each new IB and RT module will be similar to the modules in a switch with no virtual channel. However, these two modules must be replicated by the number of virtual channels, thus increasing the area of the entire switch. Note that the increment in the area of the VA-SA stage with respect to the SA stage, or the area increment due to the increased amount of RT stages, are secondary. That can be inferred from Table A.3, where it can be seen that the RT stage and the SA modules do not highly contribute to the whole switch area. The crossbar remains identical to the case where no virtual channels are used as the policy used in the VA-SA stage is to reduce the requests from an input port down to one (independently from the number of virtual channels implemented). Finally, the latency of the switch is affected by the increased complexity of the VA-SA stage. Remember that the VA-SA stage is divided into two phases. Table A.4 shows the difference in area and latency of the gNoC switch design with no virtual channels and with 5 virtual channels.

gNoC	area (um^2)	critical path (ns)
no VC	17651	0.75
5 VC	89976	1.01

TABLE A.4: Area and delay of the gNoC switch with no VC and five VC.

A.2.5 Virtual Cut-through and Tree-based Broadcast Support

The pipelined wormhole switch described above has been modified to support virtual cut-through (VCT) switching and tree-based broadcast support (see Chapter 5 for a description of the latter). Notice that support for the fork operations described in that chapter is also provided. In order to migrate a wormhole switch to a VCT switch, some changes at different levels must be applied. The most important consideration is that packets must fit into switch buffers, because VCT is leveraged. Thus, the changes required are: (1) packetization at the network interfaces; (2) adjusting buffer size to packet size; (3) changes in flow control, in order to adapt it to the nature of VCT. In addition, to support a tree-based broadcast and fork operations, we need also to: (4) change the arbiter logic to support replicating packets; (5) remove stale copies of packets (being replicated/forked).

The main changes between a wormhole switch with packet-level arbitration and a VCT switch relies on the flow control mechanism, since a VCT switch must guarantee that a packet that is crossing the switch will be completely transmitted without interruption. Furthermore, the changes in the flow control mechanism applied to the VCT switch will incur in modifications in the structure of both the IB and the SA stages. In addition, packetization is performed at the node interfaces when required (thus not covered by this discussion on switch design). Nevertheless, some short explanation on packet size is necessary. Messages in CMPs (using a coherence protocol) can be either short messages containing a coherency command and a memory address or long messages containing a cache line. In both cases, the length of those messages is known. Additionally, the percentage of short messages is usually much larger than that for large ones. Thus, the maximum packet size can be seized to that of short messages. In that case, long and less frequent messages should be packetized.

On the other hand, the Stop&Go flow control protocol implemented in the wormhole switch can be relaxed in order to make it more efficient. Instead of leveraging a flow control at the flit level, the nature of VCT allows implementing it at the packet level. The reason is that when a packet is granted an output port in the VCT, all the flits of that packet will be forwarded. Therefore, a stop or go signal may be asserted per packet. Note that we assume links with one cycle delay, and thus round-trip time is set to three cycles. In this case buffers must be set to four flits to avoid introducing bubbles. However, for messages with sizes shorter than packet size and round-trip time (e.g., one-flit packets), bubbles between packets are generated if the buffer size does not ful-

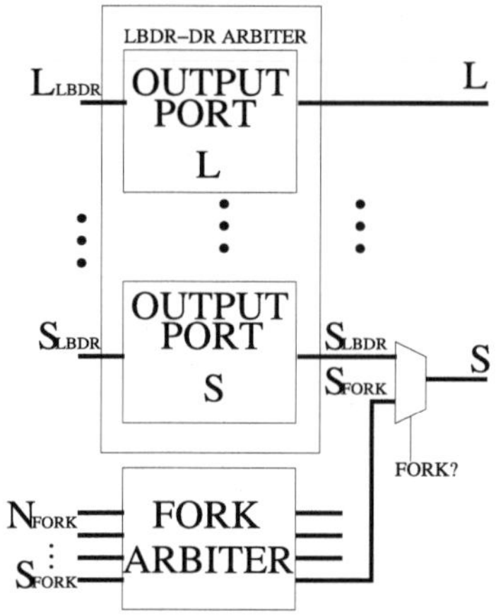

(a) New arbiter for the VCT switch with fork requests.

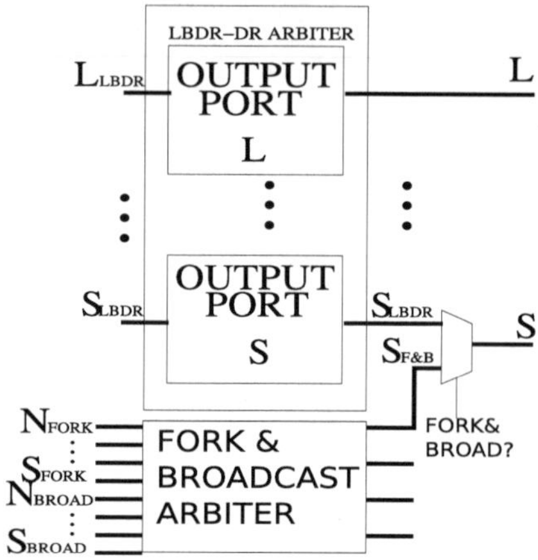

(b) New arbiter for the VCT switch with fork and broadcast requests.

FIGURE A.11: Arbiter for the VCT switch with fork or broadcast requests.

area / Freq	Wormhole	VCT
area ($um2$)	17651	13643
freq (GHz)	1.33	1.54

TABLE A.5: Area and frequency for the wormhole and VCT switches.

fill the round trip-time restriction. To avoid bubbles we decided to pad short packets to the input buffer depth. Obviously, this may affect performance.

Changes in the SA stage should be minimum since the SA stage is the most critical stage in our design (explained previously). To address this concern we have implemented the arbiters shown in Figure A.11. Figure A.11(a) describes the arbiter in charge of unicast fork operations, while Figure A.11(b) shows the arbiter for broadcast operations. Both arbiters follow the same design strategy used in the wormhole switch. Both arbiters add a new module performed in parallel with the unicast arbiter previously described (see Figure A.11). This new module arbitrates between fork and broadcast requests similarly to the unicast requests. The grant signals of this module enable (or disable) the unicast grants. Higher priority is given to fork/broadcast requests. Finally, a multiplexer decides if a unicast request or a broadcast/fork request is winning the crossbar. By doing this, minimum impact on the SA stage latency is expected.

Finally, Table A.5 summarizes the frequency and area of the wormhole and VCT implementations with no virtual channels. Both switches have the same input buffer size and flit width. Note that VCT is faster than the wormhole switch. This is due to the fact that VCT switching allows the switch to relax the Stop&Go signaling because in VCT the Stop&Go flow control is performed per-packet rather than per-flit as in the wormhole switching. Then, relaxing the Stop&Go signal reduces the critical path of the VCT switch. Furthermore, VCT is designed with an IB size equal to the packet size. This design constraint, in addition to the flow control decreasing complexity, allows the IB stage to be simpler than that of the wormhole switch. For that reason, the area of the VCT switch design is smaller. Notice, however, the improvement when moving to VCT is mostly due to the constant packet size assumed when designing the switch. For variable packet sizes, a truly credit-based flow control mechanism would be required, probably incurring the overheads reported for the xpipesLite switch design.

Acknowledgment

This work was supported by the European Union in part through the NaNoC project (project label 248972) and in part through the Galaxy project (project label 214364).

Bibliography

[1] R Development Core Team (2005). R: A language and environment for statistical computing. Available at `http://www.R-project.org`.

[2] C. Ababei, Y. Feng, B. Goplen, H. Mogal, T. Zhang, K. Bazargan, and S. Sapatnekar. Placement and routing in 3-D integrated circuits. *IEEE Design and Test of Computers*, 22(6):520–531, 2005.

[3] P. Abad, V. Puente, and J. A. Gregorio. MRR: Enabling fully adaptive multicast routing for CMP interconnection networks. In *Proceedings of the 15th International Symposium on High-Performance Computer Architecture*, pages 355–366, 2009.

[4] D. Abts and D. Weisser. Age-based packet arbitration in large-radix k-ary n-cubes. In *Proceedings of the ACM/IEEE Conference on Supercomputing*, pages 1–11, New York, NY, USA, 2007. ACM.

[5] A. Adriahantenaina, H. Charlery, A. Greiner, L. Mortiez, and C. A. Zeferino. SPIN: A scalable, packet switched, on-chip micro-network. In *Proceedings of the Design, Automation and Test in Europe Conference*, Washington, DC, USA, 2003. IEEE Computer Society.

[6] V. Agarwal, M. S. Hrishikesh, S. W. Keckler, and D. Burger. Clock rate versus IPC: The end of the road for conventional microarchitectures. In *Proceedings of the 27th International Symposium on Computer Architecture*, pages 248–259, 2000.

[7] F. Akopyan, C. Otero, D. Fang, S. J. Jackson, and R. Manohar. Variability in 3-D integrated circuits. In *Proceedings of the Custom Integrated Circuits Conference*, pages 659–662, September 2008.

[8] A. R. Alameldeen and D. A. Wood. Technical report. Frequent pattern compression: A significance-based compression scheme for l2 caches, University of Wisconsin, Madison, 2004.

[9] M. J. Alexander, J. P. Cohoon, J. L. Colflesh, J. Karro, E. L. Peters, and G. Robins. Placement and routing for three-dimensional FPGAs. In *Proceedings of the Canadian Workshop on Field-Programmable Devices*, pages 11–18, May 1996.

[10] M. Ali, M. Welzl, and S. Hellebrand. A dynamic routing mechanism for network on chip. In *Proceedings of the 23rd NORCHIP Conference*, pages 70–73, 2005.

[11] T. E. Anderson, S. S. Owicki, J. B. Saxe, and C. P. Thacker. High speed switch scheduling for local area networks. *ACM Transactions on Computer Systems*, 11(4):319–352, 1993.

[12] M. Arjomand and H. Sarbazi-Azad. Performance evaluation of butterfly on-chip network for MPSoCs. In *Proceedings of the International SoC Design Conference*, pages 296–299, Washington, DC, USA, 2008. IEEE Computer Society.

[13] A. Asenov, S. Kaya, and J. H. Davies. Intrinsic threshold voltage fluctuations in decanano MOSFETs due to local oxide thickness variations. *IEEE Transactions on Electron Devices*, 49(1):112–119, January 2002.

[14] V. Aslot, M. J. Domeika, R. Eigenmann, G. Gaertner, W. B. Jones, and B. Parady. SPEComp: A new benchmark suite for measuring parallel computer performance. In *Proceedings of the International Workshop on OpenMP Applications and Tools*, pages 1–10, London, UK, 2001. Springer-Verlag.

[15] P. Avasare, V. Nollet, J. Y. Mignolet, D. Verkest, and H. Corporaal. Centralized end-to-end flow control in a best-effort network-on-chip. In *Proceedings of the 5th ACM International Conference on Embedded Software*, pages 17–20, 2005.

[16] M. Azimi, D. Dai, A. Kumar, A. Mejía, D. Park, S. Saharoy, and A. S. Vaidya. Flexible and adaptive on-chip interconnect for tera-scale architectures. *Intel Technology Journal*, 13(4):62–79, 2009.

[17] J. Balfour and W. J. Dally. Design tradeoffs for tiled CMP on-chip networks. In *Proceedings of the 20th Annual International Conference on Supercomputing*, pages 187–198, New York, NY, USA, 2006. ACM.

[18] A. Banerjee, R. Mullins, and S. Moore. A power and energy exploration of network-on-chip architectures. In *Proceedings of the 1st International Symposium on Networks-on-Chip*, pages 163–172, Washington, DC, USA, 2007. IEEE Computer Society.

[19] P. Barford and M. Crovella. Generating representative web workloads for network and server performance evaluation. *Proceedings of the 1998 ACM SIGMETRICS International Conference on Measurement and Modeling of Computer Systems*, 26(1):151–160, 1998.

[20] L. A. Barroso, K. Gharachorloo, R. McNamara, A. Nowatzyk, S. Qadeer, B. Sano, S. Smith, R. Stets, and B. Verghese. Piranha: A scalable architecture based on single-chip multiprocessing. *Proceedings of the*

30th International Symposium on Computer Architecture, pages 12–14, 2000.

[21] E. Baydal and P. López. A robust mechanism for congestion control: INC. In *Proceedings of the Euro-Par Conference*, pages 958–968, 2003.

[22] D. U. Becker and W. J. Dally. Allocator implementations for network-on-chip routers. In *Proceedings of the Conference on High Performance Computing Networking, Storage and Analysis*, pages 1–12, New York, NY, USA, 2009. ACM.

[23] B. M. Beckmann and D. A. Wood. Managing wire delay in large chip-multiprocessor caches. In *Proceedings of the 37th Annual IEEE/ACM International Symposium on MicroArchitecture*, pages 319–330, Washington, DC, USA, 2004. IEEE Computer Society.

[24] E. Beigné, F. Clermidy, P. V. A. Clouard, and M. Renaudin. An asynchronous NoC architecture providing low latency service and its multi-level design framework. In *Proceedings of the International Symposium on Asynchronous Circuits and Systems*, pages 54–63. IEEE, 2005.

[25] E. Beigné and P. Vivet. Design of on-chip and off-chip interfaces for a GALS NoC architecture. In *Proceedings of the International Symposium on Asynchronous Circuits and Systems*, page 172. IEEE, 2006.

[26] L. Benini. Application specific NoC design. In *Proceedings of the Design, Automation and Test in Europe Conference*, pages 491–495, 2006.

[27] D. Bertozzi, G. De Micheli, and L. Benini. Network interface architecture and design issues. In Giovanni De Micheli and Luca Benini, editors, *Networks on Chips: Technology and Tools (Systems on Silicon)*, chapter 6. Morgan Kaufmann Publishers Inc., San Francisco, CA, USA, 2006.

[28] D. Bertozzi, A. Jalabert, S. Murali, R. Tamhankar, S. Stergiou, L. Benini, and G. De Micheli. NoC synthesis flow for customized domain specific multiprocessor systems-on-chip. *IEEE Transactions on Parallel and Distributed Systems*, 16(2):113–129, 2005.

[29] L. N. Bhuyan and D. P. Agrawal. Generalized hypercube and hyperbus structures for a computer network. *IEEE Transactions on Computers*, 33(4):323–333, 1984.

[30] R. Bianchini et al. Alleviating memory contention in matrix computations on large-scale shared-memory multiprocessors. Technical Report 449, Department of Computer Science, Rochester University, 1993.

[31] C. Bienia, S. Kumar, J. P. Singh, and K. Li. The PARSEC benchmark suite: Characterization and architectural implications. In *Proceedings of*

the 17th International Conference on Parallel Architectures and Compilation Techniques, pages 72–81, New York, NY, USA, 2008. ACM.

[32] D. Bing, P. Joseph, M. Bakir, T. Spencer, P. Kohl, and J. Meindl. Wafer-level microfluidic cooling interconnects for GSI. In *Proceedings of the Interconnect Technology Conference*, pages 180–182, 2005.

[33] B. Black, D. W. Nelson, C. Webb, and N. Samra. 3D processing technology and its impact on IA32 microprocessors. In *Proceedings of the IEEE International Conference on Computer Design*, pages 316–318, Washington, DC, USA, 2004. IEEE Computer Society.

[34] C. Bobda, A. Ahmadinia, M. Majer, J. Teich, S. P. Fekete, and J. Van der Veen. DyNoC: A dynamic infrastructure for communication in dynamically reconfigurable devices. *CoRR*, abs/cs/0510039, 2005.

[35] E. Bolotin, I. Cidon, R. Ginosar, and A. Kolodny. Routing table minimization for irregular mesh NoCs. In *Proceedings of the Conference on Design, Automation and Test in Europe*, pages 942–947, San Jose, CA, USA, 2007. EDA Consortium.

[36] R. V. Bopanna and S. Chalasani. Fault-tolerant wormhole routing algorithms for mesh networks. *IEEE Transactions on Computers*, 44(7):848–864, July 1995.

[37] S. Borkar. Thousand core chips: A technology perspective. In *Proceedings of the Design Automation Conference*, pages 746–749. ACM/IEEE, 2007.

[38] S. Borkar. Design perspectives on 22 nm CMOS and beyond. In *Proceedings of the Design Automation Conference*, pages 93–94. ACM/IEEE, 2009.

[39] S. Borkar, R. Cohn, G. Cox, S. Gleason, and T. Gross. iWarp: An integrated solution of high-speed parallel computing. In *Proceedings of the ACM/IEEE Conference on Supercomputing*, pages 330–339, Los Alamitos, CA, USA, 1988. IEEE Computer Society Press.

[40] Y. M. Boura and C. R. Das. Performance analysis of buffering schemes in wormhole routers. *IEEE Transactions on Computers*, 46(6):687–694, 1997.

[41] K. A. Bowman, S. G. Duvall, and J. D. Meindl. Impact of die-to-die and within-die parameter fluctuations on the maximum clock frequency distribution for gigascale integration. *IEEE Journal of Solid-State Circuits*, 37(2):183–190, February 2002.

[42] E. A. Brewer, F. T. Chong, L. T. Liu, S. D. Sharma, and J. D. Kubiatowicz. Remote queues: Exposing message queues for optimization

and atomicity. In *Proceedings of 7th ACM Symposium on Parallel Algorithms and Architectures*, pages 42–53, June 1995.

[43] H. O. Bugge. An evaluation of Intel's core i7 architecture using a comparative approach. *Computer Science—R&D*, 23(3-4):203–209, 2009.

[44] D. Burger, S. W. Keckler, K. S. McKinley, M. Dahlin, L. K. John, C. Lin, C. R. Moore, J. Burrill, R. G. McDonald, W. Yoder, and the TRIPS Team. Scaling to the end of silicon with EDGE architectures. *IEEE Computer*, 37(7):44–55, July 2004.

[45] G. T. Byrd and M. J. Flynn. Producer-consumer communication in distributed shared memory multiprocessors. *Proceedings of the IEEE*, 87(3):456–466, March 1999.

[46] G. Campobello et al. GALS networks on chip: A new solution for asynchronous delay-insensitive links. In *Proceedings of the Design Automation and Test in Europe Conference*, pages 160–165, 2006.

[47] A. P. Chandrakasan et al. Low power CMOS digital design. *IEEE Journal of Solid State Circuits*, 27:473–484, 1992.

[48] T. Chelcea and S. M. Nowick. Robust interfaces for mixed-timing systems. *IEEE Transactions on Very Large Integration Scale Systems*, 12(8):857–873, 2004.

[49] G. Chen and E. G. Friedman. Low-power repeaters driving RC and RLC interconnects with delay and bandwidth constraints. *IEEE Transactions on Very Large Scale Integration Systems*, 14(2):161–172, February 2006.

[50] Y. Chen, A. Bilas, S. N. Damianakis, C. Dubnicki, and K. Li. UTLB: A mechanism for address translation on network interfaces. In *Proceedings of the 8th International Conference on Architectural Support for Programming Languages and Operating Systems*, pages 193–204, New York, NY, USA, 1998. ACM.

[51] L. Cherkasova, V. Kotov, and T. Rokicki. Designing fibre channel fabrics. In *Proceedings of the 1995 International Conference on Computer Design*, page 346, Washington, DC, USA, 1995. IEEE Computer Society.

[52] G. M. Chiu. The odd-even turn model for adaptive routing. *IEEE Transactions on Parallel and Distributed Systems*, 11(7):729–738, July 2000.

[53] F. M. Chiussi and A. Francini. Scalable electronic packet switches. *IEEE Journal on Selected Areas in Communications*, 21(4):486–500, May 2003.

[54] N. Chrysos and G. Dimitrakopoulos. Practical high-throughput crossbar scheduling. *IEEE Micro*, 29:22–35, 2009.

[55] J. Cong and Y. Zhang. Thermal via planning for 3-D ICs. In *Proceedings of the IEEE/ACM International Conference on Computer-Aided Design*, pages 745–752, Washington, DC, USA, 2005. IEEE Computer Society.

[56] Faraday Technology Corp. UMC free library—90nm IPs. Available at http://freelibrary.faraday-tech.com/ips/90library.html.

[57] Intel Corporation. http://www.intel.com.

[58] Transaction Processing Performance Council. TPC-W (Web Commerce) benchmark. Available at `http://cs.nyu.edu/totok/professional/software/tpcw/tpcw.html`.

[59] C. E. Cummings. Simulation and synthesis techniques for asynchronous FIFO design. In *Synopsys Users Group (SNUG)*, 2002.

[60] C. E. Cummings and P. Alfke. Simulation and synthesis techniques for asynchronous FIFO design with asynchronous pointer comparison. In *Synopsys Users Group (SNUG)*, 2002.

[61] D. Dai, A. S. Vaidya, S. Saharoy, S. Park, D. Park, H. L. Thantry, R. Plate, E. Maas, A. Kumar, and M. Azimi. FPGA-based prototyping of a 2D mesh/torus on-chip interconnect. In *Proceedings of the ACM/SIGDA International Symposium on Field-Programmable Gate Arrays (FPGA)*, page 293, February 2010.

[62] W. J. Dally. Express cubes: Improving the performance of k-ary n-cube interconnection networks. *IEEE Transactions on Computers*, 40(9):1016–1023, September 1991.

[63] W. J. Dally. Virtual-channel flow control. *IEEE Transactions on Parallel and Distributed Systems*, 3(3):194–205, March 1992.

[64] W. J. Dally and H. Aoki. Deadlock-free adaptive routing in multicomputer networks using virtual channels. *IEEE Transactions on Parallel and Distributed Systems*, 4(4):466–475, 1993.

[65] W. J. Dally, J. A. S. Fiske, J. S. Keen, R. A. Lethin, M. D. Noakes, P. R. Nuth, R. E. Davison, and G. A. Fyler. The message-driven processor: A multicomputer processing node with efficient mechanisms. *IEEE Micro*, 12(2):23–39, 1992.

[66] W. J. Dally and C. L. Seitz. The torus routing chip. *Journal of Parallel and Distributed Computing*, 1(3):187–196, 1986.

[67] W. J. Dally and C. L. Seitz. Deadlock free message routing in multiprocessor interconnection networks. *IEEE Transactions on Computers*, 36(5):547–553, May 1987.

[68] W. J. Dally and B. Towles. Route packets, not wires: On-chip interconnection networks. In *Proceedings of the 38th Design Automation Conference*, pages 684–689, New York, NY, USA, 2001. ACM.

[69] W. J. Dally and B. Towles. *Principles and Practices of Interconnection Networks.* Morgan Kaufmann Publishers Inc., San Francisco, CA, USA, 2003.

[70] R. Das, O. Mutlu, T. Moscibroda, and C. R. Das. Application-aware prioritization mechanisms for on-chip networks. In *Proceedings of the Annual IEEE/ACM International Symposium on Microarchitecture*, pages 280–291, December 2009.

[71] S. Das, A. Fan, K. Chen, C. S. Tan, N. Checka, and R. Reif. Technology, performance, and computer-aided design of three-dimensional integrated circuits. In *Proceedings of the International Symposium on Physical design*, pages 108–115, New York, NY, USA, 2004. ACM.

[72] W. R. Davis, J. Wilson, S. Mick, J. Xu, H. Hua, C. Mineo, A. M. Sule, M. Steer, and P. D. Franzon. Demystifying 3D ICs: The pros and cons of going vertical. *IEEE Design & Test of Computers*, 22(6):498–510, 2005.

[73] G. De Micheli. *Synthesis and Optimization of Digital Circuits.* McGraw-Hill, 1994.

[74] M. Dehyadgari, M. Nickray, A. Afzali-Kusha, and Z. Navabi. Evaluation of pseudo adaptive XY routing using an object oriented model for NoC. In *Proceedings of the 17th International Conference on Microelectronics*, 2005.

[75] G. Della Veccia and C. Sanges. Recursively scalable networks for message passing architectures. In E. Chiricozzi and A. D'Amico, editors, *Parallel Processing and Applications*, pages 33–40. Elsevier Science Publishers, North-Holland, 1987.

[76] A. Demers et al. Epidemic algorithms for replicated database maintenance. In *Proceedings of the Symposium on Principles of Distributed Computing*, pages 1–12. ACM, 1987.

[77] R. Dick. Embedded System Synthesis Benchmarks Suites (E3S): `http://ziyang.eecs.umich.edu/~dickrp/e3s/`.

[78] S. Dighe, S. Vangal, P. Aseron, S. Kumar, T. Jacob, K. Bowman, J. Howard, J. Tschanz, V. Erraguntla, N. Borkar, V. De, and S. Borkar. Within-die variation-aware dynamic-voltage-frequency scaling core mapping and thread hopping for an 80-core processor. In *Proceedings of the IEEE International Solid-State Circuits Conference*, pages 174–175, February 2010.

[79] G. Dimitrakopoulos. Multiple port arbiter. Technical report, Institute of Computer Science (ICS), Foundation for Reseach and Technology–Hellas (FORTH), 2010.

[80] G. Dimitrakopoulos, N. Chrysos, and K. Galanopoulos. Fast arbiters for on-chip network switches. In *Proceedings of the International Conference on Computer Design*, pages 644–670. IEEE, October 2008.

[81] G. Dimitrakopoulos, K. Galanopoulos, C. Mavrokefalidis, and D. Nikolos. Low-power leading-zero counting and anticipation logic for high-speed floating point units. *IEEE Transactions on Very Large Scale Integration Systems*, 16(7):837–850, 2008.

[82] J. Duato. A new theory of deadlock-free adaptive routing in wormhole networks. *IEEE Transactions on Parallel Distributed Systems*, 4(12):1320–1331, 1993.

[83] J. Duato. A theory of fault-tolerant routing in wormhole networks. In *International Conference on Parallel and Distributed Systems*, pages 600–607, 1994.

[84] J. Duato. A necessary and sufficient condition for deadlock-free adaptive routing in wormhole networks. *IEEE Transactions on Parallel and Distributed Systems*, 6(10):1055–1067, 1995.

[85] J. Duato, S. Yalamanchili, and Ni. L. M. *Interconnection Networks: An Engineering Approach.* Morgan Kaufmann Publishers Inc., San Francisco, CA, USA, 2002.

[86] M. Dubois, C. Scheurich, and F. Briggs. Memory access buffering in multiprocessors. In *ISCA '98: 25 Years of the International Symposia on Computer Architecture (Selected Papers)*, pages 320–328, New York, NY, USA, 1998. ACM.

[87] T. Dumitras, S. Kerner, and R. Marculescu. Towards on-chip fault-tolerant communication. In *Proceedings of the Asia and South Pacific Design Automation Conference*, pages 225–232, New York, NY, USA, 2003. ACM.

[88] N. D. Enricht Jerger, L. Peh, and M. H. Lipasti. Virtual circuit tree multicasting: A case for on-chip hardware multicast support. In *Proceedings of the 35th International Symposium on Computer Architecture*, pages 229–240, 2008.

[89] N. D. Enricht Jerger, L. Peh, and M. H. Lipasti. Virtual tree coherence: Leveraging regions and in-network multicast trees for scalable cache coherence. In *Proceedings of the 41st Annual IEEE/ACM International Symposium on Microarchitecture*, pages 35–46, Washington, DC, USA, 2008. IEEE Computer Society.

[90] B. S. Feero and P. P. Pande. Networks-on-chip in a three-dimensional environment: A performance evaluation. *IEEE Transactions on Computers*, 58(1):32–45, January 2009.

[91] E. Flamand. Strategic directions towards multicore application specific computing. In *Proceedings of the Design Automation and Test in Europe Conference*, page 1266, 2009.

[92] J. Flich, M. P. Malumbres, P. López, and J. Duato. Performance evaluation of a new routing strategy for irregular networks with source routing. In *Proceedings of the 14th International Conference on Supercomputing*, pages 34–43, New York, NY, USA, 2000. ACM.

[93] J. Flich, A. Mejía, P. López, and J. Duato. Region-based routing: An efficient routing mechanism to tackle unreliable hardware in network on chips. In *Proceedings of the First International Symposium on Networks-on-Chip*, pages 183–194, Washington, DC, USA, 2007. IEEE Computer Society.

[94] J. Flich, S. Rodrigo, and J. Duato. An efficient implementation of distributed routing algorithms for NoCs. In *Proceedings of the 2nd ACM/IEEE International Symposium on Networks-on-Chip*, pages 87–96, Washington, DC, USA, 2008. IEEE Computer Society.

[95] International Technology Roadmap for Semiconductors. ITRS 2007 Online Edition. Available at `http://www.itrs.net/Links/2007ITRS/Home2007.htm`.

[96] M. Galles. Scalable pipelined interconnect for distributed endpoint routing: The SGI SPIDER chip. In *Proceedings of the Hot Interconnects*, 1996.

[97] M. Galles. Spider: A high-speed network interconnect. *IEEE Micro*, 17(1):34–39, January/February 1997.

[98] A. Gara, M. A. Blumrich, D. Chen, G. L.-T. Chiu, P. Coteus, M. E. Giampapa, R. A. Haring, P. Heidelberger, D. Hoenicke, G. V. Kopcsay, T. A. Liebsch, M. Ohmacht, B. D. Steinmacher-Burow, T. Takken, and P. Vranas. Overview of the Blue Gene/L system architecture. *IBM J. Res. Dev.*, 49(2):195–212, 2005.

[99] P. J. García, F. J. Quiles, J. Flich, J. Duato, I. Johnson, and F. Naven. Efficient, scalable congestion management for interconnection networks. *IEEE Micro*, 26(5):52–66, 2006.

[100] M. Ghoneima, Y. Ismail, M. Khellah, and V. De. Variation-tolerant and low-power source-synchronous multicycle on-chip interconnection scheme. *VLSI Design*, 2007:1–12, 2007.

[101] F. Gilabert, M. E. Gómez, S. Medardoni, and D. Bertozzi. Improved utilization of NoC channel bandwidth by switch replication for cost-effective multi-processor systems-on-chip. In *Proceedings of the 4th ACM/IEEE International Symposium on Networks-on-Chip*, pages 165–172, Washington, DC, USA, 2010. IEEE Computer Society.

[102] F. Gilabert, D. Ludovici, S. Medardoni, D. Bertozzi, L. Benini, and G. N. Gaydadjiev. Designing regular network-on-chip topologies under technology, architecture and software constraints. In *Proceedings of the International Conference on Complex, Intelligent and Software Intensive Systems*, pages 681–687, 2009.

[103] F. Gilabert, S. Medardoni, D. Bertozzi, L. Benini, M. E. Gómez, P. López, and J. Duato. Exploring high-dimensional topologies for NoC design through an integrated analysis and synthesis framework. In *Proceedings of the 2nd ACM/IEEE International Symposium on Networks-on-Chip*, pages 107–116, Washington, DC, USA, 2008. IEEE Computer Society.

[104] R. Ginosar. Fourteen ways to fool your synchronizer. In *Proceedings of the International Symposium on Asynchronous Circuits and Systems*, pages 89–97, 2003.

[105] C. J. Glass and L. M. Ni. The turn model for adaptive routing. *Journal of the ACM*, 41(5):874–902, 1994.

[106] M. E. Gómez, P. López, and J. Duato. A memory-effective routing strategy for regular interconnection networks. In *Proceedings of the 19th IEEE International Parallel and Distributed Processing Symposium*, page 41.2, Washington, DC, USA, 2005. IEEE Computer Society.

[107] M. E. Gómez, N. A. Nordbotten, J. Flich, P. López, A. Robles, J. Duato, T. Skeie, and O. Lysne. A routing methodology for achieving fault tolerance in direct networks. *IEEE Transactions on Computers*, 55(4):400–415, 2006.

[108] B. Goplen and S. Sapatnekar. Efficient thermal placement of standard cells in 3D ICs using a force directed approach. In *Proceedings of the IEEE/ACM International Conference on Computer-Aided Design*, page 86, Washington, DC, USA, 2003. IEEE Computer Society.

[109] P. Gratz, B. Grot, and S. W. Keckler. Regional congestion awareness for load balance in networks-on-chip. In *Proceedings of the 14th International Conference on High-Performance Computer Architecture*, pages 203–214, 2008.

[110] P. Gratz, C. Kim, R. McDonald, S. W. Keckler, and D. Burger. Implementation and evaluation of on-chip network architectures. In *Proceedings of the IEEE International Conference on Computer Design*, 2006.

[111] P. Gratz, K. Sankaralingam, H. Hanson, P. Shivakumar, R. McDonald, S. W. Keckler, and D. Burger. Implementation and evaluation of a dynamically routed processor operand network. In *Proceedings of the 1st IEEE International Symposium on Networks-on-Chips*, pages 7–17, 2007.

[112] F. Gray. Pulse code communication. U.S. Patent 2632058, 1953.

[113] P. Gupta and N. McKeown. Designing and implementing a fast crossbar scheduler. *IEEE Micro*, 19(1):20–28, 1999.

[114] N. Hardavellas, M. Ferdman, B. Falsafi, and A. Ailamaki. Reactive NUCA: near-optimal block placement and replication in distributed caches. In *Proceedings of the 36th Annual International Symposium on Computer Architecture*, pages 184–195, New York, NY, USA, 2009. ACM.

[115] B. Hargreaves, H. Hult, and S. Reda. Within-die process variations: How accurately can they be statistically modeled? In *Proceedings of the Design Automation Conference*, pages 524–530, March 2008.

[116] J. Heinlein, K. Gharachorloo, S. Dresser, and A. Gupta. Integration of message passing and shared memory in the stanford FLASH multiprocessor. *SIGOPS Operating Systems Review*, 28(5):38–50, 1994.

[117] J. Held, J. Bautista, and S. Koehl. A few cores to many: A tera-scale computing research overview. Technical report, Intel Corporation, Santa Clara, CA, 2006.

[118] J. L. Henning. SPEC CPU2000: Measuring CPU performance in the new millennium. *IEEE Computer*, 33(7):28–35, 2000.

[119] S. Herbert and D. Marculescu. Analysis of dynamic voltage/frequency scaling in chip multiprocessors. In *Proceedings of the International Symposium on Low Power Electronics and Design*, pages 38–43, 2007.

[120] C. Hernández, A. Roca, F. Silla, J. Flich, and J. Duato. Improving the performance of GALS-based NoCs in the presence of process variation. In *Proceedings of 4th International Symposium on Networks-on-Chip*, May 2010.

[121] C. Hernández, F. Silla, and J. Duato. A methodology for the characterization of process variation in NoC links. In *Proceedings of the Design, Automation and Test in Europe Conference*, pages 685–690, March 2010.

[122] G. Hinton, D. Sager, M. Upton, D. Boggs, D. Carmean, A. Kyker, and P. Roussel. The microarchitecture of the Pentium 4 processor. *Intel Technology Journal*, 5(1):1–13, February 2001.

[123] H. Hluchyj and M. Karol. Queueing in high-performance packet switching. *IEEE Journal on Selected Areas in Communications*, 6(9):1587–1597, December 1988.

[124] R. Ho and M. A. Horowitz. *On-Chip Wires: Scaling and Efficiency.* PhD thesis, Stanford University, 2003.

[125] W. S. Ho and D. L. Eager. A novel strategy for controlling hot spot contention. In *Proceedings of the International Conference on Parallel Processing*, pages 14–18, 1989.

[126] Y. Hoskote, S. Vangal, A. Singh, N. Borkar, and S. Borkar. A 5-ghz mesh interconnect for a teraflops processor. *IEEE Micro*, 27(5):51–61, 2007.

[127] J. Howard, S. Dighe, Y. Hoskote, S. Vangal, S. Finan, G. Ruhl, D. Jenkins, H. Wilson, N. Borka, G. Schrom, F. Pailet, S. Jain, T. Jacob, S. Yada, S. Marella, P. Salihundam, V. Erraguntla, M. Konow, M. Riepen, G. Droege, J. Lindemann, M. Gries, T. Apel, K. Henriss, T. Lund-Larsen, S. Steibl, S. Borkar, V. De, R. Van Der Wijngaart, and T. Mattson. A 48-core IA-32 message-passing processor with DVFS in 45nm CMOS. In *International Solid-State Circuits Conference*, pages 58–59, 2010.

[128] M. S. Hrishikesh, D. Burger, N. P. Jouppi, S. W. Keckler, K. I. Farkas, and P. Shivakumar. The optimal logic depth per pipeline stage is 6 to 8 FO4 inverter delays. In *Proceedings of the 29th Annual International Symposium on Computer Architecture*, pages 14–24, 2002.

[129] J. Hu and R. Marculescu. DyAD—Smart routing for networks-on-chip. In *Proceedings of the Design Automation Conference*, 2004.

[130] J. Hu and R. Marculescu. Communication and task scheduling of application-specific networks-on-chip. In *Proceedings on Computers and Digital Techniques*, pages 643–651. IEE, 2005.

[131] J. Huh, C. Kim, H. Shafi, L. Zhang, D. Burger, and S. W. Keckler. A NUCA substrate for flexible CMP cache sharing. *IEEE Transactions on Parallel and Distributed Systems*, 18:1028–1040, 2007.

[132] E. Humenay, D. Tarjan, and K. Skadron. Impact of process variations on multicore performance symmetry. In *Proceedings of the Design, Automation and Test in Europe Conference*, pages 1–6, April 2007.

[133] J. Hurt, A. May, X. Zhu, and B. Lin. Design and implementation of high-speed symmetric crossbar schedulers. In *Proceedings of the International Conference on Communication*, pages 253–258, June 1999.

[134] Synopsys Inc. TSMC manuals.

[135] T. N. K. Jain, P. V. Gratz, A. Sprintson, and G. Choi. Asynchronous bypass channels: Improving performance for multi-synchronous NoCs. In *Proceedings of the 4th ACM/IEEE International Symposium on Networks-on-Chip*, pages 51–58, 2010.

[136] J. W. Joyner, P. Zarkesh-Ha, , and J. D. Meindl. A stochastic global net-length distribution for a three-dimensional system-on-a-chip (3D-SoC). In *Proceedings of the International ASIC/SOC Conference*, 2001.

[137] J. W. Joyner, P. Zarkesh-Ha, and J. D. Meindl. Global interconnect design in a three-dimensional system-on-a-chip. *IEEE Transactions on Very Large Scale Integration Systems*, 12(4):367–372, April 2004.

[138] J. A. Kahle, M. N. Day, H. P. Hofstee, C. R. Johns, T. R. Maeurer, and D. Shippy. Introduction to the Cell multiprocessor. *IBM Journal of Research & Development.*, 49(4/5):589–604, 2005.

[139] M.R. Kakoee, I. Loi, and L. Benini. A new physical routing approach for robust bundled signaling on NoC links. In *Proceedings of the 20th Great Lakes Symposium on VLSI*, pages 3–8. ACM, 2010.

[140] R. Kalla, B. Sinharoy, W. Starke, and M. Floyd. POWER7TM: IBM's next generation server processor. *IEEE Micro*, 99(PrePrints), 2010.

[141] G. Kalokerinos, V. Papaefstathiou, G. Nikiforos, S. Kavadias, M. Katevenis, D. Pnevmatikatos, and X. Yang. FPGA implementation of a configurable cache/scratchpad memory with virtualized user-level RDMA capability. *Proceedings of IEEE International Conference on Embedded Computer Systems: Architectures, Modeling, and Simulation*, July 2009.

[142] U. Kapasi, W. J. Dally, S. Rixner, J. D. Owens, and B. Khailany. The Imagine Stream Processor. In *Proceedings of the IEEE International Conference on Computer Design*, pages 282–288, September 2002.

[143] M. Karol, M. Hluchyj, and S. Morgan. Input versus output queueing on a space-division packet switch. *IEEE Transactions on Communications*, COM-35(12):1374–1356, December 1987.

[144] M. Katevenis, V. Papaefstathiou, S. Kavadias, D. Pnevmatikatos, F. Silla, and D. S. Nikolopoulos. Explicit communication and synchronization in SARC. *IEEE Micro*, 2010. Accepted for publication.

[145] M. Katevenis, D. Serpanos, and E. Spyridakis. Credit-flow-controlled ATM for MP Interconnection: The ATLAS I Single-chip ATM switch. In *Proceedings of the 4th International Symposium on High-Performance Computer Architecture*, page 47, Washington, DC, USA, 1998. IEEE Computer Society.

[146] M. Katevenis, P. Vatsolaki, and A. Efthymiou. Pipelined memory shared buffer for VLSI switches. In *Proceedings of the Conference on Applications, Technologies, Architectures, and Protocols for Computer Communication*, pages 39–48, New York, NY, USA, 1995. ACM.

[147] S. Kavadias, M. G. H. Katevenis, M. Zampetakis, and D. S. Nikolopoulos. On-chip communication and synchronization with cache-integrated network interfaces. In *Proceedings of the 7th ACM Conference on Computing Frontiers*, New York, NY, USA, May 2010. ACM.

[148] S. W. Keckler, W. J. Dally, D. Maskit, N. P. Carter, A. Chang, and W. S. Lee. Exploiting fine-grain thread level parallelism on the MIT Multi-ALU processor. In *Proceedings of the 25th International Symposium on Computer Architecture*, pages 306–317, 1998.

[149] C. Kenyon, A. Kornfeld, K. Kuhn, M. Liu, A. Maheshwari, W. Shih, S. Sivakumar, G. Taylor, P. VanDerVoorn, and K. Zawadzki. Managing process variation in Intel's 45nm CMOS technology. *Intel Technology Journal. http://www.intel.com/technology/itj/2008/v12i2/3-managing/1-abstract.htm*, June 2008.

[150] P. Kermani and L. Kleinrock. Virtual cut-through: a new computer communication switching technique. *Computer Networks*, 3(4):276–286, 1979.

[151] R. E. Kessler. The alpha 21264 microprocessor. *IEEE Micro*, 19(2):24–36, 1999.

[152] T. Kgil, S. D'Souza, A. Saidi, N. Binkert, R. Dreslinski, T. Mudge, S. Reinhardt, and K. Flautner. Picoserver: using 3D stacking technology to enable a compact energy efficient chip multiprocessor. *ACM SIGPLAN Notices*, 41(11):117–128, 2006.

[153] C. Kim, D. Burger, and S. W. Keckler. An adaptive, non-uniform cache structure for wire-delay dominated on-chip caches. *ACM SIGPLAN Notices*, 37(10):211–222, 2002.

[154] J. Kim. Low-cost router microarchitecture for on-chip networks. In *Proceedings of the 42nd Annual IEEE/ACM International Symposium on Microarchitecture*, pages 255–266, New York, NY, USA, 2009. ACM.

[155] J. Kim, J. Balfour, and W. J. Dally. Flattened butterfly topology for on-chip networks. In *Proceedings of the 40th Annual IEEE/ACM International Symposium on Microarchitecture*, pages 172–182, Washington, DC, USA, 2007. IEEE Computer Society.

[156] J. Kim, W. J. Dally, B. Towles, and A. K. Gupta. Microarchitecture of a high-radix router. In *Proceedings of the 32nd Annual International Symposium on Computer Architecture*, pages 420–431, Washington, DC, USA, 2005. IEEE Computer Society.

[157] J. Kim, C. Nicopoulos, D. Park, R. Das, Y. Xie, V. Narayanan, M. S. Yousif, and C. R. Das. A novel dimensionally-decomposed router for on-chip communication in 3D architectures. In *Proceedings of the 34th Annual International Symposium on Computer Architecture*, ISCA '07, pages 138–149, New York, NY, USA, 2007. ACM.

[158] J. Kim, C. A. Nicopoulos, and D. Park. A gracefully degrading and energy-efficient modular router architecture for on-chip networks. In *Proceedings of the 33rd Annual International Symposium on Computer Architecture*, pages 4–15, Washington, DC, USA, 2006. IEEE Computer Society.

[159] J. Kim, C. A. Nicopoulos, D. Park, R. Das, Y. Xie, M. S. Narayanan, V.and Yousif, and C. R. Das. A novel dimensionally-decomposed router for on-chip communication in 3D architectures. *SIGARCH Computer Architecture News*, 35(2):138–149, 2007.

[160] J. Kim, D. Park, T. Theocharides, N. Vijaykrishnan, and C. R. Das. A low latency router supporting adaptivity for on-chip interconnects. In *Proceedings of the 42nd Annual Design Automation Conference*, pages 559–564, New York, NY, USA, 2005. ACM.

[161] J. H. Kim, Z. Liu, and A. A. Chien. Compressionless routing: A framework for adaptive and fault-tolerant routing. *IEEE Transactions on Parallel and Distributed Systems*, 8(3):229–244, March 1997.

[162] T. Kim and T. Kim. Clock tree embedding for 3D ICs. In *Proceedings of the 15th Asia and South Pacific Design Automation Conference*, pages 486–491, January 2010.

[163] A. J. KleinOsowski and D. J. Lilja. MinneSPEC: A new SPEC benchmark workload for simulation-based computer architecture research. *Computer Architecture Letters*, 1, 2002.

[164] A. Kodi, A. Sarathy, and A. Louri. Design of adaptive communication channel buffers for low-power area-efficient network-on-chip architecture. In *Proceedings of the 3rd ACM/IEEE Symposium on Architecture for Networking and Communications Systems*, pages 47–56, New York, NY, USA, 2007. ACM.

[165] A. Kohler and M. Radetzki. Fault-tolerant architecture and deflection routing for degradable NoC switches. In *Proceedings of the 3rd ACM/IEEE International Symposium on Networks-on-Chip*, pages 22–31, Washington, DC, USA, 2009. IEEE Computer Society.

[166] M. Koibuchi, A. Funahashi, A. Jouraku, and H. Amano. L-turn routing: An adaptive routing in irregular networks. In *Proceedings of the International Conference on Parallel Processing*, pages 383–392, Washington, DC, USA, 2001. IEEE Computer Society.

[167] M. Koibuchi, A. Jouraku, K. Watanabe, and H. Amano. Descending layers routing: A deadlock-free deterministic routing using virtual channels in system area networks with irregular topologies. In *Proceedings of the International Conference on Parallel Processing*, pages 527–536, Los Alamitos, CA, USA, 2003. IEEE Computer Society.

[168] M. Koibuchi, H. Matsutani, H. Amano, and T. M. Pinkston. A lightweight fault-tolerant mechanism for network-on-chip. In *Proceedings of the 2nd ACM/IEEE International Symposium on Networks-on-Chip*, pages 13–22, Washington, DC, USA, 2008. IEEE Computer Society.

[169] P. Kongetira, K. Aingaran, and K. Olukotun. Niagara: A 32-way multithreaded Sparc processor. *IEEE Micro*, 25(2):21–29, 2005.

[170] S. Kottapalli and J. Baxter. Nehalem-EX CPU architecture. In *Proceedings of the Hot Chips*, Stanford, CA, August 2009.

[171] M. Krstic, E. Grass, F. K. Gurkaynak, and P. Vivet. Globally asynchronous, locally synchronous circuits: Overview and outlook. *IEEE Design and Test of Computers*, 24:430–441, 2007.

[172] A. Kumar, P. Kundu, A. Singh, L. Peh, and N. K. Jha. A 4.6tbits/s 3.6 GHz single-cycle NoC router with a novel switch allocator in 65 nm CMOS. In *Proceedings of the International Conference on Computer Design*, pages 63–70. IEEE, October 2007.

[173] S. Kumar, A. Jantsch, M. Millberg, J. Öberg, J. Soininen, M. Forsell, K. Tiensyrja, and A. Hemani. A network on chip architecture and design methodology. In *Proceedings of the IEEE Computer Society Annual Symposium on VLSI*, page 117, Washington, DC, USA, 2002. IEEE Computer Society.

[174] R. O. LaMaire and D. N. Serpanos. Two-dimensional round-robin schedulers for packet switches with multiple input queues. *ACM/IEEE Transactions on Networking*, 2(5):471–482, October 1994.

[175] L. Lamport. How to make a multiprocessor computer that correctly executes multiprocess programs. *IEEE Transactions on Computers*, 28(9):690–691, 1979.

[176] J. M. Lebak. Polymorphous computing architectures (PCA) example application 4: Corner-turn. Technical report, MIT Lincoln Laboratory, October 2001.

[177] D. L. Lewis and H.-H. S. Lee. A scanisland based design enabling prebond testability in die-stacked microprocessors. In *Proceedings of the IEEE International Test Conference*, pages 1–8, October 2007.

[178] F. Li, C. A. Nicopoulos, T. Richardson, Y. Xie, V. Narayanan, and M. Kandemir. Design and management of 3D chip multiprocessors using network-in-memory. In *Proceedings of the 33rd Annual International Symposium on Computer Architecture*, pages 130–141, Washington, DC, USA, 2006. IEEE Computer Society.

[179] The Nangate Open Cell Library. 45 nm FreePDK. Available at `https://www.si2.org/openeda.si2.org/projects/nangatelib/`.

[180] C. C. Liu, I. Ganusov, M. Burtscher, and S. Tiwari. Bridging the processor-memory performance gapwith 3D IC technology. *IEEE Design & Test of Computers*, 22(6):556–564, 2005.

[181] Y. Liu, Y. Ma, E. Kursun, G. Reinman, and J. Cong. Fine grain 3D integration for microarchitecture design through cube packing exploration. In *Proceedings of the International Conference on Computer Design*, 2007.

[182] I. Loi, F. Angiolini, and L. Benini. Supporting vertical links for 3D networks-on-chip: Toward an automated design and analysis flow. In *Proceedings of the 2nd International Conference on Nano-Networks*, pages 1–5, ICST, Brussels, Belgium, 2007. ICST (Institute for Computer Sciences, Social-Informatics and Telecommunications Engineering).

[183] I. Loi, F. Angiolini, and L. Benini. Developing mesochronous synchronizers to enable 3D NoCs. In *Proceedings of the Design, Automation and Test in Europe Conference*, pages 1414–1419, New York, NY, USA, 2008. ACM.

[184] I. Loi, F. Angiolini, and L. Benini. Synthesis of low-overhead configurable source routing tables for network interfaces. In *Proceedings of the Design Automation and Test in Europe Conference*, pages 262–267. ACM/IEEE, 2009.

[185] I. Loi, S. Mitra, T. H. Lee, S. Fujita, and L. Benini. A low-overhead fault tolerance scheme for TSV-based 3D network on chip links. In *Proceedings of the IEEE/ACM International Conference on Computer-Aided Design*, pages 598–602, Piscataway, NJ, USA, 2008. IEEE Press.

[186] Z. Lu, M. Liu, and A. Jantsch. Layered switching for networks on chip. In *Proceedings of the 44th Annual Design Automation Conference*, pages 122–127, New York, NY, USA, 2007. ACM.

[187] D. Ludovici, D. Bertozzi, L. Benini, and G. N. Gaydadjiev. Capturing topology-level implications of link synthesis techniques for nanoscale networks-on-chip. In *Proceedings of the 19th ACM/IEEE Great Lakes Symposium on VLSI*, pages 125–128, 2009.

[188] D. Ludovici, F. Gilabert, S. Medardoni, C. Gómez, M. E. Gómez, P. López, G. H. Gaydadjiev, and D. Bertozzi. Assessing fat-tree topologies for regular network-on-chip design under nanoscale technology constraints. In *Proceedings of the Design, Automation and Test in Europe Conference*, pages 562–565. IEEE, 2009.

[189] D. Ludovici, A. Strano, and D. Bertozzi. Architecture design principles for the integration of synchronization interfaces into network-on-chip switches. In *Proceedings of the 2nd ACM/IEEE International Workshop on Network-on-Chip Architectures*, pages 31–36, December 2009.

[190] D. Ludovici, A. Strano, D. Bertozzi, L. Benini, and G.N. Gaydadjiev. Comparing tightly and loosely coupled mesochronous synchronizers in a NoC switch architecture. In *Proceedings of the 3rd International Symposium on Networks-on-Chip*, pages 244–249. ACM/IEEE, 2009.

[191] D. Ludovici, A. Strano, G. N. Gaydadjiev, L. Benini, and D. Bertozzi. Design space exploration of a mesochronous link for cost-effective and flexible GALS NoCs. In *Proceedings of the Design, Automation and Test in Europe Conference*, pages 679–684, March 2010.

[192] O. Lysne, T. Skeie, S. A. Reinemo, and I. Theiss. Layered routing in irregular networks. *IEEE Transactions on Parallel and Distributed Systems*, 17(1):51–65, 2006.

[193] K. Mackenzie, J. Kubiatowicz, M. Frank, W Lee, V Lee, A. Agarwal, and M. Kaashoek. Exploiting two-case delivery for fast protected messaging. In *Proceedings of the 4th International Symposium on High-Performance Computer Architecture*, page 231, Washington, DC, USA, 1998. IEEE Computer Society.

[194] N. Madan and R. Balasubramonian. Leveraging 3D technology for improved reliability. In *Proceedings of the 40th Annual IEEE/ACM International Symposium on MicroArchitecture*, pages 223–235, Washington, DC, USA, 2007. IEEE Computer Society.

[195] R. A. Maddox, G. Singh, and R. Safranek. *Weaving High Performance Multiprocessor Fabric: Architectural Insights to the Intel (R) QuickPath Interconnect.* Intel Press, Beaverton, OR, 2009.

[196] P. S. Magnusson, M. Christensson, J. Eskilson, D. Forsgren, G. Hållberg, J. Högberg, F. Larsson, A. Moestedt, and B. Werner. Simics: A full system simulation platform. *IEEE Computer*, 35(2):50–58, 2002.

[197] S. Mahadevan, F. Angiolini, M. Storoaard, R. G. Olsen, J. Sparso, and J. Madsen. Network traffic generator model for fast network-on-chip simulation. In *Proceedings of the Design, Automation and Test in Europe Conference*, pages 780–785, 2005.

[198] S. Manolache, P. Eles, and Z. Peng. Fault and energy-aware communication mapping with guaranteed latency for applications implemented on NoC. In *Proceedings of the 42nd Annual Design Automation Conference*, pages 266–269, New York, NY, USA, 2005. ACM.

[199] R. Marculescu, U. Y. Ogras, L. S. Peh, N. E. Jerger, and Hoskote Y. Outstanding research problems in NoC design: System, microarchitecture, and circuit perspectives. *IEEE Transactions on Computer-Aided Design of Integrated Circuits and Systems*, 28(1):3–21, 2009.

[200] M. M. Martin, M. D. Hill, and D. A Wood. Token coherence: Decoupling performance and correctness. In *Proceedings of the 30th International Symposium on Computer Architecture*, pages 182–193, 2003.

[201] M. M. Martin, D. Sorin, B. Beckmann, M. R. Marty, M. Xu, A. R. Alameldeen, K. E. Moore, M. Hill, and D. A. Wood. Multifacet's general execution-driven multiprocessor simulator (GEMS) toolset. *Computer Architecture News (CAN)*, pages 96–108, 2005.

[202] H. Matsutani, M. Koibuchi, H. Amano, and T. Yoshinaga. Prediction router: Yet another low latency on-chip router architecture. In *Proceedings of the International Symposium on High-Performance Computer Architecture*, pages 367–378. IEEE, February 2009.

[203] T. Mattner and F. Olbrich. FPGA based tera-scale IA prototyping system. In *Workshop on Architectural Research Prototyping*, June 2008.

[204] N. McKeown. The iSLIP scheduling algorithm for input-queued switches. *IEEE/ACM Transactions on Networking*, 7(2):188–201, 1999.

[205] N. McKeown, M. Izzard, A. Mekkittikul, W. Ellersick, and M. Horowitz. Tiny tera: A packet switch core. *IEEE Micro*, 17(1):26–33, January/February 1997.

[206] A. Mejía, J. Flich, J. Duato, S. A. Reinemo, and T. Skeie. Segment-based routing: an efficient fault-tolerant routing algorithm for meshes and tori. In *Proceedings of the International Parallel and Distributed Processing Symposium*, pages 84–94, Los Alamitos, CA, USA, 2006. IEEE Computer Society.

[207] Message Passing Interface Forum. *MPI: A Message-Passing Interface Standard*, May 1994. http://www.mpi-forum.org/docs/docs.html.

[208] G. Micheli and L. Benini. Networks on chips: Technology and tools. In Giovanni De Micheli and Luca Benini, editors, *Networks on Chips: Technology and Tools (Systems on Silicon)*, chapter 2. Morgan Kaufmann Publishers Inc., San Francisco, CA, USA, 2006.

[209] G. Michelogiannakis, J. Balfour, and W. J. Dally. Elastic-buffer flow control for on-chip networks. In *Proceedings of the International Symposium on High Performance Computer Architecture*, pages 151–162. IEEE, February 2009.

[210] C. A. Mineo. Clock tree insertion and verification for 3D integrated circuits. Master's thesis, North Carolina State University at Raleigh, September 2005.

[211] C. Minkenberg, F. Abel, P. Muller, R. Krishnamurthy, M. Gusat, P. Dill, I. Iliadis, R. Luijten, B. R. Hemenway, R. Grzybowski, and E. Schiattarella. Designing a crossbar scheduler for HPC applications. *IEEE Micro*, 26(3):58–71, May/June 2006.

[212] C. Minkenberg and M. Gusat. Design and performance of speculative flow control for high-radix datacenter interconnect switches. *Journal of Parallel and Distributed Computing*, 69(8):680–695, 2009.

[213] C. Minkenberg, I. Iliadis, and F. Abel. Low-latency pipelined crossbar arbitration. In *Proceedings of Globecom*, December 2004.

[214] I. Miro-Panades, F. Clermidy, P. Vivet, and A. Greiner. Physical implementation of the DSPIN network-on-chip in the FAUST architecture. In *Proceedings of the 2nd ACM/IEEE International Symposium on Networks-on-Chip*, pages 139–148, Washington, DC, USA, 2008. IEEE Computer Society.

[215] I. Miro-Panades and A. Greiner. Bi-synchronous FIFO for synchronous circuit communication well suited for network-on-chip in GALS architectures. In *Proceedings of the International Symposium on Networks-on-Chip*, pages 83–94. IEEE, 2007.

[216] M. Mirza-Aghatabar, S. Koohi, S. Hessabi, and M. Pedram. An empirical investigation of mesh and torus NoC topologies under different routing algorithms and traffic models. In *Proceedings of the 10th Euromicro Conference on Digital System Design Architectures, Methods and Tools*, pages 19–26, Washington, DC, USA, 2007. IEEE Computer Society.

[217] M. Mondal, T. Ragheb, X. Wu, A. Aziz, and Y. Massoud. Provisioning on-chip networks under buffered RC interconnect delay variations. In *Proceedings of the 8th International Symposium on Quality Electronic Design*, pages 873–878, March 2007.

[218] M. Mondal, A. J. Ricketts, S. Kirolos, T. Ragheb, G. Link, N. Vijaykrishnan, and Y. Massoud. Thermally robust clocking schemes for 3D integrated circuits. In *Proceedings of the Design, Automation and Test in Europe Conference*, pages 1–6, April 2007.

[219] G. Mora, J. Flich, J. Duato, E. Baydal, P. López, and O. Lysne. Towards an efficient switch architecture for high-radix switches. In *Proceedings of the ACM/IEEE Symposium on Architectures for Networking and Communications Systems*, December 2006.

[220] P. Morrow, M. Kobrinsky, S. Ramanathan, C. M. Park, M. Harmes, V. Ramachandrarao, H. Park, G. Kloster, S. List, and S. Kim. Wafer-level 3D interconnects via Cu bonding. In *Proceedings of the 21st Advanced Metallization Conference*, 2004.

[221] D. Mosberger. Memory consistency models. *SIGOPS Operating Systems Review*, 27(1):18–26, 1993.

[222] S. S. Mukherjee, B. Falsafi, M. D. Hill, and D. A. Wood. Coherent network interfaces for fine-grain communication. In *Proceedings of the 23rd International Symposium on Computer Architecture*, pages 247–258, May 1996.

[223] S. S. Mukherjee, F. Silla, P. Bannon, J. Emer, S. Lang, and D. Webb. A comparative study of arbitration algorithms for the Alpha 21364 pipelined router. In *Proceedings of the International Symposium on Architectural Support for Programming Languages and Operating Systems*, May 2002.

[224] R. Mullins and S. Moore. Demystifying data-driven and pausible clocking schemes. In *Proceedings of the International Symposium on Asynchronous Circuits and Systems*, pages 175–185, 2007.

[225] R. Mullins, A. West, and S. Moore. Low-latency virtual-channel routers for on-chip networks. In *Proceedings of the 31st International Symposium on Computer Architecture*, pages 188–197. IEEE, June 2004.

[226] S. Murali, C. Seiculescu, L. Benini, and G. De Micheli. Synthesis of networks on chips for 3D systems on chips. In *Proceedings of the Asia and South Pacific Design Automation Conference*, pages 242–247, Piscataway, NJ, USA, 2009. IEEE Press.

[227] C. J. Myers. *Asynchronous Circuit Design.* Wiley, 2001.

[228] T. Nachiondo, J. Flich, and J. Duato. Efficient reduction of HoL blocking in multistage networks. In *Proceedings of the International Parallel and Distributed Processing Symposium*, page 211.2. IEEE, 2005.

[229] T. Nachiondo, J. Flich, and J. Duato. Destination-based HoL blocking elimination. In *Proceedings of the International Conference on Parallel and Distributed Systems*, pages 213–222. IEEE Computer Society, July 2006.

[230] C. A. Nicopoulos, D. Park, J. Kim, N. Vijaykrishnan, M. S. Yousif, and C. R. Das. ViChaR: A dynamic virtual channel regulator for network on-chip routers. In *International Symposium on Microarchitecture*, pages 333–344. IEEE, December 2006.

[231] C. A. Nicopoulos, S. Srinivasan, A. Yanamandra, D. Park, V. Narayanan, C. Das, and M. Irwin. On the effects of process variation in network-on-chip architectures. *IEEE Transactions on Dependable and Secure Computing*, 7(3):240–254, 2010.

[232] E. Nilsson, M. Millberg, J. Öberg, and A. Jantsch. Load distribution with the proximity congestion awareness in a network on chip. In *Proceedings of the Design, Automation and Test in Europe Conference*, pages 1126–1127, 2003.

[233] J. L. Nuñez Yanez, D. Edwards, and A. M. Coppola. Adaptive routing strategies for fault-tolerant on-chip networks in dynamically reconfigurable systems. *IET Computers & Digital Techniques*, 2(3):184–198, 2008.

[234] U. Y. Ogras and R. Marculescu. Prediction-based flow control for network-on-chip traffic. In *Proceedings of the Design Automation Conference*, 2006.

[235] U. Y. Ogras and R. Marculescu. Analysis and optimization of prediction-based flow control in networks-on-chip. *ACM Transactions on Design Automation of Electronic Systems*, 13:1–28, January 2008.

[236] U. Y. Ogras, R. Marculescu, P. Choudhary, and D. Marculescu. Voltage-frequency island partitioning for GALS-based networks-on-chip. In *Proceedings of the Design Automation Conference*, pages 110–115, June 2007.

[237] M. Orshansky, S. Nassif, and D. Boning. *Design for Manufacturability and Statistical Design: A Comprehensive Approach.* Springer-Verlag New York, Inc., Secaucus, NJ, USA, 2006.

[238] J. D. Owens, W. J. Dally, R. Ho, D. N. Jayasimha, S. W. Keckler, and L. S. Peh. Research challenges for on-chip interconnection networks. *IEEE Micro*, pages 96–108, 2007.

[239] J. M. Owner, M. D. Hummel, D. R. Meyer, and J. B. Keller. System and method of maintaining coherency in a distributed communication system. U.S. Patent 7069361, 2006.

[240] M. Palesi, S. Kumar, and R. Holsmark. A method for router table compression for application specific routing in mesh topology NoC architectures. In *Proceedings of the International Symposium on Sistems, Architectures, Modeling and Simulation VI Workshop*, pages 373–384, 2006.

[241] P. P. Pande, C. Grecu, A. Ivanov, and R. Saleh. Design of a switch for network on chip applications. In *Proceedings of the 2003 IEEE International Symposium on Circuits and Systems*, volume 5, pages 217–220, Bangkok, Thailand, 2003. IEEE.

[242] D. Park, S. Eachempati, R. Das, A. K. Mishra, Y. Xie, N. Vijaykrishnan, and C. R. Das. MIRA: A multi-layered on-chip interconnect router architecture. In *Proceedings of the 35th Annual International Symposium on Computer Architecture*, pages 251–261, Washington, DC, USA, 2008. IEEE Computer Society.

[243] D. Park, C. A. Nicopoulos, J. Kim, N. Vijaykrishnan, and C. R. Das. Exploring fault-tolerant network-on-chip architectures. In *Proceedings of the International Conference on Dependable Systems and Networks*, pages 93–104, Washington, DC, USA, 2006. IEEE Computer Society.

[244] S. Pasricha. Exploring serial vertical interconnects for 3D ICs. In *Proceedings of the Design Automation Conference*, pages 581–586, 2009.

[245] G. Passas, M. Katevenis, and D. Pnevmatikatos. A 128 x 128 x 24Gb/s crossbar interconnecting 128 tiles in a single hop and occupying 6% of their area. In *Proceedings of the 4th International Symposium on Networks-on-Chip*, pages 87–95, Los Alamitos, CA, USA, 2010. IEEE Computer Society.

[246] R. S. Patti. Three-dimensional integrated circuits and the future of system-on-chip designs. *Proceedings of the IEEE*, 94(6), June 2006.

[247] V. F. Pavlidis and E. G. Friedman. 3-D topologies for networks-on-chip. *IEEE Transactions on Very Large Scale Integration Systems*, 15(10):1081–1090, October 2007.

[248] V. F. Pavlidis and E. G. Friedman. Interconnect-based design methodologies for three-dimensional integrated circuits. *Proceedings of the IEEE*, 97(1):123–140, January 2009.

[249] V. F. Pavlidis, I. Savidis, and E. G. Friedman. Clock distribution networks for 3-D integrated circuits. In *Proceedings of the Custom Integrated Circuits Conference*, pages 651–654, September 2008.

[250] L. Peh and W. J. Dally. Flit-reservation flow control. In *Proceedings of the 6th International Symposium on High-Performance Computer Architecture*, pages 73–84, January 2000.

[251] L. Peh and W. J. Dally. A delay model and speculative architecture for pipelined routers. In *Proceedings of 7th International Symposium on High-Performance Computer Architecture*, pages 255–266, January 2001.

[252] F. Petrini and M. Vanneschi. K-ary N-trees: High performance networks for massively parallel architectures. Technical report, University of Pisa, 1995.

[253] T. M. Pinkston and J. Shin. Trends toward on-chip networked microsystems. *International Journal on High Performance Computing and Networking*, 3(1):3–18, 2005.

[254] M. Pirretti, G. M. Link, R. R. Brooks, N. Vijaykrishnan, M. Kandemir, and M. J. Irwin. Fault tolerant algorithms for network-on-chip interconnect. *IEEE Computer Society Annual Symposium on VLSI*, 0:46, 2004.

[255] European IP Project. SARC: Scalable computer ARChitecture. Available at http://www.sarc-ip.org/, 2005–2009.

[256] V. Puente, J. A. Gregorio, F. Vallejo, and R. Beivide. Immunet: A cheap and robust fault-tolerant packet routing mechanism. *SIGARCH Computer Architecture News*, 32(2):198, 2004.

[257] A. Pullini, F. Angiolini, D. Bertozzi, and L. Benini. Fault tolerance overhead in network-on-chip flow control schemes. In *SBCCI '05: Proceedings of the 18th Annual Symposium on Integrated Circuits and System Design*, pages 224–229, New York, NY, USA, 2005. ACM.

[258] A. Pullini, F. Angiolini, S. Murali, D. Atienza, G. De Micheli, and L. Benini. Bringing NoCs to 65 nm. In *IEEE Micro 27*, Washington, DC, USA, 2007. IEEE Computer Society.

[259] K. Puttaswamy and G. H. Loh. Implementing caches in a 3D technology for high performance processors. In *Proceedings of the 2005 International Conference on Computer Design*, pages 525–532, Washington, DC, USA, 2005. IEEE Computer Society.

[260] K. Puttaswamy and G. H. Loh. Thermal herding: Microarchitecture techniques for controlling hotspots in high-performance 3D-integrated processors. In *Proceedings of the IEEE 13th International Symposium on High Performance Computer Architecture*, pages 193–204, Washington, DC, USA, 2007. IEEE Computer Society.

[261] W. Qiao and L. M. Ni. Adaptive routing in irregular networks using cut-through switches. In *Proceedings of the International Conference on Parallel Processing*, pages 52–60, 1996.

[262] B. R. Quinton et al. Asynchronous IC interconnect network design and implementation using a standard ASIC flow. In *Proceedings of the International Conference of Computer Design*, pages 267–274, 2005.

[263] J. M. Rabaey. *Digital Integrated Circuits: A Design Perspective.* Prentice-Hall, 2003.

[264] D. Rahmati, A. E. Kiasari, S. Hessabi, and H. Sarbazi-Azad. A performance and power analysis of WK-recursive and mesh networks for network-on-chips. In *Proceedings of the International Conference on Computer Design*, pages 142–147, 2006.

[265] S. Ramabhadran and J. Pasquale. Stratified round robin: A low complexity packet scheduler with bandwidth fairness and bounded delay. In *Proceedings of the Conference on Applications, Technologies, Architectures, and Protocols for Computer Communications*, pages 239–250, New York, NY, USA, 2003. ACM.

[266] J. Rattner. Single-chip cloud computer: An experimental many-core processor from intel labs. Available online at http://download.intel.com/pressroom/pdf/rockcreek/SCC_Announcement_JustinRattner.pdf.

[267] E. Rijpkema, K. G. W. Goossens, A. Radulescu, J. Dielissen, J. van Meerbergen, P. Wielage, and E. Waterlander. Trade offs in the design of a router with both guaranteed and best-effort services for networks on chip. *Computers and Digital Techniques*, 150(5):294–302, 2003.

[268] A. Roca, J. Flich, F. Silla, and J. Duato. VCTlite: Towards an efficient implementation of virtual cut-through switching in on-chip networks. In *Proceedings of the 17th Annual International Conference on High Performance Computing*, 2010.

[269] S. Rodrigo, J. Flich, J. Duato, and M. Hummel. Efficient unicast and multicast support for CMPs. In *Proceedings of the 41st Annual IEEE/ACM International Symposium on Microarchitecture*, pages 364–375, Washington, DC, USA, 2008. IEEE Computer Society.

[270] S. Rodrigo, J. Flich, A. Roca, S. Medardoni, D. Bertozzi, J. Camacho, F. Silla, and J. Duato. Addressing manufacturing challenges with cost-efficient fault tolerant routing. In *Proceedings of the 4th ACM/IEEE International Symposium on Networks-on-Chip*, pages 25–32, 2010.

[271] S. Rodrigo, C. Hernández, J. Flich, F. Silla, J. Duato, S. Medardoni, D. Bertozzi, A. Mejía, and D. Dai. Yield-oriented evaluation methodology of network-on-chip routing implementations. In *Proceedings of the 11th International Conference on System-on-chip*, pages 100–105, Piscataway, NJ, USA, 2009. IEEE Press.

[272] S. Rodrigo, S. Medardoni, J. Flich, D. Bertozzi, and J. Duato. Efficient implementation of distributed routing algorithms for NoCs. *IET Computers and Digital Techniques*, 3:460–475, 2009.

[273] F. A. Samman, T. Hollstein, and M. Glesner. Multicast parallel pipeline router architecture for network-on-chip. In *Proceedings of the Confer-*

ence on Design, Automation and Test in Europe, pages 1396–1401, New York, NY, USA, 2008. ACM.

[274] D. Sanchez, R. M. Yoo, and C. Kozyrakis. Flexible architectural support for fine-grain scheduling. In *Proceedings of the 15th International Conference on Architectural Support for Programming Languages and Operating Systems*, pages 311–322, New York, NY, USA, 2010. ACM.

[275] J. C. Sancho, A. Robles, and J. Duato. A flexible routing scheme for networks of workstations. In *Proceedings of the International Symposium on High Performance Computing*, pages 260–267. Springer Berlin/Heidelberg, 2000.

[276] J. C. Sancho, A. Robles, and J. Duato. A new methodology to compute deadlock-free routing tables for irregular networks. In *Proceedings of the 4th International Workshop on Communication, Architecture and Applications for Network-based Parallel Computing*, 2000.

[277] J. C. Sancho, A. Robles, J. Flich, P. López, and J. Duato. Effective methodology for deadlock-free minimal routing in Infiniband networks. In *Proceedings of the International Conference on Parallel Processing*, page 409, Washington, DC, USA, 2002. IEEE Computer Society.

[278] K. Sankaralingam, R. Nagarajan, R. McDonald, R. Desikan, S. Drolia, M. S. Govindan, P. Gratz, D. Gulati, H. Hanson, C. Kim, H. Liu, N. Ranganathan, S. Sethumadhavan, S. Sharif, P. Shivakumar, S. W. Keckler, and D. Burger. The distributed microarchitecture of the TRIPS prototype processor. In *Proceedings of the 39th ACM/IEEE International Symposium on Microarchitecture*, pages 480–491, 2006.

[279] K. Sankaralingam, V. Ajay Singh, S. W. Keckler, and D. Burger. Routed inter-ALU networks for ILP scalability and performance. In *Proceedings of the IEEE International Conference on Computer Design*, pages 170–177, 2003.

[280] S. R. Sarangi, B. Greskamp, R. Teodorescu, J. Nakano, A. Tiwari, and J. Torrellas. VARIUS: A model of process variation and resulting timing errors for microarchitects. *IEEE Transactions on Semiconductor Manufacturing*, 21(1):3–13, February 2008.

[281] I. Savidis and E. G. Friedman. Closed-form expressions of 3-D via resistance, inductance, and capacitance. *IEEE Transactions on Electron Devices*, 56(9):1873–1881, September 2009.

[282] I. Schoinas and M. D. Hill. Address translation mechanisms in network interfaces. In *Proceedings of the 4th International Symposium on High-Performance Computer Architecture*, page 219, Washington, DC, USA, 1998. IEEE Computer Society.

[283] M. D. Schroeder, A. D. Birrell, M. Burrows, H. Murray, R. M. Needham, T. L. Rodeheffer, E. H. Satterthwaite, and C. P. Thacker. Autonet: a high-speed, self-configuring local area network using point-to-point links. *IEEE Journal on Selected Areas in Communications*, 9, 1991.

[284] S. L. Scott and G. S. Sohi. The use of feedback in multiprocessors and its application to tree saturation control. *IEEE Transactions on Parallel and Distributed Systems*, 1(4):385–398, October 1990.

[285] C. Seiculescu, S. Murali, L. Benini, and G. De Micheli. Noc topology synthesis for supporting shutdown of voltage islands in SoCs. In *Proceedings of the Design Automation Conference*, pages 822–825. ACM/IEEE, 2009.

[286] D. Seo, A. Ali, W. Lim, N. Rafique, and M. Thottethodi. Near-optimal worst-case throughput routing for two-dimensional mesh networks. In *Proceedings of the 32nd Annual International Symposium on Computer Architecture*, pages 432–443, Washington, DC, USA, 2005. IEEE Computer Society.

[287] L. Shang, L. Peh, A. Kumar, and N. K. Jha. Thermal modeling, characterization and management of on-chip networks. In *Proceedings of the 37th Annual IEEE/ACM International Symposium on Microarchitecture*, pages 67–78, Washington, DC, USA, 2004. IEEE Computer Society.

[288] T. Shimizu, T. Horie, and H. Ishihata. Low-latency message communication support for the AP1000. In *Proceedings of the 19th Annual International Symposium on Computer Architecture*, pages 288–297, New York, NY, USA, 1992. ACM.

[289] E. S. Shin, V. J. Mooney III, and G. F. Riley. Round-robin arbiter design and generation. In *Proceedings of the 15th International Symposium on System Synthesis*, pages 243–248, 2002.

[290] K. G. Shin and S. W. Daniel. Analysis and implementation of hybrid switching. *IEEE Transactions on Computers*, 45(6):684–692, 1996.

[291] D. Sigüenza-Tortosa and J. Nurmi. Proteo: A new approach to network-on-chip. In *Proceedings of the IASTED International Conference on Communication Systems and Networks*, 2002.

[292] F. Silla and J. Duato. Improving the efficiency of adaptive routing in networks with irregular topology. In *In Proceedings of the International Conference on High Performance Computing*, 1997.

[293] K. Siozios, K. Sotiriadis, V. F. Pavlidis, and D. Sondris. Exploring alternative 3D FPGA architectures: Design methodology and CAD tool support. In *Proceedings of the International Conference on Field Programmable Logic and Applications*, pages 652–655, August 2007.

[294] K. Skadron, M. R. Stan, W. Huang, S. Velusamy, K. Sankaranarayanan, and D. Tarjan. Temperature-aware microarchitecture. *SIGARCH Computer Architecture News*, 31(2):2–13, 2003.

[295] T. Skeie, O. Lysne, and I. Theiss. Layered shortest path (LASH) routing in irregular system area networks. In *Proceedings of the 16th International Parallel and Distributed Processing Symposium*, page 194, Washington, DC, USA, 2002. IEEE Computer Society.

[296] T. Skeie, F. O. Sem-Jacobsen, S. Rodrigo, J. Flich, D. Bertozzi, and S. Medardoni. Flexible DOR routing for virtualization of multicore chips. In *Proceedings of the 11th International Conference on System-on-chip*, pages 73–76, Piscataway, NJ, USA, 2009. IEEE Press.

[297] W. Song, D. Edwards, J. L. Nuñez Yanez, and S. Dasgupta. Adaptive stochastic routing in fault-tolerant on-chip networks. In *Proceedings of the 3rd ACM/IEEE International Symposium on Networks-on-Chip*, pages 32–37, Washington, DC, USA, 2009. IEEE Computer Society.

[298] V. Soteriou and L. Peh. Dynamic power management for power optimization of interconnection networks using on/off links. In *Proceedings of the Symposium on High-Performance Interconnects*, pages 15–20, Los Alamitos, CA, USA, 2003. IEEE Computer Society.

[299] J. Sparso and S. Furber. *Principles of Asynchronous Circuit Design: A System Perspective.* Kluwer, 2001.

[300] SPEC. SPECjbb2000 Java Benchmark. Available at `http://www.spec.org/osg/jbb2000/`.

[301] S. Spiesshoefer et al. Z-axis interconnects using fine pitch, nanoscale through-silicon vias: Process development. In *Proceedings of the Electronic Components and Technology Conference*, 2004.

[302] E. Sprangle and D. Carmean. Increasing processor performance by implementing deeper pipelines. In *Proceedings of the 30th International Symposium on Computer Architecture*, pages 25–34, 2002.

[303] B. Stackhous et al. A 65 nm 2-billion transistor quad-core Itanium processor. *IEEE Journal of Solid State Circuits*, 44:18–31, 2009.

[304] B. Stefano, D. Bertozzi, L. Benini, and E. Macii. Process variation tolerant pipeline design through a placement-aware multiple voltage island design style. In *Proceedings of the Design, Automation and Test in Europe*, pages 967–972, March 2008.

[305] S. Stergiou, F. Angiolini, S. Carta, L. Raffo, D. Bertozzi, and G. De Micheli. xPipes Lite: A synthesis oriented design library for networks on chips. In *Proceedings of the Design Automation and Test in Europe Conference*, pages 1188–1193. IEEE, 2005.

[306] S. Strickland, E. Ergin, D. R. Kaeli, and P. Zavracky. VLSI design in the 3rd dimension. *Journal on Integrated VLSI*, 25(1):1–16, 1998.

[307] S. Suboh, M. Bakhouya, and T. El-Ghazawi. Simulation and evaluation of on-chip interconnect architectures: 2D mesh, spidergon, and WK-recursive network. In *Proceedings of the 2nd ACM/IEEE International Symposium on Networks-on-Chip*, pages 205–206, Washington, DC, USA, 2008. IEEE Computer Society.

[308] H. Sullivan and T. R. Bashkow. A large scale, homogeneous, fully distributed parallel machine. In *Proceedings of the 4th Annual Symposium on Computer Architecture*, pages 105–117, New York, NY, USA, 1977. ACM.

[309] R. Sunkam Ramanujam and B. Lin. A novel 3D layer-multiplexed on-chip network. In *Proceedings of the ACM/IEEE Symposium on Architectures for Networking and Communications Systems*, 2009.

[310] S. Swanson, A. Putnam, M. Mercaldi, K. Michelson, A. Petersen, A. Schwerin, M. Oskin, and S. Eggers. Area-performance trade-offs in tiled dataflow architectures. In *Proceedings of the 33rd International Symposium on Computer Architecture*, pages 314–326, 2006.

[311] Y. Tamir and H. C. Chi. Symmetric crossbar arbiters for VLSI communication switches. *IEEE Transactions on Parallel and Distributed Systems*, 4:13–27, 1993.

[312] Y. Tamir and G. L. Frazier. Dynamically-allocated multi-queue buffers for VLSI communication switches. *IEEE Transactions on Computers*, 41(6):725–737, June 1992.

[313] A. S. Tanenbaum. *Computer Networks*. Prentice-Hall, Inc., Upper Saddle River, NJ, USA, 2003.

[314] D. Tarjan, S. Thoziyoor, and N. P. Jouppi. CACTI 4.0, technical report, hpl-2006-86, 2006.

[315] M. B. Taylor, W. Lee, S. P. Amarasinghe, and A. Agarwal. Scalar operand networks: On-chip interconnect for ILP in partitioned architecture. In *Proceedings of the 9th International Symposium on High-Performance Computer Architecture*, pages 341–353, 2003.

[316] M. B. Taylor, W. Lee, S. P. Amarasinghe, and A. Agarwal. Scalar operand networks: Design, implementation, and analysis, 2004.

[317] M. B. Taylor, W. Lee, J. Miller, D. Wentzlaff, I. Bratt, B. Greenwald, H. Hoffman, P. Johnson, J. Kim, J. Psota, A. Saraf, N. Shnidman, V. Strumpen, M. Frank, S. P. Amarasinghe, and A. Agarwal. Evaluation of the RAW microprocessor: An exposed-wire-delay architecture

for ILP and streams. In *Proceedings of the International Symposium on Computer Architecture*, pages 2–13, June 2004.

[318] T. Thorolfsson, S. Melamed, G. Charles, and P. D. Franzon. Comparative analysis of two 3D integration implementations of a SAR processor. In *Proceedings of the International Conference on 3D System Integration*, pages 1–4, September 2009.

[319] M. Thottethodi, A. R. Lebeck, and S. S. Mukherjee. Self-tuned congestion control for multiprocessor networks. In *Proceedings of the International Symposium on High-Performance Computer Architecture*, pages 107–120, 2001.

[320] TILERA. TILE-Gx processors family. Available at `http://www.tilera.com/products/TILE-Gx.php`.

[321] R. Tomasulo. An efficient algorithm for exploring multiple arithmetic units. *IBM Journal of Research and Development*, 11(1):25–33, January 1967.

[322] A. T. Tran, D. N. Truong, and B. M. Baas. A reconfigurable source-synchronous on-chip network for GALS many-core platforms. *IEEE Transactions on Computer-Aided Design of Integrated Circuits and Systems*, 29:897–910, June 2010.

[323] Y. Tsai, Y. Xie, N. Vijaykrishnan, and M. J. Irwin. Three-dimensional cache design exploration using 3DCacti. In *Proceedings of the International Conference on Computer Design*, pages 519–524, Washington, DC, USA, 2005. IEEE Computer Society.

[324] J. Turner. Design and evaluation of a practical, high performance crossbar scheduler. Technical report, Computer Science and Engineering Department, Washington University in St.Louis, December 2009.

[325] Arizona State University. Predictive technology model. Available at `http://ptm.asu.edu`.

[326] St. Louis University. MediaBench II. Available at `http://euler.slu.edu/fritts/mediabench/`.

[327] A. S. Vaidya, A. Sivasubramaniam, and C. R. Das. LAPSES: A recipe for high performance adaptive router design. In *Proceedings of the 5th International Symposium on High Performance Computer Architecture*, page 236, Washington, DC, USA, 1999. IEEE Computer Society.

[328] J. W. van den Brand, C. Ciordas, K. Goosens, and T. Basten. Congestion-controlled best-effort communication for networks-on-chip. In *Proceedings of the Design, Automation and Test in Europe Conference*, pages 948–953, 2007.

[329] J. Van Leeuwen and R. B. Tan. Interval Routing. *The Computer Journal*, 30(4):298–307, 1987.

[330] S. Vangal, J. Howard, G. Ruhl, S. Dighe, H. Wilson, J. Tschanz, D. Finan, P. Iyer, A. Singh, T. Jacob, S. Jain, S. Venkataraman, Y. Hoskote, and N. Borkar. An 80-tile 1.28 TFLOPS network-on-chip in 65 nm CMOS. In *Proceedings of the International Solid-State Circuits Conference*, pages 5–7, February 2007.

[331] F. Vitullo et al. Low-complexity link microarchitecture for mesochronous communication in networks-on-chip. *IEEE Transactions on Computers*, 57(9):1196–1201, 2008.

[332] T. von Eicken, D. E. Culler, S. C. Goldstein, and K. E. Schauser. Active messages: a mechanism for integrated communication and computation. In *Proceedings of the 19th Annual International Symposium on Computer Architecture*, pages 256–266, New York, NY, USA, 1992. ACM.

[333] E. Waingold, M. B. Taylor, D. Srikrishna, V. Sarkar, W. Lee, V. Lee, J. Kim, M. Frank, P. Finch, R. Barua, J. Babb, S. P. Amarasinghe, and A. Agarwal. Baring it all to software: Raw machines. *IEEE Computer*, 30(9):86–93, September 1997.

[334] H. Wang, L. Peh, and S. Malik. Power-driven design of router microarchitectures in on-chip networks. In *Proceedings of the 36th Annual IEEE/ACM International Symposium on MicroArchitecture*, page 105, Washington, DC, USA, 2003. IEEE Computer Society.

[335] H. Wang, X. Zhu, L. Peh, and S. Malik. Orion: a power-performance simulator for interconnection networks. In *Proceedings of the 35th Annual ACM/IEEE International Symposium on MicroArchitecture*, pages 294–305, Los Alamitos, CA, USA, 2002. IEEE Computer Society Press.

[336] L. Wang, Y. Jin, H. Kim, and E. J. Kim. Recursive partitioning multicast: A bandwidth-efficient routing for networks-on-chip. In *Proceedings of the 3rd ACM/IEEE International Symposium on Networks-on-Chip*, pages 64–73, Washington, DC, USA, 2009. IEEE Computer Society.

[337] R. Weerasekera, L. Zheng, D. Pamunuwa, and H. Tenhunen. Extending systems-on-chip to the third dimension: Performance, cost and technological tradeoffs. In *Proceedings of the IEEE/ACM International Conference on Computer-Aided Design*, pages 212–219, Piscataway, NJ, USA, 2007. IEEE Press.

[338] A. R. Weiss. The standardization of embedded benchmarking: Pitfalls and opportunities. In *Proceedings of the IEEE International Conference on Computer Design*, pages 492–498, 1999.

[339] D. Wentzlaff, P. Griffin, H. Hoffmann, L. Bao, B. Edwards, C. Ramey, M. Mattina, C. Miao, J. F. Brown III, and A. Agarwal. On-chip interconnection architecture of the tile processor. *IEEE Micro*, 27(5):15–31, September/October 2007.

[340] N. H. Weste and D. Harris. *CMOS VLSI Design: A Circuits and Systems Perspective*. Addison Wesley, 4th edition, 2010.

[341] Intel Whitepaper. *64-bit Intel Xeon Processor MP with up to 8 MB L3 cache*. Santa Clara, CA, 2005.

[342] M. V. Wilkes. *The Best Way to Design an Automatic Calculating Machine*. MIT Press, Cambridge, MA, USA, 1989.

[343] S. C. Woo, M. Ohara, E. Torrie, J. P. Singh, and A. Gupta. The SPLASH-2 programs: Characterization and methodological considerations. In *Proceedings of the 22nd Annual International Symposium on Computer Architecture*, pages 24–36, New York, NY, USA, 1995. ACM.

[344] H. Xu, V. F. Pavlidis, and G. De Micheli. Process-induced skew variation in scaled 2-D and 3-D ICs. In *SLIP '10: Proceedings of the 12th ACM/IEEE International Workshop on System Level Interconnect Prediction*, pages 17–24, New York, NY, USA, 2010. ACM.

[345] Y. Xu, B. Zhao, X. Zhou, Y. Zhang, and J. Yang. A low-radix and low-diameter 3D interconnection network design. In *Proceedings of the 15th International Symposium on High Performance Computer Architecture*, pages 30–42, 2009.

[346] Y. S. Yeh, M. Hiuchyj, and A. Acampora. The knockout switch: A simple modular architecture for high performance packet switching. *IEEE Journal in Selected Areas in Communications*, SAC-5(8):1274–1283, 1987.

[347] P. Yew, N. Tzeng, and D. H. Lawrie. Distributing hot-spot addressing in large-scale multiprocessors. *IEEE Transactions on Computers*, 36(4):388–395, April 1987.

[348] Z. Yu and B. M. Baas. Implementing tile-based chip multiprocessors with GALS clocking styles. In *Proceedings of the International Conference of Computer Design*, pages 174–179, 2006.

[349] K. Y. Yun and R. P. Donohue. Pausible clocking: A first step toward heterogeneous systems. In *Proceedings of the International Conference of Computer Design*, pages 118–123, 1996.

[350] A. Y. Zeng, J. J. Lu, K. Rose, and R. J. Gutmann. First-order performance prediction of cache memory with wafer-level 3D integration. *IEEE Design & Test of Computers*, 22(6):548–555, 2005.

[351] ZEUS. ZEUS Web Server. Available at `http://www.zeus.com/products/zws/`.

[352] M. Zhang and K. Asanovic. Victim replication: maximizing capacity while hiding wire delay in tiled chip multiprocessors. In *Proceedings of the 32nd International Symposium on Computer Architecture*, pages 336–345, June 2005.

[353] L. Zhao, R. Iyer, S. Makineni, J. Moses, R. Illikkal, and D. Newell. Performance, area and bandwidth implications on large-scale CMP cache design. In *Proceedings of the Workshop on Chip Multiprocessor Memory Systems and Interconnects*, 2007.

[354] X. Zhao, D. L. Lewis, H.-H. S. Lee, and S. K. Lim. Pre-bond testable low-power clock tree design for 3D stacked ICs. In *Proceedings of the IEEE/ACM International Conference on Computer-Aided Design*, pages 184–190, November 2009.

[355] X. Zhao and S. K. Lim. Power and slew-aware clock network design for through-silicon-via (TSV) based 3D ICs. In *Proceedings of the 15th Asia and South Pacific Design Automation Conference*, pages 175–180, January 2010.

Index